Advanced Composites

Advanced Composites:
From Materials Characterization to Structural Application

Editor

Viktor Gribniak

MDPI • Basel • Beijing • Wuhan • Barcelona • Belgrade • Manchester • Tokyo • Cluj • Tianjin

Editor
Viktor Gribniak
Vilnius Gediminas Technical University
Lithuania

Editorial Office
MDPI
St. Alban-Anlage 66
4052 Basel, Switzerland

This is a reprint of articles from the Special Issue published online in the open access journal *Materials* (ISSN 1996-1944) (available at: https://www.mdpi.com/journal/materials/special_issues/ Advanced_Application).

For citation purposes, cite each article independently as indicated on the article page online and as indicated below:

LastName, A.A.; LastName, B.B.; LastName, C.C. Article Title. *Journal Name* **Year**, *Volume Number*, Page Range.

ISBN 978-3-0365-0724-8 (Hbk)
ISBN 978-3-0365-0725-5 (PDF)

Contents

About the Editor

Viktor Gribniak is the Head of the Laboratory of Innovative Building Structures and Professor with the Department of Steel and Composite Structures at Vilnius Gediminas Technical University (Vilnius Tech), Lithuania. From 1989 to 1994, he studied Applied Mathematics at Vilnius University. From 1994 to 2009, he received B.S., M.S., and Ph.D. degrees in Structural Engineering at Vilnius Gediminas Technical University. During that time, Dr. V. Gribniak worked as an engineer for several building companies in Lithuania, with a total working experience of over ten years. In 2011–2012, Dr. V. Gribniak expanded his core scientific competencies during his postdoctoral internship in the civil engineering centers in Austria, Germany, Spain, Switzerland, and the Czech Republic. Dr. V. Gribniak is the co-author of several patents and more than 170 scientific publications, more than 70 of which in journals references in the ISI Web of Science (WoS) database. Since 2010, he has given more than 60 presentations at conferences in the USA, Argentina, South Africa, South Korea, Japan, Hong Kong, Australia, Austria, France, Italy, Spain, and the Czech Republic. Dr. V. Gribniak's research interests are related to material characterization and mechanics of composites, with noticeable results achieved in constitutive and numerical modeling of composite structures. In 2013, Dr. V. Gribniak, in a co-authorship, obtained the Moisseiff Award by the American Society of Civil Engineers (ASCE). Since 1948, the best article annual nomination has been awarded for essential inputs in structural design and theoretical analysis. Dr. V. Gribniak is a top peer reviewer, with more than 340 records verified in WoS Publons and a member of editorial/advisory Boards of Composites Part B: Engineering, Nanotechnology Reviews, and Építőanyag—*Journal of Silicate-Based and Composite Materials*.

Preface to "Advanced Composites: From Materials Characterization to Structural Application"

Modern industry can produce composite materials with a wide range of mechanical properties applicable in the medicine, aviation, and automotive industries, etc. However, the construction industry is responsible for a substantial part of the budgets worldwide and uses vast amounts of materials. Engineering practice has revealed that innovative technologies' structural application requires new design concepts related to developing materials with mechanical properties tailored to construction purposes. This is the opposite to the current practice, where design solutions are associated with applying existing materials, the physical characteristics of which, in general, are imperfectly suited to the application requirements. The resolution of the above problem should ensure the efficient consumption of engineering materials when the efficiency is understood in a simplified and heuristic manner as the optimization of the performance and the proper combination of structural components. The optimization, based on the materials' adequate characterization, enables implementing environmentally- friendly engineering principles when the efficient use of advanced materials guarantees the required structural safety.

The progress of the research project Industrialised material-oriented engineering for eco-optimised structures, funded by the Research Council of Lithuania, has inspired the emergence of this book, exploring sustainable composites with valorized manufacturability corresponding to the Industrial Revolution 4.0 ideology. Identifying fundamental relationships between the structure of advanced composites and the related physical properties is the focus of the collected articles. Among others, these publications reveal that the application of nano-particles improves the mechanical performance of composite materials; heat-resistant aluminum composites ensure the safety of overhead power transmission lines; chemical additives can detect the impact of temperature on concrete structures. Several publications investigate the connection problems of high-performance composites and expand the application of fiber-reinforced polymer materials to strengthen concrete structures. The Democritus University of Thrace research team achieved remarkable results in developing and analyzing fibrous reinforcements, improving the structural components' mechanical performance and sustainability. New experimental results of cyclic tests of fiber-reinforced concrete beams with bar reinforcement provide a valuable reference for further analysis and development of advanced cement-based composites. The articles collected in this book demonstrate that the choice of construction materials has considerable room for improvement from a scientific viewpoint, following heuristic approaches. The Editor expresses sincere gratitude to all authors for their valuable contributions, the numerous eminent referees, the editorial team of Materials, and especially Elsa Qiu for making this publication successful.

Viktor Gribniak
Editor

Editorial

Special Issue "Advanced Composites: From Materials Characterization to Structural Application"

Viktor Gribniak [1,2]

[1] Department of Steel and Composite Structures, Vilnius Gediminas Technical University (Vilnius Tech), Saulėtekio av. 11, LT-10223 Vilnius, Lithuania; Viktor.Gribniak@vgtu.lt; Tel.: +370-6134-6759

[2] Laboratory of Innovative Building Structures, Vilnius Tech, Saulėtekio av. 11, LT-10223 Vilnius, Lithuania

Received: 5 December 2020; Accepted: 17 December 2020; Published: 21 December 2020

The modern industry allows synthesizing and manufacturing composite materials with a wide range of mechanical properties applicable in medicine, aviation, automotive industry, etc. Construction production generates a substantial part of budgets worldwide and utilizes vast amounts of materials. Nowadays, the practice has revealed that structural applications of innovative engineering technologies require new design concepts related to the development of materials with mechanical properties tailored for construction purposes. It is the opposite way to the current design practice where design solutions are associated with the application of existing materials, the physical characteristics of which, in general, are imperfectly suiting the application requirements.

The efficient utilization of engineering materials is the result of achieving the solution of the above problem. The efficiency can be understood in a simplified and heuristic manner as the optimization of the performance and the proper combination of structural components, leading to the utilization of the minimum volume of materials; consumption of the least amount of natural resources should ensure the development of more durable and sustainable structures. The solution of the eco-optimization problem, based on the adequate characterization of the materials, enables implementing principles of environment-friendly engineering when the efficient utilization of advanced materials guarantees the structural safety required.

The identification of fundamental relationships between the structure of advanced composites and the related physical properties was the focus of the completed Special Issue. The research team from the Democritus University of Thrace achieved outstanding results in the development and analysis of fibrous reinforcement, improving the mechanical performance of structural components [1–3]. The manuscripts [1,2] present new experimental results of cyclic tests of fiber-reinforced concrete beams with bar reinforcement. The reported outcomes are a valuable reference for further analysis and development of advanced cement-based composites. The application of the steel fibers improved the cracking resistance and failure ductility of concrete beams. However, the success of this solution requires a suitable proportioning of the concrete. An extensive review of the literature is a beneficial attribute of the article [2]. The listed publications adequately describe the current state of the art and determine the unsolved problems for further research. This paper contributes to the limited existing literature on cyclic tests of reinforced concrete beams with steel fibers, providing detailed experimental data of beams with a relatively high amount of steel fibers (3%) that has not been examined in structural applications. The work [3] develops a numerical (finite element) model, investigating short-term deformation behavior of structural concrete elements reinforced with a combination of steel fibers and bar reinforcement. The comparison of the numerical and experimental results reveals reliability and computational-efficiency of the developed model, capturing well the critical aspects of the mechanical response of fiber-reinforced cement-based composites and the favorable contribution of the steel fibers on the residual stiffness and ductility of such structures.

The articles [4,5] expand the application of fiber-reinforced polymer (FRP) reinforcement to the strengthening of concrete structures. The paper [4] investigates the local bond stress–slip effects on the global response of the near-surface-mounted (NSM) FRP-strengthening systems in terms of load-bearing capacity, bonding mechanisms, slip resistance, and distribution of shear stresses and strains along the bonded length. A numerical procedure suitable for the design of strengthening systems was developed as well. The calculations employ the finite difference method to solve the governing equations describing the mechanical resistance of the FRP-to-concrete joint. The model was verified experimentally using pull-out shear tests of FRP strips. The article [5] investigates material behavior within the multiple-component system of a structural concrete element strengthened in flexure with an externally bonded fiber-reinforced polymer plate. The specific feature of the proposed approach is the possibility to track the evolution of the damage accumulation in concrete, which is related directly to the initiation of micro-cracks and further propagation of cracks at the macroscopic scale.

The articles included in the Special Issue explore the development of sustainable composites with valorized manufacturability for a breakthrough from conventional practices and corresponding to the Industrial Revolution 4.0 ideology. The publications revealed that the application of nano-particles improves the mechanical performance of composite materials [6]; advanced woven fabrics efficiently reinforce soft body-armor [7]; heat-resistant aluminum composites ensure the safety of overhead power transmission lines [8]; chemical additives can help in detecting temperature impact on concrete structures [9]. The publications [10,11] investigated connection problems of high-performance composites. The sound radiation of shape memory alloy (SMA) composite laminates is the focus of interest of the paper [12]. The study demonstrates that composite laminates embedded with SMA wires have an excellent ability to alter sound radiation characteristics and adjusting modal frequencies. That provides a solid rationale for diminishing the resonance problems and reducing the vibrations and sound radiation of composite laminates from the point of the view of materials engineering. The article [13] analyzes tribological properties and resonance problems of polymeric composites utilized in the aviation industry. The research confirmed several hypotheses related to the abrasive wear process of new materials.

The publication [14] investigates the application of lightweight aluminum honeycombs for developing an efficient system for blast protection. The review article [15] is a valuable reference of design solutions of acoustic coatings. Based on the requirements of underwater acoustic stealth, the classification and research background of acoustic coatings are discussed.

The application of advanced manufacturing technologies is the research focus of articles [16,17]. Supersaturated Cu-Cr solid solution was formed in the reference [16] using the mechanical alloying method. Extension of the high-energy milling procedure presents new insights in the field of research where many articles have been published in the past ten years. A prefabricated Ti-based alloy plate was processed in the study [17] using laser surface remelting technology to obtain an in situ, strongly binding amorphous coating. Dental implants and orthopedic prostheses are the common application objects of such alloys because of high corrosion resistance and good biocompatibility.

The articles [18–20] investigate problems related to the development of advanced steel structures. It was shown that the refinement of grain size improves the low-temperature toughness of high-strength pipeline steel [18]. The study [19] focuses on the evaluation and analysis of the fracture mechanisms (fracture toughness and microstructural and cracking behavior) of T2 copper-45 steel. The reported results provide a basis for improvement in the performance of welded joints. Although the current trends in the construction industry are related to the use of high-strength steel, the increase in the load-bearing capacity of the member is often not proportional to the rise of the strength of the material. The study [20] employed flexural test results of two groups of structural tubular profiles produced from high- and normal-strength steel to illustrate the above observation. The tests demonstrated that the strength condition controlled the load-bearing capacity of the normal-strength specimens. On the contrary, the stiffness condition governed the failure behavior of the high-strength samples. Thus, identification of the optimum configuration of the cross-section should be the subject for further research.

The published works demonstrate that the choice of construction materials has considerable room for improvement from a scientific viewpoint, following heuristic approaches. At the closure of the Special Issue, altogether, the manuscripts included in the Special Issue were cited 51 times. That highlights the essential impact of the reported outcomes on the research community and their valuable contributions to materials science. The onward Special Issue "Advanced Composites: From Materials Characterization to Structural Application (Second Volume)" of *Materials* continues the successful publication series, aiming to combine the innovative achievements of experts in the fields of materials and structural engineering to raise the scientific and practical value of the gathered results of the interdisciplinary research.

Funding: This project received funding from the European Regional Development Fund (Project "Industrialised material-oriented engineering for eco-optimised structures" No 01.2.2-LMT-K-718-03-0010) under grant agreement with the Research Council of Lithuania (LMTLT).

Acknowledgments: The Guest Editor expresses sincere gratitude to the valuable contributions by all authors, numerous eminent referees, the editorial team of *Materials*, and especially Elsa Qiu, for making this Special Issue successful.

Conflicts of Interest: The author declares no conflict of interest.

References

1. Kytinou, V.K.; Chalioris, C.E.; Karayannis, C.G.; Elenas, A. Effect of steel fibers on the hysteretic performance of concrete beams with steel reinforcement—Tests and analysis. *Materials* **2020**, *13*, 2923. [CrossRef] [PubMed]
2. Chalioris, C.E.; Kosmidou, P.-M.K.; Karayannis, C.G. Cyclic response of steel fiber reinforced concrete slender beams: An experimental study. *Materials* **2019**, *12*, 1398. [CrossRef] [PubMed]
3. Kytinou, V.K.; Chalioris, C.E.; Karayannis, C.G. Analysis of residual flexural stiffness of steel fiber-reinforced concrete beams with steel reinforcement. *Materials* **2020**, *13*, 2698. [CrossRef] [PubMed]
4. Gómez, J.; Torres, L.; Barris, C. Characterization and simulation of the bond response of NSM FRP reinforcement in concrete. *Materials* **2020**, *13*, 1770. [CrossRef] [PubMed]
5. Zhelyazov, T. Structural materials: Identification of the constitutive models and assessment of the material response in structural elements strengthened with externally-bonded composite material. *Materials* **2020**, *13*, 1272. [CrossRef] [PubMed]
6. Kumar, A.; Kumar, S.; Mukhopadhyay, N.K.; Yadav, A.; Winczek, J. Effect of SiC reinforcement and its variation on the mechanical characteristics of AZ91 composites. *Materials* **2020**, *13*, 4913. [CrossRef] [PubMed]
7. Abtew, M.A.; Boussu, F.; Bruniaux, P.; Liu, H. Fabrication and mechanical characterization of dry three-dimensional warp interlock para-aramid woven fabrics: Experimental methods toward applications in composite reinforcement and soft body armor. *Materials* **2020**, *13*, 4233. [CrossRef] [PubMed]
8. Qiao, K.; Zhu, A.; Wang, B.; Di, C.; Yu, J.; Zhu, B. Characteristics of heat resistant aluminum alloy composite core conductor used in overhead power transmission lines. *Materials* **2020**, *13*, 1592. [CrossRef] [PubMed]
9. Rajadurai, R.S.; Lee, J.-H. High temperature sensing and detection for cementitious materials using manganese violet pigment. *Materials* **2020**, *13*, 993. [CrossRef] [PubMed]
10. Puchala, K.; Szymczyk, E.; Jachimowicz, J.; Bogusz, P.; Salacinski, M. The influence of selected local phenomena in CFRP laminate on global characteristics of bolted joints. *Materials* **2019**, *12*, 4139. [CrossRef] [PubMed]
11. Gautam, B.G.; Xiang, Y.; Liao, X.; Qiu, Z.; Guo, S. Experimental investigation of a slip in high-performance steel-concrete small box girder with different combinations of group studs. *Materials* **2019**, *12*, 2781. [CrossRef] [PubMed]
12. Huang, Y.; Zhang, Z.; Li, C.; Wang, J.; Li, Z.; Mao, K. Sound radiation of orthogonal antisymmetric composite laminates embedded with pre-strained SMA wires in thermal environment. *Materials* **2020**, *13*, 3657. [CrossRef] [PubMed]
13. Krzyzak, A.; Kosicka, E.; Borowiec, M.; Szczepaniak, R. Selected tribological properties and vibrations in the base resonance zone of the polymer composite used in the aviation industry. *Materials* **2020**, *13*, 1364. [CrossRef] [PubMed]

14. Li, X.; Lin, Y.; Lu, F. Numerical simulation on in-plane deformation characteristics of lightweight aluminum honeycomb under direct and indirect explosion. *Materials* **2019**, *12*, 2222. [CrossRef] [PubMed]
15. Bai, H.; Zhan, Z.; Liu, J.; Ren, Z. From local structure to overall performance: An overview on the design of an acoustic coating. *Materials* **2019**, *12*, 2509. [CrossRef] [PubMed]
16. Shan, L.; Wang, X.; Wang, Y. Extension of solid solubility and structural evolution in nano-structured Cu-Cr solid solution induced by high-energy milling. *Materials* **2020**, *13*, 5532. [CrossRef]
17. Li, P.; Meng, L.; Wang, S.; Wang, K.; Sui, Q.; Liu, L.; Zhang, Y.; Yin, X.; Zhang, Q.; Wang, L. In situ formation of Ti47Cu38Zr7.5Fe2.5Sn2Si1Nb2 amorphous coating by laser surface remelting. *Materials* **2019**, *12*, 3660. [CrossRef]
18. Niu, Y.; Jis, S.; Liu, Q.; Tong, S.; Li, B.; Ren, Y.; Wang, B. Influence of effective grain size on low temperature toughness of high-strength pipeline steel. *Materials* **2019**, *12*, 3672. [CrossRef]
19. Ding, H.; Huang, Q.; Liu, P.; Bao, Y.; Chai, G. Fracture toughness, breakthrough morphology, microstructural analysis of the T2 copper-45 steel welded joints. *Materials* **2020**, *13*, 488. [CrossRef] [PubMed]
20. Misiūnaitė, I.; Gribniak, V.; Rimkus, A.; Jakubovskis, R. The efficiency of utilisation of high-strength steel in tubular profiles. *Materials* **2020**, *13*, 1193. [CrossRef] [PubMed]

Publisher's Note: MDPI stays neutral with regard to jurisdictional claims in published maps and institutional affiliations.

Article

Effect of Steel Fibers on the Hysteretic Performance of Concrete Beams with Steel Reinforcement—Tests and Analysis

Violetta K. Kytinou, Constantin E. Chalioris *, Chris G. Karayannis and Anaxagoras Elenas

Laboratory of Reinforced Concrete and Seismic Design of Structures, Civil Engineering Department, Faculty of Engineering, Democritus University of Thrace (D.U.Th.), 67100 Xanthi, Greece; vkytinou@civil.duth.gr (V.K.K.); karayan@civil.duth.gr (C.G.K.); elenas@civil.duth.gr (A.E.)
* Correspondence: chaliori@civil.duth.gr; Tel.: +30-2541-079-632

Received: 5 June 2020; Accepted: 29 June 2020; Published: 29 June 2020

Abstract: The use of fibers as mass reinforcement to delay cracking and to improve the strength and the post-cracking performance of reinforced concrete (RC) beams has been well documented. However, issues of common engineering practice about the beneficial effect of steel fibers to the seismic resistance of RC structural members in active earthquake zones have not yet been fully clarified. This study presents an experimental and a numerical approach to the aforementioned question. The hysteretic response of slender and deep steel fiber-reinforced concrete (SFRC) beams reinforced with steel reinforcement is investigated through tests of eleven beams subjected to reversal cyclic loading and numerical analysis using 3D finite element (FE) modeling. The experimental program includes flexural and shear-critical SFRC beams with different ratios of steel reinforcing bars (0.55% and 1.0%), closed stirrups (from 0 to 0.5%), and fibers with content from 0.5 to 3% per volume. The developed nonlinear FE numerical simulation considers well-established relationships for the compression and tensional behavior of SFRC that are based on test results. Specifically, a smeared crack model is proposed for the post-cracking behavior of SFRC under tension, which employs the fracture characteristics of the composite material using stress versus crack width curves with tension softening. Axial tension tests of prismatic SFRC specimens are also included in this study to support the experimental project and to verify the proposed model. Comparing the numerical results with the experimental ones it is revealed that the proposed model is efficient and accurately captures the crucial aspects of the response, such as the SFRC tension softening effect, the load versus deformation cyclic envelope and the influence of the fibers on the overall hysteretic performance. The findings of this study also reveal that SFRC beams showed enhanced cyclic behavior in terms of residual stiffness, load-bearing capacity, deformation, energy dissipation ability and cracking performance, maintaining their integrity through the imposed reversal cyclic tests.

Keywords: steel fiber-reinforced concrete (SFRC); cyclic tests; reinforced concrete; direct tension tests; hysteretic response; tension softening; smeared crack model; residual stiffness; finite element (FE) analysis; shear; flexure; numerical analysis

---

1. Introduction

Short discrete fibers are used as mass reinforcement in concrete structural members to enhance tensile characteristics and to control crack width by the crack-bridging phenomenon observed at a local crack. The structural advances of fiber-reinforced cement-based members depend mainly on the content and the geometrical and mechanical properties of the fibers. Each fiber type influences in some particular function acting as crack arrestors due to their strength, bond, and pullout mechanisms across a crack surface. The addition of fibers to concrete has been shown to increase its toughness significantly and to promote more ductile material behavior [1–6]. The incorporation of randomly distributed short steel fibers has been developed as an alternative reinforcement technique to improve the brittle tensional failure and poor cracking performance of concrete by the debonding and pullout process of the fibers [7–9]. For this reason, deformed steel fibers with hooked ends exhibit anchoring action, providing increased bond characteristics that leads to a crucial enhancement of the post-cracking response, improving the energy-absorbing capability [10–12].

Reinforced concrete (RC) structural members with a low shear span-to-depth ratio, such as deep beams, are shear-vulnerable exhibiting brittle failure due to the low tensile properties of concrete [13]. The favorable influence of fiber-reinforced concrete inspired researchers to study the application of fibers in shear-critical RC beams instead of conventional transverse steel reinforcement [14–16]. In beams, the addition of steel fibers improves the concrete's diagonal tension capacity, leading to increased shear resistance, which can promote flexural failure and ductility [17,18]. Although the full replacement of conventional steel stirrups with fibers proved to be rather difficult, at least a partial replacement of closed stirrups with steel fibers leading to desirable reduce of conventional reinforcement congestion could be possible under certain circumstances [19–21]. The replacement of steel stirrups is important in shear-deficient RC joints [22–25], columns [26,27], deep [28] and torsional beams [29], where design criteria usually require a high amount of transverse reinforcement leading to dense placing of stirrups or the installation of cumbersome reinforcing systems. In this direction, a significant contribution is the recent analytical models that have been proposed to predict the shear capacity of steel fiber-reinforced concrete (SFRC) beam with [30,31] or without stirrups [32–34].

Reinforced concrete is capable of bearing some tensile stresses between cracks; this effect is called tension stiffening and is responsible for increased tensile stiffness in the RC member before the yielding of reinforcement [35]. Rigidity, deflections and crack widths may be greatly affected by this effect under the service limit state. The inclusion of steel fibers in the concrete matrix can reduce control crack splitting and significantly enhances residual stiffness, as SFRC can bear tensile stresses along the cracks [36–40]. However, the very restricted usage of SFRC in structural applications is mostly related to the difficulties in determining accurate and logical approaches that represent and estimate the performance of the material in either service or ultimate limit state. At flexure ultimate limit state, the tensile strength of SFRC members is usually disregarded [41,42]. Nevertheless, the member is stiffer as SFRC can bear tensile stresses both across cracks and among them. Thus, as strain increases and cracks coalesce, adequate residual tensile stresses are still developed since SFRC utilizes the fiber crack bridging effect, the tension stiffening attributed to steel reinforcement bond with concrete and the fracture mechanics of SFRC. This residual stiffness must be included in the calculations for the design of SFRC structural members, as it has been observed that the residual stresses get increased by adding higher amounts of steel fibers in the concrete mixture [43,44].

It is known that steel fibers and conventional steel reinforcing bars with stirrups are usually combined. In such cases, the tensile stresses that are being developed at a crack are distributed to the reinforcing bars and steel fibers bridging the crack. This way, the added steel fibers enhance the residual stiffness, provide crack control and enables the usage of higher strength steel reinforcement while retaining the control of crack widths based on the type and amount of steel fibers added [45,46]. These effects have become vital in determining cracking processes at service loads and in developing accurate constitutive models of cracked SFRC, which can be used in an analysis to predict member behavior [47,48].

It is recognized that the effective design of SFRC structures against complex external actions, such as earthquake, fatigue, explosion, and other types of loads, require a clear understanding of SFRC's mechanical behavior under monotonic and cyclic loads. Cyclic loading is one of the most complex loading conditions, both in structural performance and from a modeling point of view. Researchers have studied the fatigue life of fiber-reinforced high-strength concretes to determine the number of cycles that the specimen can withstand [49,50]. Others have studied cyclic tensile behavior of SFRC to reveal the underlying damage mechanism and SFRC specimens subjected to cyclic compressive load to a variety of high-stress range and high strain rate [51–53]. There are only a few cyclic experimental tests on SFRC beams under flexure [54–59]. Besides, while there is extensive research on the shear behavior of SFRC subject to monotonic loading, there is minimal research on the behavior of shear-critical SFRC structural members subject to reverse cyclic loading [60–62].

From the above literature review, it can be summarized that although some attempts have been made to capture the cyclic response of SFRC, broad experimental studies on the reverse cyclic behavior of SFRC have still been limited. The cyclic stress–strain relation and the damage evolution law of SFRC materials are some of the critical aspects in uncovering the realistic failure process as well as probing the variation of the structural responses of SFRC structures.

The present study aims to contribute to the ongoing research on SFRC by examining the significance of considering the tension softening and residual stiffness effect in SFRC beams subjected to reverse cyclic loading failing in flexure or shear. The precise definition of how this effect progresses with the number of cycles has not yet been adequately investigated as far as SFRC is concerned. For this purpose, a research program has been carried out herein, combining experimental investigation and numerical modeling. Precise finite element (FE) simulation that has been established with test data enables to explore even more the parameters affecting and to solve difficult structural problems [63–65].

The experimental part of this study includes two series of concrete beams with conventional steel reinforcing bars, stirrups, and short steel fibers. The first series consists of two slender SFRC beams tested under displacement control cyclic loading conducted by Chalioris et al. [66]. The second series includes nine deep SFRC beams subjected to a force control cyclic reversal loading for the purposes of this study. This test project brings new data concerning the improvement of the hysteretic behavior of realistic beams under reversal loading due to the addition of steel fibers. That might broaden the application of SFRC to structures in regions with high seismic activity.

Furthermore, a computationally efficient simulation using ABAQUS [67] software that takes into account the nonlinearities of the SFRC is also presented. The developed nonlinear FE analysis allows for an accurate prediction of the overall hysteretic response of realistic SFRC members. Comparisons between test and numerical results showed the feasibility of the proposed approach to simulate the response of flexural and shear-critical SFRC beams subjected to cyclic deformations. The effect of steel fibers on the overall performance and cracking behavior is also presented and discussed.

2. Materials and Methods

2.1. Experimental Investigation

The experimental part of this study includes two series of concrete beams with conventional steel reinforcing bars, stirrups, and short steel fibers. The first series consists of two slender SFRC beams tested under displacement control cyclic loading that was conducted by Chalioris et al. [66]. The second series includes nine deep SFRC beams subjected to a force control cyclic reversal loading for the purposes of this study. Details of the beam specimens are presented in this section.

2.1.1. Characteristics of the Beam Specimens

Beams are sorted in three groups ("FL", "SH-s" and "SH") as shown in Table 1 and described below:

Table 1. Properties of the tested beams.

| Group | Beam Name | Geometrical Data | | | | | Steel Reinforcement (Bars and Stirrups) | | | | | Steel Fiber Characteristics * | | |
		b (mm)	h (mm)	d (mm)	a_s (mm)	a_s/d	$A_{s1} = A_{s2}$ (Ø in mm)	$\rho_{l1} = \rho_{l2}$ (%)	$\varnothing_w/s$ (mm/mm)	ρ_w (%)	f_y (MPa)	V_{SF} (%)	l_{SF}/d_{SF} (mm/mm)	F
"FL"	FL0.3	200	200	170	1000	5.9	3Ø12	1.00	Ø8/200	0.25	590	1.00	44/1.0	0.3
	FL1.0	200	200	170	1000	5.9	3Ø12	1.00	Ø8/200	0.25	590	3.00	44/1.0	1.0
"SH-s"	SH0-s37	100	300	275	550	2.0	3Ø8	0.55	Ø8/275	0.37	575	–	–	–
	SH0-s50	100	300	275	550	2.0	3Ø8	0.55	Ø8/200	0.50	575	–	–	–
	SH0.3-s37	100	300	275	550	2.0	3Ø8	0.55	Ø8/275	0.37	575	0.50	60/0.8	0.3
	SH0.3-s50	100	300	275	550	2.0	3Ø8	0.55	Ø8/200	0.50	575	0.50	60/0.8	0.3
"SH"	SH0	100	300	275	550	2.0	3Ø8	0.55	–	–	575	–	–	–
	SH0.3	100	300	275	550	2.0	3Ø8	0.55	–	–	575	0.50	60/0.8	0.3
	SH0.4	100	300	275	550	2.0	3Ø8	0.55	–	–	575	0.75	60/0.8	0.4
	SH0.6	100	300	275	550	2.0	3Ø8	0.55	–	–	575	1.00	60/0.8	0.6
	SH0.8	100	300	275	550	2.0	3Ø8	0.55	–	–	575	1.50	60/0.8	0.8

* Hooked steel fibers with $f_{SF} = 1000$ MPa.

Group "FL" includes the two slender beams (2500 mm long) failed in flexure (specimens FL0.3 and FL1.0) containing 1% and 3% steel fibers per volume fraction, V_{SF}, respectively, that correspond to 80 kg and 240 kg per cubic meter of concrete, respectively. The added steel fibers are hooked-ended with a length to diameter ratio (aspect ratio) of l_{SF}/d_{SF} = 44 mm/1 mm = 44, a bond factor of β = 0.75 and, therefore, the fiber factor of beams FL0.3 and FL1.0 is F = 0.3 and 1.0, respectively. The geometrical and the reinforcement characteristics of these flexural beams are presented in Table 1 and further details can also be found in the recent study of Chalioris et al. [66].

All deep beams (nine shear-critical specimens) are 1600 mm long, have the same width to height ratio b/h = 100/300 mm, effective depth d = 275 mm, shear span a_s = 550 mm, shear span to effective depth ratio a_s/d = 2 and three longitudinal steel reinforcing bars with a diameter of 8 mm at the top and at the bottom with geometrical longitudinal reinforcement ratio ρ_{l1} = ρ_{l2} = 0.55% and f_y = 575 MPa. Deep beams are sorted in two groups "SH-s" and "SH", with and without closed steel stirrups, respectively.

Group "SH-s" includes four deep beams (specimens SH0-s37, SH0-s50, SH0.3-s37 and SH0.3-s50) with stirrups of 8 mm diameter at a uniform spacing of 275 and 200 mm that corresponds to a geometrical web (transverse) reinforcement ratio of ρ_w = 0.37% (specimens SH0-s37 and SH0.3-s37) and ρ_w = 0.50% (specimens SH0-s50 and SH0.3-s50), respectively. Beams SH0-s37 and SH0-s50 are made of plain concrete (reference specimens), whereas beams SH0.3-s37 and SH0.3-s50 contain steel fibers with V_{SF} = 0.5%.

Group "SH" consists of five deep beams without stirrups (specimens SH0, SH0.3, SH0.4, SH0.6 and SH0.8) that contain steel fibers with V_{SF} = 0 (plain concrete reference beam), 0.5%, 0.75%, 1.0% and 1.5%, respectively.

The steel fibers added in the deep beams are hooked ended with a length to diameter ratio (aspect ratio) of l_{SF}/d_{SF} = 60 mm/0.8 mm = 75, bond factor β = 0.75 and, therefore, the fiber factor of the SFRC beams of group "SH-s" is F = 0.3 and of group "SH" is F = 0.3, 0.4, 0.6 and 0.8, as shown in Table 1. Geometrical and reinforcement characteristics of all the examined beams are also summarized in Table 1 and illustrated in Figure 1. Details concerning the preparation stages, the mixture procedure and the curing of steel fiber reinforced concrete specimens can be found in [66].

2.1.2. Test Rig and Loading Histories

Beams were imposed to full-cycle deformations and subsequently to increasing load till failure. The test setup is shown in Figure 1a,b for the slender and the deep beams, respectively. A four-point-bending test setup was implemented for the cyclic reversal loading of the beams that were simply edge-supported on roller supports 2200 mm (slender beams) and 1450 mm (deep beams) apart in a rigid laboratory frame.

The imposed load was applied in two points in the mid-span of the beams and therefore the shear span, a_s, is equal to 1000 mm (slender beams) and 550 mm (deep beams) with span-to-depth ratio: a_s/d = 5.9 (slender beams) and 2.0 (deep beams). The load was measured by a load cell with 0.05 kN accuracy and net midspan deformations by LVDTs with 0.01 mm accuracy that were placed in the midspan of the beams and the supports. The load and corresponding deflection measurements were recorded continuously during the performed cyclic tests.

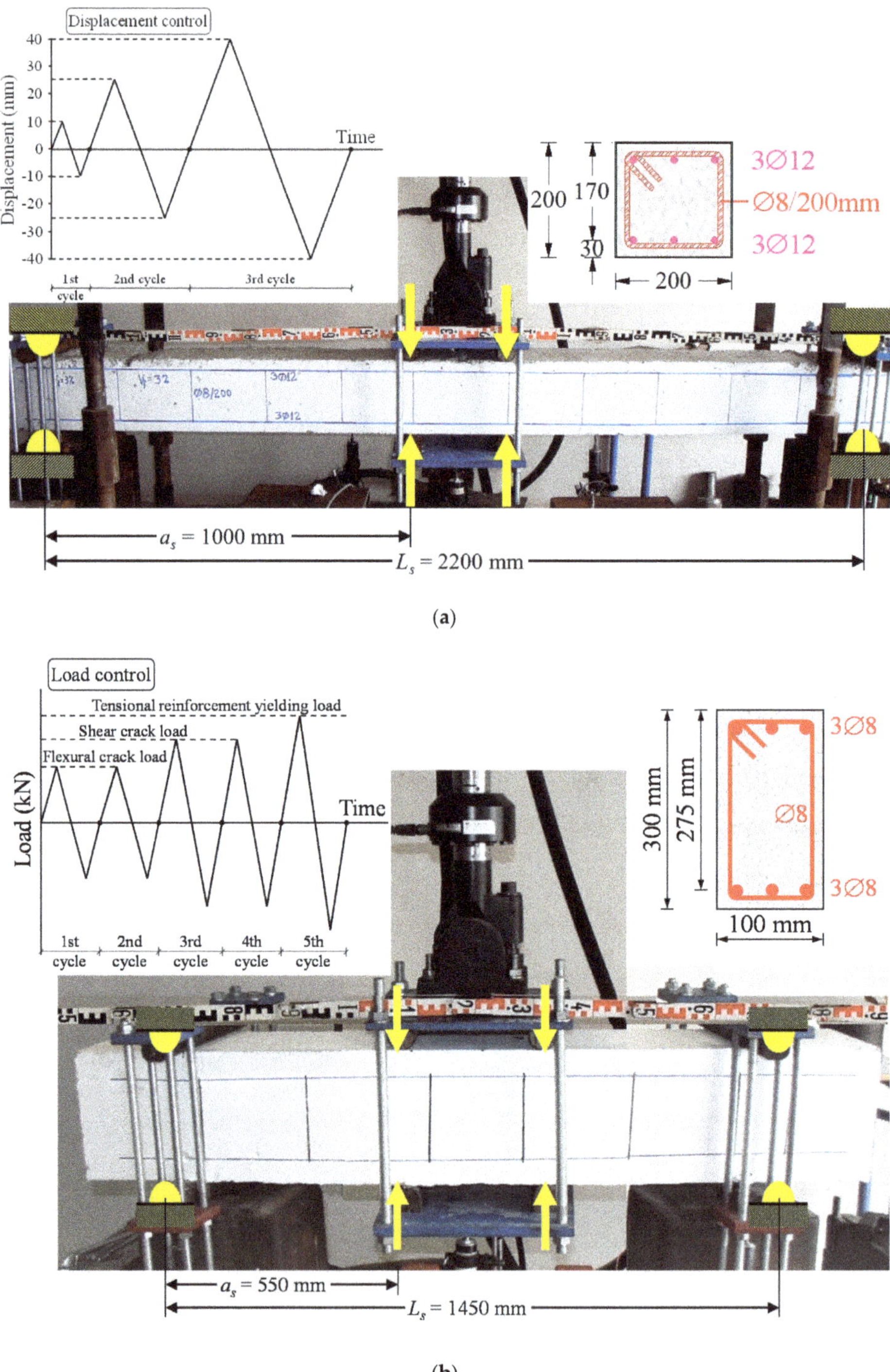

(**a**)

(**b**)

Figure 1. Geometry, reinforcement and cyclic reversal loading sequence of the tested specimens (dimensions in mm): (**a**) slender (flexural) beams; (**b**) deep (shear-critical) beams.

Beams of group "FL" (flexural specimens) were tested under increasing cyclic loading history that includes three loading steps as shown in Figure 1a. The loading history of the deep beams of group "SH" (specimens without stirrups) and the beams of group "SH-s" (specimens with stirrups) includes three and five full loading cycles, respectively, as shown in Figure 1b. The first two loading cycles correspond to the load for the initiation of the flexural crack, the next two to the load for the onset of the inclined shear cracking and the last one almost to the load for the steel yielding of the tension bars of the reference beams SH0, SH0-s37 and SH0-s50 (specimens without steel fibers).

2.1.3. Properties of the SFRC

The plain concrete and SFRC mixture used consisted of an ordinary general-purpose Portland-type cement (type CEM II 32.5 N, Greek type pozzolan cement with 10% fly ash), high fineness modulus sand, crushed stone aggregates with a maximum size of 16 mm (for the slender beams of group "FL") and 9.5 mm (for the deep beams of groups "SH-s" and "SH") and a water-to-cement ratio of 0.55 and 0.57, respectively. Uniaxial compression and tension tests were also performed to measure the compressive strength and the full tensile behavioral curves of the SFRC mixtures at the day of the cyclic tests of the beams (after 28 days of curing).

Three standard concrete cylinders with dimensions diameter to height = 150/300 mm were cast from each plain concrete and SFRC mixture and tested under axial compression using a universal testing machine (UTM) with an ultimate capacity of 3000 kN. The average compressive strength of the concrete without steel fibers is 27 MPa, whereas the average compressive strength, $f_{c,SF}$, of each SFRC beam is presented in Table 2.

Table 2. Compressive and tensile properties of the steel fiber-reinforced concrete (SFRC) of each tested beam.

Group	Beam Name	$E_c = E_{t,SF}$ (GPa)	$f_{c,SF}$ (MPa)	$\varepsilon_{cu,SF}$ (mm/m)	$f_{t,SF}$ (MPa)	k_f	$G_{f,SF}$ (N/mm)	$\varepsilon_{to,SF}$ (mm/m)	k_w
"FL"	FL0.3	29.336	25.51	3.59	3.30	0.19	2.001	0.113	0.01
	FL1.0	30.442	27.25	8.47	4.39	0.43	8.561	0.144	0.01
"SH-s"	SH0.3-s37	30.055	28.76	3.43	2.39	0.22	0.655	0.080	0.04
	SH0.3-s50	30.055	28.76	3.43	2.39	0.22	0.655	0.080	0.04
"SH"	SH0.3	30.055	28.76	3.43	2.39	0.22	0.601	0.080	0.04
	SH0.4	30.264	29.64	4.33	2.44	0.33	0.892	0.081	0.04
	SH0.6	30.474	30.52	5.33	2.69	0.40	1.397	0.088	0.04
	SH0.8	30.893	32.27	7.64	2.79	0.57	2.084	0.090	0.04

Concerning the tensile behavior of SFRC, three prismatic specimens were cast from each SFRC mixture and tested under axial tension. Figure 2 illustrates the geometry of the specimens and the direct tension test setup. The extension rate was 0.02 mm/m/sec and tensile deformations and crack width were monitored through four LVDTs with 0.001 mm accuracy that were placed symmetrically on two steel hoops fixed on the wide edges of the specimen. It is noted that tensile cracking and final failure occurred in the middle of the gauge length area of the SFRC notched specimens, away from the edge-mounted clamps, as shown in Figure 2.

The experimentally measured stress versus crack width behavior of each SFRC mixture under direct tension are presented in Figure 3a,b for the slender and the deep beams, respectively. For comparison reasons, the predictions of the proposed smeared crack tensional model in stress versus crack width curves for each SFRC beam are also illustrated in Figure 3a,b and compared the test results. The initial tensile behavior before cracking was elastic and linear to the point of the SFRC tensile strength, $f_{t,SF}$, with elastic modulus under tension, $E_{t,SF}$ (see also Table 2). The main SFRC variables derived from the tests and the proposed model are also presented in Table 2 for each beam.

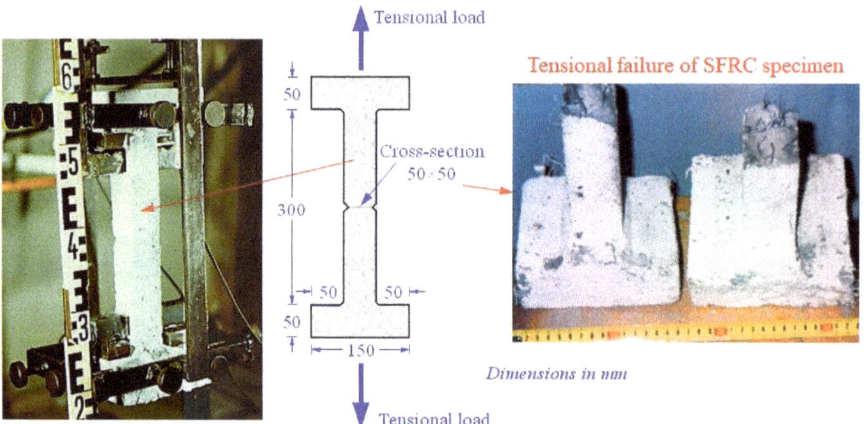

Figure 2. Prismatic SFRC specimen tested and failed under axial tension test.

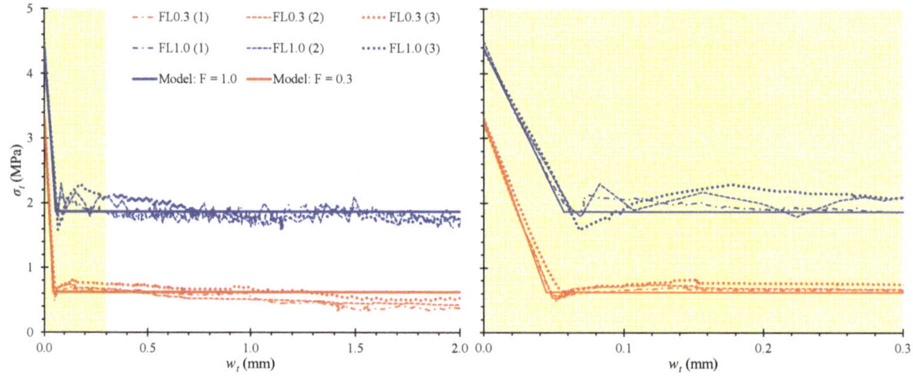

(a)

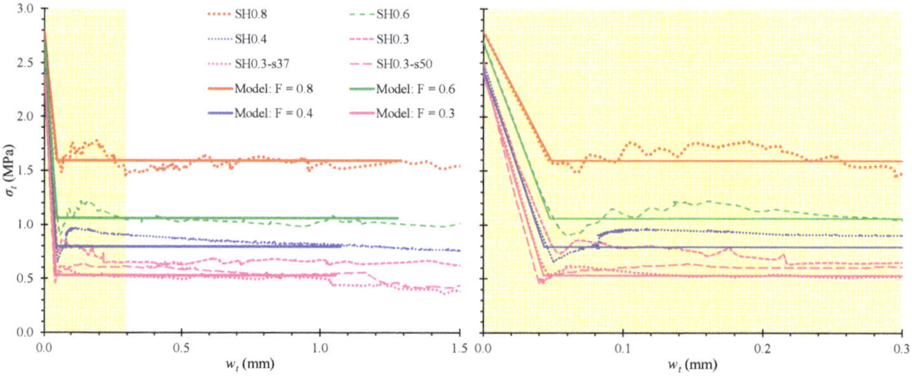

(b)

Figure 3. Tensional stress versus crack width (σ_t - w_t) behavioral curves obtained from axial direct tension tests and predicted from the proposed model: (**a**) SFRC mixtures of the slender beams (three tensional specimens from each SFRC mixture); (**b**) SFRC mixtures of the deep beams.

The experimental results of the direct tension tests indicate that steel fibers substantially influence the tensile behavior after cracking. The descending post-peak part of the tensile response, especially the residual stress versus crack width curve, depends on the fiber factor, F, of the fibers added in the mixtures. The test curves indicate that SFRC with a higher content of steel fibers (see for example beams FL1.0 and SH0.8 with V_{SF} = 3% and 1.5%, respectively, and F = 1.0 and 0.8, respectively) demonstrate higher post-cracking stress with regard to the corresponding SFRC mixtures with lower steel fiber content (see for example beams FL0.3 and SH0.3 with V_{SF} = 1% and 0.5%, respectively and F = 0.3, both of them).

It is also deduced that the fiber factor, F, is a more efficient parameter than the volume fraction, V_{SF}, for the evaluation of the steel fiber contribution on the residual stress since SFRC mixtures with the same fiber factor F = 0.3 and different steel fiber content, such as beam FL0.3 with V_{SF} = 1% and beams SH0.3, SH0.3-s37 and SH0.3-s50 with V_{SF} = 0.5% exhibit more or less the same post-cracking stress (approximately 0.6 MPa), as shown in Figure 3a,b.

Furthermore, the comparison between the experimental and the analytical diagrams of Figure 3a,b reveals that the stress versus crack width curves derived from the proposed smeared crack analysis that takes into account the tension softening phenomenon are in very good compliance with the test results.

2.2. Proposed Model, Constitutive Relationships of the Materials and Nonlinear FE Analysis

A nonlinear 3D FE analysis has been performed to predict the response of slender and deep SFRC beams with steel reinforcement subjected to reversal cyclic loading. This way, the efficacy and accuracy of the developed FE model are checked using experimental data of SFRC beams failing in flexure and in shear. The performed analysis adopts properly modified constitutive laws of the materials that consider the influence of the added steel fibers to the compressive and to the tensile behavior of SFRC with tension softening and residual stiffness effect.

The constitutive laws of SFRC under reversal compression and tension are based on test results and models that have been addressed by the authors in previous relative studies. Especially for the simulation of the tensional response of SFRC, a smeared crack analysis with tension softening is adopted. This model uses the fracture characteristics of the material taking into account stress versus crack width constitutive laws with post-cracking descending part (softened tensional behavior). The post-cracking response of SFRC under tension near the reinforcing bars is simulated by a residual stiffness approach that combines the interaction of steel fibers in the cracked concrete regions, the bond performance of steel reinforcing bars corresponding to the tension stiffening effect, the reinforcement characteristics and the fracture mechanics of SFRC. Furthermore, the performed FE analysis utilizes the concrete damaged plasticity (CDP) approach [68] to simulate the behavior of SFRC.

2.2.1. SFRC under Reversal Compression

It is known that steel fibers become more effective after cracking and, therefore, they mainly improve the post-peak compressive response of SFRC elements. The content, the geometrical and the bond properties of the added fibers are the main parameters that influence the enhanced post-cracking behavior of SFRC under compression [69–71]. The ultimate stress is slightly increased due to the presence of steel fibers since their progressive debonding failure improves the crack growth resistance of the material that undergoes after the compressive strength. Most of the proposed models of the literature that simulate the SFRC under compression are based on test results and proper regression analysis [72–76].

The proposed model for SFRC under reversal compression is illustrated in Figure 4. It can simulate the observed properties of the material behavior, such as the cracking and crushing of SFRC, the accumulation of damage (fracture), and the degradation of stiffness under cyclic loading. The SFRC constitutive law can be defined using multiple points on the compressive stress-strain (σ_c - ε_c). A user-defined damage curve is implemented in the proposed FE analysis to account for the gradual SFRC stiffness degradation as the cracks spread. When the unloading takes place after the first crack formulation, the unloading branch stiffness is equivalent to the elastic stiffness, decreased by a factor, d_c, which considers the degradation due to damage. A substantial decrease in the stiffness is observed up to the full closure of the crack. The crack closure stage ends when the elastic displacement is restored. At this point, the elastic stiffness is also restored, and tension is performed.

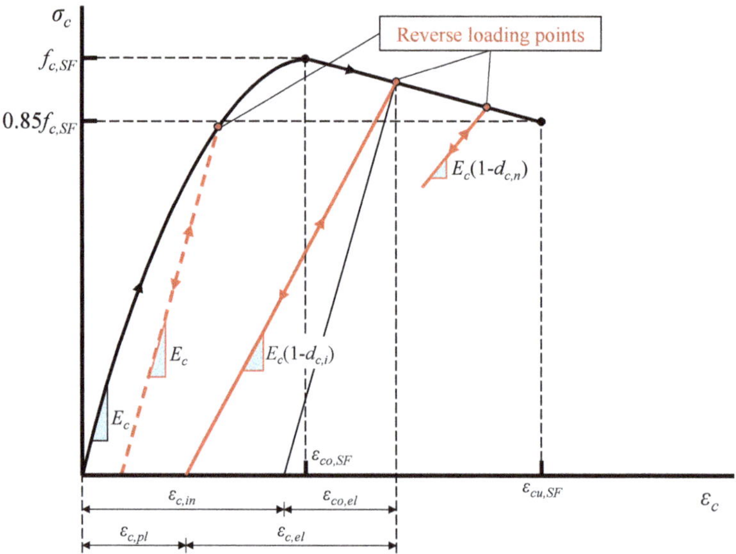

Figure 4. Model of SFRC under reversal compression.

The uniaxial compressive behavior of SFRC in relation to the CDP model can be formulated using stress versus plastic strain curves. The stress and the strain parameters of the proposed reversal compressive response shown in Figure 4 are calculated as follows:

$$\sigma_c = (1 - d_c) E_c \left(\varepsilon_c - \varepsilon_{c,pl} \right), \tag{1}$$

$$\varepsilon_{c,pl} = \varepsilon_{c,in} - \frac{d_c}{1 - d_c} \varepsilon_{co,el}, \tag{2}$$

$$\varepsilon_{c,in} = \varepsilon_c - \varepsilon_{co,el}, \tag{3}$$

$$\varepsilon_{co,el} = \sigma_c / E_c, \tag{4}$$

where E_c is the initial elastic modulus of SFRC under compression and d_c is the compressive plastic damage factor that takes values $0 \leq d_c \leq 1$: 0 for the undamaged SFRC and 1 for the complete loss of the SFRC compressive strength, $f_{c,SF}$:

$$d_c = 1 - \sigma_c / f_{c,SF}, \tag{5}$$

The analytical formulation of the compressive stress–strain behavior of SFRC adopted in this study has been derived from test data of 125 stress versus strain curves and 257 strength values [77]. The ascending part of the compressive behavior until the ultimate strength of commonly used SFRC with $f_{c,SF} \leq 50$ MPa is expressed by (see also Figure 4):

$$\sigma_c = f_{c,SF} \left[1 - \left(1 - \frac{\varepsilon_c}{\varepsilon_{co,SF}} \right)^2 \right], \tag{6}$$

where $\varepsilon_{co,SF}$ is the strain corresponding to the maximum compressive stress, $f_{c,SF}$, of the SFRC.

Concerning the post-peak compressive behavior, a linear descending part is considered from the ultimate strength, $f_{c,SF}$, until the value of $0.85f_{c,SF}$ is obtained. The SFRC compressive strength, $f_{c,SF}$, the corresponding strain, $\varepsilon_{co,SF}$, and the strain, $\varepsilon_{cu,SF}$, corresponding to the stress value $0.85f_{c,SF}$ are estimated by [77]:

$$f_{c,SF} = f_c(0.2315F + 1), \tag{7}$$

$$\varepsilon_{co,SF} = \varepsilon_{co}(0.95F + 1), \tag{8}$$

$$\varepsilon_{cu,SF} = \varepsilon_{co,SF}(1.40F + 1), \tag{9}$$

where f_c is the compressive strength of plain concrete, ε_{co} is the corresponding strain that is usually equal to 0.002, F is the fiber factor $F = \beta V_{SF}(l_{SF}/d_{SF})$, where β is a bond factor (taken 0.50 for round, 0.75 for deformed and 1.0 for indented fibers), V_{SF} is the volume fraction of the steel fibers and l_{SF} and d_{SF} are their length and diameter, respectively.

2.2.2. SFRC under Reversal Tension

The performance of concrete under tension can be substantially improved by the addition of steel fibers since SFRC exhibits increased tensile strength and mainly post-peak deformation capability showing pseudo-ductile behavior due to the gradual debonding failure of the fibers [78]. Various analytical stress versus strain expressions have been proposed to simulate the SFRC tensile behavior [79–81]. In this study, a smeared crack model for plain concrete with tension softening that has been addressed and experimentally verified by the authors [82,83] is adopted. This model has properly been modified to simulate the favorable influence of steel fibers in SFRC under tension. Smeared crack approaches have also been used to evaluate the uncertainty of crack width in large-scale RC beams [84]. Furthermore, the uniaxial tensile response of SFRC in relation to the CDP model can be formulated using stress versus plastic strain curves. The parameters used in the proposed reversal tensile behavior shown in Figure 5 are calculated as follows:

$$\sigma_t = (1 - d_t)E_{t,SF}\left(\varepsilon_t - \varepsilon_{t,pl}\right), \tag{10}$$

$$\varepsilon_{t,pl} = \varepsilon_{t,cr} - \frac{d_t}{1 - d_t}\varepsilon_{to,el}, \tag{11}$$

$$\varepsilon_{t,cr} = \varepsilon_t - \varepsilon_{to,el}, \tag{12}$$

$$\varepsilon_{to,el} = \sigma_t/E_{t,SF}, \tag{13}$$

where $E_{t,SF}$ is the initial elastic modulus of SFRC under tension and d_t is the tensile damage factor, which takes values $0 \leq d_t \leq 1$: 0 for the undamaged SFRC and 1 for the complete loss of the SFRC tensile strength, $f_{t,SF}$:

$$d_t = 1 - \sigma_t/f_{t,SF}, \tag{14}$$

The analytical formulation of the proposed smeared crack approach for the post-cracking tensile behavior of SFRC utilizes stress versus crack width relationships. SFRC cracking takes place within a fracture process zone that is initiated at the tensile strength of SFRC, $f_{t,SF}$. The boundary of the strain-softening region and the SFRC characteristics define this zone, assuming that less damaged or even elastic parts coexist between the cracks of this fracture process zone. The total tensional strain, ε_t, is estimated as the sum of an elastic, $\varepsilon_{to,el}$, and a fracture component, $\varepsilon_{t,fr}$ (see also Figure 5):

$$\varepsilon_t = \varepsilon_{to,el} + \varepsilon_{t,fr}, \tag{15}$$

$$\varepsilon_{t,fr} = w_t/L_{fr,SF}, \tag{16}$$

where σ_t is the tensile stress, w_t is the crack width and $L_{fr,SF}$ is the fracture process zone length that can be taken equal to $3l_{sf}$ [85].

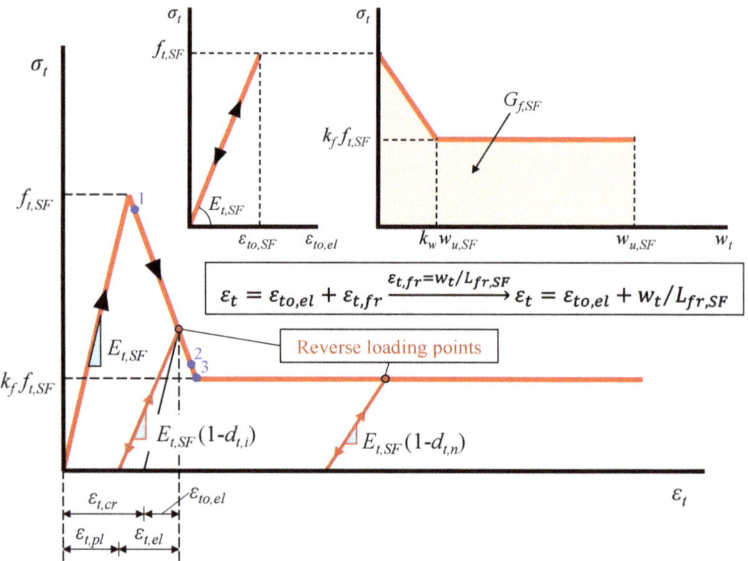

Figure 5. Model of SFRC under reversal tension.

The following equations describe the elastic component of the model (see also the linear σ_t - $\varepsilon_{to,el}$ diagram of Figure 5) [86]:

$$f_{t,SF} = \varepsilon_{to,SF}E_{t,SF},\tag{17}$$

$$\varepsilon_{to,SF} = 0.167n_{l,el}V_{SF}(f_{SF}/E_{SF} - f_t/E_t) + f_t/E_t,\tag{18}$$

$$E_{t,SF} = \frac{3}{8}[E_t(1 - V_{SF}) + E_{SF}V_{SF}] + \frac{5}{8}\left[\frac{E_{SF}E_t}{E_{SF}(1 - V_{SF}) + E_tV_{SF}}\right],\tag{19}$$

where $E_{t,SF}$ is the modulus of elasticity under tension of SFRC, f_t and E_t are the ultimate tensile strength and the elastic modulus under tension of the plain concrete, respectively, f_{SF} and E_{SF} are the tensile strength and the elastic modulus of the steel fiber, respectively, and $n_{l,el}$ is the ratio of the average elastic stress to the strength of the steel fiber that is usually equal to 0.5 [86].

The properties of the SFRC softening and fracture response define the parameters of the fracture components of the proposed smeared crack approach [87,88]. The fracture energy, $G_{f,SF}$, is the energy required for the cracking formation within the fracture process zone and for the full opening of one single crack [89]:

$$G_{f,SF} = \int_{f_{t,SF}}^{0} \sigma_t dw_t \xrightarrow{w_t=L_{fr,SF}\varepsilon_{t,fr}} G_{f,SF} = L_{fr,SF}\int_{f_{t,SF}}^{0} \sigma_t d\varepsilon_{t,fr},\tag{20}$$

The post-peak behavior shown in the σ_t - w_t curve of Figure 5 is defined by a linear descending part until the point of the maximum post-cracking tensional stress, $k_f f_{t,SF}$, and the corresponding crack width, $k_w w_{u,SF}$. After this point, the stress is assumed to have the following constant value until the ultimate crack width, $w_{u,SF}$: $\sigma_t = k_t f_{t,SF}$ ($w_t > k_w w_{u,SF}$). The fracture energy is the area under the curve of SFRC stress versus crack width (see also the bilinear σ_t - w_t diagram of Figure 5):

$$G_{f,SF} = f_{t,SF}w_{u,SF}(k_f + 0.5k_w - 0.5k_f k_w),\tag{21}$$

where the values of the coefficients k_f and k_w depend on the SFRC characteristics as [85]:

$$k_f = \frac{0.405 n_l \sigma_{fu} V_{SF}}{\varepsilon_{to,SF} E_{t,SF}},\tag{22}$$

$$k_w = \frac{(3 \div 8) L_{fr,SF} \varepsilon_{to,SF}}{w_{u,SF}}\tag{23}$$

where σ_{fu} is the maximum stress of the fiber when a uniform bond stress, τ_u, is assumed at the interface between the steel fiber and the concrete:

$$\sigma_{fu} = \left\{ \begin{array}{ll} 2\tau_u l_{SF}/d_{SF} & l_{SF} \leq l_{cr} \\ f_{SF} & l_{SF} > l_{cr} \end{array} \right\},\tag{24}$$

$$n_l = \left\{ \begin{array}{ll} 0.50 & l_{SF} \leq l_{cr} \\ 1 - \frac{l_{SF}}{2 l_{cr}} & l_{SF} > l_{cr} \end{array} \right\},\tag{25}$$

where l_{cr} is the length of the fiber in which the ultimate fiber stress is developed:

$$l_{cr} = 0.5 f_{SF} d_{SF}/\tau_u,\tag{26}$$

The fracture energy, $G_{f,SF}$, can been estimated by tension tests as a function of the known steel fiber factor, F, and the value of the fracture energy of the plain concrete, G_f [85]:

$$G_{f,SF} = G_f (104F + 1),\tag{27}$$

where the fracture energy of the plain concrete, G_f, can be calculated using linear relation from the tensile strength of concrete, f_t, to zero at the maximum crack width, w_u:

$$G_f = 0.5 f_t w_u,\tag{28}$$

$$w_u = \varepsilon_{tu,fr} L_{fr} \overset{\varepsilon_{tu,fr} = a_{fr}\varepsilon_{to}}{\longrightarrow} w_u = a_{fr} \varepsilon_{to} L_{fr} \overset{\varepsilon_{to} = f_t/E_t}{\longrightarrow} w_u = a_{fr} L_{fr} \frac{f_t}{E_t}\tag{29}$$

where a_{fr} is a coefficient that takes values from 5 to 8 for maximum aggregate size d_g = 32 to 8 mm, respectively [82], and L_{fr} is the plain concrete fracture process that can be taken as $3d_g$ [90]). This way, Equation (21) can be written as:

$$f_{t,SF} w_{u,SF} (k_f + 0.5 k_w - 0.5 k_f k_w) = 0.5 f_t a_{fr} L_{fr} \frac{f_t}{E_t} (104F + 1) \overset{L_{fr} = 3 d_g}{\longrightarrow},\tag{30}$$

$$w_{u,SF} = \frac{1.5 f_t^2 a_{fr} d_g (104F + 1)}{f_{t,SF} E_t (k_f + 0.5 k_w - 0.5 k_f k_w)}\tag{31}$$

The stress at each stage can be calculated as:

$$\sigma_t = \left\{ \begin{array}{ll} \varepsilon_t E_{t,SF} & \text{if } 0 < \varepsilon_t \leq \varepsilon_{to,SF} \\ f_{t,SF} \left(1 - \frac{1 - k_f}{k_w w_{u,SF}} w_t \right) & \text{if } 0 < w_t \leq k_w w_{u,SF} \\ k_f f_{t,SF} & \text{if } k_w w_{u,SF} < w_t \leq w_{u,SF} \end{array} \right\},\tag{32}$$

Furthermore, the following CDP-material-associated parameters define the inelastic behavior of SFRC [91] and their values used herein are presented in Table 3:

Table 3. Concrete damaged plasticity (CDP) model input parameters.

Parameter	Value
ψ	40°
K_c	2/3
σ_{b0}/σ_{c0}	1.16
$\in$	0.10
μ	0.0001

Where ψ is the dilatation angle affecting the plastic deformation, $\in$ is the flow potential eccentricity that defines the rate of the plastic potential hyperbolic to its asymptote, σ_{b0}/σ_{c0} is the ratio of the strength in the biaxial state to the strength in the uniaxial state, K_c is the ratio of the tensile to the compressive meridian and μ is the viscosity parameter.

2.2.3. Modeling of Steel Reinforcement

The cyclic response of the steel reinforcing bars and stirrups is derived by a superposition of several elastic and perfectly plastic models in parallel. This takes account of a nonlinear kinematic positive strain-hardening since plastic behavior is defined by the values of the yield strength and the corresponding plastic strain (f_y, ε_y), the ultimate strength and the corresponding strain (f_u, ε_u) and is characterized by permanent deformations. The values of the steel modulus of elasticity, E_s, and Poisson's ratio are also used in the FE analysis according to the test data of the steel reinforcement.

2.2.4. Element Types

SFRC was simulated by using 8-node 3-dimensional solid elements with reduced integration (C3D8R) to avoid the effect of shear locking. The 3-dimensional 2-node truss elements (T3D2) were selected for the simulation of steel reinforcement (longitudinal and stirrups). Every element's node has three degrees of freedom with x, y, and z (global coordinate system) translation, as depicted in Figure 6. The bond between reinforcement and concrete is modeled using the embedded process, and precisely the Abaqus feature "truss in solid" [85].

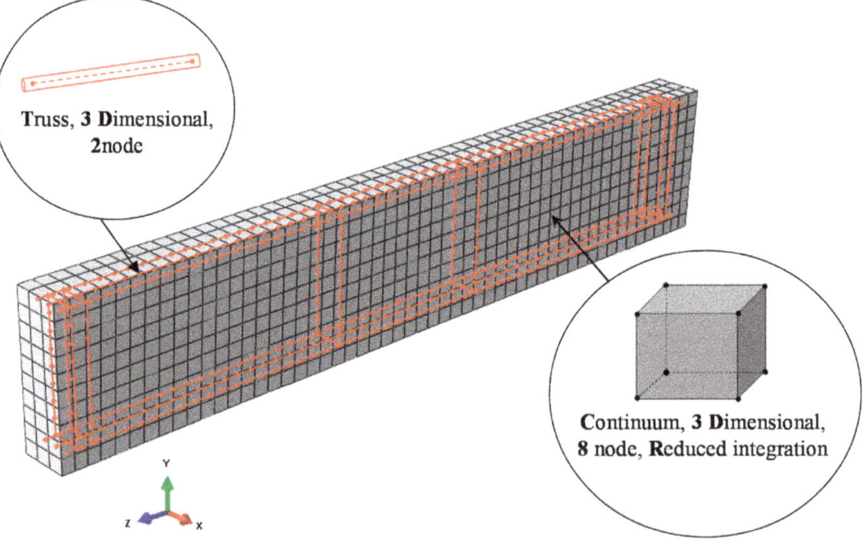

Figure 6. Finite element (FE) mesh of the SFRC and the steel reinforcement of the beam.

2.2.5. Boundary Conditions

The simulated beam's boundary conditions were adopted according to the experimental setup shown in Figure 7. The supports were positioned at a particular distance from each edge, while the edges remained free. At the left side, a line of nodes was constrained in the Ux, Uy, Uz directions, while at the right side only the Uy direction was constrained.

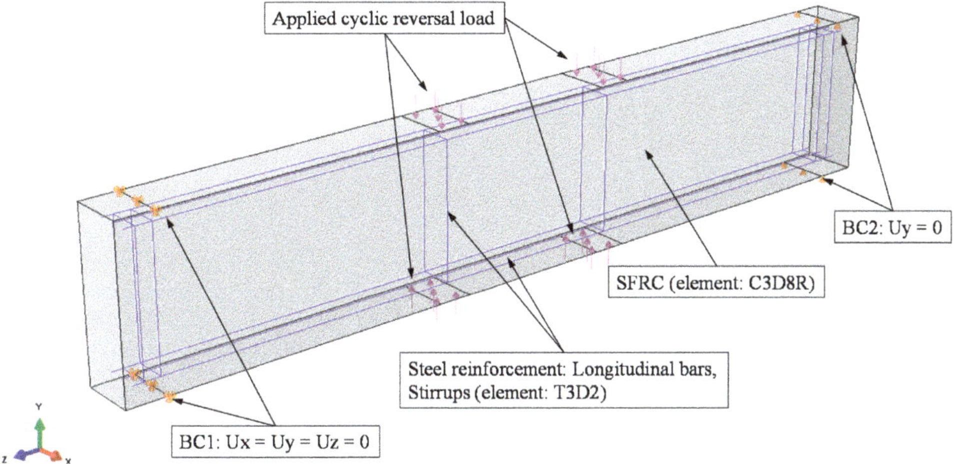

Figure 7. Boundary conditions and loading of the FE simulation of the beam.

2.2.6. Loading, Mesh and Convergence

The load was applied to the beam specimens in a quasi-static manner as a reverse cyclic loading, which implies that each cycle consisted of loading in both directions, as shown in Figure 7. Analysis up to load failure capacity of the tested SFRC beams was beyond the scope of this study. The slender beams were subjected to displacement control while the deep beams to load control cyclic histories. Furthermore, the load was applied constantly and smoothly to achieve a quasi-static solution and prevent any essential acceleration alteration through each iteration, which further ensures that the stress and displacement changes remain smooth.

Mesh size was selected based on the assumption that the distribution of SFRC cracking typically includes spatial scales between two to three dominant aggregate sizes of the concrete mixture [90]. The maximum aggregate size for the slender and the deep beams was 16 and 9.5 mm, respectively. Mesh sizes of 40 and 30 mm have been applied to all types of elements (truss and solid).

2.2.7. FE Simulation of the Tested Beams and Material Input

The simulations of the examined SFRC beam specimens were developed according to the geometrical and mechanical properties of the tested beams and the boundary conditions of the experimental setup. The aspects of the developed FE modeling are described in Sections 2.2.1–2.2.7 of this paper. Furthermore, the main characteristics of the materials for each specimen are presented in Tables 1 and 2. Supplementary input parameters are presented in Table 4. Stress, σ, crack width, w, and strain, ε, values at points 1, 2 and 3, and the corresponding plastic damage factors, d_t, of Table 4 are defined in the tensional behavioral model of the SFRC shown in Figure 5.

Table 4. Supplementary input variables for the simulation of the tested SFRC beams.

Beam Name	v	σ_{t1} (MPa)	w_{t1} (mm)	ε_{t1} (mm/m)	σ_{t2} (MPa)	w_{t2} (mm)	ε_{t2} (mm/m)	σ_{t3} (MPa)	w_{t3} (mm)	ε_{t3} (mm/m)	d_{t1}	d_{t2}	d_{t3}
FL0.3	0.212	2.56	0.012	0.09	1.07	0.034	0.26	0.62	0.045	0.34	0.225	0.676	0.811
FL1.0	0.237	3.69	0.013	0.10	2.29	0.041	0.31	1.87	0.053	0.40	0.159	0.478	0.573
SH0.3-s37	0.210	1.87	0.011	0.06	0.84	0.032	0.18	0.53	0.045	0.25	0.219	0.656	0.787
SH0.3-s50	0.210	1.87	0.011	0.06	0.84	0.032	0.18	0.53	0.045	0.25	0.219	0.656	0.787
SH0.3	0.210	1.87	0.011	0.06	0.84	0.032	0.18	0.53	0.043	0.24	0.216	0.648	0.778
SH0.4	0.216	1.98	0.011	0.06	1.07	0.032	0.18	0.80	0.041	0.23	0.187	0.561	0.673
SH0.6	0.215	2.24	0.011	0.06	1.33	0.034	0.19	1.06	0.045	0.25	0.168	0.504	0.605
SH0.8	0.231	2.46	0.011	0.06	1.79	0.032	0.18	1.59	0.043	0.24	0.119	0.357	0.428

3. Results and Discussion

3.1. Verification of the Model

The numerical results yielded from the developed nonlinear FE simulations are compared with the experimental data using load versus hysteretic deformation curves. Figures 8 and 9 clearly demonstrate the analytical and the test curves of the slender beams FL0.3 and FL1.0 with steel fiber factor $F = 0.3$ ($V_{SF} = 1\%$) and $F = 1.0$ ($V_{SF} = 3\%$), respectively, for each loading cycle (see the three diagrams of the 1st, 2nd and 3rd cycle in Figures 8 and 9).

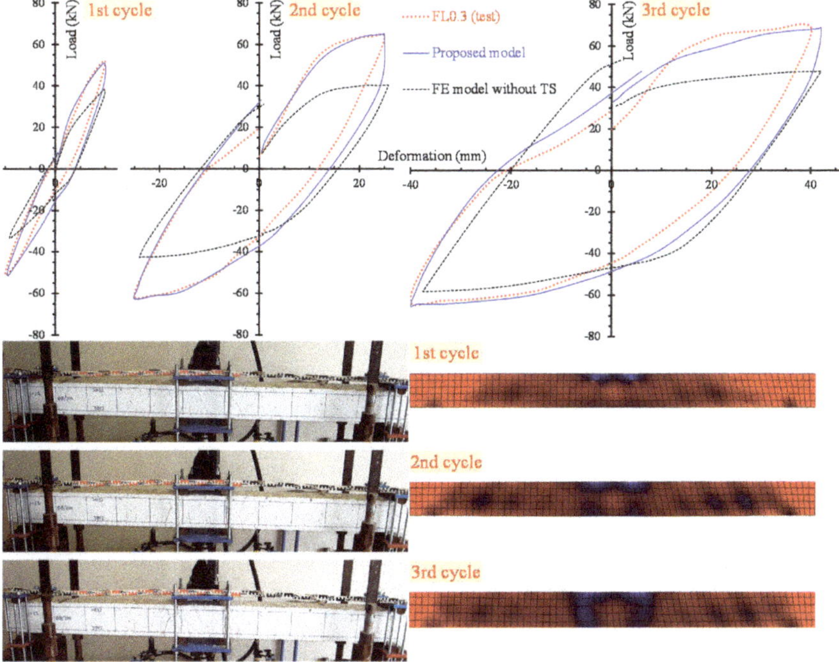

Figure 8. Comparisons between the analytical and experimental results of slender beam FL0.3 in load versus deformation curves and cracking patterns for the 1st, 2nd and 3rd loading cycle.

In order to comprehend the effect of the tension softening and the residual stiffness on the flexural hysteretic response of the tested SFRC beams, two analytical curves are compared with the test curve in Figures 8 and 9. The first curve (continuous blue line) has been derived from the proposed nonlinear FE smeared crack analysis with tension softening and residual stiffness approach and the second curve (black thin dotted line) from the FE analysis without taking into account this effect (denoted as "FE model without TS" in Figures 8 and 9).

Figures 8 and 9 also present the cracking patterns of the flexural beams FL0.3 and FL1.0, respectively, at each loading cycle obtained from the tests and compared to the corresponding cracking pattern at the same loading level derived from the proposed analysis using stress distribution data. Experimental and numerical crack propagation due to flexure at the end of each hysteretic loading cycle are in good compliance.

The ability of the proposed model to calculate accurately the entire hysteretic load versus the deformation behavior of SFRC beams with different failure modes and various steel fiber volumetric fractions is examined in Figures 10–12. In these Figures, the analytical and the experimental hysteretic response and cracking patterns at the failure of the shear-critical beams are compared. Each load versus deformation diagram includes an experimental and two numerical curves yielded from the proposed FE analysis with tension softening and a residual stiffness effect (continuous blue line) and form the FE analysis without taking into account this effect (black thin dotted line). Figure 10 presents the diagrams of shear-critical beams without stirrups SH0.3 and SH0.4, and Figure 11 shows the diagrams of shear-critical beams SH0.6 and SH0.8, also without stirrups. Figure 12 presents the diagrams of shear-critical beams with stirrups SH0.3-s37 and SH0.3-s50.

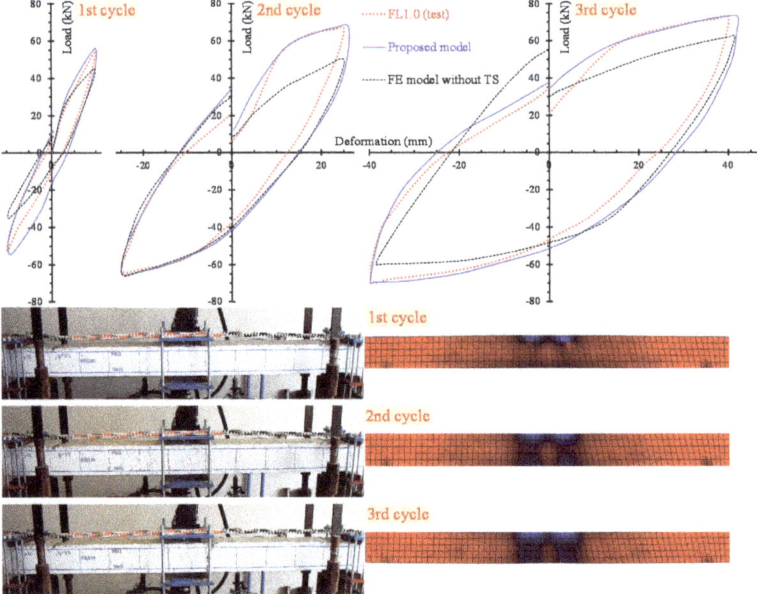

Figure 9. Comparisons between analytical and experimental results of slender beam FL1.0 in load versus deformation curves and cracking patterns for the 1st, 2nd and 3rd loading cycle.

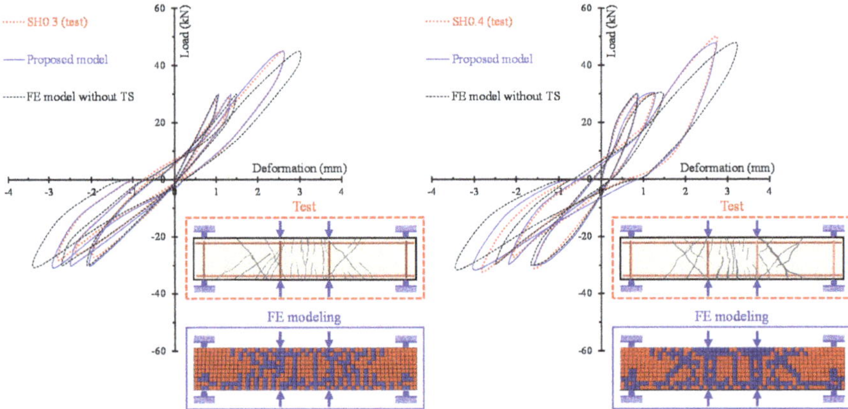

Figure 10. Comparisons between the analytical and experimental hysteretic response and cracking patterns of shear-critical beams without stirrups SH0.3 and SH0.4.

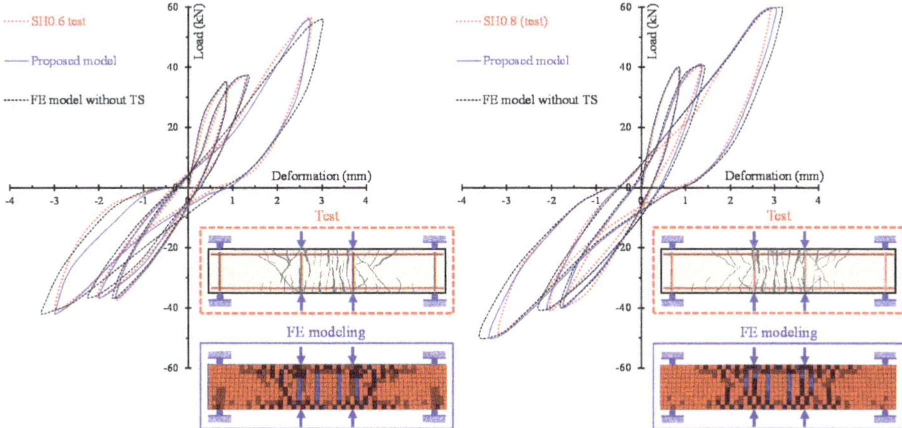

Figure 11. Comparisons between the analytical and experimental hysteretic response and cracking patterns of shear-critical beams without stirrups SH0.6 and SH0.8.

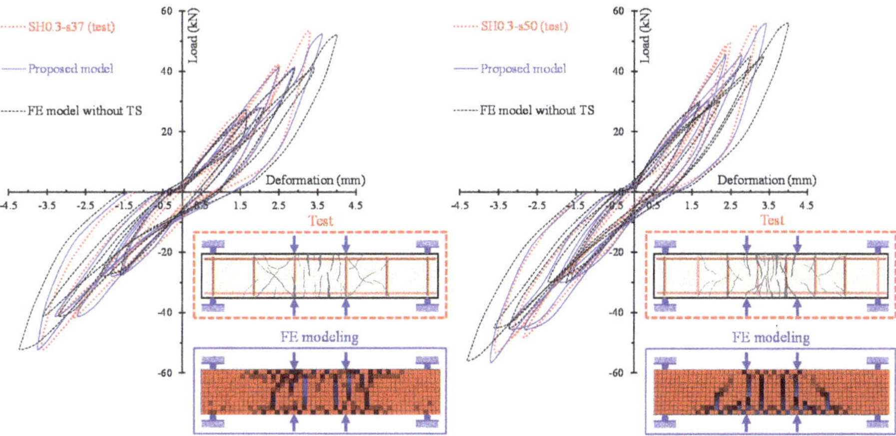

Figure 12. Comparisons between the analytical and experimental hysteretic response and cracking patterns of shear-critical beams with stirrups SH0.3-s37 and SH0.3-s50.

The comparisons of the hysteretic response obtained from the tests and derived from the proposed model, as illustrated in Figures 8–12, indicate that in most beams the predictions of the developed FE analysis that takes into account the smeared crack model with tension softening and residual stiffness effect are in closer agreement with the experimental data than the predictions without this effect. It is noted that numerical curves derived from the proposed model fit well to the test results in both types of beams that failed in flexure (specimens of group "FL") and in shear (specimens of groups "SH-s" and "SH"). Furthermore, the formation of cracks during the performed cyclic tests and the corresponding cracking patterns derived from the FE model exhibit many similarities.

3.2. Analysis of the Hysteric Behavior and Accuracy of the Proposed Model

The load, P, versus displacement, δ, curves of the structural members under cyclic reversal loading are the basis of their hysteretic performance. Based on the test results as well as the FE analysis, the P-δ hysteretic curves for the SFRC beams have been created.

3.2.1. Simplification of the hysteretic Loop

The hysteretic loop (or else full loading cycle) can be interpreted either by the actual path of the loop's curve, or, by parameters that define its general form. One of the essential properties of the hysteretic loop is its inclination. A complete loop comprises a loading and an unloading curve. After the first crack formation on the member, the slope of the loading curve decreases with the increase in displacement, indicating that the members' stiffness has decreased during each repeated loading cycle. Like the loading curve, the inclination of the unloading curve also declines as the number of cycles increases, and the members' unloading stiffness gradually degrades.

When the loading of the members starts, the member's stiffness is equal to the initial elastic deformation stiffness, K_{in}, as shown in Figure 13a, this is also declared to be the maximum stiffness of the element. As the loading cycle proceeds, the loop inclination depends on the stiffness of the member, which can be estimated by the tangent stiffness, K_{tan}, at any point throughout the loading phase (Figure 13b). The value of tangent stiffness varies throughout the cycle of loading, but its average value over the entire loop can be approximated by the cyclic stiffness, K_{cyclic}. The average value of tangent stiffness, K_{tan}, for a half loading cycle can be approximated by the secant stiffness, K_{sec}. When the loop is symmetrical the average values of $K_{sec}^{(+)}$ and $K_{sec}^{(-)}$ equal to the value of cyclic stiffness, K_{cyclic}. Referring to Figure 13a,b, in the linear elastic load range, $K_{in} = K_{sec} = K_{tan}$. The use of K_{sec} is preferred rather than K_{tan} in the processing of test data because it is an order-of-magnitude less influenced by random errors. Nevertheless, K_{tan} is preferred in numerical procedures that require the assembly of an incremental stiffness matrix. In this study K_{sec}, K_{tan} and K_{cyclic} are calculated and compared for each cycle.

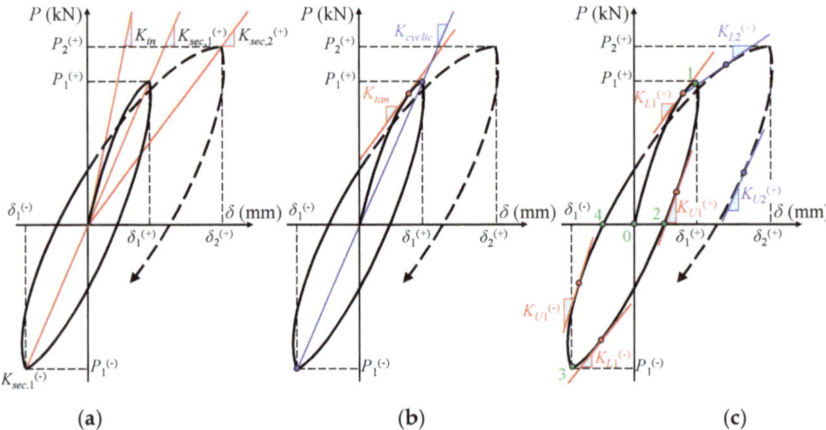

Figure 13. Definitions and variables of the hysteretic loop-loading cycle: (**a**) initial elastic deformation stiffness and secant stiffness; (**b**) tangent and cyclic stiffness; (**c**) loading and unloading stiffness during cyclic reversal imposed load.

3.2.2. Degradation Analysis of Strength and Stiffness

Stiffness is an important assessment index to evaluate the cyclic response of an SFRC member. Thus, stiffness at different points has been calculated and then statistically analyzed to evaluate the member's performance and the effectiveness of the proposed model. Tangent stiffness, K_{tan}, has been calculated at multiple points in each loading cycle, according to Figure 13b. As shown in Figure 13c, the loading starts from point 0 to the unloading point 1, and then reverse loading starts at point 2. During this process, along the path 0-1-2 the tangent stiffness of the member is changed from initial stiffness, K_{in}, to tangent stiffness, $K_{L1}^{(+)}$, as the member's behavior changes from elastic to post-cracking, and then again after reaching the maximum loading point 1 of unloading starts and the tangent stiffness changes again from $K_{L1}{}^{(+)}$ to $K_{U1}{}^{(+)}$. A half cycle is completed from point 0 to point 2, the reverse loading starts from point 2 and increases until the reach of maximum point 3, at which point, unloading starts again to the reverse unloading point 4. During this reverse loading process, stiffness changes again from $K_{L1}{}^{(-)}$ to $K_{U1}{}^{(-)}$. This process continuous in every next cycle until the end of the experimental testing.

3.2.3. Accuracy of the Model

In order to establish the validity and check the accuracy of the proposed nonlinear FE analysis, Tables 5–8 summarize the differences between the numerical calculations and the experimental results in terms of "calculation errors". The discrepancy of the load, the deformation and the stiffness (tangent, secant and cyclic) between the predictions and the tests along the entire hysteretic diagrams of each beam have been calculated in order to evaluate the accuracy of the examined models. The following know expression is used to calculate the discrepancy as a percentage error:

$$Error\ of\ VARiable\ (\%) = \left| \frac{VAR_{model} - VAR_{exp}}{V_{exp}} \right| \times 100, \tag{33}$$

where VAR_{model} and VAR_{exp} are the values of the examined variable derived from the numerical models (A: using the proposed model or B: using the FE model without tension softening and residual stiffness effect) and the experiments, respectively.

Table 5. Error calculation in the prediction of load, P (slender beams), or deformation, δ (deep beams).

Beam Name		MAE at Each Cycle					Overall MAE	SE	CV
		Cycle 1	Cycle 2	Cycle 3	Cycle 4	Cycle 5			
FL0.3	A	1.6%	1.8%	1.7%	–	–	1.7%	0.4%	26%
	B	30.3%	33.8%	19.2%	–	–	27.8%	4.3%	16%
FL1.0	A	2.6%	1.9%	2.2%	–	–	2.2%	0.3%	12%
	B	25.4%	13.2%	12.7%	–	–	17.1%	4.6%	27%
SH0.3-s37	A	1.6%	9.5%	3.9%	7.7%	6.8%	5.9%	1.7%	29%
	B	3.2%	4.2%	17.6%	25.7%	19.1%	13.9%	3.7%	26%
SH0.3-s50	A	0.1%	5.6%	1.1%	8.1%	6.2%	4.2%	1.5%	36%
	B	3.1%	22.2%	22.4%	26.9%	24.2%	19.8%	3.7%	19%
SH0.3	A	5.1%	5.1%	1.8%	–	–	4.0%	0.9%	23%
	B	6.3%	15.1%	17.5%	–	–	13.0%	2.2%	17%
SH0.4	A	3.9%	3.6%	2.6%	–	–	3.4%	1.0%	27%
	B	4.7%	18.9%	18.9%	–	–	14.2%	3.0%	20%
SH0.6	A	0.1%	3.2%	0.0%	–	–	1.1%	0.7%	64%
	B	1.6%	8.9%	10.6%	–	–	7.0%	1.8%	26%
SH0.8	A	3.2%	3.1%	5.8%	–	–	4.0%	1.0%	24%
	B	4.6%	9.9%	11.5%	–	–	8.7%	1.5%	18%

A: Proposed model with tension softening and residual stiffness effect. B: FE model without TS (without tension softening and residual stiffness effect).

The variables examined herein are the applied load, P, or the deformation, δ (Table 5), the tangent stiffness, K_{tan} (Table 6), the secant stiffness, K_{sec} (Table 7) and the cyclic stiffness, K_{cyclic} (Table 8). Each table summarizes the mean absolute error (MAE), the standard error (SE) and the coefficient of variation (CV) of the examined variable for each beam.

Table 6. Error calculation in the prediction of tangent stiffness, K_{tan}.

| Beam Name | | MAE at Each Cycle | | | | | | | | | | Overall MAE | SE | CV |
| | | Cycle 1 | | Cycle 2 | | Cycle 3 | | Cycle 4 | | Cycle 5 | | | | |
		$K_{tan}^{(+)}$	$K_{tan}^{(-)}$	$K_{tan}^{(+)}$	$K_{tan}^{(-)}$	$K_{tan}^{(+)}$	$K_{tan}^{(-)}$	$K_{tan}^{(+)}$	$K_{tan}^{(-)}$	$K_{tan}^{(+)}$	$K_{tan}^{(-)}$			
FL0.3	A	1.8%	3.5%	8.3%	14.2%	6.0%	6.1%	–	–	–	–	6.7%	1.6%	24.3%
	B	31.8%	41.5%	30.5%	28.4%	35.8%	24.5%	–	–	–	–	32.1%	2.0%	6.1%
FL1.0	A	5.9%	4.4%	21.6%	13.3%	4.1%	11.7%	–	–	–	–	10.2%	2.4%	24.0%
	B	25.6%	30.9%	17.4%	13.2%	16.1%	19.0%	–	–	–	–	20.4%	2.5%	12.5%
SH0.3-s37	A	5.4%	6.4%	3.2%	12.5%	1.1%	8.0%	9.2%	6.6%	6.6%	8.3%	6.7%	1.3%	18.8%
	B	5.7%	8.8%	4.7%	12.0%	16.8%	6.7%	23.0%	14.0%	23.8%	20.6%	13.6%	1.7%	12.6%
SH0.3-s50	A	8.8%	11.5%	4.8%	11.7%	9.3%	5.1%	9.4%	6.2%	13.3%	1.3%	8.1%	1.0%	12.6%
	B	8.0%	8.0%	7.2%	10.4%	23.0%	18.6%	26.7%	16.7%	23.4%	19.9%	16.2%	1.9%	12.0%
SH0.3	A	4.5%	9.3%	7.2%	23.1%	10.8%	6.1%	–	–	–	–	10.1%	2.9%	28.7%
	B	3.5%	7.4%	9.2%	14.1%	3.8%	9.9%	–	–	–	–	8.0%	2.1%	26.2%
SH0.4	A	8.0%	5.4%	14.2%	6.2%	7.5%	24.3%	–	–	–	–	10.9%	2.5%	22.5%
	B	8.0%	6.1%	22.7%	9.8%	20.2%	30.3%	–	–	–	–	16.2%	3.1%	19.2%
SH0.6	A	14.8%	10.4%	17.6%	7.0%	6.5%	8.4%	–	–	–	–	10.8%	1.5%	14.1%
	B	13.4%	10.4%	23.0%	11.1%	12.4%	21.5%	–	–	–	–	15.3%	2.0%	13.1%
SH0.8	A	12.9%	7.4%	11.0%	18.2%	10.4%	6.9%	–	–	–	–	11.2%	2.3%	20.4%
	B	12.9%	9.2%	16.2%	6.2%	11.7%	7.2%	–	–	–	–	10.6%	2.2%	20.9%

A: Proposed model with tension softening and residual stiffness effect. B: FE model without TS (without tension softening and residual stiffness effect).

Table 7. Error calculation in the prediction of secant stiffness, K_{sec}.

Beam Name		MAE at Each Cycle					Overall MAE	SE	CV
		Cycle 1	Cycle 2	Cycle 3	Cycle 4	Cycle 5			
FL0.3	A	4.7%	1.8%	4.0%	–	–	3.5%	1.0%	28.2%
	B	25.9%	36.1%	29.7%	–	–	30.6%	1.9%	6.2%
FL1.0	A	8.3%	12.5%	4.1%	–	–	8.3%	1.8%	21.5%
	B	22.1%	1.9%	12.0%	–	–	12.0%	4.3%	35.9%
SH0.3-s37	A	4.8%	0.4%	4.0%	8.1%	11.1%	5.7%	1.7%	30.1%
	B	5.8%	8.0%	15.9%	19.4%	20.5%	13.9%	2.8%	19.9%
SH0.3-s50	A	7.4%	1.9%	6.2%	12.8%	10.0%	7.7%	1.7%	21.8%
	B	4.0%	12.0%	23.0%	26.0%	27.9%	18.6%	3.4%	18.3%
SH0.3	A	3.9%	4.4%	2.8%	–	–	3.7%	1.4%	37.6%
	B	2.6%	5.4%	11.0%	–	–	6.3%	2.1%	33.4%
SH0.4	A	2.9%	3.0%	5.4%	–	–	3.8%	0.7%	19.0%
	B	4.4%	11.4%	18.4%	–	–	11.4%	2.8%	24.3%
SH0.6	A	0.9%	2.9%	2.9%	–	–	2.3%	0.7%	30.7%
	B	2.3%	7.7%	7.7%	–	–	5.9%	1.5%	25.1%
SH0.8	A	3.3%	3.0%	3.5%	–	–	3.3%	0.5%	15.3%
	B	2.2%	9.0%	8.5%	–	–	6.6%	1.6%	23.8%

A: Proposed model with tension softening and residual stiffness effect. B: FE model without TS (without tension softening and residual stiffness effect).

Table 8. Error calculation in the prediction of cyclic stiffness, K_{cyclic}.

Beam Name		MAE at Each Cycle					Overall MAE	SE	CV
		Cycle 1	Cycle 2	Cycle 3	Cycle 4	Cycle 5			
FL0.3	A	4.6%	1.8%	1.0%	–	–	2.5%	1.1%	44.2%
	B	25.8%	33.7%	19.2%	–	–	26.2%	4.2%	15.9%
FL1.0	A	8.2%	12.4%	0.3%	–	–	7.0%	3.5%	50.6%
	B	22.0%	3.0%	11.2%	–	–	12.1%	5.5%	45.4%
SH0.3-s37	A	3.7%	7.5%	0.0%	12.1%	7.3%	6.1%	2.0%	33.1%
	B	2.1%	2.8%	15.4%	22.9%	22.5%	13.1%	4.6%	34.8%
SH0.3-s50	A	7.0%	1.9%	5.2%	12.5%	9.9%	7.3%	1.8%	25.2%
	B	3.9%	12.0%	22.8%	25.8%	23.7%	17.7%	4.2%	23.7%
SH0.3	A	3.8%	3.4%	1.7%	–	–	3.0%	0.7%	22.0%
	B	2.2%	5.2%	11.9%	–	–	6.4%	2.9%	44.9%
SH0.4	A	2.0%	6.3%	5.8%	–	–	4.7%	1.3%	28.5%
	B	4.0%	7.1%	18.7%	–	–	9.9%	4.5%	44.9%
SH0.6	A	0.4%	0.5%	1.6%	–	–	0.8%	0.4%	45.4%
	B	2.3%	7.9%	8.1%	–	–	6.1%	1.9%	31.1%
SH0.8	A	0.0%	2.9%	3.7%	–	–	2.2%	1.1%	50.2%
	B	1.7%	9.5%	8.7%	–	–	6.6%	2.5%	37.4%

A: Proposed model with tension softening and residual stiffness effect. B: FE model without TS (without tension softening and residual stiffness effect).

The overall MAE values in Tables 5–8 indicate that the proposed model predictions (Model A) lead to an error below 5.9%, 11.2%, 8.3% and 7.3% for P or δ, K_{tan}, K_{sec} and K_{cyclic}, respectively, for each tested SFRC beam and the average MAE of all tested beams is 3.3%, 9.3%, 4.8% and 4.2%, respectively, for the aforementioned variables. The corresponding average MAE of all tested beams for the same variables using the predictions of the FE model without tension softening; the stiffening effect (model B) is much higher and equal to 15.2%, 16.5%, 13.2% and 12.3%, respectively. These average values of MAE clearly indicate that the developed FE analysis that takes into account the proposed model with tension softening for the tensional behavior of SFRC and residual stiffness effect yields to accurate predictions of the hysteretic response of concrete members reinforced with conventional reinforcement (bars and stirrups) and steel fibers.

3.3. Effect of Steel Fibers on the Hysteretic Response

The influence of steel fibers on the cyclic response of flexural and shear-critical beams is demonstrated in this section through the hysteretic response of the tested specimens in terms of load versus deformation curves and cracking patterns. The experimental and the numerical hysteretic curves and the cracking patterns of the slender beams FL0.3 and FL1.0 are presented and compared in Figure 14. It is emphasized that SFRC beam FL1.0 with a higher content of steel fibers ($F = 1.0$ and $V_{SF} = 3\%$) demonstrates higher strength, increased energy absorption capacity and improved cracking performance with less diagonal cracks formed at the shear spans, near the supports of the beam with regard to the SFRC beam FL0.3 with $F = 0.3$ and $V_{SF} = 1\%$.

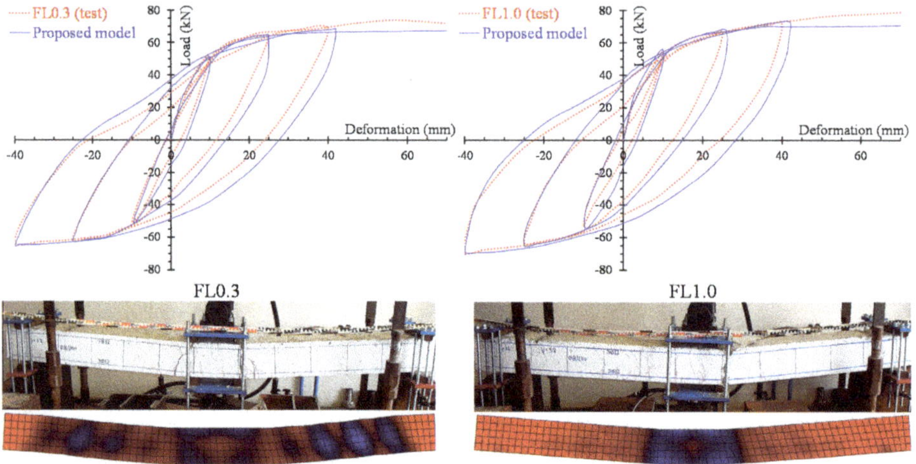

Figure 14. Full hysteretic response and cracking pattern at failure of the slender beams (group "FL" specimens).

The beneficial effect of the used steel fibers on the hysteretic performance is even more revealed and emphasized in the experimental results of the shear-critical beams. Figure 15 presents the load versus deformation curves per loading cycle of the beams of group "SH" (deep beams without stirrups). It is obvious that SFRC beams with a higher amount of steel fibers exhibit improved strength, stiffness and energy absorption capacity.

Furthermore, the gradual increase in the amount of the steel fibers added in the SFRC beams causes a consistent enhancement of the overall hysteretic response and the cracking performance of the deep beams without stirrups, as shown in Figure 16. Furthermore, the cracking patterns shown in Figure 16 indicate that less shear diagonal cracks have been formed in the SFRC beams with a higher amount of steel fibers. In particular, SFRC specimen SH0.8 with $F = 0.8$ and $V_{SF} = 1.5\%$ exhibited more and wide flexural cracks, whereas only slight diagonal cracks have been developed in the shear spans of the beam. On the contrary, plain concrete beam SH0 and SFRC beam SH0.3 with a low amount of steel fibers ($F = 0.3$ and $V_{SF} = 0.5\%$) demonstrated severe shear diagonal cracks and quite brittle behavior.

More or less, similar concluding remarks are also deduced from the comparisons of the hysteretic and cracking performance of the tested deep beams with stirrups (Figure 17). Beams with steel fibers, even in low amounts ($F = 0.3$ and $V_{SF} = 0.5\%$) demonstrated improved overall behavior and especially cracking patterns with less severe shear diagonal cracks than the corresponding plain concrete beams.

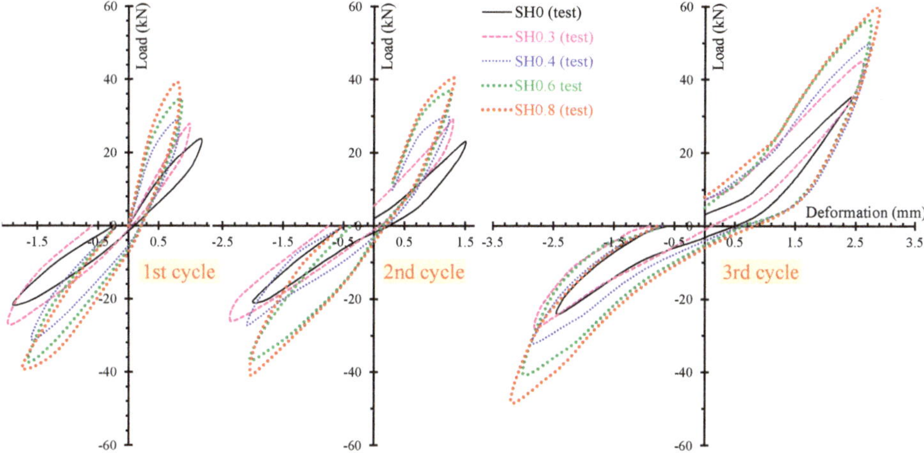

Figure 15. Hysteretic response at each loading cycle of the shear-critical beams without stirrups (group "SH" specimens).

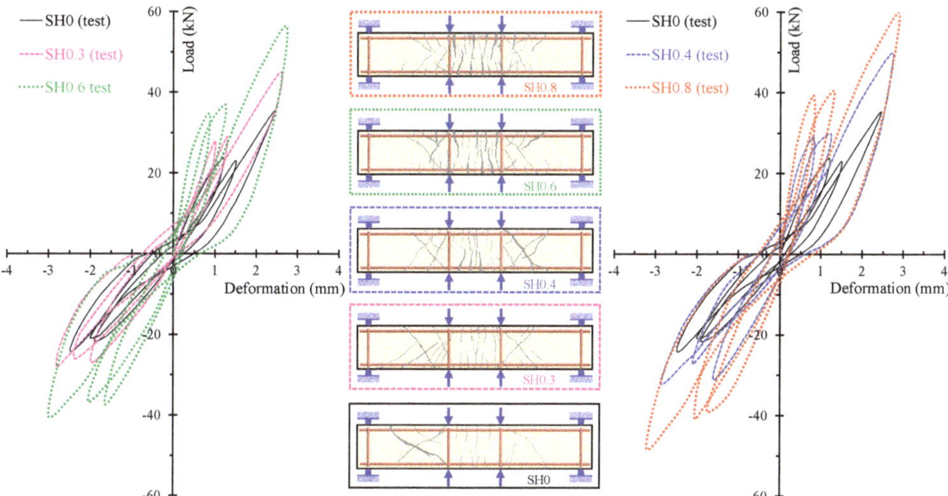

Figure 16. Influence of the steel fibers on the hysteretic and cracking behavior of the shear-critical beams without stirrups (group "SH" specimens).

Furthermore, the improvement of the hysteretic response due to the addition of steel fibers is also indicated in the ratios of the cumulative total energy absorbed per loading cycle of each SFRC beam summarized in Table 9. These ratios have been evaluated from the area enclosed within a full loading cycle of every cycle of the SFRC beam divided by the area of the same cycle of the corresponding reference plain concrete beam (without steel fibers). The absorbed energy reflects the capacity and toughness of the beam.

The results presented in Table 9 indicate that SFRC deep beams absorbed a substantially larger amount of energy than the beams without steel fibers since the ratios are much greater than 1.0. Although this improvement seems to be lower in the slender beams, the ability of fibers to enhance cyclic loading conditions is profound. Thus, SFRC beams maintain their integrity through a potential seismic excitation exhibiting higher energy dissipation capacities than the corresponding beams without steel fibers.

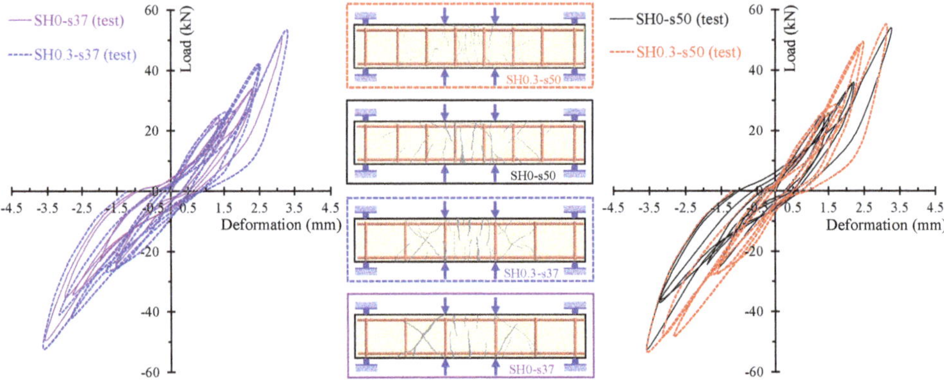

Figure 17. Influence of the steel fibers on the hysteretic and cracking behavior of the shear-critical beams with stirrups (group "SH-s" specimens).

Table 9. Energy dissipation ratios.

Group	Beam Name	F	V_{SF} (%)	ρ_w (%)	Cycle 1	Cycle 2	Cycle 3	Cycle 4	Cycle 5
"FL"	FL0.3	0.3	1.00	0.25	1.00	1.17	1.09	–	–
	FL1.0	1.0	3.00	0.25	1.00	1.28	1.20	–	–
"SH-s"	SH0.3-s37	0.3	0.50	0.37	1.33	1.80	1.59	1.26	1.49
	SH0.3-s50	0.3	0.50	0.50	1.42	1.73	1.40	1.21	1.27
"SH"	SH0.3	0.3	0.50	–	1.38	1.60	1.60	–	–
	SH0.4	0.4	0.75	–	1.40	1.67	2.40	–	-
	SH0.6	0.6	1.00	–	1.73	2.08	3.09	–	–
	SH0.8	0.8	1.50	–	1.98	2.55	3.47	–	–

The effectiveness of steel fibers as the only shear reinforcement, as an alternative of conventional steel stirrups is examined in Figure 18. This figure illustrates and compares the hysteretic and cracking behavior of the SFRC deep beam SH0.8 (V_{SF} = 1.5%) without stirrups and the plain concrete deep beams SH0-s37 and SH0-s50 with stirrups ratio ρ_w = 0.37% and 0.50%, respectively. From the comparison of the load versus deformation curves, it is deduced that the SFRC beam without stirrups exhibited, more or less, a comparable hysteretic response with the RC beams with stirrups. Thus, a potential replacement of stirrups with steel fibers could be achieved under certain circumstances.

These specific conditions depend on the ability of an SFRC beam with longitudinal bars to satisfy pre-set strength and ductility requirements that are defined by design criteria. The optimum amount of the steel fibers that should be added in the mixture of the beam can be evaluated using a recently proposed analytical methodology by Chalioris [92]. Based on this approach, steel fibers as the only shear reinforcement or a desirable combination of steel fibers and stirrups can be used to achieve the above requirement. The methodology is based on the fact that the desirable flexural failure mode occurs when the shear resistance of the examined SFRC beam is higher than its ultimate flexural strength. Analytical expressions to calculate the flexural and the shear strength of SFRC structural members are implemented. A formula to evaluate the minimum steel fibers required in terms of the fiber factor, F, has also been addressed in order for the examined SFRC beam to demonstrate pure flexural response with adequate strength and ductility, whereas its validity has been checked by the test data of 256 SFRC beams under monotonic loading from the literature [92]. However, more cyclic tests are required to provide sound conclusions concerning this important issue.

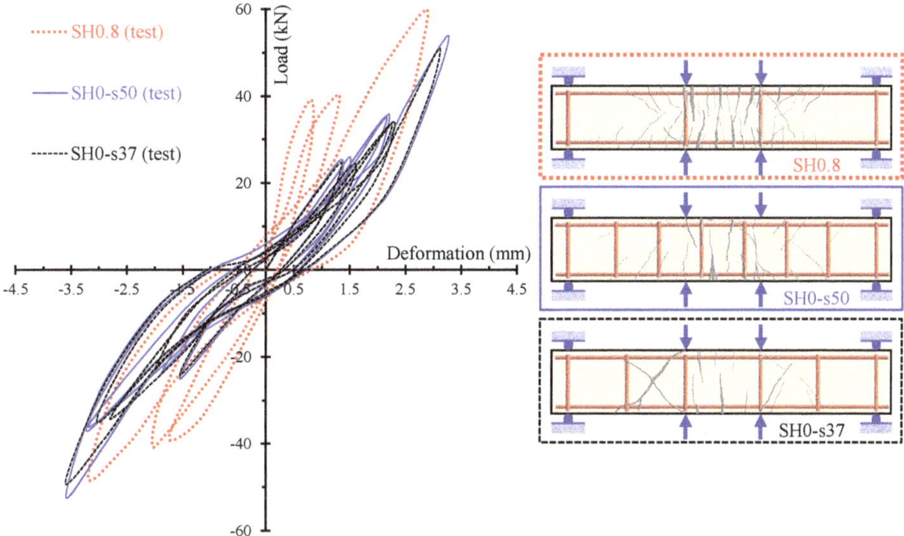

Figure 18. Potential replacement of common closed steel stirrups with short steel fibers.

4. Conclusions

The efficiency of steel fibers on the hysteretic performance of realistic flexural and shear-critical steel fiber-reinforced concrete (SFRC) beams reinforced with steel reinforcements has been investigated. An experimental program of eleven beam specimens subjected to reversal cyclic loading and a numerical nonlinear finite element (FE) analysis have been presented. Based on the results of this study, the following concluding remarks can be drawn:

- The performed cyclic loading tests of slender and deep beams indicate that SFRC beams with increased values of the fiber factor, F, exhibit an improved hysteretic response in terms of stiffness, load-bearing capacity, deformation, energy dissipation ability and cracking behavior. The favorable effect of the used steel fibers on the overall seismic response has been highlighted since SFRC specimens maintain their integrity through the imposed reversal cyclic tests exhibiting higher values of load-bearing capacity and cumulative energy absorbed per loading cycle than the corresponding plain concrete beams without fibers. It is noted that steel fibers with a volume fraction of 1% and 3% provided a 17% and 28% increase in the energy dissipation, respectively, for the case of the flexural beams. This increase was much higher in the shear-critical beams without stirrups. In particular, the ratio of the cumulative energy absorbed of the last loading cycle of the SFRC deep beams to the corresponding energy of the reference plain concrete beams was 1.60, 2.40, 3.09 and 3.47 for beams with fiber factor $F = 0.3, 0.4, 0.6$ and 0.8, respectively.
- Shear-critical beams reinforced with longitudinal bars and steel fibers without stirrups exhibited comparable hysteretic response in terms of strength and absorbed energy with the corresponding deep beams reinforced with bars and stirrups without steel fibers. Although more tests are still required to provide wide-ranging conclusions, it is indicated that a potential replacement of stirrups with steel fibers could be achieved under certain circumstances that depend on the ability of the SFRC beam to satisfy pre-set strength and ductility requirements.
- The developed FE simulation considers the nonlinearities of the materials by a smeared crack approach with tension softening and residual stiffness effect. The favorable influence of the steel

fibers is evaluated according to their characteristics and content in order to achieve a more realistic prediction of SFRC behavior under compression and tension.

- The direct tension experimental results of SFRC specimens carried out in this study verify the analytical predictions of the proposed tensional model. These tests indicate that steel fibers substantially improve the post-cracking tensile behavior and the residual stress versus crack width curve according to the values of the fiber factor, F. Specifically, the value of the maximum post-cracking residual tensile stress was found to be 0.22, 0.33, 0.40 and 0.57 times the value of the tensile strength for SFRC mixtures with fiber factor $F = 0.3, 0.4, 0.6$ and 0.8, respectively. Furthermore, this fiber factor is a more efficient parameter than the volume fraction for the evaluation of the steel fiber contribution.

- The developed nonlinear FE analysis accurately predicts the overall hysteretic response and points out the beneficial effect of the added fibers. Comparisons between the test and numerical results reveal that the developed nonlinear FE analysis with a smeared crack model that takes into account the tension softening and residual stiffness effect accurately predicts the hysteretic response of realistic SFRC beams with steel reinforcement. Furthermore, its validity and accuracy have been checked by calculating the discrepancies between test data and numerical predictions for various variables, such as load, deformation, and stiffness. The mean absolute errors of these variables were found to be satisfactorily low for the tested beams.

Author Contributions: All authors contributed extensively to this study, discussed the results and reviews, and agreed to the amendments at all stages of the paper. C.E.C. and C.G.K. designed and performed the tests. V.K.K. and C.E.C. analyzed the test results, developed the aspects of the experimental investigation and prepared the manuscript. V.K.K. developed the aspects of the numerical investigation and performed the FE analyses under the supervision of C.E.C., C.G.K. and A.E. All authors have read and agreed to the published version of the manuscript.

Funding: This research received no external funding.

Acknowledgments: Author Violetta K. Kytinou gratefully acknowledges the financial support received from Eugenides Foundation towards doctoral studies.

Conflicts of Interest: The authors declare no conflict of interest.

References

1. Conforti, A.; Zerbino, R.; Plizzari, G.A. Influence of steel, glass and polymer fibers on the cracking behavior of reinforced concrete beams under flexure. *Struct. Concr.* **2019**, *20*, 133–143. [CrossRef]
2. Smarzewski, P. Effect of curing period on properties of steel and polypropylene fibre reinforced ultra-high performance concrete. *Mater. Sci. Eng.* **2017**, *245*, 032059. [CrossRef]
3. Bencardino, F.; Nisticò, M.; Verre, S. Experimental investigation and numerical analysis of bond behavior in SRG-strengthened masonry prisms using UHTSS and stainless-steel fibers. *Fibers* **2020**, *8*, 8. [CrossRef]
4. Cuenca, E.; Ferrara, L. Self-healing capacity of fiber reinforced cementitious composites. State of the art and perspectives. *KSCE J. Civ. Eng.* **2017**, *21*, 2777–2789. [CrossRef]
5. Guerini, V.; Conforti, A.; Plizzari, G.; Kawashima, S. Influence of steel and macro-synthetic fibers on concrete properties. *Fibers* **2018**, *6*, 47. [CrossRef]
6. Martinelli, E.; Lima, C.; Pepe, M.; Caggiano, A.; Faella, C. Post-cracking response of hybrid recycled/industrial steel fiber-reinforced concrete. *ACI Spec. Publ.* **2018**, *326*, 64.1–64.10.
7. Belletti, B.; Cerioni, R.; Meda, A.; Plizzari, G. Design aspects on steel fiber-reinforced concrete pavements. *J. Mater. Civ. Eng.* **2008**, *20*, 599–607. [CrossRef]
8. Bencardino, F. Mechanical parameters and post-cracking behaviour of HPFRC according to three-point and four-point bending test. *Adv. Civ. Eng.* **2013**, *2013*, 179712. [CrossRef]
9. Caggiano, A.; Gambarelli, S.; Martinelli, E.; Nisticò, N.; Pepe, M. Experimental characterization of the post-cracking response in hybrid steel/polypropylene fiber-reinforced concrete. *Constr. Build. Mater.* **2016**, *125*, 1035–1043. [CrossRef]
10. Tsonos, A.-D.G. Steel fiber high-strength reinforced concrete: A new solution for earthquake strengthening of old R/C structures. *WIT Trans. Built Env.* **2009**, *104*, 153–164.

11. Vougioukas, E.; Papadatou, M. A model for the prediction of the tensile strength of fiber-reinforced concrete members, before and after cracking. *Fibers* **2017**, *5*, 27. [CrossRef]
12. Abambres, M.; Lantsoght, E.O.L. ANN-Based Shear Capacity of Steel Fiber-Reinforced Concrete Beams without Stirrups. *Fibers* **2019**, *7*, 88. [CrossRef]
13. Campione, G.; Minafo, G. Behaviour of concrete deep beams with openings and low shear span-to-depth ratio. *Eng. Struct.* **2012**, *41*, 294–306. [CrossRef]
14. Ma, K.; Qi, T.; Liu, H.; Wang, H. Shear behavior of hybrid fiber reinforced concrete deep beams. *Materials* **2018**, *11*, 2023. [CrossRef]
15. Smarzewski, P. Analysis of failure mechanics in hybrid fibre-reinforced high-performance concrete deep beams with and without openings. *Materials* **2019**, *12*, 101. [CrossRef]
16. Cuenca, E.; Echegaray-Oviedo, J.; Serna, P. Influence of concrete matrix and type of fiber on the shear strength behavior of self-compacting fiber reinforced concrete. *Compos. Part B Eng.* **2015**, *75*, 135–147. [CrossRef]
17. Tsonos, A.-D.G. Ultra-high-performance fiber reinforced concrete: An innovative solution for strengthening old R/C structures and for improving the FRP strengthening method. *WIT Trans. Eng. Sci.* **2009**, *64*, 273–284.
18. Leone, M.; Centonze, G.; Colonna, D.; Micelli, F.; Aiello, M.A. Fiber-reinforced concrete with low content of recycled steel fiber: Shear behavior. *Constr. Build. Mater.* **2018**, *161*, 141–155. [CrossRef]
19. Cucchiara, C.; Mendola, L.; Papia, M. Effectiveness of stirrups and steel fibres as shear reinforcement. *Cem. Concr. Compos.* **2004**, *26*, 777–786. [CrossRef]
20. Zhao, J.; Liang, J.; Chu, L.; Shen, F. Experimental study on shear behavior of steel fiber reinforced concrete beams with high-strength reinforcement. *Materials* **2018**, *11*, 1682. [CrossRef]
21. Torres, J.A.; Lantsoght, E.O.L. Influence of fiber content on shear capacity of steel fiber-reinforced concrete beams. *Fibers* **2019**, *7*, 102. [CrossRef]
22. Tsonos, A.G.; Stylianidis, K. Seismic retrofit of beam-to-column joints with high-strength fiber jackets. *Europ. Earthq. Eng.* **2002**, *16*, 56–72.
23. Amato, G.; Campione, G.; Cavaleri, L.; Minafo, G. Flexural behaviour of external R/C steel fibre reinforced beam-column joints. *Europ. J. Env. Civ. Eng.* **2011**, *15*, 1253–1276. [CrossRef]
24. Kabir, M.R.; Alam, M.S.; Said, A.M.; Ayad, A. Performance of hybrid reinforced concrete beam column joint: A critical review. *Fibers* **2016**, *4*, 13. [CrossRef]
25. Abbas, A.A.; Mohsin, S.M.S.; Cotsovos, D.M. Seismic response of steel fiber reinforced concrete beam-column joints. *Eng. Struct.* **2014**, *59*, 261–283. [CrossRef]
26. Colajanni, P.; De Domenico, F.; Recupero, A.; Spinella, N. Concrete columns confined with fibre reinforced cementitious mortars: Experimentation and modelling. *Constr. Build. Mater.* **2014**, *52*, 375–384. [CrossRef]
27. Cascardi, A.; Longo, F.; Micelli, F.; Aiello, M.A. Compressive strength of confined column with fiber reinforced mortar (FRM): New design-oriented-models. *Constr. Build. Mater.* **2017**, *156*, 387–401. [CrossRef]
28. Chalioris, C.E.; Sfiri, E.F. Shear performance of steel fibrous concrete beams. *Procedia Eng.* **2011**, *14*, 2064–2068. [CrossRef]
29. Chalioris, C.E.; Karayannis, C.G. Effectiveness of the use of steel fibers on the torsional behavior of flanged concrete beams. *Cem. Concr. Compos.* **2009**, *31*, 331–341. [CrossRef]
30. Spinella, N.; Colajanni, P.; Recupero, A. Simple plastic model for shear critical SFRC beams. *J. Struct. Eng.* **2010**, *136*, 390–400. [CrossRef]
31. Colajanni, P.; Recupero, A.; Spinella, N. Generalization of shear truss model to the case of SFRC beams with stirrups. *Comput. Concr.* **2012**, *9*, 227–244. [CrossRef]
32. Spinella, N. Shear strength of full-scale steel fibre-reinforced concrete beams without stirrups. *Comput. Concr.* **2013**, *11*, 365–382. [CrossRef]
33. Lantsoght, E.O.L. Database of shear experiments on steel fiber reinforced concrete beams without stirrups. *Materials* **2019**, *12*, 917. [CrossRef] [PubMed]
34. Bernat, A.M.; Spinella, N.; Recupero, A.; Cladera, A. Mechanical model for the shear strength of steel fiber reinforced concrete (SFRC) beams without stirrups. *Mater. Struct.* **2020**, *53*, 1–20.
35. Bischoff, P.H. Tension stiffening and cracking of steel fiber-reinforced concrete. *ASCE J. Mat. Civ. Eng.* **2003**, *15*, 174–182. [CrossRef]
36. Meskenas, A.; Kaklauskas, G.; Daniunas, A.; Bacinskas, D.; Jakubovskis, R.; Gribniak, V.; Gelazius, V. Determination of the stress-crack opening relationship of SFRC by an inverse analysis. *Mech. Compos. Mater.* **2014**, *49*, 685–690. [CrossRef]

37. Morelli, F.; Amico, C.; Salvatore, W.; Squeglia, N.; Stacul, S. Influence of tension stiffening on the flexural stiffness of reinforced concrete circular sections. *Materials* **2017**, *10*, 669. [CrossRef]

38. Gribniak, V.; Arnautov, A.K.; Norkus, A.; Kliukas, R.; Tamulenas, V.; Gudonis, E.; Sokolov, A.V. Steel fibers: Effective way to prevent failure of the concrete bonded with FRP sheets. *Adv. Mater. Sci. Eng.* **2016**, *2016*, 10. [CrossRef]

39. Gribniak, V.; Tamulenas, V.; Ng, P.-L.; Arnautov, A.K.; Gudonis, E.; Misiunaite, I. Mechanical behavior of steel fiber-reinforced concrete beams bonded with external carbon fiber sheets. *Materials* **2017**, *10*, 666. [CrossRef]

40. Gribniak, V.; Ng, P.-L.; Tamulenas, V.; Misiunaite, I.; Norkus, A.; Šapalas, A. Strengthening of fibre reinforced concrete elements: Synergy of the fibres and external sheet. *Sustainability* **2019**, *11*, 4456. [CrossRef]

41. Meda, A.; Minelli, F.; Plizzari, G.A. Flexural behaviour of RC beams in fibre reinforced concrete. *Compos. Part B Eng.* **2012**, *43*, 2930–2937. [CrossRef]

42. Smarzewski, P. Flexural toughness of high-performance concrete with basalt and polypropylene short fibres. *Adv. Civ. Eng.* **2018**, *2018*, 5024353. [CrossRef]

43. Gribniak, V.; Kaklauskas, G.; Kwan, H.; Bacinskas, D.; Ibinas, D. Deriving stress-strain relationships for steel fiber concrete in tension from tests of beams with ordinary reinforcement. *Eng. Struct.* **2012**, *42*, 387–395. [CrossRef]

44. Gribniak, V.; Kaklauskas, G.; Torres, L.; Daniunas, A.; Timinskas, E.; Gudonis, E. Comparative analysis of deformations and tension-stiffening in concrete beams reinforced with GFRP or steel bars and fibers. *Compos. Part B Eng.* **2013**, *50*, 158–170. [CrossRef]

45. Kaklauskas, G.; Gribniak, V.; Meskenas, A.; Bacinskas, D.; Juozapaitis, A.; Sokolov, A.; Ulbinas, D. Experimental investigation of the deformation behavior of SFRC beams with an ordinary reinforcement. *Mech. Compos. Mater.* **2014**, *50*, 417–426. [CrossRef]

46. Meskenas, A.; Gribniak, V.; Kaklauskas, G.; Sokolov, A.; Gudonis, E.; Rimkus, A. Experimental investigation of cracking behaviour of concrete beams reinforced with steel fibres produced in Lithuania. *Baltic J. Road Bridge. Eng.* **2017**, *12*, 82–87. [CrossRef]

47. Gribniak, V.; Arnautov, A.K.; Norkus, A.; Tamulenas, V.; Gudonis, E.; Sokolov, A. Experimental investigation of the capacity of steel fibers to ensure the structural integrity of reinforced concrete specimens coated with CFRP sheets. *Mech. Compos. Mater.* **2016**, *52*, 401–410. [CrossRef]

48. Ng, P.; Gribniak, V.; Jakubovskis, R.; Rimkus, A. Tension stiffening approach for deformation assessment of flexural reinforced concrete members under compressive axial load. *Struct. Concr.* **2019**, *20*, 2056–2068. [CrossRef]

49. Plizzari, G.A.; Cangiano, S.; Cere, N. Postpeak behavior of fiber-reinforced concrete under cyclic tensile loads. *ACI Mater. J.* **2000**, *97*, 182–192.

50. Graeff, A.G.; Pilakoutas, K.; Neocleous, K.; Peres, M.V.N.N. Fatigue resistance and cracking mechanism of concrete pavements reinforced with recycled steel fibres recovered from post-consumer tyres. *Eng. Struct.* **2012**, *45*, 385–395. [CrossRef]

51. Li, B.; Xu, L.; Chi, Y.; Huang, B.; Li, C. Experimental investigation on the stress-strain behavior of steel fiber reinforced concrete subjected to uniaxial cyclic compression. *Constr. Build. Mater.* **2017**, *140*, 109–118. [CrossRef]

52. Li, B.; Chi, Y.; Xu, L.; Li, C.; Shi, Y. Cyclic tensile behavior of SFRC: Experimental research and analytical model. *Constr. Build. Mater.* **2018**, *190*, 1236–1250. [CrossRef]

53. Gonzalez, D.C.; Vicente, M.A.; Ahmad, S. Effect of cyclic loading on the residual tensile strength of steel fiber-reinforced high-strength concrete. *ASCE J. Mater. Civ. Eng.* **2015**, *27*, 04014241. [CrossRef]

54. Daniel, L.; Loukili, A. Behavior of high-strength fiber-reinforced concrete beams under cyclic loading. *ACI Struct. J.* **2002**, *99*, 248–256.

55. Kotsovos, G.; Zeris, C.; Kotsovos, M. The effect of steel fibers on the earthquake-resistant design of reinforced concrete structures. *Mater. Struct.* **2007**, *40*, 175–188. [CrossRef]

56. Harajli, M.H.; Gharzeddine, O. Effect of steel fibers on bond performance of steel bars in NSC and HSC under load reversals. *ASCE J. Mater. Civ. Eng.* **2007**, *19*, 864–873. [CrossRef]

57. Campione, G.; Mangiavillano, M.L. Fibrous reinforced concrete beams in flexure: Experimental investigation, analytical modelling and design considerations. *Eng. Struct.* **2008**, *30*, 2970–2980. [CrossRef]

58. Tavallali, H.; Lepage, A.; Rautenberg, J.M.; Pujol, S. Concrete beams reinforced with high-strength steel subjected to displacement reversals. *ACI Struct. J.* **2014**, *111*, 1037–1048. [CrossRef]

59. Caratelli, A.; Meda, A.; Rinaldi, Z. Monotonic and cyclic behaviour of lightweight concrete beams with and without steel fiber reinforcement. *Constr. Build. Mater.* **2016**, *122*, 23–35. [CrossRef]

60. Parra-Montesinos, G.J.; Chompreda, P. Deformation capacity and shear strength of fiber-reinforced cement composite flexural members subjected to displacement reversals. *ASCE J. Struct. Eng.* **2007**, *133*, 421–431. [CrossRef]

61. Luo, J.W.; Vecchio, F.J. Behavior of steel fiber-reinforced concrete under reversed cyclic shear. *ACI Struct. J.* **2016**, *113*, 75–84. [CrossRef]

62. Chalioris, C.E. Steel fibrous RC beams subjected to cyclic deformations under predominant shear. *Eng. Struct.* **2013**, *49*, 104–118. [CrossRef]

63. Rousakis, T.C.; Manolitsi, G.; Karabinis, A.I. FRP strengthening of RC columns: Parametric finite element analyses of bar quality effect. In Proceedings of the 1st Asia-Pacific Conference on FRP in Structures (APFIS 2007), Hong-Kong, China, 12–14 December 2007; Smith, S.T., Ed.; International Institute for FRP in Construction: Winnipeg, MB, Canada, 2007.

64. Gómez, J.; Torres, L.; Barris, C. Characterization and simulation of the bond response of NSM FRP reinforcement in concrete. *Materials* **2020**, *13*, 1770. [CrossRef]

65. Zhelyazov, T. Structural Materials: Identification of the constitutive models and assessment of the material response in structural elements strengthened with externally-bonded composite material. *Materials* **2020**, *13*, 1272. [CrossRef] [PubMed]

66. Chalioris, C.E.; Kosmidou, P.-M.K.; Karayannis, C.G. Cyclic response of steel fiber reinforced concrete slender beams: An experimental study. *Materials* **2019**, *12*, 1398. [CrossRef]

67. Dassault Systèmes Simulia System Information. In *Abaqus 2017 User's Manual*; Version 6.12.1; SIMULIA: Providence, RI, USA, 2017.

68. Lubliner, J.; Oliver, J.; Oller, S.; Oñate, E. A plastic-damage model for concrete. *Int. J. Solids Struct.* **1989**, *25*, 299–326. [CrossRef]

69. Karayannis, C.G. A numerical approach to steel-fibre reinforced concrete under torsion. *Struct. Eng. Rev.* **1995**, *7*, 83–91.

70. Lantsoght, E.O.L. How do steel fibers improve the shear capacity of reinforced concrete beams without stirrups? *Compos. Part B Eng.* **2019**, *175*, 107079. [CrossRef]

71. Yun, H.-D.; Lim, S.H.; Choi, W.-C. Effects of reinforcing fiber strength on mechanical properties of high-strength concrete. *Fibers* **2019**, *7*, 93. [CrossRef]

72. Neves, R.; Almeida, J. Compressive behaviour of steel fibre reinforced concrete. *Struct. Concr.* **2005**, *6*, 1–8. [CrossRef]

73. Bencardino, F.; Rizzuti, L.; Spadea, G. Stress-strain behavior of steel fiber-reinforced concrete in compression. *ASCE J. Mater. Civ. Eng.* **2008**, *20*, 255–263. [CrossRef]

74. Chalioris, C.E.; Liotoglou, F.A. Tests and simplified behavioral model for steel fibrous concrete under compression. In *Advances in Civil Engineering and Building Materials IV.*; Chang, S.-Y., Al Bahar, S.K., Husain, A.-A.M., Zhao, J., Eds.; CRC Press/Balkema: Leiden, The Netherlands, 2015; pp. 195–199.

75. Lee, S.-C.; Oh, J.-H.; Cho, J.-Y. Compressive behavior of fiber-reinforced concrete with end-hooked steel fibers. *Materials* **2015**, *8*, 1442–1458. [CrossRef] [PubMed]

76. Zhao, M.; Zhang, B.; Shang, P.; Fu, Y.; Zhang, X.; Zhao, S. Complete stress–strain curves of self-compacting steel fiber reinforced expanded-shale lightweight concrete under uniaxial compression. *Materials* **2019**, *12*, 2979. [CrossRef] [PubMed]

77. Chalioris, C.E.; Panagiotopoulos, T.A. Flexural analysis of steel fibre-reinforced concrete members. *Comput. Concr.* **2018**, *22*, 11–25.

78. Choi, W.-C.; Jung, K.-Y.; Jang, S.-J.; Yun, H.-D. The influence of steel fiber tensile strengths and aspect ratios on the fracture properties of high-strength concrete. *Materials* **2019**, *12*, 2105. [CrossRef]

79. Wang, Z.L.; Wu, J.; Wang, J.G. Experimental and numerical analysis on effect of fiber aspect ratio on mechanical properties of SFRC. *Constr. Build. Mater.* **2010**, *24*, 559–565. [CrossRef]

80. Choi, S.W.; Choi, J.; Lee, S.C. Probabilistic analysis for strain-hardening behavior of high-performance fiber-reinforced concrete. *Materials* **2019**, *12*, 2399. [CrossRef]

81. Bezerra, A.; Maciel, P.S.; Corrêa, E.; Soares Junior, P.R.R.; Aguilar, M.T.P.; Cetlin, P.R. Effect of high temperature on the mechanical properties of steel fiber-reinforced concrete. *Fibers* **2019**, *7*, 100. [CrossRef]

82. Karayannis, C.G. Smeared crack analysis for plain concrete in torsion. *ASCE J. Struct. Eng.* **2000**, *126*, 638–645. [CrossRef]

83. Karayannis, C.G.; Chalioris, C.E. Experimental validation of smeared analysis for plain concrete in torsion. *ASCE J. Struct. Eng.* **2000**, *126*, 646–653. [CrossRef]

84. Rimkus, A.; Cervenka, V.; Gribniak, V.; Cervenka, J. Uncertainty of the smeared crack model applied to RC beams. *Eng. Fract. Mech.* **2020**, *233*, 107088. [CrossRef]

85. Kytinou, K.V.; Chalioris, C.E. Analysis of residual flexural stiffness of steel fiber-reinforced concrete beams with steel reinforcement. *Materials* **2020**, *13*, 2698. [CrossRef] [PubMed]

86. Karayannis, C.G. Nonlinear analysis and tests of steel-fiber concrete beams in torsion. *Struct. Eng. Mech.* **2000**, *9*, 323–338. [CrossRef]

87. Mudadu, A.; Tiberti, G.; Plizzari, G.A.; Morbi, A. Post-cracking behavior of polypropylene fiber reinforced concrete under bending and uniaxial tensile tests. *Struct. Concr.* **2019**, *20*, 1411–1424. [CrossRef]

88. Mudadu, A.; Tiberti, G.; Germano, F.; Plizzari, G.A.; Morbi, A. The effect of fiber orientation on the post-cracking behavior of steel fiber reinforced concrete under bending and uniaxial tensile tests. *Cem. Concr. Compos.* **2018**, *93*, 274–288. [CrossRef]

89. Bencardino, F.; Rizzuti, L.; Spadea, G.; Swamy, R.N. Experimental evaluation of fiber reinforced concrete fracture properties. *Compos. Part B Eng.* **2010**, *41*, 17–24. [CrossRef]

90. Bazant, Z.P.; Oh, B.H. Crack band theory for fracture of concrete. *Mater. Constr.* **1983**, *16*, 155–177. [CrossRef]

91. Genikomsou, A.S.; Polak, M.A. Finite element analysis of punching shear of concrete slabs using damaged plasticity model in ABAQUS. *Eng. Struct.* **2015**, *98*, 38–48. [CrossRef]

92. Chalioris, C.E. Analytical approach for the evaluation of minimum fiber factor required for steel fibrous concrete beams under combined shear and flexure. *Constr. Build. Mater.* **2013**, *43*, 317–336. [CrossRef]

Article

Cyclic Response of Steel Fiber Reinforced Concrete Slender Beams: An Experimental Study

Constantin E. Chalioris *, Parthena-Maria K. Kosmidou and Chris G. Karayannis

Reinforced Concrete and Seismic Design of Structures Laboratory, Civil Engineering Department, School of Engineering, Democritus University of Thrace, 67100 Xanthi, Greece; pkosmido@civil.duth.gr (P.-M.K.K.); karayan@civil.duth.gr (C.G.K.)
* Correspondence: chaliori@civil.duth.gr; Tel.: +30-25410-79632

Received: 6 April 2019; Accepted: 25 April 2019; Published: 29 April 2019

Abstract: Reinforced concrete (RC) beams under cyclic loading usually suffer from reduced aggregate interlock and eventually weakened concrete compression zone due to severe cracking and the brittle nature of compressive failure. On the other hand, the addition of steel fibers can reduce and delay cracking and increase the flexural/shear capacity and the ductility of RC beams. The influence of steel fibers on the response of RC beams with conventional steel reinforcements subjected to reversal loading by a four-point bending scheme was experimentally investigated. Three slender beams, each 2.5 m long with a rectangular cross-section, were constructed and tested for the purposes of this investigation; two beams using steel fibrous reinforced concrete and one with plain reinforced concrete as the reference specimen. Hook-ended steel fibers, each with a length-to-diameter ratio equal to 44 and two different volumetric proportions (1% and 3%), were added to the steel fiber reinforced concrete (SFRC) beams. Accompanying, compression, and splitting tests were also carried out to evaluate the compressive and tensile splitting strength of the used fibrous concrete mixtures. Test results concerning the hysteretic response based on the energy dissipation capabilities (also in terms of equivalent viscous damping), the damage indices, the cracking performance, and the failure of the examined beams were presented and discussed. Test results indicated that the SFRC beam demonstrated improved overall hysteretic response, increased absorbed energy capacities, enhanced cracking patterns, and altered failure character from concrete crushing to a ductile flexural one compared to the RC beam. The non-fibrous reference specimen demonstrated shear diagonal cracking failing in a brittle manner, whereas the SFRC beam with 1% steel fibers failed after concrete spalling with satisfactory ductility. The SFRC beam with 3% steel fibers exhibited an improved cyclic response, achieving a pronounced flexural behavior with significant ductility due to the ability of the fibers to transfer the developed tensile stresses across crack surfaces, preventing inclined shear cracks or concrete spalling. A report of an experimental database consisting of 39 beam specimens tested under cyclic loading was also presented in order to establish the effectiveness of steel fibers, examine the fiber content efficiency and clarify their role on the hysteretic response and the failure mode of RC structural members.

Keywords: steel fiber reinforced concrete (SFRC); slender beams; cyclic loading; hysteretic response; failure mode; tests

---

1. Introduction

Reinforced concrete (RC) beams with inadequate transverse reinforcement exhibit a lack of ductility and fail in a rather brittle manner due to the weak tensile resistance and the reduced deflection capacity in the presence of cracks. These weaknesses can be overwhelmed by the addition of steel fibers as shear resistance mass reinforcement, which can reduce and delay cracking and ameliorate the overall performance of RC members. It is known that concrete reinforced with discrete, short, and randomly distributed fibers is a composite material with considerable cracking resistance due to the ability of the fibers to transfer the developed tensile stresses across crack surfaces (crack-bridging). The properties and the structural behavior of steel fiber reinforced concrete (SFRC) under compression, tension, flexure, shear, and torsion have been studied usually under monotonic tests, as next presented in the following subsections.

1.1. Compression and Tensile Behavior of Steel Fiber Reinforced Concrete (SFRC)

SFRC under compression exhibits increased strength only in mixtures with a high amount and adequate aspect ratio of fibers, whereas it exhibits a rather marginal contribution of the fibers on the compressive strength in most of the examined cases [1–6]. Nevertheless, significant improvement of the post-cracking stress–strain compressive behavior with noticeable toughness and a ductile response even in low volumetric proportions of fibers has been revealed [7–10]. A significant increase in the SFRC compressive strength is achieved in mixtures with at least 3% volume fraction of steel fibers [11,12].

The flexural tensile behavior of SFRC has been studied by using three- or four-point bending tests of small-scaled notched prismatic specimens. The majority of the experimental studies pointed out the favorable influence of the added steel fibers on the post-cracking regime [13]. Further, fibrous concrete mixtures with a high percentage of fibers showed an enhanced overall performance and a re-hardening response after cracking [2,14–16]. Furthermore, long steel fibers proved more effective than the short ones to ameliorate the flexural response at large deflections in terms of strength, deformation capacity, toughness, and cracking behavior [17]. Similar concluding remarks concerning the splitting tensile strength of SFRC have also been derived from extensive data of splitting tests performed on cylinders [18–20] and cubes [5,15]. The favorable influence of steel fiber orientation on tensile strength increase has been highlighted through splitting and flexural tensile tests on magnetically driven concrete and mortar cube specimens [21].

The orientation distribution effect of the added steel fibers has also been examined in the post-cracking tensile behavior of SFRC through tests of specially fabricated specimens subjected to direct tension and by applied different casting methods of the fresh mixtures [22]. It has been found that distribution and orientation of fibers greatly influence the overall performance of fibrous concrete. For this purpose, X-ray computed tomography techniques have recently been developed and experimentally verified, which can evaluate the microstructural parameters along with the orientation distribution and the concentration of the short fibers added in cementitious matrices as mass reinforcement [23,24]. The results of these novel techniques also confirmed that the post-cracking tensile response of fibrous concrete significantly depends on the elastic behavior of the uncracked regions of the composite material located between the cracking areas [25,26]. Micro-cracks are initially formed, propagating into localized macro-cracks after overcoming the elastic response. Hence, although fibers practically do not affect the pre-cracking behavior, it has long been recognized that SFRC exhibits increased strength and an ameliorated response after the formation of cracks due to direct tensile loading [27].

Uniaxial tests of prismatic specimens under direct tension revealed that the addition of an adequate amount of short steel fibers mainly increase the tensile load capacity, whereas longer steel fibers enlarge the ultimate tensile deformation [28]. However, SFRC with inadequate dosage of steel fibers demonstrated a negligible increase in the tensile strength and limited improvement in the post-cracking behavior [27]. Thus, a critical volume fraction of steel fibers has been proposed in order to design high-performance fibrous concrete mixtures that achieve strain hardening under direct tension with advanced ductility and energy absorption capacity [29–31], such as ultra-high performance SFRC [32]. Nevertheless, dispersion of long steel fibers with higher volumetric proportions was found to be problematic [33] with regard to the fine workability properties of fibrous mixtures with short lengths [34] or even microfilament steel fibers that were 13 mm long [35].

Recent analytical studies associate the tensile performance and the fracture properties of SFRC with the fiber-to-concrete interface behavior, which can be calibrated by pullout tests [36,37].

Unified formulations and simplified analytical approaches for the simulation of the overall bond behavior of fibers embedded in concrete and the prediction of the tensile response of SFRC members after cracking have also been proposed [38,39], which have been calibrated, validated or based on various pullout tests and numerical models [40–42]. However, due to the various available shapes, types, materials, and dimensions of fibers, along with the different mechanical properties of the cementitious mixtures, their bond characteristics also vary. Thus, there are no widely accepted or reliable constitutive tensile models that can be applied broadly in SFRC members [43]. Consequently, an interesting alternative approach has been developed by Gribniak et al. [44], which uses an inverse technique to derive average stress–strain relationships from a wide range of experimental moment–curvature curves of flexural SFRC members.

Further, the mechanical recovery of high-performance steel fiber reinforced concrete under uniaxial tensile loading has been investigated by [45,46]. It was found that fibrous concrete mixtures subjected to direct tension present multiple micro-cracks with fine crack widths, which offers an important self-healing capability [46]. The width of the developed cracks along with the type of sand and the added fibers are the main parameters that influence the self-healing capacity by generating improved interfacial bond conditions [45]. Further, a hybrid fiber reinforcing system containing polyethylene and steel fibers showed more enhanced recovery characteristics since, even in cracks with rather large width, the bond in the crack region around the longer steel fiber was bridged, controlled, and densified by the shorter polyethylene fiber [46].

1.2. Reinforced Concrete (RC) Structural Members with Steel Fibers in Shear and/or Flexure

Shear-critical RC structural members usually exhibit brittle catastrophic failure due to the apparent weakness of concrete in tension, which is commonly compensated by the presence of an adequate amount of steel stirrups. The advantageous characteristics of SFRC inspired researchers to investigate the use of short fibers as mass reinforcement against shear instead of conventional transverse steel reinforcement. The full or even partial replacement of stirrups is crucial in RC joints and deep, torsional, and coupling beams, where design criteria require a high ratio of shear reinforcement that leads to the extremely short spacing of stirrups and/or to the use of cumbersome reinforcement systems such as spirals, cross inclined bars, diagonal reinforcement, etc. [47–50]. Thus, at least in the critical sections of these shear-vulnerable RC members, the use of steel fibers could lead to reduced reinforcement congestion [19,51–53].

Several experimental works demonstrated the feasibility of substituting a significant amount of stirrups for an adequate volume fraction of short steel fibers, achieving comparable load capacities at the first shear cracking and at the maximum shear strength in RC beams [54–58]. Smarzewski [59,60] examined the synergetic positive effect of steel and polypropylene fibers in deep RC beams in order to replace conventional steel reinforcement. Zhao et al. [61] highlighted the favorable influence of steel fibers on the stiffness, the shear capacity, and the deformation of shear-critical RC beams, along with the reduction in crack width, height, and strains of concrete and stirrup across the diagonal section. Further, Chalioris [62] proposed an efficient analytical method to evaluate the proper dosage and type of steel fibers that should be added to concrete to replace a desirable number of stirrups and to satisfy pre-set strength and ductility requirements in shear-critical RC members. Furthermore, analytical models have also been developed to evaluate the total shear strength of concrete beams reinforced with steel fibers and longitudinal reinforcing bars, with or without stirrups [63–65]. Moreover, the well-known softened truss model was properly modified to implement the contribution of steel fibers on the shear strength of SFRC members [66,67].

For many years, researchers have been studying the effectiveness of steel fibers on the flexural behavior of RC structural members and it has long been acknowledged that they improved bending moment strength, ductility, failure toughness, and energy absorption capacity. Flexural cracking performance of SFRC beams with longitudinal steel reinforcing bars also appears to be enhanced with increased crack number, reduced crack width and height, restricted crack propagation, and delayed concrete spalling [68–70]. It has also been demonstrated that deformed or hook-ended steel fibers with higher aspect ratios can increase cracking resistance, energy dissipation, and ductility index most effectively [71,72]. Further, a recent experimental study revealed that even straight and short mill-cut steel fibers with a rather low aspect ratio equal to 40 added to a volume fraction up to 2% were capable of enhancing sectional flexural stiffness and increasing ductility related to the lateral deformations of RC columns under eccentric compression [73].

The implementation of feasible constitutive stress–strain relationships of SFRC under compression and tension in numerical analysis of concrete cross-sections reinforced with steel reinforcing bars and fibers provides rational and accurate predictions of the flexural response in terms of bending moment versus curvature analytical curves [31]. Further, the ability of steel fibers to control flexural cracking and to prevent concrete spalling inspired researchers to use SFRC in deficient and in blast-damaged RC beams that had been strengthened using externally bonded carbon fiber-reinforced polymer sheets, in order to avoid rip-off failure of the concrete cover, to prevent premature failures, to increase ductility, and to ensure structural integrity [74–77].

1.3. Cyclic Response of Reinforced Concrete (RC) Structural Members with Steel Fibers

The behavior of RC members under cyclic loading is significantly influenced by the fact that each concrete layer is subjected to alternate tension and compression stresses. Tests have shown that RC specimens under reversal loading suffer from degraded aggregate interlock and intensive cracking at both tension and compression zones that reduce concrete shear strength and weaken compression zones, resulting in a premature, catastrophic brittle failure [78–80]. The addition of steel fibers into the concrete mixture of flexural beams [72,79,81–85], columns [69,80]), and joints [52,86] has been proposed and examined as an alternative or additional reinforcement for seismic resistance of structures. Due to the tensile stress transfer ability of the fibers, SFRC under cyclic deformations exhibits satisfactory resistance to the formation, propagation, and widening of cracks, improved post-cracking behavior, and an ameliorated energy dissipation capability.

Daniel and Loukili [81] suggested that the use of steel fibers in flexural fibrous concrete beams with different longitudinal reinforcement ratios can be efficient to prevent an early development of macro-cracks during the pre-peak stage and enhances the energy absorption over both the elastic and inelastic stages. Tests by Campione and Mangiavillano [72] demonstrated that the addition of fibers increases the load-bearing capacity of the examined specimens, ensures more ductile behavior, and reduces degradation effects under cyclic deformations, especially in beams with increased concrete cover thickness.

Further, Harajli and Gharzeddine [82] investigated the influence of steel fibers on the bond performance of spliced steel bars in normal and high strength concrete. The experimental results showed that the existence of steel fibers delayed the formation and propagation of splitting cracks along the spliced region and increased the absorbed energy of the SFRC beams, resulting in a less brittle failure. Cyclic tests in concrete beams with high-strength steel reinforcement by Tavallali et al. [85] showed that the addition of steel fibers significantly reduced the maximum crack width of both flexural and shear cracks regarding the non-fibrous concrete beams.

Parra-Montesinos and Chompreda [83] investigated the experimental behavior of SFRC beams with and without steel transverse reinforcement. Results from the tests showed that the fibrous concrete specimens presented a pure strain-hardening behavior with advanced damage tolerance developing multiple flexural and diagonal cracks. Chalioris [84] investigated the influence of steel fibers in shear-critical RC beams with and without steel transverse reinforcement under reversal loading. Steel fibrous beams demonstrated improved overall shear performance with increased shear strength, ameliorated pre-crack and post-crack behavior, and enhanced energy dissipation capabilities compared to the non-fibrous specimens.

Cyclic experiments in a two-span continuous column performed by Kotsovos et al. [69] showed that specimens with steel fibers satisfied the performance requirements of Eurocodes for concrete strength up to 60 MPa, as the existence of steel fibers enhanced the overall performance of the specimens regarding ultimate strength and the observed failure mode. The addition of steel fibers in concrete columns with advanced high-steel longitudinal reinforcement was experimentally investigated by Lepage et al. [80]. Results from these tests showed the efficiency of steel fibers by reducing the amount of spalling of the concrete cover and altering the mode of failure from buckling of the compression bars to fracture of the tension bars.

1.4. Research Significance

The aforementioned literature review reveals that the majority of the conducted research is focused on the behavior of SFRC specimens under monotonic loading, whereas the cyclic response of SFRC members has been rarely investigated with regard to the plethora of the monotonically tested ones. Cyclic testing of RC beams with steel fibers is quite limited and has preliminary and exploratory character so far. Also, the hysteretic performance of such structural members is greatly influenced by several parameters which even independently have not been thoroughly clarified yet. Various combinations of conventional reinforcement (steel bars and stirrups) with steel fibers result in different contributions of the added fibers to the flexural and shear capabilities of SFRC beams. The type, the aspect ratio, and the volumetric proportion of the fibers also affect the overall structural performance.

Thus, although there are some widely accepted findings concerning the favorable influence of the added steel fibers on the post-cracking behavior of SFRC members under reversal deformations, there are still several issues that require further and systematic investigation. The dual contribution of fibers to the increase of the flexural and the shear strength of SFRC beams under different load-bearing mechanisms and the ability of steel fibers to alter the failure mode of the structural member under certain circumstances are some examples of research gaps.

Further, the recent increased interest for the application of SFRC in retrofitting applications of deficient and/or damaged RC structural members under seismic reversal excitations and the lack of relative experimental studies are the main motives behind this work.

In this work, the influence of steel fibers on the cyclic hysteretic response of slender RC beams was experimentally investigated. Three slender beams, each 2.5 m long with a rectangular cross-section, were constructed and tested under reversal deformations.

Hook-ended steel fibers, each with a length-to-diameter ratio equal to 44 and two different contents (1% and 3%), were added to the steel fiber reinforced concrete (SFRC) beams. Accompanying compression and splitting tests were also carried out in order to acquire full stress–strain relationships of the plain and the fibrous mixtures. Test results concerning the hysteretic response based on the energy dissipation capabilities, the damage indices, the cracking performance, and the failure of the examined beams are presented and discussed in later sections. Special attention is given to comprehend the observed failure modes and to associate the experimental results of this study with the test data of relevant published works from the literature. A systematic report of an experimental database consisting of 39 beam specimens tested under cyclic loading is also presented herein in order to clarify the effectiveness of steel fibers and their role on the hysteretic response and the failure mode of RC structural members.

This paper contributes to the limited existing literature on cyclic tests of RC beams with steel fibers, providing detailed experimental data of beams with a rather high amount of steel fibers (3%) that has not been examined before. An additional innovation of this paper regarding the existing experimental studies is the thorough demonstration of the available relevant tests in order to compare the experimental results and to derive new quantitative concluding remarks concerning the cyclic performance of SFRC beams.

2. Steel Fiber Reinforced Concrete

2.1. Materials

The concrete mixture used in this study consisted of a general-purpose ordinary Portland type cement (type CEM II 32.5 N, Greek type pozzolan cement containing 10% fly ash), crushed and natural river sand with a high fineness modulus, crushed stone aggregates with a maximum size of 16 mm, and water, in a mass proportion of 1:3.62:2.67:0.55, respectively. Further, 1 L retarder (Pozzolith 134 CF) per 1 m^3 concrete was added to the mix in order to slow the rate of the concrete setting.

The steel fibers added to the fibrous concrete mixtures (see Figure 1) were hook-ended fibers to enhance the anchorage of the fiber in the matrix. The dimensions of the used fibers are an aspect ratio (length-to-diameter ratio) equal to $\ell_f / d_f = 44$ mm/1 mm = 44, as shown in Figure 2a. Two different steel fiber volumetric proportions, V_f, were chosen; 1% or 80 kg per 1 m^3 concrete and 3% or 240 kg per 1 m^3 concrete. The nominal yield tensile strength of the fibers was 1000 MPa.

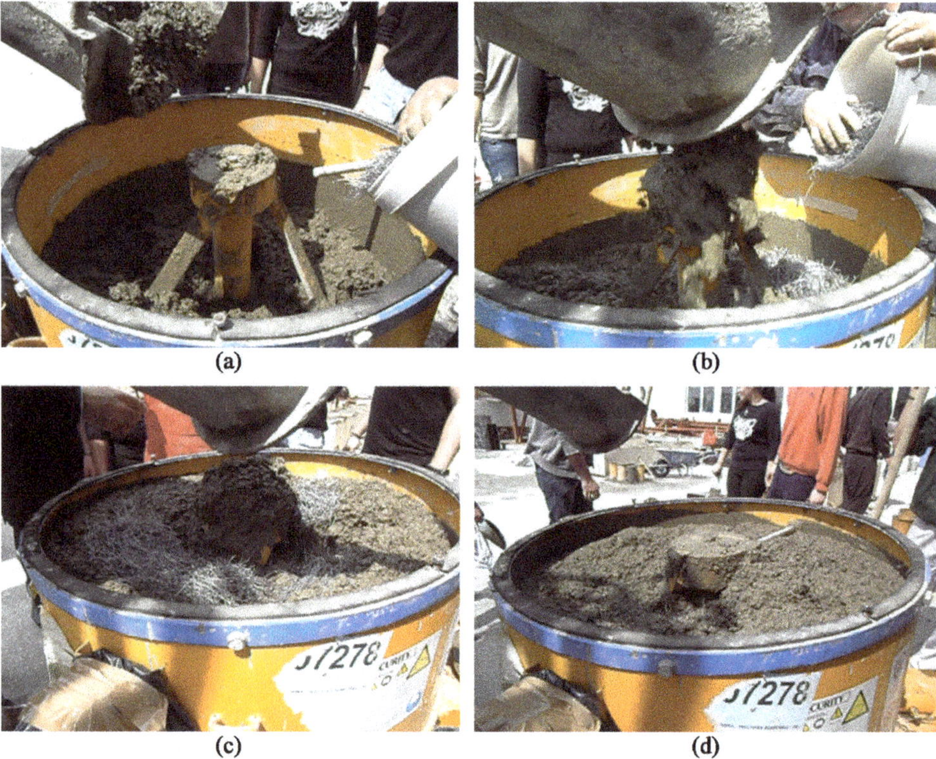

Figure 1. Preparation stages of steel fiber reinforced concrete: (**a**) Pouring of ready-mix concrete gradually into the pan type mixer; (**b**) and (**c**) gradual addition of the steel fibers into the fresh concrete mixture during stirring; (**d**) final homogeneous wet steel fiber reinforced concrete mixture.

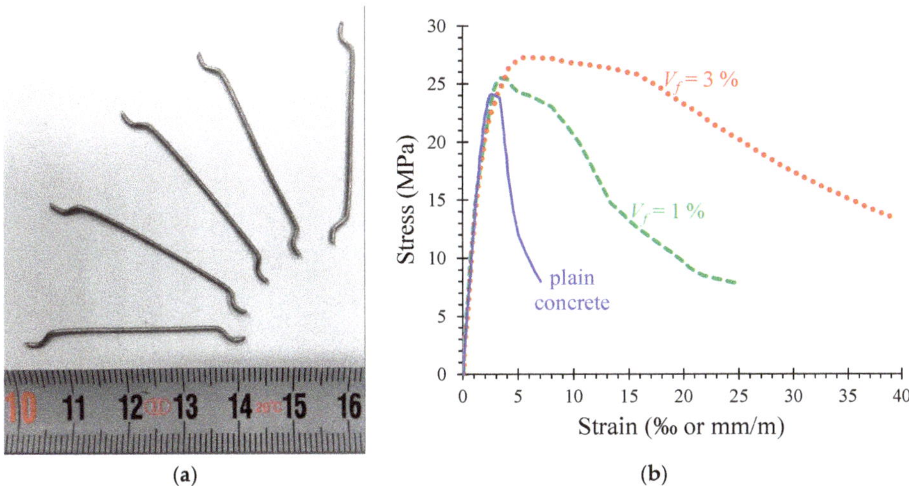

Figure 2. (**a**) Dimensions of the used hook-ended steel fibers; (**b**) experimental compressive stress–strain relationships of plain and steel fibrous concrete.

2.2. Mix Preparation

The mix preparation of the steel fiber reinforced concrete (SFRC) was carried out using a pan type concrete mixer (see Figure 1). Two different SFRC mixtures were produced with steel fiber volume fractions, V_f, equal to 1% and 3%, respectively. Special attention was given to the 3% SFRC mixture in order to ensure flowability of the fresh mixture and uniformity of fiber distribution, since it is known that high fiber content ($\geq$1.5%) can decrease mechanical properties of concrete [35].

The ready-mix concrete was poured into the pan type mixer in three stages (Figure 1a–d). The steel fibers were first dispersed clump-free by hand and added steadily in small amounts into the fresh concrete mixture during stirring in order to prevent clump formation (Figure 1b,c). Stirring continued gradually, ensuring that the produced mixture would obtain uniform material consistency, adequate workability, and homogeneous fiber distribution (Figure 1d). The freshly prepared SFRC mixtures were placed in the cylinders and the molds of the specimens and adequately vibrated. During mixing and casting of the fresh mixtures, no steel fiber segregation was observed.

2.3. Compression and Splitting Tests

Standard concrete cylinders with a diameter to height ratio of 150/300 mm were cast from the plain concrete (PC) batch and from the batches of each SFRC mixture ($V_f = 1\%$ and $V_f = 3\%$). Three specimens from each batch were tested under axial compression and three specimens under splitting tension on the day of the tests of the beams using a universal testing machine (UTM, ELE International, Leighton Buzzard, UK) with an ultimate capacity of 3000 kN. The compression tests were carried out under displacement control mode (about 2 mm/min constant rate of strain) to obtain the post-peak behavior of the SFRC mixtures. During the tests, two linear variable differential transducers (LVDTs, Kyowa, Tokyo, Japan) with 0.01 mm accuracy were installed to measure the axial strain.

The mean and standard deviation values (in parentheses) of the compressive and tensile splitting strength of the examined PC and the SFRC mixtures are given in Table 1. Figure 2b illustrates the experimental compressive behavior of the examined cases in terms of stress–strain curves. It is clear that the addition of the steel fibers to the concrete results in a low increase in the concrete compression strength, but it improves the post-peak behavior and increases the absorbed energy of the fibrous concrete. For SFRC with $V_f = 1\%$, the increase of the compression strength can be considered negligible.

The slope of the descending part of the compressive stress–strain curves is increased as the fiber volume fraction increases (see Figure 2b).

Table 1. Experimental results of compression and splitting tests.

Concrete Mixture	Cylinder Compressive Strength, f_c (MPa)	Splitting Tensile Strength, $f_{ct,spl}$ (MPa)
Plain concrete	24.13 (0.67)	2.62 (0.30)
SFRC with $V_f = 1\%$	25.51 (0.71)	3.32 (0.59)
SFRC with $V_f = 3\%$	27.25 (1.05)	4.96 (0.69)

3. Experimental Program of the Cyclic Tests

The current experimental work was carried out to investigate the hysteretic response of steel fiber reinforced concrete (SFRC) beams with conventional steel reinforcements subjected to cyclic four-point bending load. The influence of two different dosages of steel fibers ($V_f = 1\%$ and 3%) was also studied.

3.1. Specimens' Characteristics

The experimental program included three (3) slender beams, each 2.5 m long, tested in cyclic loading. One beam was constructed using plain concrete as the reference beam (non-fibrous specimen denoted as "B-P") and two beams using SFRC containing 1% and 3% steel fibers (specimens "B-F1" and "B-F3", respectively).

The geometry, the cross-sectional dimensions, and the reinforcement layout of the tested beams are presented in Figure 3. All beams had the same dimensions and conventional reinforcement. Their cross-sections have a width to height ratio of $b/h = 200/200$ mm, an effective depth of $d = 170$ mm, and a shear span of a = 1 m, with the shear span to the effective depth ratio being a/d = 5.9 (slender beams).

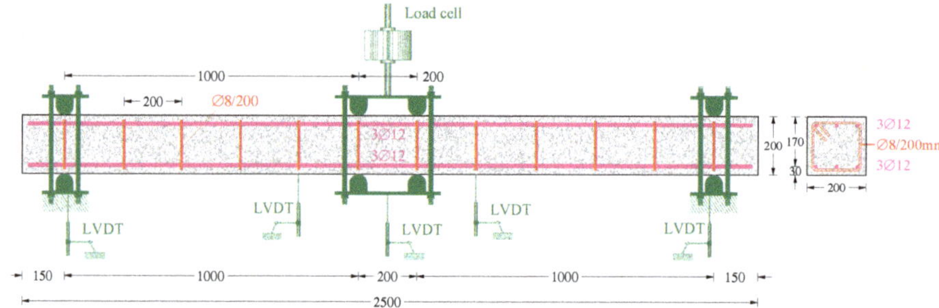

Figure 3. Geometry and reinforcement layout of the tested beams (dimensions in mm).

The top and the bottom longitudinal reinforcement consisted of three common deformed steel bars with a diameter of 12 mm (3Ø12 top and 3Ø12 bottom bars) that corresponded to a geometrical longitudinal reinforcement ratio equal to 1%. Further, the transverse shear reinforcement included closed steel stirrups with a diameter of 8 mm at a uniform spacing of $s = 200$ mm (Ø8/200 mm) that corresponded to a geometrical web reinforcement ratio of 0.25%. The yield tensile strength of the bars and stirrups was $f_{yl} = 590$ MPa.

3.2. Experimental Setup and Instrumentation

A four-point-bending experimental setup was used for the cyclic loading of the RC beams, as presented in Figure 4a. The beam specimens were simply edge-supported on roller supports 2.2 m apart in a rigid laboratory frame. The imposed load was applied using a steel spreader beam in two points 200 mm apart in the mid-span of the beams in order for the shear span to be equal to 1 m and, consequently, the span-to-depth ratio to be equal to 5.9. This way, the adopted loading scheme and apparatus simulate slender beams, although they were designed with a rather low ratio of transverse reinforcement. Thus, the design shear capacity of the reference non-fibrous beam is slightly higher than its flexural strength at yielding, but less than its ultimate flexural capacity. The specimens were subjected to increasing reversal cyclic deformation with a loading history of three loading steps with maximum deflections ±10 mm, ±25mm, and ±40 mm, respectively (Figure 4b).

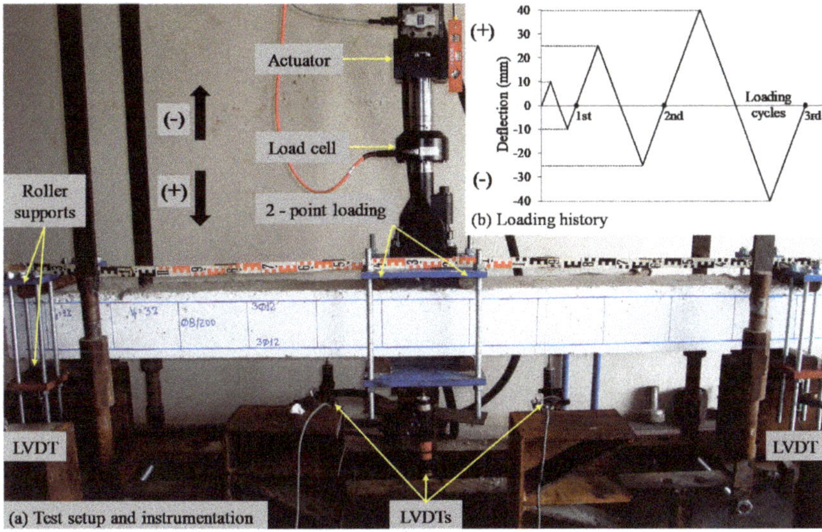

Figure 4. (a) Experimental setup and instrumentation; (b) loading sequence.

The aforementioned quasi-static cyclic loading history was chosen to simulate the seismic effect and to capture the critical issues of the beams' performance, as well as their seismic demands. Further, it is known that every excursion in the inelastic range causes cumulative damage in the structural elements. The amplitude of the inelastic excursions increases with a decrease in the period of the structural system, the rate of increase being very high for short period systems. Therefore, it is obvious that in the adopted loading program, emphasis is on the rapid increase of the rate of the inelastic excursions representing the common cases of very short period systems as the low rise buildings with dual structural systems that include RC frames and walls.

The load was imposed consistently by a pinned-end hydraulic actuator and measured by a load cell with an accuracy of 0.05 kN. The deflections of the beams were measured using five linear variable differential transducers (LVDTs, Kyowa, Tokyo, Japan) with 0.01 mm accuracy. One of the installed LVDTs was placed at the mid-span of the beams, two at a distance of 0.8 m from the supports within the left and the right shear span and two at the supports (see also Figure 4a). Load and corresponding deflection measurements were recorded continuously during the performed tests until the failure of the beams.

4. Test Results

The hysteretic responses of the tested specimens are presented and compared in Figure 5 in terms of the applied load versus mid-span deflection observed curves (Figure 5a) and the envelope curves of the observed maximum loads (Figure 5b). Calculated results of the flexural response of the tested beams according to the methodology proposed by Chalioris and Panagiotopoulos [31] are also presented in Figure 5b. Table 2 summarizes the test results of each beam related to the maximum cycle loads, the hysteretic energy dissipation in terms of the area enclosed within a full cycle of the load versus mid-span deflection curves, and the damage indices per each loading cycle. All specimens completed the testing protocol (three full loading cycles, see Figure 4b) and failed after the third loading cycle during the final downward loading direction. Figures 6–9 illustrates the experimental behavior and the cracking patterns of the tested beams per each loading cycle until the total failure. During the tests, cracks that developed at the downward and upward loading directions were marked using red and green colors, respectively, as shown in the photographs of Figures 6b, 7b, 8b and 9b.

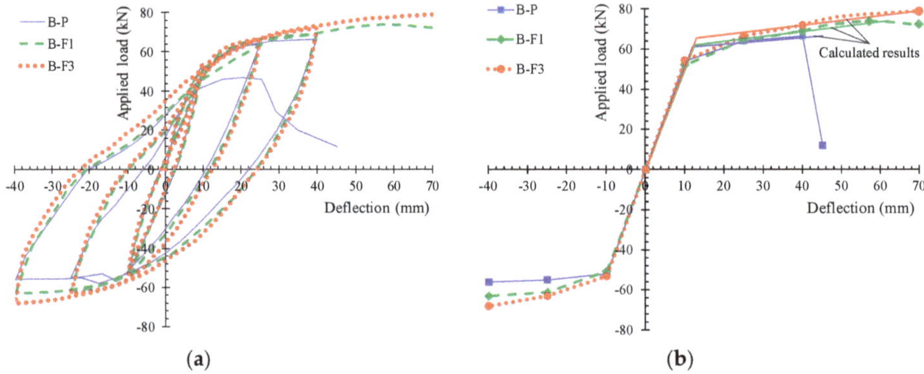

Figure 5. (**a**) Experimental hysteretic responses of the tested beams subjected to cyclic loading; (**b**) envelope curves of the observed maximum loads of the hysteretic response of the tested beams.

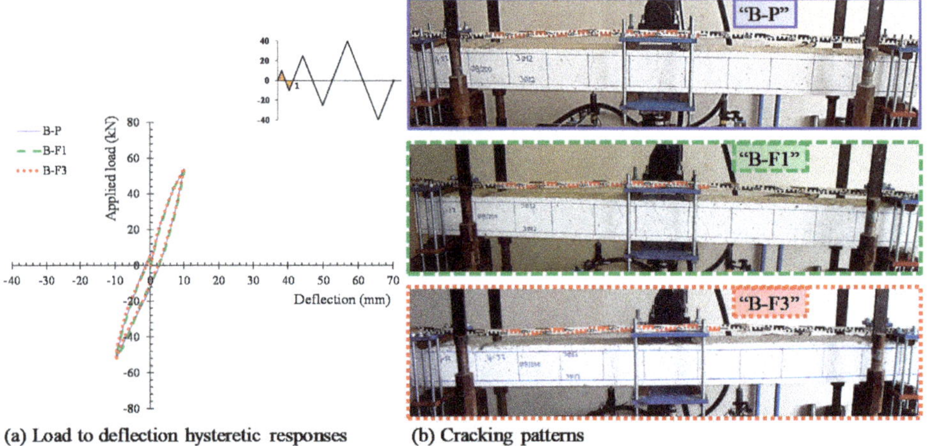

(a) Load to deflection hysteretic responses   **(b) Cracking patterns**

Figure 6. (**a**) Load to mid-span deflection observed curves and (**b**) cracking patterns of the tested beams at the end of the first loading cycle.

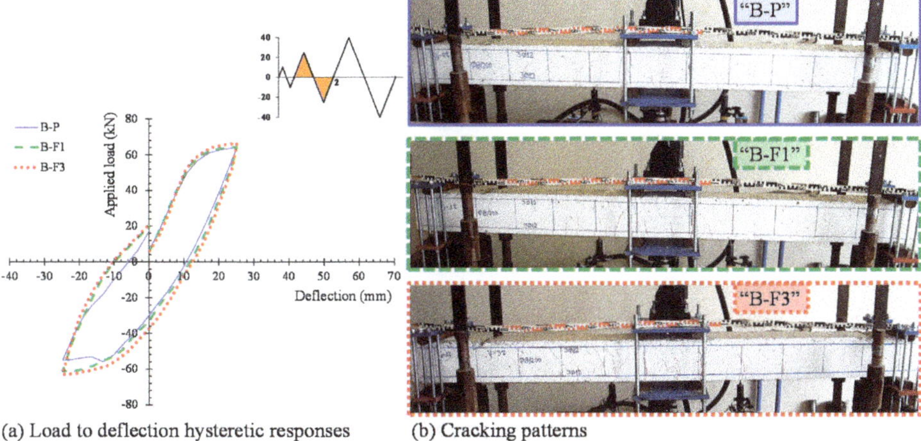

(a) Load to deflection hysteretic responses (b) Cracking patterns

Figure 7. (**a**) Load to mid-span deflection observed curves and (**b**) cracking patterns of the tested beams at the end of the second loading cycle.

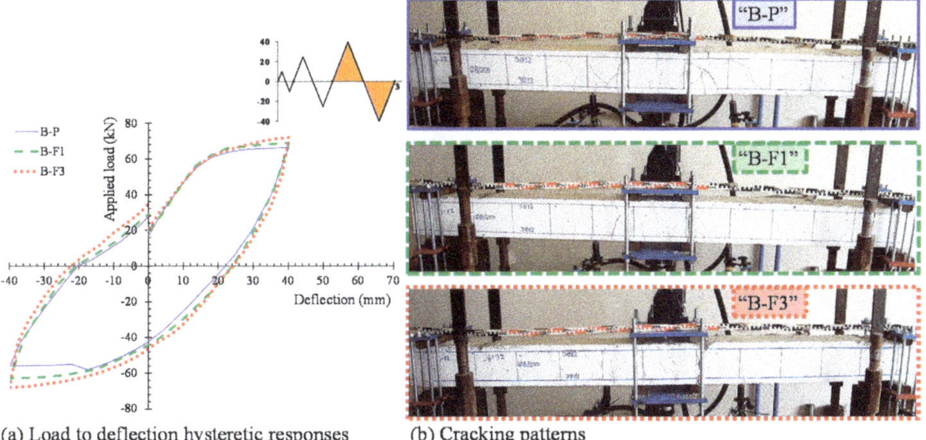

(a) Load to deflection hysteretic responses (b) Cracking patterns

Figure 8. (**a**) Load to mid-span deflection observed curves and (**b**) cracking patterns of the tested beams at the end of the third loading cycle.

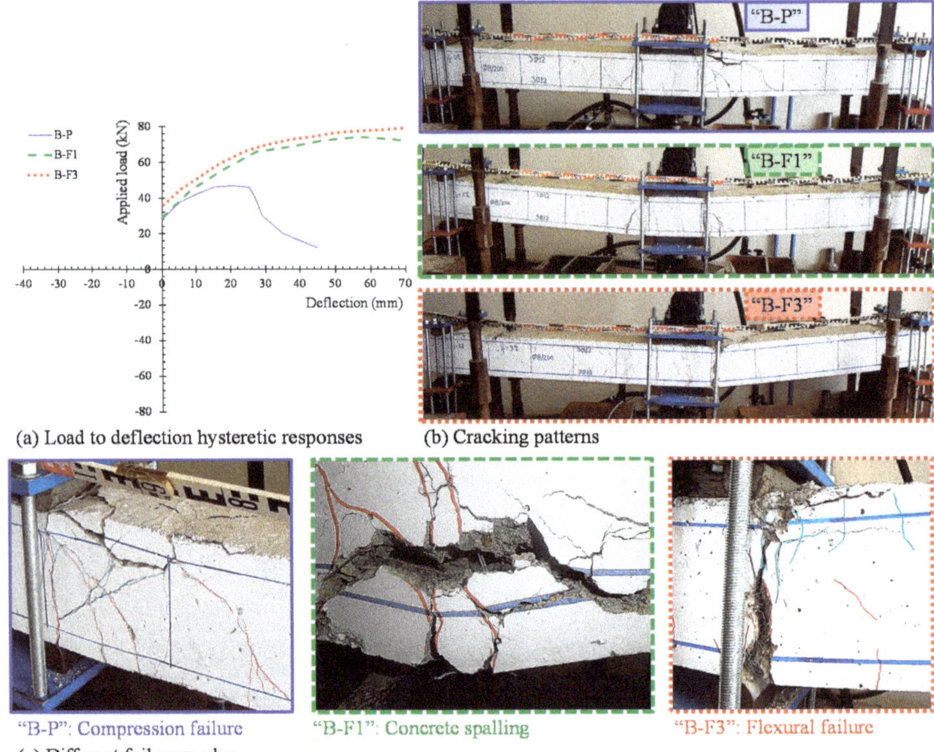

(a) Load to deflection hysteretic responses   (b) Cracking patterns

"B-P": Compression failure   "B-F1": Concrete spalling   "B-F3": Flexural failure

(c) Different failure modes

Figure 9. (**a**) Load to mid-span deflection observed curves, (**b**) cracking patterns, and (**c**) failure modes of the tested beams.

Table 2. Experimental results of the tested beams.

Cycle	Maximum Deflection (mm)	B-P			B-F1			B-F3		
		Load (kN)	Energy (kNmm)	Damage Index	Load (kN)	Energy (kNmm)	Damage Index	Load (kN)	Energy (kNmm)	Damage Index
1st	+10 −10	54.0 −52.0	282.5	0.29	52.0 −50.7	271.4	0.20	54.4 −53.2	269.1	0.16
2nd	+25 −25	64.0 −55.0	1704.5	0.86	64.2 −61.1	2002.2	0.63	66.5 −63.0	2186.6	0.52
3rd	+40 −40	66.5 −56.0	4091.1	1.58	69.0 −63.0	4465.3	1.14	72.0 −68.0	4915.9	0.94
Cracking load:		24.0 kN			30.0 kN			35.2 kN		
Failure mode:		Concrete crushing			Concrete spalling			Flexural failure		

4.1. Beam "B-P"

The cracking patterns per each loading cycle of the reference beam "B-P" (non-fibrous specimen) are presented in Figures 6b, 7b, 8b and 9b. During the first loading cycle, typical vertical flexural cracks were initiated within and in the vicinity of the constant bending moment area. During the second loading cycle, new flexural cracks continued to form while the already existed cracks propagated vertically. In the upward loading direction and after the yielding of the tensional reinforcement at a load of about −55.9 kN, two flexural cracks in both shear spans developed inclined branches at their top and at their bottom ends (see Figure 7b). During the third cycle, more inclined cracks and severe flexural cracks developed without further significant increase of the applied load, as shown in Figure 8. In the upward loading direction, the critical flexural-shear crack at the right shear span continued to propagate in a stable manner.

The developed inclined cracks had shear character that decreased the concrete strength at the compression zone in both loading directions. Thus, in the final downward loading direction (after the third loading cycle, see Figures 5a and 9) the beam failed, due to crushing of the weakened concrete compression zone in a rather brittle manner. The load and the mid-span deflection at that point were equal to +47.0 kN and 20.3 mm, respectively, which correspond to 71% of the maximum observed load and 51% of the maximum deflection of the third loading cycle, respectively.

4.2. Beam "B-F1"

Specimen "B-F1" was constructed using SFRC mixture with steel fiber volume fraction equal to 1%. Up until the second loading cycle, typical flexural vertical and thinner cracks developed within and very close to the constant bending moment area. The cracks were distributed uniformly along this region of the beam in both loading directions in comparison with the reference non-fibrous concrete beam "B-P", as shown in Figures 6b and 7b. After yielding of the tensional reinforcements at loads of about +62.0 kN and −58.0 kN, one severe flexural crack under the right point of the applied load propagated vertically and grew wider. The existence of steel fibers restrained the crack widening and splitting cracks developed at the level of the longitudinal bars located at the constant bending moment area due to the excessive vertical tensile stresses (Figure 8b).

At the final downward loading direction after the third loading cycle (see Figures 5a and 9), the concrete at the tension zone was severely cracked and degraded by tensile stresses and the beam failed after concrete spalling, demonstrating satisfactory ductility. The load and the mid-span deflections at that point were equal to +72.0 kN and 70.0 mm, respectively. The maximum observed applied load was equal to +74.0 kN at a mid-span deflection of about 57.0 mm.

4.3. Beam "B-F3"

Specimen "B-F3" was constructed using SFRC mixture with steel fiber volume fraction equal to 3%. Up until the second loading cycle. a greater number of vertical, thinner, uniformly distributed flexural cracks were developed within and very close to the constant bending moment area in comparison with the fibrous concrete beam "B-F1", as shown in Figures 6b and 7b. Due to the higher volume fraction of the steel fibers, the number of cracks increased while the crack spacing decreased. After yielding of the tensional reinforcements at loads of about +65.5 kN and −60.8 kN, no severe flexural cracks were observed along the flexure-dominated zone of the beam due to the favorable contribution of the fibers. During the third loading cycle, one severe flexural crack under the right point of the applied load propagated vertically and grew wider. The beam exhibited a flexural response at the final downward loading direction (after the third loading cycle, see Figures 5a and 9) and noteworthy ductility until failure. The load and the mid-span deflections at that point were equal to +79.0 kN and 70.0 mm, respectively.

It should be mentioned that the existence of high steel fiber volume fraction enhanced the post-peak behavior of the beam and no inclined cracks or concrete spalling were observed along the length of the beam.

5. Comparisons and Discussion of Test Results

To enable a better understanding of the behavioral characteristics of the fibrous concrete specimens, data in terms of the hysteretic energy absorption, the damage index, and the equivalent viscous damping (Figures 10 and 11) of the tested specimens were acquired and examined in comparison with the data of the reference non-fibrous specimen.

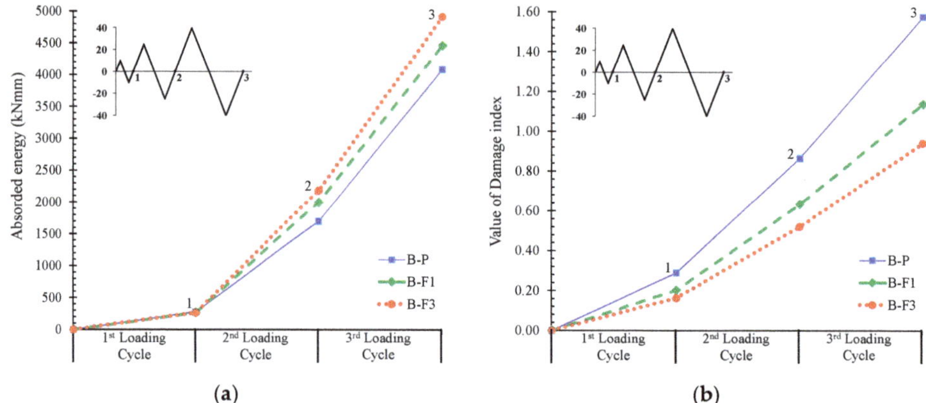

Figure 10. (a) Absorbed energy values per loading cycle of the tested beams; (b) comparisons of the damage indices according to Park and Ang.

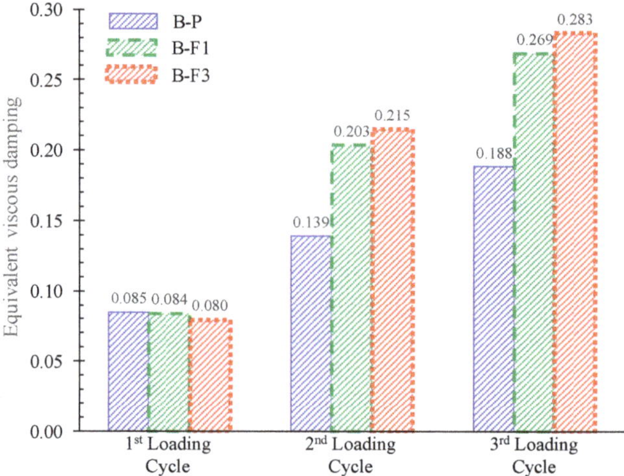

Figure 11. Comparisons of the equivalent viscous damping values of the tested beams.

The value of the damage index in Figure 10b was calculated according to the Park and Ang model [87]. This model is based on the idea that the seismic structural damage is expressed as a linear combination of the damage caused by excessive deformation and the damage accumulated by a repeated cyclic loading effect. Thus, the value of the damage index per loading cycle is given by the following sum:

$$D = \frac{\delta_M}{\delta_u} + \frac{\beta}{Q_y \delta_u} \int dE, \tag{1}$$

where δ_M is the maximum deformation under an earthquake; δ_u is the ultimate deformation under monotonic loading; Q_y is the calculated yield strength; dE is the incremental absorbed hysteretic energy; and β is a non-negative parameter representing the effect of cyclic loading on structural damage given by the following expression:

$$\beta = \left(-0.447 + 0.073\frac{a}{d} + 0.24n_o + 0.314\rho_\ell \right) 0.7^{\rho_w}, \tag{2}$$

where a/d is the shear span to depth ratio; n_o is the normalized axial stress (replaced by 0.2 if $n_o < 0.2$); ρ_ℓ is the longitudinal steel reinforcement ratio as a percentage (replaced by 0.75% if $\rho_\ell < 0.75\%$); and ρ_w is the confinement ratio.

From the observed load versus mid-span deflection envelopes of the tested specimens in Figure 5b, it can be deduced that the addition of steel fibers improves the overall behavior of the specimens "B-F1" and "B-F3" with regard to the non-fibrous reference beam "B-P". Due to the existence of steel fibers, the first cracking load increased (see Table 2), confirming the efficiency of the fibers to delay the propagation of the macro-cracks. Further, the addition of the fibers prevented the formation of inclined cracks along the length of the fibrous specimens in both loading directions by increasing the tensile strength of concrete and the shear capacity of the beam and, consequently, by altering the brittle failure to a ductile flexural failure. Further increase of the amount of steel fibers from 1% to 3% resulted in a noteworthy ductility until failure (see Figures 5–9 and Table 2).

The lower amount of the steel fibers used in the "B-F1" beam seemed to be inefficient in bridging the tensile cracks developed at the level of the longitudinal reinforcing bars, due to the degradation of the steel–concrete bond performance and the beam failure due to concrete spalling. Nevertheless, the addition of a higher volume fraction of steel fibers in "B-F3" prevented the propagation of horizontal cracking, resulting in a typical flexural failure (see also Figure 9).

The deterioration of the steel–concrete bond performance in RC members under cyclic loading was first investigated and discussed by Popov [88]. When reverse load was applied, internal tensile cracks developed around the reinforcing bars and at both sides of their ribs. During the repeated application of the cyclic load, these cracks opened and closed according to the loading direction. At a higher level of the applied load, a gap between the ribs of the steel bars and the surrounding concrete formed and the bars began to move back and forth through a small distance rather freely, which revealed the deterioration of the steel–concrete bond. The existence of the steel fibers improved the steel–concrete bond performance, as also highlighted by Daniel and Loukili [81] and Hameed et al. [79].

Figure 10a shows the absorbed hysteretic energy per loading cycle in terms of the area enclosed within a full cycle of the load versus mid-span deflection curves of the tested beams in Figure 5a. From the improved ductility of the fibrous concrete beams, it is clear that "B-F1" and "B-F3" demonstrated a higher energy dissipation capacity compared to the non-fibrous beam "B-P", as also shown in Figure 10a and Table 2. This is also confirmed by the values of the equivalent viscous damping in Figure 11.

The calculated values of damage indices according to Park and Ang [87] are presented and compared in Figure 10b and Table 2 for the tested specimens. Values higher than 1.0 are related to the collapse of the system corresponding to the situation when the structure is no longer able to absorb additional energy. From these results it can be deduced that the fibrous beams present lower damage index factors compared to the reference beam "B-P". Beam "B-F3", containing a higher percentage of steel fibers, exhibited a considerable lower damage level regarding the reference beam even from the first loading cycle.

6. Experimental Database and Comparisons

There are only a few experimental works available in the literature investigating the hysteretic behavior of RC beams with steel fibers under cyclic loading. The parameters examined so far have been the volume fraction of the steel fibers, with a maximum amount of 1.5%, the thickness of the concrete cover, the concrete strength, and the ratio and the strength of the longitudinal reinforcement. The experimental database presented in Table 3 consists of 39 non-fibrous and fibrous RC beams subjected to reversal deformations derived from the present study and five existing works from the literature [72,79,81,82,85].

Table 3. Experimental data and results of beam specimens under cyclic loading derived from the literature and from the current study.

Beam Codified Name	Geometrical and Mechanical Characteristics								Reinforcements			Experimental Results			
	b/h (mm/mm)	d (mm)	c (mm)	L (mm)	a/d	f_c[1] (MPa)	ρ_{sl} (%)	ρ_{st} (%)	L_f (mm)	D_f (mm)	V_f (%)	P_{max} (kN)	d_{Pmax} (mm)	ΔP_{max}	Failure Mode[2]
Present study															
B-P	200/200	170	16	2200	5.9	24	1.00	0.25	-	-	-	67	40	-	CC
B-F1	200/200	170	16	2200	5.9	26	1.00	0.25	44	1.00	1.0	74	57	11%	CS
B-F3	200/200	170	16	2200	5.9	27	1.00	0.25	44	1.00	3.0	79	70	19%	Fl
Daniel and Loukili [81]															
L-ref	150/300	270	16	2400	4.1	97	0.55	0.34	-	-	-	59	9.2	-	Fl
L-30	150/300	270	16	2400	4.1	110	0.55	0.34	30	0.38	1.0	86	9.9	47%	Fl
L-60	150/300	270	16	2400	4.1	116	0.55	0.34	60	0.75	1.0	93	9.9	59%	Fl
M-ref	150/300	270	14	2400	4.1	95	0.97	0.34	-	-	-	89	10.3	-	Fl
M-30	150/300	270	14	2400	4.1	112	0.97	0.34	30	0.38	1.0	116	11.0	30%	Fl
M-60	150/300	270	14	2400	4.1	117	0.97	0.34	60	0.75	1.0	124	12.1	40%	Fl
H-ref	150/300	270	12	2400	4.1	94	1.52	0.34	-	-	-	151	15.1	-	Fl
H-30	150/300	270	12	2400	4.1	114	1.52	0.34	30	0.38	1.0	159	13.7	5%	Fl
H-60	150/300	270	12	2400	4.1	117	1.52	0.34	60	0.75	1.0	179	15.0	18%	Fl
Harajli and Gharzeddine [82]															
NB20F0.0	240/300	250	40	2000	2.8	43	1.05	0.65	-	-	-	176	5.8	-	CS
NB20F0.5	240/300	250	40	2000	2.8		1.05	0.65	30	0.50	0.5	220	7.0	25%	CS
NB20F1.0	240/300	250	40	2000	2.8		1.05	0.65	30	0.50	1.0	227	12.5	29%	Fl
NB20F1.5	240/300	250	40	2000	2.8		1.05	0.65	30	0.50	1.5	228	15.0	30%	Fl
NB25F0.0	240/300	253	35	2000	2.8	43	1.62	0.65	-	-	-	217	5.5	-	CS
NB25F0.5	240/300	253	35	2000	2.8		1.62	0.65	30	0.50	0.5	242	6.0	12%	CS
NB25F1.0	240/300	253	35	2000	2.8		1.62	0.65	30	0.50	1.0	302	7.5	39%	CS
NB20F1.5	240/300	253	35	2000	2.8		1.62	0.65	30	0.50	1.5	255	7.0	18%	CS
HSC25F0.0	240/300	253	35	2000	2.8	68	1.62	0.65	-	-	-	244	5.5	-	CS
HSC25F0.5	240/300	253	35	2000	2.8		1.62	0.65	30	0.50	0.5	310	7.0	27%	CS
HSC25F1.0	240/300	253	35	2000	2.8		1.62	0.65	30	0.50	1.0	342	7.0	40%	CS
HSC20F1.5	240/300	253	35	2000	2.8		1.62	0.65	30	0.50	1.5	372	9.0	52%	CS
Campione and Mangiavillano [72]															
Beam I	150/150	133	5	550	2.1	31	1.13	0.75	-	-	-	95	6.0	-	Sh
Beam II	150/150	133	5	550	2.1	35	1.13	0.75	30	0.50	1.0	120	9.5	26%	CS
Beam III	150/150	123	15	550	2.2	31	1.23	0.75	-	-	-	105	4.0	-	Sh
Beam IV	150/150	123	15	550	2.2	35	1.23	0.75	30	0.50	1.0	126	9.5	20%	Fl
Beam V	150/150	113	25	550	2.4	31	1.33	0.75	-	-	-	112	9.5	-	Fl
Beam VI	150/150	113	25	550	2.4	35	1.33	0.75	30	0.50	1.0	100	9.5	−11%	Fl
Hameed et al. [79]															
Beam-cont	150/200	176	15	1000	2.8	41–45	0.21	0.38	-	-	-	24	8.0	-	CS
Beam-DF40	150/200	176	15	1000	2.8	41–45	0.21	0.38	30	0.50	0.5	29	8.0	19%	Fl
Tavallali [85]															
CC4-X	406/254	203	31	914	3.0	41	1.84	0.69	-	-	-	245	24.4	-	CC
UC4-X	406/254	203	33	914	3.0	43	1.24	0.69	-	-	-	236	16.0	-	CC
UC4-F	406/254	203	33	914	3.0	44	1.24	0.69	30	0.38	1.5	285	18.3	21%	Fl
UC2-F	406/254	203	33	914	3.0	44	1.24	0.34	30	0.38	1.5	271	24.4	15%	Fl
CC2-F	406/254	203	31	914	3.0	40	1.84	0.34	30	0.38	1.5	255	29.3	4%	Fl
CC4-X$	406/254	203	31	914	3.0	43	1.84	0.69	-	-	-	220	18.3	-	CC
UC2-F$	406/254	203	33	914	3.0	43	1.24	0.34	30	0.38	1.5	267	17.1	13%	Fl

[1] Cylinder compressive strength for plain concrete or for steel fiber reinforced concrete (SFRC). [2] Failure mode notation: Fl: Flexural failure; CS: Concrete spalling; CC: Crushing of concrete compression zone; Sh: Shear failure.

Table 3 presents the geometrical, the mechanical, and the reinforcement data and compares them with the experimental results yielded from the examined beams. The values of ΔP_{max} express the increase of the ultimate load capacity, P_{max}, due to the addition of steel fibers. From these values it is deduced that, although there is an obvious tendency of the SFRC beams to demonstrate increased strength with respect to the reference specimens without fibers, this increase is not consistent, neither is it analogous to the increase of the steel fiber content. This aspect was more or less expected since the provided longitudinal reinforcing bars greatly influence flexural strength and absorbed energy at a more significant level than the added steel fibers do. However, the favorable contribution of the steel fibers is focused on the increase of the shear capacity of the beams, which is higher than the corresponding increase of the flexural strength and, under certain circumstances, causes an important modification of the failure mode from brittle-shear to flexural-ductile.

Further, fibrous beams present higher values of d_{Pmax}, which defines the displacement of the beam at the maximum load capacity, than the corresponding non-fibrous beams. This is attributed to the ability of steel fibers to increase the residual tensile stresses after cracking [7,44,89] and to delay cracking and, consequently, increase the deformation capability of the beams, even in reversal loading conditions.

Furthermore, according to the experimental results presented in Table 3, it can also be concluded that the thickness of concrete cover influences the efficiency of the steel fibers and the nature of the observed failure mode. The majority of the non-fibrous beams failed in a rather brittle manner dominated by shear failure or by concrete crushing or failed due to concrete spalling. The failure mode of the fibrous specimens with sufficient thickness of concrete cover and adequate steel fiber volume fraction (greater than 1.5%, which could be considered as a recommended lower content level of fibers) is flexural with satisfactory ductility. Otherwise, the beams tend to fail due to concrete spalling. It is also noted that, in the majority of the examined cases, the addition of steel fibers in a dosage equal to or less than 1% seems to be insufficient to improve the failure mode of the SFRC beams and to alter it from brittle to ductile. Further, the enhanced flexural behavior with noticeable ductility of concrete beams reinforced with bars and steel fibers is also attributed to the ability of steel fibers to significantly increase tension-stiffening stresses [90].

An additional conclusion that can also be deduced from the comparison of the test results of Table 3 is that the usage of steel fibers could be more efficient by taking extra care to choose the right thickness of concrete cover with the appropriate combination of a rather high steel fiber volumetric proportion.

7. Concluding Remarks

The experimental investigation has been conducted in this paper to investigate the influence of steel fibers on the cyclic hysteretic response of slender RC beams. The following conclusions can be drawn within the scope of this study:

- Based on the hysteretic responses, the cracking, and the failure modes of the tested beams, it can be deduced that the overall performance of the RC beams with steel fibers was improved with respect to the behavior of the reference specimen without fibers, confirming most of the known aspects.
- The non-fibrous reference specimen demonstrated shear diagonal cracking failing in a rather brittle manner, whereas the SFRC beam with 1% steel fibers failed after concrete spalling with satisfactory ductility. Moreover, it is stressed that the SFRC beam with 3% steel fibers exhibited an improved cyclic response, since no inclined cracks of shear nature or concrete spalling were observed along the length of the beam that failed due to flexure with significant ductility.
- The lower amount of the steel fibers seemed to be inefficient to bridge the tensile cracks developed at the level of the longitudinal reinforcing bars due to the degradation of the steel–concrete bond performance and, consequently, SFRC demonstrated concrete spalling. The addition of a higher volume fraction of steel fibers (3%) prevented the propagation of horizontal cracking, resulting in a pronounced flexural failure with enhanced post-peak hysteretic behavior in terms of strength, ductility, cracking performance, absorbed energy capability, and equivalent viscous damping.
- The obtained increase of the first cracking load due to the addition of steel fibers in dosage 1% and 3% was found to be 25% and 47%, respectively. Further, although the increase of the ultimate load during the third cycle of the SFRC beams with 1% and 3% fibers was only 4% and 8%, respectively (downward loading direction), and 13% and 21%, respectively (upward loading direction), compared to the reference beam, the increase of the load-bearing capacity at failure was 61% and 72% for the SFRC beams with 1% and 3% steel fibers, respectively.
- Based on the calculated values of damage indices, it can be deduced that the fibrous beams presented lower damage index factors than the corresponding non-fibrous reference specimen. Further, the SFRC beam containing 3% steel fibers exhibited a considerable lower damage level regarding the reference specimen even from the first loading cycle.
- A systematic report of an experimental database consisting of 39 beams tested under cyclic loading was also presented in order to clarify the effectiveness of steel fibers and their role on

the hysteretic response and the failure mode of RC structural members. Although the amount of the examined specimens was rather limited to derive sound conclusions, it was found that the favorable influence of steel fibers on the overall performance can be achieved by taking extra care to choose the right thickness of concrete cover with a rather high amount of steel fibers (1.5% seems an appropriate lower volume content level of steel fibers).

Author Contributions: All authors contributed extensively to this study, discussed the results and reviews, prepared the manuscript, and agreed to the amendments at all stages of the paper. P.-M.K.K. and C.E.C. performed the tests under the supervision of C.G.K., who designed the experiments.

Acknowledgments: The contribution of A. Droutsas to the construction of the specimens and to the experimental procedure during his undergraduate diploma thesis, along with the technical support of S. Kellis and V. Kanakaris, Laboratory and Technical Employees of the Reinforced Concrete and Seismic Design of Structures Laboratory, is greatly appreciated.

Conflicts of Interest: The authors declare no conflict of interest.

References

1. Wang, R.; Gao, X. Relationship between flowability, entrapped air content and strength of UHPC mixtures containing different dosage of steel fiber. *Appl. Sci.* **2016**, *6*, 216. [CrossRef]
2. Caggiano, A.; Gambarelli, S.; Martinelli, E.; Nisticò, N.; Pepe, M. Experimental characterization of the post-cracking response in hybrid steel/polypropylene fiber-reinforced concrete. *Constr. Build. Mater.* **2016**, *125*, 1035–1043. [CrossRef]
3. Caggiano, A.; Folino, P.; Lima, C.; Martinelli, E.; Pepe, M. On the mechanical response of hybrid fiber reinforced concrete with recycled and industrial steel fibers. *Constr. Build. Mater.* **2017**, *147*, 286–295. [CrossRef]
4. Simoes, T.; Octavio, C.; Valença, J.; Costa, H.; Dias-da-Costa, D.; Júlio, E. Influence of concrete strength and steel fiber geometry on the fiber/matrix interface. *Compos. Part B Eng.* **2017**, *122*, 156–164. [CrossRef]
5. Smarzewski, P. Effect of curing period on properties of steel and polypropylene fiber reinforced ultra-high performance concrete. *Mater. Sci. Eng.* **2017**, *245*. [CrossRef]
6. Shon, C.-S.; Mukashev, T.; Lee, D.; Zhang, D.; Kim, J.R. Can common reed fiber become an effective construction material? Physical, mechanical, and thermal properties of mortar mixture containing common reed fiber. *Sustainability* **2019**, *11*, 903. [CrossRef]
7. Rizzuti, L.; Bencardino, F. Effects of fiber volume fraction on the compressive and flexural experimental behavior of SFRC. *Contemp. Eng. Sci.* **2014**, *7*, 379–390. [CrossRef]
8. Lee, S.-C.; Oh, J.-H.; Cho, J.-Y. Compressive behavior of fiber-reinforced concrete with end-hooked steel fibers. *Materials* **2015**, *8*, 1442–1458. [CrossRef]
9. Perceka, W.; Liao, W.-C.; Wang, Y.-D. High strength concrete columns under axial compression load: Hybrid confinement efficiency of high strength transverse reinforcement and steel fibers. *Materials* **2016**, *9*, 264. [CrossRef]
10. Song, W.; Yin, J. Hybrid effect evaluation of steel fiber and carbon fiber on the performance of the fiber reinforced concrete. *Materials* **2016**, *9*, 704. [CrossRef] [PubMed]
11. Chalioris, C.E.; Liotoglou, F.A. Tests and simplified behavioral model for steel fibrous concrete under compression. In *Advances in Civil Engineering and Building Materials IV*; Chang, S.-Y., Al Bahar, S.K., Husain, A.-A.M., Zhao, J., Eds.; CRC Press/Balkema: Leiden, The Netherlands, 2015; pp. 195–199.
12. Saidani, M.; Saraireh, D.; Gerges, M. Behavior of different types of fiber reinforced concrete without admixture. *Eng. Struct.* **2016**, *113*, 328–334. [CrossRef]
13. Bencardino, F.; Rizzuti, L.; Spadea, G.; Swamy, R.N. Implications of test methodology on post-cracking and fracture behavior of steel fiber reinforced concrete. *Compos. Part B Eng.* **2013**, *46*, 31–38. [CrossRef]
14. Caggiano, A.; Cremona, M.; Faella, C.; Lima, C.; Martinelli, E. Fracture behavior of concrete beams reinforced with mixed long/short steel fibers. *Constr. Build. Mater.* **2012**, *37*, 832–840. [CrossRef]
15. Smarzewski, P.; Barnat-Hunek, D. Fracture properties of plain and steel-polypropylene fiber reinforced high-performance concrete. *Mater. Technol.* **2015**, *49*, 563–571.

16. Simoes, T.; Costa, H.; Dias-da-Costa, D.; Júlio, E. Influence of fibers on the mechanical behavior of fiber reinforced concrete matrixes. *Constr. Build. Mater.* **2017**, *137*, 548–556. [CrossRef]
17. Yoo, D.-Y.; Banthia, N.; Lee, J.-Y.; Yoon, Y.-S. Effect of fiber geometric property on rate dependent flexural behavior of ultra-high-performance cementitious composite. *Cem. Concr. Compos.* **2018**, *86*, 57–71. [CrossRef]
18. Song, P.S.; Hwang, S. Mechanical properties of high-strength steel fiber-reinforced concrete. *Constr. Build. Mater.* **2004**, *18*, 669–673. [CrossRef]
19. Chalioris, C.E.; Karayannis, C.G. Effectiveness of the use of steel fibers on the torsional behavior of flanged concrete beams. *Cem. Concr. Compos.* **2009**, *31*, 331–341. [CrossRef]
20. Lin, W.-T.; Wu, Y.-C.; Cheng, A.; Chao, S.-J.; Hsu, H.-M. Engineering properties and correlation analysis of fiber cementitious materials. *Materials* **2014**, *7*, 7423–7435. [CrossRef]
21. Xue, W.; Chen, J.; Xie, F.; Feng, B. Orientation of steel fibers in magnetically driven concrete and mortar. *Materials* **2018**, *11*, 170. [CrossRef] [PubMed]
22. Choi, M.S.; Kang, S.-T.; Lee, B.Y.; Koh, K.-T.; Ryu, G.-S. Improvement in predicting the post-cracking tensile behavior of ultra-high performance cementitious composites based on fiber orientation distribution. *Materials* **2016**, *9*, 829. [CrossRef]
23. Herrmann, H.; Pastorelli, E.; Kallonen, A.; Suuronen, J.-P. Methods for fiber orientation analysis of X-ray tomography images of steel fiber reinforced concrete (SFRC). *J. Mater. Sci.* **2016**, *51*, 3772–3783. [CrossRef]
24. Trofimov, A.; Mishurova, T.; Lanzoni, L.; Radi, E.; Bruno, G.; Sevostianov, I. Microstructural analysis and mechanical properties of concrete reinforced with polymer short fibers. *Int. J. Eng. Sci.* **2018**, *133*, 210–218. [CrossRef]
25. Eik, M.; Puttonen, J.; Herrmann, H. The effect of approximation accuracy of the orientation distribution function on the elastic properties of short fiber reinforced composites. *Compos. Struct.* **2016**, *148*, 12–18. [CrossRef]
26. Mishurova, T.; Rachmatulin, N.; Fontana, P.; Oesch, T.; Bruno, G.; Radi, E.; Sevostianov, I. Evaluation of the probability density of inhomogeneous fiber orientations by computed tomography and its application to the calculation of the effective properties of a fiber-reinforced composite. *Int. J. Eng. Sci.* **2018**, *122*, 14–29. [CrossRef]
27. Bentur, A.; Mindess, S. *Fiber Reinforced Cementitious Composites: Modern Concrete Technology Series*, 2nd ed.; Taylor & Francis: New York, NY, USA, 2007; p. 624.
28. Olivito, R.S.; Zuccarello, F.A. An experimental study on the tensile strength of steel fiber reinforced concrete. *Compos. Part B Eng.* **2010**, *41*, 246–255. [CrossRef]
29. Karayannis, C.G. Nonlinear analysis and tests of steel-fiber concrete beams in torsion. *Struct. Eng. Mech.* **2000**, *9*, 323–338. [CrossRef]
30. Naaman, A.E. Engineered steel fibers with optimal properties for reinforcement of cement composites. *J. Adv. Concr. Technol.* **2003**, *1*, 241–252. [CrossRef]
31. Chalioris, C.E.; Panagiotopoulos, T.A. Flexural analysis of steel fiber-reinforced concrete members. *Comput. Concr.* **2018**, *22*, 11–25.
32. Wille, K.; El-Tawil, S.; Naaman, A.E. Properties of strain hardening ultra-high performance fiber reinforced concrete (UHP-FRC) under direct tensile loading. *Cem. Concr. Compos.* **2014**, *48*, 53–66. [CrossRef]
33. Ding, X.; Li, C.; Han, B.; Lu, Y.; Zhao, S. Effects of different deformed steel-fibers on preparation and properties of self-compacting SFRC. *Constr. Build. Mater.* **2018**, *168*, 471–481. [CrossRef]
34. Guerini, V.; Conforti, A.; Plizzari, G.A.; Kawashima, S. Influence of steel and macro-synthetic fibers on concrete properties. *Fibers* **2018**, *6*, 47. [CrossRef]
35. Zhou, Y.; Xiao, Y.; Gu, A.; Lu, Z. Dispersion, workability and mechanical properties of different steel-microfiber-reinforced concretes with low fiber content. *Sustainability* **2018**, *10*, 2335. [CrossRef]
36. Caggiano, A.; Xargay, H.; Folino, P.; Martinelli, E. Experimental and numerical characterization of the bond behavior of steel fibers recovered from waste tires embedded in cementitious matrices. *Cem. Concr. Compos.* **2015**, *62*, 146–155. [CrossRef]
37. Montero-Chacón, F.; Cifuentes, H.; Medina, F. Mesoscale characterization of fracture properties of steel fiber-reinforced concrete using a lattice–particle model. *Materials* **2017**, *10*, 207. [CrossRef]
38. Caggiano, A.; Martinelli, E. A unified formulation for simulating the bond behavior of fibers in cementitious materials. *Mater. Des.* **2012**, *42*, 204–213. [CrossRef]

39. Vougioukas, E.; Papadatou, M. A model for the prediction of the tensile strength of fiber-reinforced concrete members, before and after cracking. *Fibers* **2017**, *5*, 27. [CrossRef]

40. Shannag, M.J.; Brincker, R.; Hansen, W. Pullout behavior of steel fibers from cement-based composites. *Cem. Concr. Res.* **1997**, *27*, 925–936. [CrossRef]

41. Karayannis, C.G. Analysis and experimental study for steel fiber pullout from cementitious matrices. *Adv. Compos. Lett.* **2000**, *9*, 243–256. [CrossRef]

42. Georgiadi-Stefanidi, K.; Mistakidis, E.; Pantousa, D.; Zygomalas, M. Numerical modeling of the pull-out of hooked steel fibers from high-strength cementitious matrix, supplemented by experimental results. *Constr. Build. Mater.* **2010**, *24*, 2489–2506. [CrossRef]

43. Kaklauskas, G.; Gribniak, V.; Bacinskas, D. Inverse technique for deformational analysis of concrete beams with ordinary reinforcement and steel fibers. *Procedia Eng.* **2011**, *14*, 1439–1446. [CrossRef]

44. Gribniak, V.; Kaklauskas, G.; Kwan, H.; Bacinskas, D.; Ulbinas, D. Deriving stress–strain relationships for steel fiber concrete in tension from tests of beams with ordinary reinforcement. *Eng. Struct.* **2012**, *42*, 387–395. [CrossRef]

45. Kim, D.J.; Kang, S.H.; Ahn, T.-H. Mechanical characterization of high-performance steel-fiber reinforced cement composites with self-healing effect. *Materials* **2014**, *7*, 508–526. [CrossRef] [PubMed]

46. Nishiwaki, T.; Kwon, S.; Homma, D.; Yamada, M.; Mihashi, H. Self-healing capability of fiber-reinforced cementitious composites for recovery of watertightness and mechanical properties. *Materials* **2014**, *7*, 2141–2154. [CrossRef]

47. Tsonos, A.G. Steel fiber high-strength reinforced concrete: A new solution for earthquake strengthening of old R/C structures. *WIT Trans. Built Environ.* **2009**, *104*, 153–164.

48. Chalioris, C.E.; Karayannis, C.G. Experimental investigation of RC beams with rectangular spiral reinforcement in torsion. *Eng. Struct.* **2013**, *56*, 286–297. [CrossRef]

49. Chalioris, C.E.; Bantilas, K.E. Shear strength of reinforced concrete beam-column joints with crossed inclined bars. *Eng. Struct.* **2017**, *140*, 241–255. [CrossRef]

50. Chalioris, C.E.; Kosmidou, P.-M.K.; Papadopoulos, N.A. Investigation of a new strengthening technique for RC deep beams using carbon FRP ropes as transverse reinforcements. *Fibers* **2018**, *6*, 52. [CrossRef]

51. Canbolat, B.A.; Parra-Montesinos, G.J.; Wight, J.K. Experimental study on the seismic behavior of high-performance fiber-reinforced cement composite coupling beams. *ACI Struct. J.* **2005**, *102*, 159–166.

52. Abbas, A.A.; Mohsin, S.M.S.; Cotsovos, D.M. Seismic response of steel fiber reinforced concrete beam-column joints. *Eng. Struct.* **2014**, *59*, 261–283. [CrossRef]

53. Ma, K.; Qi, T.; Liu, H.; Wang, H. Shear behavior of hybrid fiber reinforced concrete deep beams. *Materials* **2018**, *11*, 2023. [CrossRef]

54. Furlan, S., Jr.; Bento de Hanai, J. Shear behavior of fiber reinforced concrete beams. *Cem. Concr. Compos.* **1997**, *19*, 359–366. [CrossRef]

55. Cucchiara, C.; Mendola, L.; Papia, M. Effectiveness of stirrups and steel fibers as shear reinforcement. *Cem. Concr. Compos.* **2004**, *26*, 777–786. [CrossRef]

56. Juárez, C.; Valdez, P.; Durán, A.; Sobolev, K. The diagonal tension behavior of fiber reinforced concrete beams. *Cem. Concr. Compos.* **2007**, *29*, 402–408. [CrossRef]

57. Chalioris, C.E.; Sfiri, E.F. Shear performance of steel fibrous concrete beams. *Procedia Eng.* **2011**, *14*, 2064–2068. [CrossRef]

58. Lee, J.-Y.; Shin, H.-O.; Yoo, D.-Y.; Yoon, Y.-S. Structural response of steel-fiber-reinforced concrete beams under various loading rates. *Eng. Struct.* **2018**, *156*, 271–283. [CrossRef]

59. Smarzewski, P. Hybrid fibers as shear reinforcement in high-performance concrete beams with and without openings. *Appl. Sci.* **2018**, *8*, 2070. [CrossRef]

60. Smarzewski, P. Analysis of failure mechanics in hybrid fiber-reinforced high-performance concrete deep beams with and without openings. *Materials* **2019**, *12*, 101. [CrossRef]

61. Zhao, J.; Liang, J.; Chu, L.; Shen, F. Experimental study on shear behavior of steel fiber reinforced concrete beams with high-strength reinforcement. *Materials* **2018**, *11*, 1682. [CrossRef] [PubMed]

62. Chalioris, C.E. Analytical approach for the evaluation of minimum fiber factor required for steel fibrous concrete beams under combined shear and flexure. *Constr. Build. Mater.* **2013**, *43*, 317–336. [CrossRef]

63. Kim, K.S.; Lee, D.H.; Hwang, J.H.; Kuchma, D.A. Shear behavior model for steel fiber-reinforced concrete members without transverse reinforcements. *Compos. Part B Eng.* **2012**, *43*, 2324–2334. [CrossRef]

64. Zhang, F.; Ding, Y.; Xu, J.; Zhang, Y.; Zhu, W.; Shi, Y. Shear strength prediction for steel fiber reinforced concrete beams without stirrups. *Eng. Struct.* **2016**, *127*, 101–116. [CrossRef]
65. Lee, D.H.; Han, S.-J.; Kim, K.S.; LaFave, J.M. Shear capacity of steel fiber-reinforced concrete beams. *Struct. Concr.* **2017**, *18*, 278–291. [CrossRef]
66. Colajanni, P.; Recupero, A.; Spinella, N. Generalization of shear truss model to the case of SFRC beams with stirrups. *Comput. Concr.* **2012**, *9*, 227–244. [CrossRef]
67. Hwang, J.-H.; Lee, D.H.; Ju, H.; Kim, K.S.; Seo, S.-Y.; Kang, J.-W. Shear behavior models of steel fiber reinforced concrete beams modifying softened truss model approaches. *Materials* **2013**, *6*, 4847–4867. [CrossRef]
68. Bafghi, M.A.B.; Amini, F.; Nikoo, H.S.; Sarkardeh, H. Effect of steel fiber and different environments on flexural behavior of reinforced concrete beams. *Appl. Sci.* **2017**, *7*, 1011. [CrossRef]
69. Kotsovos, G.; Zeris, C.; Kotsovos, M. The effect of steel fibers on the earthquake-resistant design of reinforced concrete structures. *Mater. Struct.* **2007**, *40*, 175–188. [CrossRef]
70. Soutsos, M.N.; Le, T.T.; Lampropoulos, A.P. Flexural performance of fiber reinforced concrete made with steel and synthetic fibers. *Constr. Build. Mater.* **2012**, *36*, 704–710. [CrossRef]
71. Barros, J.A.; Figueiras, J.A. Flexural behavior of SFRC: Testing and modeling. *ASCE J. Mater. Civ. Eng.* **1999**, *11*, 331–339. [CrossRef]
72. Campione, G.; Mangiavillano, M.L. Fibrous reinforced concrete beams in flexure: Experimental investigation, analytical modelling and design considerations. *Eng. Struct.* **2008**, *30*, 2970–2980. [CrossRef]
73. Li, C.; Geng, H.; Deng, C.; Li, B.; Zhao, S. Experimental investigation on columns of steel fiber reinforced concrete with recycled aggregates under large eccentric compression load. *Materials* **2019**, *12*, 445. [CrossRef]
74. Gribniak, V.; Arnautov, A.K.; Norkus, A.; Tamulenas, V.; Gudonis, E.; Sokolov, A. Experimental investigation of the capacity of steel fibers to ensure the structural integrity of reinforced concrete specimens coated with CFRP sheets. *Mech. Compos. Mater.* **2016**, *52*, 401–410. [CrossRef]
75. Gribniak, V.; Arnautov, A.K.; Norkus, A.; Kliukas, R.; Tamulenas, V.; Gudonis, E.; Sokolov, A.V. Steel fibers: Effective way to prevent failure of the concrete bonded with FRP sheets. *Adv. Mater. Sci. Eng.* **2016**, *2016*. [CrossRef]
76. Gribniak, V.; Tamulenas, V.; Ng, P.-L.; Arnautov, A.K.; Gudonis, E.; Misiunaite, I. Mechanical behavior of steel fiber-reinforced concrete beams bonded with external carbon fiber sheets. *Materials* **2017**, *10*, 666. [CrossRef] [PubMed]
77. Lee, J.-Y.; Shin, H.-O.; Min, K.-H.; Yoon, Y.-S. Flexural assessment of blast-damaged RC beams retrofitted with CFRP sheet and steel fiber. *Int. J. Polym. Sci.* **2018**, *2018*, 2036436. [CrossRef]
78. Kotsovos, M.D.; Baka, A.; Vougioukas, E. Earthquake-resistant design of reinforced concrete structures: Shortcomings of current methods. *ACI Struct. J.* **2003**, *100*, 11–18.
79. Hameed, R.; Duprat, F.; Turatsinze, A.; Sellier, A.; Siddiqi, Z.A. Behavior of reinforced fibrous concrete beams under reversed cyclic loading. *Pak. J. Eng. Appl. Sci.* **2011**, *9*, 1–12.
80. Lepage, A.; Tavallali, H.; Pujol, S.; Rautenberg, J.M. High-performance steel bars and fibers as concrete reinforcement for seismic-resistant frames. *Adv. Civ. Eng.* **2012**, *2012*, 450981. [CrossRef]
81. Daniel, L.; Loukili, A. Behavior of high-strength fiber-reinforced concrete beams under cyclic loading. *ACI Struct. J.* **2002**, *99*, 248–256.
82. Harajli, M.H.; Gharzeddine, O. Effect of steel fibers on bond performance of steel bars in NSC and HSC under load reversals. *J. Mater. Civ. Eng.* **2007**, *19*, 864–873. [CrossRef]
83. Parra-Montesinos, G.J.; Chompreda, P. Deformation capacity and shear strength of fiber-reinforced cement composite flexural members subjected to displacement reversals. *ASCE J. Struct. Eng.* **2007**, *133*, 421–431. [CrossRef]
84. Chalioris, C.E. Steel fibrous RC beams subjected to cyclic deformations under predominant shear. *Eng. Struct.* **2013**, *49*, 104–118. [CrossRef]
85. Tavallali, H.; Lepage, A.; Rautenberg, J.M.; Pujol, S. Concrete beams reinforced with high-strength steel subjected to displacement reversals. *ACI Struct. J.* **2014**, *111*, 1037–1048. [CrossRef]
86. Tsonos, A.-D.G. Ultra-high-performance fiber reinforced concrete: An innovative solution for strengthening old R/C structures and for improving the FRP strengthening method. *WIT Trans. Eng. Sci.* **2009**, *64*, 273–284.
87. Park, Y.-J.; Ang, A.H.-S. Mechanistic seismic damage model for reinforced concrete. *ASCE J. Struct. Eng.* **1985**, *111*, 722–739. [CrossRef]
88. Popov, E.P. Bond and anchorage of reinforcing bars under cyclic loading. *ACI Struct. J.* **1984**, *81*, 340–349.

89. Meskenas, A.; Kaklauskas, G.; Daniunas, A.; Bacinskas, D.; Jakubovskis, R.; Gribniak, V.; Gelazius, V. Determination of the stress-crack opening relationship of SFRC by an inverse analysis. *Mech. Compos. Mater.* **2014**, *49*, 685–690. [CrossRef]

90. Gribniak, V.; Kaklauskas, G.; Torres, L.; Daniunas, A.; Timinskas, E.; Gudonis, E. Comparative analysis of deformations and tension-stiffening in concrete beams reinforced with GFRP or steel bars and fibers. *Compos. Part B Eng.* **2013**, *50*, 158–170. [CrossRef]

Article

Analysis of Residual Flexural Stiffness of Steel Fiber-Reinforced Concrete Beams with Steel Reinforcement

Violetta K. Kytinou, Constantin E. Chalioris * and Chris G. Karayannis

Laboratory of Reinforced Concrete and Seismic Design of Structures, Department of Civil Engineering, Faculty of Engineering, Democritus University of Thrace (D.U.Th.), 67100 Xanthi, Greece; vkytinou@civil.duth.gr (V.K.K.); karayan@civil.duth.gr (C.G.K.)
* Correspondence: chaliori@civil.duth.gr; Tel.: +30-2541-079632

Received: 17 May 2020; Accepted: 10 June 2020; Published: 13 June 2020

Abstract: This paper investigates the ability of steel fibers to enhance the short-term behavior and flexural performance of realistic steel fiber-reinforced concrete (SFRC) structural members with steel reinforcing bars and stirrups using nonlinear 3D finite element (FE) analysis. Test results of 17 large-scale beam specimens tested under monotonic flexural four-point loading from the literature are used as an experimental database to validate the developed nonlinear 3D FE analysis and to study the contributions of steel fibers on the initial stiffness, strength, deformation capacity, cracking behavior, and residual stress. The examined SFRC beams include various ratios of longitudinal reinforcement (0.3%, 0.6%, and 1.0%) and steel fiber volume fractions (from 0.3 to 1.5%). The proposed FE analysis employs the nonlinearities of the materials with new and established constitutive relationships for the SFRC under compression and tension based on experimental data. Especially for the tensional response of SFRC, an efficient smeared crack approach is proposed that utilizes the fracture properties of the material utilizing special stress versus crack width relations with tension softening for the post-cracking SFRC tensile response instead of stress–strain laws. The post-cracking tensile behavior of the SFRC near the reinforcing bars is modeled by a tension stiffening model that considers the SFRC fracture properties, the steel fiber interaction in cracked concrete, and the bond behavior of steel bars. The model validation is carried out comparing the computed key overall and local responses and responses measured in the tests. Extensive comparisons between numerical and experimental results reveal that a reliable and computationally-efficient model captures well the key aspects of the response, such as the SFRC tension softening, the tension stiffening effect, the bending moment–curvature envelope, and the favorable contribution of the steel fibers on the residual response. The results of this study reveal the favorable influence of steel fibers on the flexural behavior, the cracking performance, and the post-cracking residual stress.

Keywords: reinforced concrete; steel fiber-reinforced concrete (SFRC); tension softening; tension stiffening; finite element (FE) analysis; smeared crack model; constitutive analysis; residual stresses; flexural behavior; numerical analysis

---

1. Introduction

The addition of randomly distributed discrete fibers in concrete significantly improves the overall performance of reinforced concrete (RC) structural members because of its enhanced tensile properties and cracking control. This improvement is mainly attributed to the crack-bridging phenomenon that is observed across crack surfaces due to the incorporated steel fibers. The advanced characteristics of fiber-reinforced cement-based elements depend on many factors such as size, type, elastic properties, aspect ratio, and volume fraction of fibers, and each type of fiber can be effective in some specific function [1,2].

Concrete can tolerate only a small tensile stress before it cracks in a quasi-brittle manner. Research into the use of short steel fibers as mass reinforcement in concrete over the last five decades or so has developed them as an option to mitigate the unfavorable cracking characteristics of concrete, due to the ability of the fibers to bridge and transmit tensile stresses across cracks [3,4]. Thus, cracking and, eventually, tensile failure of steel fiber-reinforced concrete (SFRC) elements requires debonding and eventually full pullout of the steel fibers crossing the developing crack. Consequently, stress situations that depend on the tensile strength of the material in the presence of steel fibers generally contribute toward the improvement of an energy-absorbing mechanism resulting in a substantially improved post-cracking behavior [5–7]. In this direction, the full or even partial replacement of traditional stirrups is crucial in shear-critical RC members because the use of SFRC reduces reinforcement congestion since in these cases design criteria require a high amount of transverse reinforcement that leads to a short spacing of stirrups [8–12].

Attempts have been made in the past to combine different types of fibers and their addition to cementitious composites to improve the cracking performance in concrete at different levels [13,14]. Small and soft fibers control the initiation and propagation of microcracks, whereas large and strong fibers control macrocracks. Such hybrid fiber-reinforced composites can also offer more attractive engineering properties because the presence of one type of fiber effectively uses the properties of the other fiber [15,16].

Tension stiffening is the ability of concrete to carry tension between cracks, which provides additional stiffness for a RC member in tension before the reinforcement yields [17]. It can significantly affect member rigidity, deflection, and width of cracks under service loads [18–20]. Increased tension stiffening was observed for the high-strength concrete specimens, and the higher reinforcement ratio decreases the tension stiffening effect [21]. The presence of steel fibers is effective in controlling splitting cracks and significantly increases the tension stiffening effect because SFRC can carry tensile stress through the crack [22–26].

Nevertheless, the rather limited utilization of SFRC in structural applications can be attributed to the difficulty in establishing reliable and rational design procedures that describe and predict the material's behavior for each design limit state. At the strength limit state in flexure, a conservative view is often taken to ignore the tensile capacity of SFRC at critical design sections [27]. This approach is often defended by the fact that the fibers are randomly orientated within a concrete matrix as opposed to aligned in the direction oriented to the principal tensile stress [28]. However, as SFRC can carry tensile stresses not only at the cracks but also between them, as a result of bond, the member is stiffer because of the steel fibers. This phenomenon termed "tension stiffening" combines the fiber interaction in cracked concrete, the bond behavior of steel reinforcing bars, and fracture mechanics of concrete, and it needs to be considered in service calculations when designing structural members made of SFRC.

Further, it is known that, in most of the practical applications of SFRC construction, steel fibers are combined with conventional steel reinforcement such as transverse stirrups and longitudinal reinforcing bars. When both forms of reinforcement bridge a crack, the transmitted tensile stress across the crack is shared between the reinforcing bars and the fibers [29]. Consequently, the average steel strains are smaller than they would be without the fibers between the cracks, leading to more closely spaced cracks [30]. Thus, the addition of steel fibers effectively increases tension stiffening effect, improves crack control, and permits the use of higher strength reinforcing steels while still maintaining control of crack widths depending on the type and dosage of fibers used. Tension stiffening effects are also useful for assessing cracking behavior at service loads and to develop suitable material models of cracked SFRC for analysis that use averaged stresses and strains to predict member behavior [31–33].

In this paper, the effectiveness of steel fibers to improve the flexural response of realistic SFRC structural members with conventional steel reinforcement using nonlinear 3D finite element (FE) analysis is investigated. For this reason, an efficient FE model implemented in ABAQUS [34] software has been developed, considering the added steel fibers according to their type, aspect ratio, and percentage of the addition so that a more realistic calculation of the behavior of SFRC members is possible. The proposed model employs the nonlinearities of the materials by constitutive relationships for the SFRC under compression and tension based on experimental data in order to simulate the flexural performance of SFRC beams with steel reinforcing bars and stirrups accurately and reach results close to reality. Experimental testing is more expensive and time consuming compared to computer analysis, thus being able to produce reliable results by studying individual RC members and even structures by FE analysis is vital. Detailed and precise FE modeling can even replace experimental testing for research purposes and enable us to explore a much more extensive range of innovative design solutions and to solve complex engineering problems [35,36]. Seventeen analyses were executed using the proposed model, and the results are discussed in this paper. Comparing the FE models' responses with the published experimental results reveals that the proposed model can successfully be used to predict the short-term behavior of SFRC beams under monotonic loading. Therefore, the favorable contribution of the steel fibers on the flexural behavior, the cracking performance, and the post-cracking residual strength and response are presented and discussed.

2. Constitutive Laws of the Materials and FE Model Formulation

The main aspects of the SFRC analyses developed in the current research are presented in this section. The constitutive laws of SFRC have been derived from experimental results and models proposed by the authors in previous studies available in the literature. Especially for the simulation of the SFRC under tension, a smeared crack approach is adopted. The proposed smeared crack model utilizes the fracture properties of the material and employs constitutive relationships of stress versus crack width with tension softening for the post-cracking SFRC tensile response instead of stress–strain laws. Further, the post-cracking tensile behavior of the SFRC near the reinforcing bars is modeled by a tension stiffening model that considers the SFRC fracture properties and the reinforcement characteristics.

2.1. Stages of SFRC Cracking and Tension Stiffening

Figure 1 illustrates an idealized moment versus curvature response of a SFRC member including four stages of cracking. In the first stage (Stage I), concrete, as well as steel, are considered to behave in a linear and elastic way. Stresses and strains are uniformly distributed along the beam, and the applied load is shared between concrete and reinforcement. Concrete is considered macroscopically uncracked, and the tensile stress in concrete does not exceed its tensile strength. Steel fibers are assumed to not contribute at this stage, as a crack must be formed for the slip between concrete and fiber to occur, and fiber to be activated.

The second stage (Stage II) begins when the first crack is formed. Once tensile stress overcomes the tensile strength of SFRC and cracking moment is exceeded, a flexural crack begins to develop in the weakest cross-section of the beam. At this stage, the concrete–steel bond mechanism is activated, and the tension stiffening effect starts contributing to the member response. Tension stiffening is the ability of concrete between the cracks to bear tensile stresses, and through the bond mechanism, tensile stresses can be transferred from reinforcement to concrete. In addition, fibers tend to "bridge" the formed cracks, improve the bond efficiency due to the ability to also carry tension across cracks and thereby improve the tension stiffening performance of the member. As a result, the average tensile stiffness of cracked RC is greater than that of the reinforcing bars alone. Therefore, if this effect is disregarded, the stiffness of the cracked RC may be underestimated. As the imposed load increases, more cracks are developed, and the process goes on until the final pattern of cracking is established. If the space between the cracks is not large enough to cultivate a sufficient bond to allow the developed stress to reach tensile strength, no further cracking will occur.

The third stage (Stage III) occurs when the crack pattern has stabilized, and the crack spacing remains constant; cracks self-propagate at this point. The tension stiffening effect decays as the load increases. In the final stage (Stage IV), as the member's average tensile stress begins to exceed the yield strain of the reinforcing bars, the member's response is determined by the behavior of the fibers and the reinforcing bars.

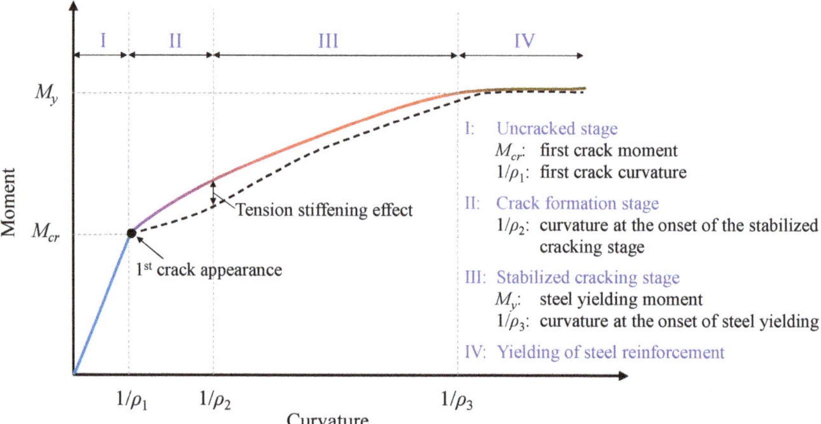

Figure 1. Stages of cracking and tension stiffening in SFRC members.

2.2. Concrete Damaged Plasticity

The well-known concrete damaged plasticity (CDP) model is usually employed in FE analyses using Abaqus to simulate the nonlinear behavior of concrete. CDP is a general-purpose commonly used model that can be applied to all types of concrete structures under monotonic, cyclic, and dynamic loading. It is a continuum, plasticity-based, damage model that assumes tensile cracking and compressive crushing are the two main failure mechanisms in concrete [37]. In the FE analyses performed in this study, the CDP model has been used for comparison reasons to evaluate the effectiveness and the accuracy of the proposed materials model. Some main aspects and equations of the CDP model are briefly reported below.

The CDP model adopts the elastic model to define the mechanical properties of the materials in the elastic stage. However, when entering the stage of damage, to describe the stiffness degradation, the modulus of elasticity is described as:

$$E = (1 - d)E_0 \tag{1}$$

where E_0 is the initial elastic modulus and d is the plastic damage factor (d_c or d_t, for compression and tension, respectively), which varies from $0 \leq d \leq 1$, with zero indicating the undamaged material, to one indicating a complete loss of strength.

According to Lubliner et al. [37], plastic degradation occurs only within the softening range and the stiffness is proportional to the material's cohesion. Solving the previous equation for the plastic damage factor, it is derived:

$$\frac{E}{E_0} = 1 - d = \frac{c}{c_{max}} \rightarrow d = 1 - \frac{c}{c_{max}} \tag{2}$$

where c is cohesion in the yield criteria, which is proportional to stress, and c_{max} is proportional to the strength of the concrete.

Adopting Equation (2) to uniaxial tension or compression converts to the following expression:

$$d = 1 - \frac{\sigma}{f} \tag{3}$$

where f is either the compressive or the tensile strength of concrete and σ is the compressive or the tensile stress, respectively.

The uniaxial compressive and tensile responses of concrete in relation to the concrete damage plasticity model subjected to compression and tension load are calculated based on the following expressions (see also Figure 2a,b for notation):

$$\sigma_c = (1 - d_c)E_0\left(\varepsilon_c - \varepsilon_{c,pl}\right) \tag{4}$$

$$\sigma_t = (1 - d_t)E_0\left(\varepsilon_t - \varepsilon_{t,pl}\right) \tag{5}$$

where $\sigma_c - \varepsilon_c$ and $\sigma_t - \varepsilon_t$ are the concrete compressive and tensile stress-strain relationships, respectively, $\varepsilon_{c,pl}$ and $\varepsilon_{t,pl}$ are the compressive and tensile plastic strain of concrete, respectively.

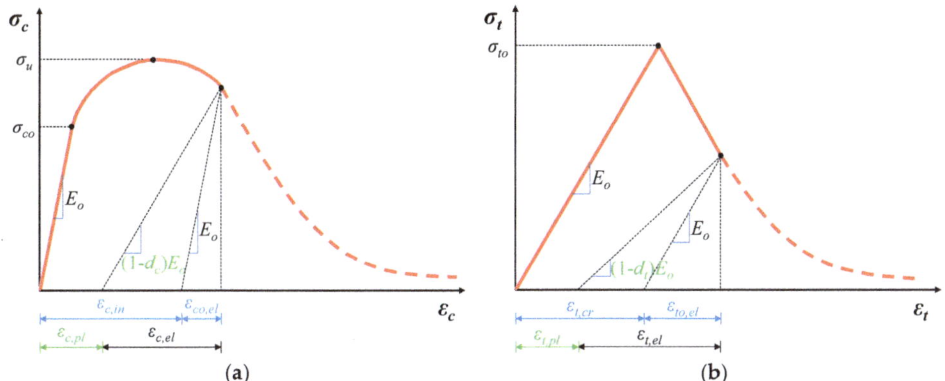

Figure 2. CDP model in Abaqus: (**a**) definition of compressive; and (**b**) definition of tensile damage.

Using the above equations, the uniaxial stress–strain curves can be converted into stress versus plastic strain curves. This conversion is carried out by Abaqus automatically when user provides in a tabular form all the variables contained in the above equations. The strain variables for the compressive behavior are calculated as follows (see also Figure 2a for notation):

$$\varepsilon_{c,in} = \varepsilon_{c,tot} - \varepsilon_{co,el}, \tag{6}$$

$$\varepsilon_{co,el} = \sigma_{co}/E_0, \tag{7}$$

The equivalent plastic strain for crushed concrete, $\varepsilon_{c,pl}$, is estimated using:

$$\varepsilon_{c,pl} = \varepsilon_{c,in} - \frac{d_c}{1 - d_c}\varepsilon_{co,el}, \tag{8}$$

where d_c is the compressive damage variable:

$$d_c = 1 - \sigma_c/\sigma_{cu}, \tag{9}$$

where and $\sigma_{cu} = f_{cm}$ (mean concrete compressive strength).

In the same manner, the cracking strain $\varepsilon_{t,cr}$ and $\varepsilon_{t0,el}$ for the tensile behavior are defined as (see also Figure 2b for notation):

$$\varepsilon_{t,cr} = \varepsilon_{t,tot} - \varepsilon_{t0,el}, \tag{10}$$

$$\varepsilon_{t0,el} = \sigma_{t0}/E_0, \tag{11}$$

where the plastic strain for damaged concrete is:

$$\varepsilon_{t,pl} = \varepsilon_{t,cr} - \frac{d_t}{1 - d_t}\varepsilon_{to,el},$$ (12)

with d_t the tensile damage variable:

$$d_t = 1 - \sigma_t/\sigma_{to},$$ (13)

and $\sigma_{to} = f_{ctm}$ (mean concrete tensile strength).

Five other CDP material associated parameters must be described to completely define the inelastic behavior of concrete [38–40]:

- ψ is the dilatation angle, which affects the amount of plastic volume deformation. Various values (20°–45°) are found in the literature. A dilation angle close to the friction angle of the material, for concrete 56.3°, results in a ductile behavior, whereas a low value near 0° leads to very brittle behavior.
- ϵ, the flow potential eccentricity, which is the rate of approach of the plastic potential hyperbolic to its asymptote, is used to define the shape of the plastic potential surface in the meridional plane.
- σ_{bo}/σ_{co} is the ratio of the strength in the biaxial state to the strength in the uniaxial state.
- K_c is the ratio of the tensile to the compressive meridian and determines the shape of the yield surface.
- μ is the viscosity parameter, which is used to help the analysis achieve a good convergence.

The values used in the current study are presented in Table 1.

Table 1. Parameters input of the CDP model.

Parameter	Value
ψ	31°
K_c	2/3
σ_{bo}/σ_{c0}	1.16
ϵ	0.10
μ	0.0001

2.3. Proposed Model of the SFRC Compressive Behavior

Steel fibers practically become effective after cracking and consequently influence post-peak compressive behavior of SFRC. The amount, geometry, and the bond characteristics of the added steel fibers are the parameters affecting the improved post-cracking ductility of SFRC under compression. This enhancement is caused by the ability of steel fibers to exhibit a progressive debonding failure that provides crack-bridging, confinement, and crack growth resisting of the developed cracks [41–44]. However, compressive strength seems to be slightly increased, especially in SFRC mixtures with a low volume fraction of steel fibers. Most analytical models proposed for the simulation of the SFRC compressive response are based on regression analysis of experimental results available in the literature and taking into account the previously mentioned parameters [45–50].

The model adopted in this paper to simulate the compressive stress–strain (σ_c–ε_c) behavior of SFRC was proposed by Chalioris and Panagiotopoulos [51], and was derived from test data of 125 stress–strain curves and 257 strength values. Based on this model, the initial ascending compressive behavior until the maximum strength of common SFRC mixtures with cylinder compressive strength, $f_{c,SF}$, less than 50 MPa is described by the expression (see also Figure 3 for notation):

$$\sigma_c = f_{c,SF}\left[1 - \left(1 - \frac{\varepsilon_c}{\varepsilon_{co,SF}}\right)^2\right],$$ (14)

where $\varepsilon_{co,SF}$ is the SFRC strain that corresponds to the ultimate compressive stress.

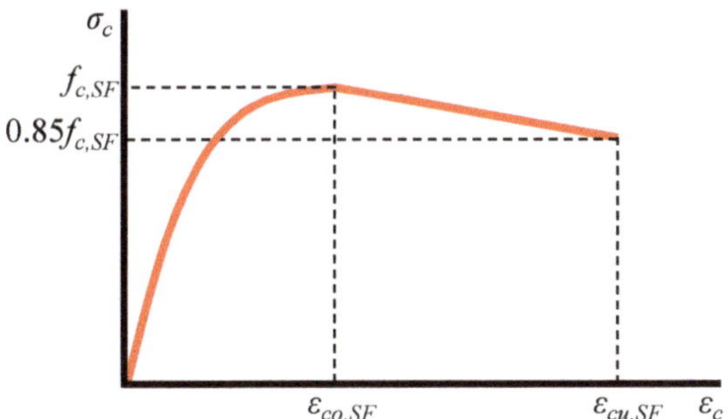

Figure 3. Stress–strain model of SFRC under compression.

The post-peak response is assumed to be linear from the value of the compressive strength, $f_{c,SF}$, until the value of $0.85 f_{c,SF}$. The parameters of the strength, $f_{c,SF}$, the strain at ultimate strength, $\varepsilon_{co,SF}$, and the ultimate strain, $\varepsilon_{cu,SF}$, that corresponds to the post-peak compressive stress $0.85 f_{c,SF}$ are calculated using the following expressions:

$$f_{c,SF} = f_c(0.2315F + 1), \tag{15}$$

$$\varepsilon_{co,SF} = \varepsilon_{co}(0.95F + 1), \tag{16}$$

$$\varepsilon_{cu,SF} = \varepsilon_{co,SF}(1.40F + 1), \tag{17}$$

where f_c is the cylinder compressive strength of plain concrete, ε_{co} is the strain corresponding to the ultimate stress of plain concrete and usually takes the value of 0.002, and F is fiber factor that depends on the characteristics of the added steel fibers in the SFRC mixture and can be evaluated as: $F = \beta V_{SF}(l_{SF}/d_{SF})$, where β is a bond factor that equals to 0.50 for round fibers, 0.75 for deformed (such as hooked, crimped, and undulated) fibers, and 1.0 for indented fibers, and V_{SF}, l_{SF}, and d_{SF} are the volume fraction, length, the diameter of the steel fibers, respectively.

The ability of this analytical model to efficiently predict the overall compressive behavior of SFRC was verified against experimental stress–strain curves of various SFRC mixtures by Chalioris and Panagiotopoulos [51].

2.4. Proposed Smeared Crack Model of the SFRC Tensile Behavior

Steel fibers significantly improve the tensile performance of SFRC increasing the strength, mainly enhancing the deformation capability of the post-peak behavior by providing pseudo-ductile response due to their gradual debonding procedure [52]. Many analytical stress–strain expressions have been developed to describe the SFRC response under tension [53,54]. In this paper, a smeared crack analysis that has previously been developed and experimentally verified by the authors [55,56] to model the behavior of plain concrete under tension is properly modified and adopted for the SFRC tensile behavior. Similar smeared crack models have also been applied recently in FE simulations to evaluate the uncertainty of crack width of RC beams [57]. The adopted smeared crack model is properly modified herein to employ the favorable contribution of the added steel fibers on the tensile response.

The proposed model employs constitutive relationships in terms of normal stress versus crack width for the post-cracking tensile response of SFRC instead of describing the crack process by stress–strain laws. Based on this aspect, crack propagation of SFRC takes place with the formation of a fracture process zone that is initiated at maximum tensile strength of SFRC, $f_{t,SF}$, and is characterized by the gradually reduction of the strength during deformation. This fracture process zone is defined as the boundary of the strain softening region, which is considered as a SFRC property and is assumed to be wider than the region of visible cracks. It is also assumed that between the cracks of this zone there are less damaged or even elastic parts. Thus, the total tensional strain, ε_t, is considered as the sum of an elastic, $\varepsilon_{t,el}$, and a fracture component, $\varepsilon_{t,fr}$, which can be estimated based on the following relationships (see also Figure 4):

$$\varepsilon_t = \varepsilon_{t,el} + \varepsilon_{t,fr}, \tag{18}$$

$$\varepsilon_{t,el} = \sigma_t / E_{t,SF}, \tag{19}$$

$$\varepsilon_{t,fr} = w_t / L_{fr,SF}, \tag{20}$$

where σ_t is the tensile stress, $E_{t,SF}$ is the modulus of elasticity under tension, w_t is the crack width, and $L_{fr,SF}$ is the length of fracture process zone of the SFRC.

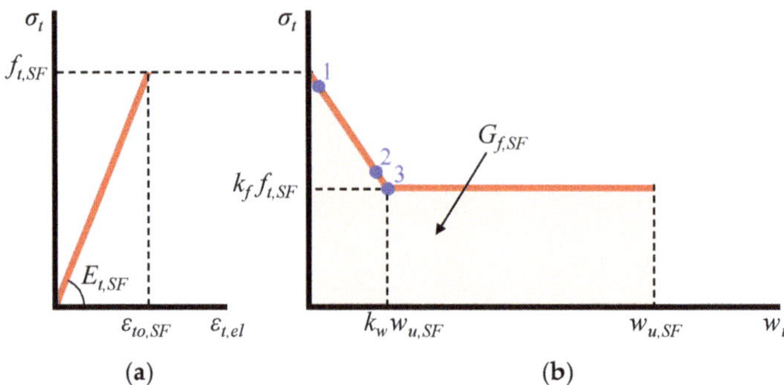

Figure 4. Tensile model of SFRC including: (**a**) stress versus strain elastic response; and (**b**) stress versus crack width post-cracking response with tension softening.

It is noted that the influence of cracks on the deformations of the beams is considered through the implementation of the aforementioned smeared crack approach to the nonlinear FE analysis using Abaqus software to provide numerical behavioral curves and cracking patterns based on the stress distribution of the simulated SFRC beams.

2.4.1. Elastic Response

The parameters of the elastic component of the proposed model for the simulation of the SFRC tensile behavior, as shown by the σ_t–$\varepsilon_{t,el}$ linear diagram in Figure 4, are calculated based on the properties of the corresponding plain concrete and the added steel fibers, as shown in the following relationships [58]:

$$f_{t,SF} = \varepsilon_{to,SF} E_{t,SF}, \tag{21}$$

$$\varepsilon_{to,SF} = 0.167 n_{l,el} V_{SF} \left(\varepsilon_{y,SF} - \varepsilon_{to} \right) + \varepsilon_{to}, \tag{22}$$

$$\varepsilon_{to} = f_t / E_t, \tag{23}$$

$$\varepsilon_{y,SF} = f_{SF} / E_{SF}, \tag{24}$$

$$E_{t,SF} = \frac{3}{8} \left[E_t (1 - V_{SF}) + E_{SF} V_{SF} \right] + \frac{5}{8} \left[\frac{E_{SF} E_t}{E_{SF}(1 - V_{SF}) + E_t V_{SF}} \right], \tag{25}$$

where $E_{t,SF}$ is the modulus of elasticity under tension of SFRC; f_t, ε_{to}, and E_t are the ultimate tensile strength, the corresponding strain, and the modulus of elasticity under tension of the plain concrete, respectively; f_{SF}, $\varepsilon_{y,SF}$, and E_{SF} are the ultimate tensile strength at yielding, the yield strain, and the modulus of elasticity of the steel fiber, respectively; and $n_{l,el}$ is the ratio of the average fiber elastic stress to the maximum fiber stress that is usually equal to 0.5.

2.4.2. Fracture Response with Tension Softening

The fracture characteristics and the tension softening response of SFRC define the parameters of the fracture component of the proposed smeared crack model. The energy required for the formation of the cracks included in the fracture process zone and for the fully opening of one single crack for a unit area crack plane is the fracture energy, $G_{f,SF}$, and can be expressed as:

$$G_{f,SF} = \int_{f_{t,SF}}^{0} \sigma_t dw_t \xrightarrow{w_t = L_{fr,SF}\varepsilon_{t,fr}} G_{f,SF} = L_{fr,SF}\int_{f_{t,SF}}^{0} \sigma_t d. \tag{26}$$

The post-peak fracture response (σ_t–w_t curve in Figure 4) is described by a linear descending line until the point of the maximum post-cracking stress, $k_f f_{t,SF}$, and the corresponding crack width, $k_w w_{u,SF}$, and then a horizontal line of constant tensile stress ($\sigma_t = k_t f_{t,SF}$ when $w_t > k_w w_{u,SF}$) until the ultimate considered crack width, $w_{u,SF}$. Thus, the fracture energy can also be expressed in terms of the area under the curve of SFRC tensile stress versus crack width, as shown by the σ_t–w_t bilinear diagram in Figure 4:

$$G_{f,SF} = k_f f_{t,SF} w_{u,SF} + 0.5(f_{t,SF} - k_\sigma f_{t,SF})k_w w_{u,SF} \rightarrow$$
$$G_{f,SF} = f_{t,SF} w_{u,SF}(k_f + 0.5k_w - 0.5k_f k_w), \tag{27}$$

The values of the maximum post-cracking stress and crack width depend on the SFRC characteristics and can be estimated using the following expressions for short deformed steel fibers [58]:

$$k_f f_{t,SF} = 0.405 n_l \sigma_{fu} V_{SF} \rightarrow k_f = \frac{0.405 n_l \sigma_{fu} V_{SF}}{\varepsilon_{to,SF} E_{t,SF}} \tag{28}$$

$$k_w w_{u,SF} = (3 \div 8)L_{fr,SF}\varepsilon_{to,SF} \rightarrow k_w = \frac{(3 \div 8)L_{fr,SF}\varepsilon_{to,SF}}{w_{u,SF}} \tag{29}$$

where σ_{fu} is the ultimate stress of the fiber when a uniform ultimate bond stress, τ_u, is assumed at the fiber–matrix interface and calculated as follows:

$$\sigma_{fu} = \left\{ \begin{array}{ll} 2\tau_u l_{SF}/d_{SF} & l_{SF} \le l_{cr} \\ f_{SF} & l_{SF} > l_{cr} \end{array} \right\} \tag{30}$$

$$n_l = \left\{ \begin{array}{ll} 0.50 & l_{SF} \le l_{cr} \\ 1 - \frac{l_{SF}}{2l_{cr}} & l_{SF} > l_{cr} \end{array} \right\} \tag{31}$$

where l_{cr} is the length in the fiber required to develop the ultimate fiber stress that can be estimated as:

$$l_{cr} = 0.5 f_{SF} d_{SF}/\tau_u \tag{32}$$

2.4.3. Fracture Energy

The value of the SFRC fracture energy is an important parameter that depends on the SFRC characteristics and can be evaluated experimentally by tension tests. In this study, the ratio of the fracture energy of the SFRC, $G_{f,SF}$, to the fracture energy of the corresponding plain concrete, G_f, was determined by 85 test results from the literature as a function of the fiber factor, F, as shown in Figure 5. The experimental results of low to high concrete strength deformed (hooked, crimped, and undulated) short steel fibers were used [59–64]. The experimental data points and fitted lines in Figure 5 include test results of low concrete strength with steel fibers of normal strength (Line A), normal to high concrete strength with steel fibers of high strength (Line B), and normal to high concrete strength with steel fibers of normal strength (Line C).

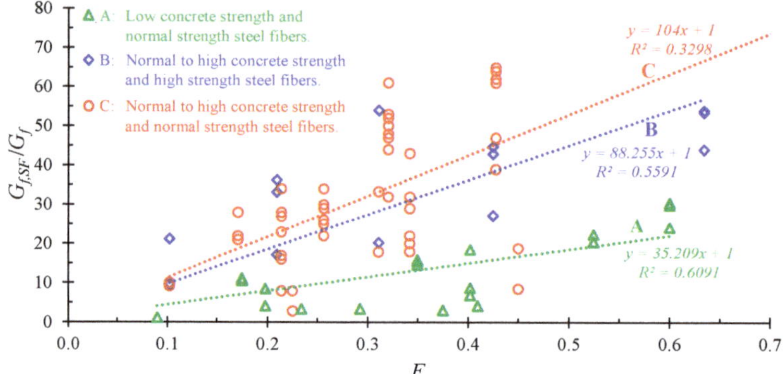

Figure 5. Estimation of SFRC fracture energy based on test results.

Thus, the expression derived from the experimental results presented in Figure 5 has the form:

$$G_{f,SF}/G_f = k_G F + 1 \rightarrow G_{f,SF} = G_f(k_G F + 1) \tag{33}$$

where k_G is a factor that can be taken based on the results in Figure 5.

The fracture energy of the plain concrete, G_f, can be expressed using the following relationship assuming a linear reduction of the tensile concrete stress from the maximum tensile strength, f_t, to zero at the ultimate crack width, w_u:

$$G_f = 0.5 f_t w_u \tag{34}$$

$$w_u = \varepsilon_{tu,fr} L_{fr} \overset{\varepsilon_{tu,fr}=a_{fr}\varepsilon_{to}}{\rightarrow} w_u = a_{fr}\varepsilon_{to}L_{fr} \overset{(23)}{\rightarrow} w_u = a_{fr}L_{fr}\frac{f_t}{E_t} \tag{35}$$

where a_{fr} is a coefficient that depends on the shape of the stress versus crack width curve and the nature and size of the concrete aggregates and takes values from 5 to 8 for concrete aggregates of maximum size d_g = 32 to 8 mm, respectively [55]. L_{fr} is the length of fracture process zone of plain concrete that can be taken as $3d_g$ [65].

Thus, using Equations (33)–(35), Equation (27) can be written as:

$$f_{t,SF}w_{u,SF}\left(k_f + 0.5k_w - 0.5k_f k_w\right) = 0.5 f_t a_{fr} L_{fr}\frac{f_t}{E_t}(k_G F + 1) \overset{L_{fr}=3d_g}{\rightarrow} \tag{36}$$

$$w_{u,SF} = \frac{1.5 f_t^2 a_{fr} d_g (k_G F + 1)}{f_{t,SF}E_t\left(k_f + 0.5k_w - 0.5k_f k_w\right)} \tag{37}$$

Based on the aspects of the proposed smeared crack model, the formulation of the stress is:

$$\sigma_t = \left\{ \begin{array}{ll} \varepsilon_t E_{t,SF} & \text{if } 0 < \varepsilon_t \leq \varepsilon_{to,SF} \\ f_{t,SF}\left(1 - \frac{1-k_f}{k_w w_{u,SF}}w_t\right) & \text{if } 0 < w_t \leq k_w w_{u,SF} \\ k_f f_{t,SF} & \text{if } k_w w_{u,SF} < w_t \leq w_{u,SF} \end{array} \right\} \tag{38}$$

2.4.4. Fracture Process Zone of the SFRC

The aforementioned formulation can provide a feasible evaluation of the entire tensile behavior of the SFRC based on the elastic and fracture characteristics of the materials used in the fibrous mixture (concrete and steel fibers). All parameters and coefficients can be defined using the tensile strength, f_t, the modulus of elasticity under tension, E_t, and the maximum size of the aggregates, d_g, of the plain concrete, along with the characteristics of the added steel fibers, such as their ultimate tensile strength at yielding, f_{SF}, their modulus of elasticity, E_{SF}, and the fiber factor, F, which depend on their characteristics (bond factor, β, volume fraction, V_{SF}, length, l_{SF}, and diameter, d_{SF}). The only parameter that requires being defined is the size of the SFRC fracture process zone length, $L_{fr,SF}$. This length cannot be assumed as mesh dependent, but it has to be attributed to the nature of the SFRC, the type of loading and the size of the specimen [55].

In this study the estimation of the crack process zone length is based on experimental data and the developed FE modeling. In the following sections, the methodology used to estimate the value of the fracture process zone length of the SFRC is presented.

2.5. Modeling of Steel Reinforcement

Steel reinforcement (longitudinal and stirrups) was modeled as a linear elastic material until the reach of yield point. Post-yielding behavior is considered to be nonlinear inelastic, referred as plastic behavior. Plastic behavior is characterized by permanent (plastic) deformations and is defined by introducing parameters of stress and plastic strain at the yield point (f_y, ε_y) as well as the stress and strain at ultimate point (f_y and ε_u). Steel modulus of elasticity, E_s, is also provided to Abaqus, according to experimental data values as well as Poisson's ratio, which is considered equal to 0.3 for all analyses performed in this study.

3. Application of the Proposed Model

3.1. Examined Beam Specimens

The current research presents a tension stiffening FE model for SFRC that properly considers the above-mentioned parameters. Seventeen SFRC beams available in the literature [66–71] were selected and a nonlinear FE analysis was performed to illustrate the capability and accuracy of the proposed model. The selected series of experimental tests had the same geometrical characteristics but differentiated in longitudinal reinforcement, concrete strength, and amount of steel fibers added to the concrete mixture.

These test series were chosen because all the necessary test data needed by the FE simulations were clearly reported. Further, the beam specimens include various ratios of longitudinal reinforcement (0.3%, 0.6%, and 1.0%) and steel fiber volume fractions (from 0.3 to 1.5%). Table 2 summarizes the geometric dimensions and reinforcement details of all examined beams as well as the material properties.

The 17 FE simulations were developed using same geometry, dimensions, material properties, and boundary conditions of the tested beams. The model is based on a 3D FE analysis implemented in Abaqus [34] software and on a smeared cracked approach, which does not require predetermined crack paths. Proper modeling and analysis of the SFRC beams involves full consideration on the material and geometry nonlinearities as well as effects related to the interaction and load sharing between steel and reinforced concrete. More details about the FE model are presented in the following sections.

3.2. Element Types Selected

SFRC was modeled using three-dimensional eight-node solid elements with reduced integration to prevent shear locking effect (C3D8R). For steel reinforcement (bars and stirrups), three-dimensional two-node truss elements were chosen (T3D2). Each node of these elements has three degrees of freedom with translation in X, Y, and Z directions (global coordinates system), as shown in Figure 6.

Table 2. Characteristics of the examined beams.

Ref.	Beam Name	Geometry		Long. Reinforcement			Material Properties			Steel Fiber Properties *				SFRC Properties (Model)					
		b (mm)	h (mm)	A_{s1} (mm^2)	A_{s2} (mm^2)	ρ_l (%)	f_{cm} (MPa)	E_s (GPa)	f_y (MPa)	V_{SF} (%)	$\frac{l_{SF}}{d_{SF}}$	F	f_{SF} (MPa)	$f_{c,SF}$ (MPa)	$\varepsilon_{cu,SF}$ (mm/m)	$f_{t,SF}$ (MPa)	k_f	$\varepsilon_{to,SF}$ (mm/m)	k_w
[66]	S3-2-7F	284	298	232 (3Ø10)	56 (2Ø6)	0.3	44.6	203	560	0.3	55/1	0.12	1020	45.88	2.98	3.37	0.18	0.097	0.56
[67]	S3-2-3	284	298	232 (3Ø10)	57 (2Ø6)	0.3	50.9	210	578	-	-	-	-	-	-	-	-	-	-
	S3-1-F05	278	302	235 (3Ø10)	56 (2Ø6)	0.3	55.6	203	560	0.5	53/1	0.20	1020	58.06	3.70	4.06	0.24	0.109	0.21
	S3-1-F10	279	300	235 (3Ø10)	56 (2Ø6)	0.3	48.0	203	560	1.0	53/1	0.40	1020	52.42	4.98	3.74	0.40	0.104	0.12
	S3-1-F15	279	300	235 (3Ø10)	56 (2Ø6)	0.3	52.2	203	560	1.5	53/1	0.60	1020	59.13	6.86	4.10	0.67	0.110	0.14
[68]	S2-4-F10	270	301	452 (4Ø12)	57 (2Ø6)	0.6	38.4	211	632	1.0	55/1	0.41	1020	42.07	4.76	3.13	0.36	0.093	0.08
[69]	S2-3	272	300	466 (3Ø14)	57 (2Ø6)	0.6	48.1	211	632	-	-	-	-	-	-	-	-	-	-
	S2-F05	273	301	477 (3Ø14)	56 (2Ø6)	0.6	55.6	205	559	0.5	53/1	0.20	1020	58.06	3.70	4.06	0.24	0.109	0.16
	S2-F10	272	301	477 (3Ø14)	56 (2Ø6)	0.6	48.0	205	559	1.0	53/1	0.40	1020	52.42	4.98	3.74	0.46	0.104	0.14
	S2-F15	272	299	477 (3Ø14)	56 (2Ø6)	0.6	52.2	205	559	1.5	53/1	0.60	1020	59.13	6.86	4.10	0.60	0.110	0.13
[70]	B1-F0	280	300	226 (2Ø12)	56 (2Ø6)	0.3	64.5	194	800	-	-	-	-	-	-	-	-	-	-
	B2-F03	280	300	226 (2Ø12)	56 (2Ø6)	0.3	65.3	194	800	0.3	50/1	0.11	1100	66.93	3.28	4.36	0.13	0.112	0.27
	B3-F06	280	300	226 (2Ø12)	56 (2Ø6)	0.3	62.3	194	800	0.6	50/1	0.23	1100	65.42	4.02	4.35	0.22	0.113	0.08
[71]	S1-3	280	300	755 (3Ø18)	57 (2Ø6)	1.0	48.1	211	632	-	-	-	-	-	-	-	-	-	-
	S1-F05	280	300	798 (3Ø18)	56 (2Ø6)	1.0	55.6	205	559	0.5	53/1	0.20	1020	58.06	3.70	4.06	0.24	0.109	0.25
	S1-F10	280	300	798 (3Ø18)	56 (2Ø6)	1.0	48.0	205	559	1.0	53/1	0.41	1020	52.58	5.10	3.74	0.42	0.104	0.13
	S1-F15	280	300	798 (3Ø18)	56 (2Ø6)	1.0	52.2	205	559	1.5	53/1	0.60	1020	59.13	6.86	4.10	0.60	0.110	0.12

* Hooked steel fibers.

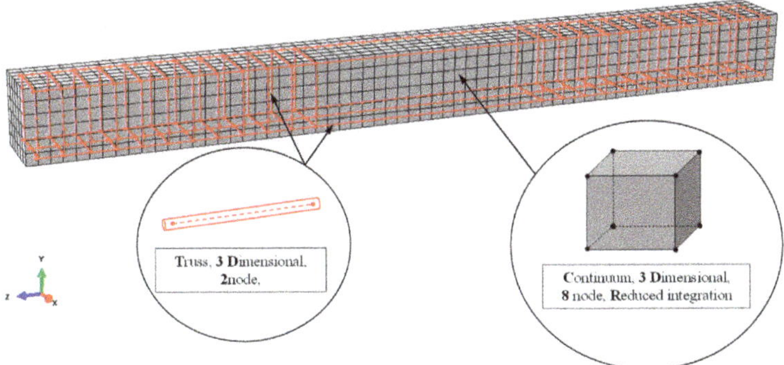

Figure 6. Concrete and steel reinforcement mesh.

Proper modeling of steel reinforcement requires that bond interaction between reinforcement and concrete is considered. In general, the steel reinforcement could be modeled using a beam element with bending stiffness. However, in this study, the bending stiffness of the reinforcement is considered to be quite low compared to the bending stiffness of the SFRC matrix, and therefore it is neglected. Thus, truss elements are used for modeling steel reinforcement. The interaction between concrete and reinforcement is considered using the embedded method and specifically "the truss in solid" type. Truss elements (longitudinal reinforcement and stirrups) are embedded in solid region (SFRC), referred as host region. Geometric relationships between host and embedded elements are explored and if an embedded element node lies within a host element, then the transactional degrees of freedom (DOFs) of the node are constrained to follow the response of the host element.

3.3. Boundary Conditions and Loading

Displacement boundary conditions are necessary to constrain the model and obtain a unique solution. To ensure that the model behaves in an equivalent way as the experimental beam, the supports and loadings must be introduced to proper boundary conditions. The entire SFRC beam was modeled and simply supported. Two supports were designed in such a way as to create one roller and one pinned support. The supports were placed at a distance of 140 mm from each end of the beam according to the experimental set up, while the ends of the beams were free. At the one support, in the Ux, Uy, and Uz directions, a single line of nodes was constrained and added as constant values of 0, while at the other support only Uy direction were restricted given value of 0, as shown in Figure 7. This allows the beam to rotate at the supports.

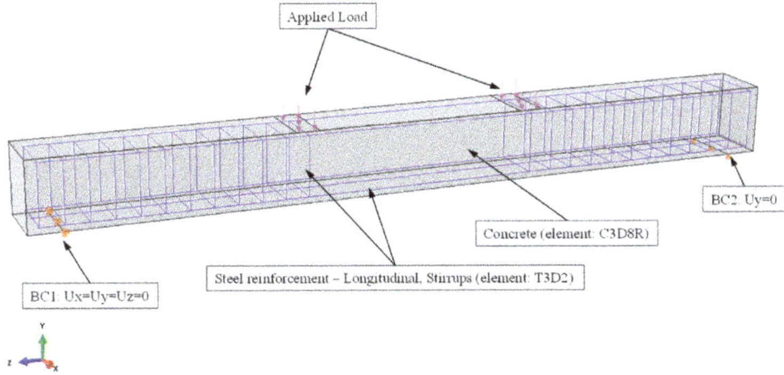

Figure 7. FE model boundary conditions and loading.

The load was applied as a distributed surface load (Figure 7), which is more realistic as in real structures forces are most likely to be transferred through contact and thus a force is likely to be distributed across a contact area. Applying the load as a concentrated force to a single node can cause troublesome effects, such as high stress concentration around the application node. When a load is applied as surface load, the total force is distributed to the face of the FE which composes the surface, and then to the nodes laying on the face, which leads to having a smaller amount of the total force at each node and excessive local stress concentration is avoided. Further, to ensure that the dynamic implicit procedure adopted in this study obtained a quasi-static solution, the load was applied steadily and gradually to avoid any significant change in acceleration from one increment to the next. This further guarantees that the alterations in stresses and displacement are smooth. The loading sequence used in the FE analysis was based on the experimental sequence and the analysis was performed using the load control technique.

3.4. Meshing and Convergence

The mesh size of 50 mm was applied to all elements (truss and solid) to make sure that different materials (steel and concrete) share common nodes. This size of mesh was chosen because concrete cracking propagation usually involves spatial scales in the range of 2–3 dominant aggregate sizes (usually between 16 and 32 mm) of the concrete [65]. A convergence study was carried out using different mesh sizes before proceeding to the main analysis and the above-mentioned mesh proved to be fine enough to generate reliable results.

3.5. Material Input

The main properties of the proposed material models for each beam are presented in Table 2. Some additional parameters also used as input in the performed nonlinear analysis of the SFRC beams are presented in Table 3. Stress and crack widths at Points 1–3 and corresponding plastic damage factors shown in Table 3 are defined in the SFRC tension model depicted in Figure 4.

Table 3. Input parameters of the examined SFRC beams.

Ref.	Beam Name	$E_{t,SF} = E_o$ (GPa)	v	$G_{f,SF}$ (N/mm)	σ_{t1} (MPa)	w_{t1} (mm)	ε_{t1} (‰)	σ_{t2} (MPa)	w_{t2} (mm)	ε_{t2} (‰)	σ_{t3} (MPa)	w_{t3} (mm)	ε_{t3} (‰)	d_{t1}	d_{t2}	d_{t3}
[66]	S3-2-7F	34.70	0.205	0.277	3.11	0.00990	0.06	0.81	0.10065	0.61	0.60	0.15675	0.95	0.076	0.761	0.822
[67]	S3-1-F05	37.23	0.207	0.534	3.63	0.01113	0.07	1.05	0.08109	0.51	0.97	0.16854	1.06	0.106	0.741	0.762
	S3-1-F10	36.06	0.215	0.893	3.32	0.01113	0.07	1.67	0.05565	0.35	1.50	0.15741	0.99	0.110	0.553	0.597
	S3-1-F15	37.38	0.222	1.394	3.85	0.01113	0.07	2.84	0.05406	0.34	2.74	0.16218	1.02	0.062	0.308	0.333
[68]	S2-4-F10	33.78	0.215	0.671	2.57	0.01155	0.07	1.45	0.03300	0.20	1.11	0.14685	0.89	0.179	0.537	0.644
[69]	S2-F05	37.23	0.207	0.534	3.48	0.01272	0.08	1.20	0.06042	0.38	0.97	0.16854	1.06	0.141	0.706	0.762
	S2-F10	36.06	0.215	0.893	3.36	0.01113	0.07	1.87	0.05406	0.34	1.72	0.15741	0.99	0.100	0.500	0.540
	S2-F15	37.38	0.222	1.394	3.80	0.01113	0.07	2.57	0.05565	0.35	2.45	0.16377	1.03	0.075	0.374	0.404
[70]	B2-F03	38.88	0.204	0.393	3.93	0.01050	0.07	0.97	0.08850	0.59	0.55	0.16650	1.11	0.097	0.777	0.875
	B3-F06	38.59	0.208	0.687	3.41	0.01350	0.09	1.54	0.03900	0.26	0.97	0.16500	1.10	0.216	0.647	0.776
[71]	S1-F05	37.23	0.207	0.564	3.72	0.01113	0.07	1.31	0.08904	0.56	0.97	0.16854	1.06	0.085	0.677	0.762
	S1-F10	36.06	0.215	0.876	3.33	0.01113	0.07	1.72	0.05406	0.34	1.56	0.15741	0.99	0.107	0.549	0.583
	S1-F15	37.38	0.222	1.430	3.80	0.01113	0.07	2.57	0.05565	0.35	2.45	0.16377	1.03	0.075	0.374	0.404

4. Results and Discussion

4.1. Validation and Accuracy of the Proposed Model

As mentioned in Section 2.4.4, the value of the SFRC fracture process zone length, $L_{fr,SF}$, cannot be assumed as mesh dependent but it has to be attributed to the nature of the SFRC, the type of loading, and the size of the specimen. In this study the estimation of the proper length size for the specific loading case is based on the examined tests using the developed FE modeling and by the comparison of the numerical results and the experimental data.

Thus, comparisons of the numerically predicted bending moment versus curvature responses of SFRC beams S3-1-F05 shown in Figure 8 for different values of fracture process zone length, $L_{fr,SF}$, with the experimental response can yield useful conclusions about the proper size of this length and for this type of loading. The nature of the material used in this procedure is represented by the adopted stress versus crack width curve. The examined values of the fracture process zone length, $L_{fr,SF}$, are taken as function of the steel fiber length, l_{SF}, and five different values are investigated: $L_{fr,SF}$ = 2.0, 2.5, 3.0, 3.5 and 4.0 l_{SF}. The comparisons of these five numerical curves (blue continuous lines) with the experimental data (red dotted line) shown in Figure 8 indicate that the best fitted numerical curve has been derived for $L_{fr,SF} = 3l_{SF}$. Further, the comparisons between the numerical predictions using this specific length of the SFRC fracture process zone the test results prove that the proposed model fits very well to the experimental behavior of the beams with higher accuracy than the simulation without considering the tension stiffening and softening effect.

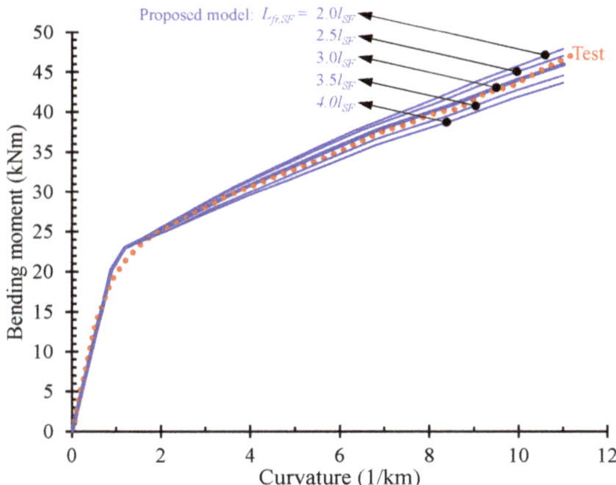

Figure 8. Evaluation of the SFRC fracture process zone length by comparisons between numerical and experimental bending moment versus curvature curves of beam S3-1-F05.

Further, the numerical results derived from the proposed FE model are presented and compared with the corresponding experimental data in terms of: (a) bending moment versus curvature; and (b) stress versus strain curves. Especially, Figures 9–12 demonstrate the FE analysis and the experimental behavioral curves of the SFRC beams S3-1-F05, S3-1-F10, S3-1-F15, and S2-4-F10 [67,68] with 0.5%, 1.0%, 1.5%, and 1.0% volume fraction of steel fibers, respectively. To comprehend the influence of the proposed smeared crack model with tension softening and the tension stiffening effect on the flexural behavior of the examined beams, two curves derived from FE analysis, in terms of bending moment versus curvature are compared with the experimental one in Figure 9a, Figure 10a, Figure 11a, and Figure 12a. The first curve (blue continuous line) was derived using the proposed material models in the nonlinear analysis with tension softening and tension stiffening approach and the second (black thin dotted line) using the FE model without considering these effects (denoted as "model without TS" where TS is tension softening and stiffening).

The presented stress versus strain curves in Figures 9b, 10b, 11b and 12b consist of the stresses corresponding to the tension stiffening effect and the residual stresses due to fiber interaction with the concrete [67,68]. The numerical curves in these diagrams were calculated based on the stress–strain values in each element, which were extracted by the FE analysis and then the average values over all elements were calculated and used. The averaged stress strain values are the overall stress versus strain for the whole beam simulation and combined in a stress–strain curve.

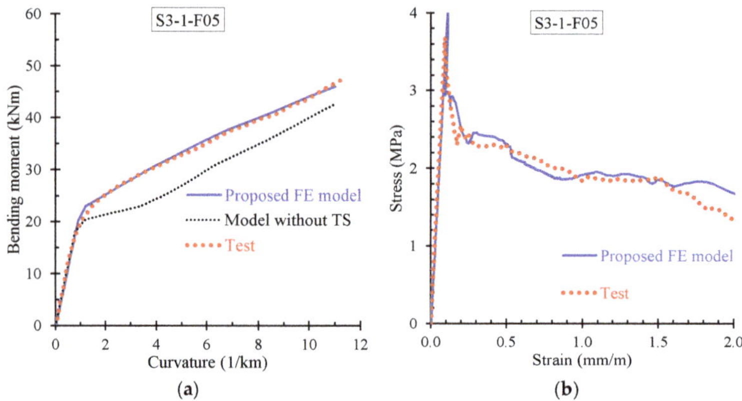

Figure 9. Comparisons between numerical curves and experimental results of beam S3-1-F05 in terms of: (**a**) bending moment versus curvature; and (**b**) residual stress versus strain curves.

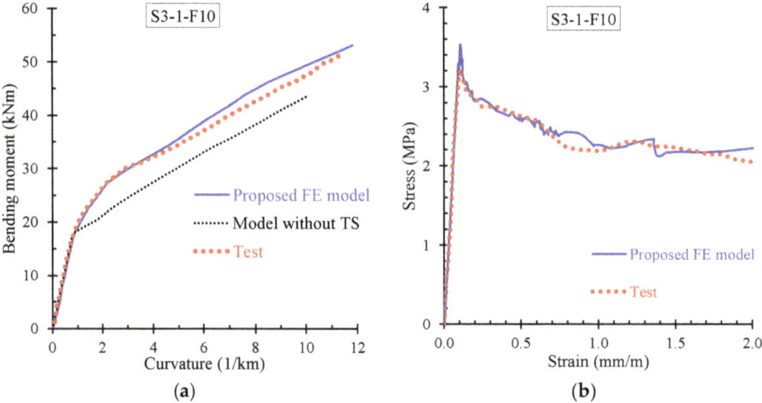

Figure 10. Comparisons between numerical curves and experimental results of beam S3-1-F10 in terms of: (**a**) bending moment versus curvature; and (**b**) residual stress versus strain curves.

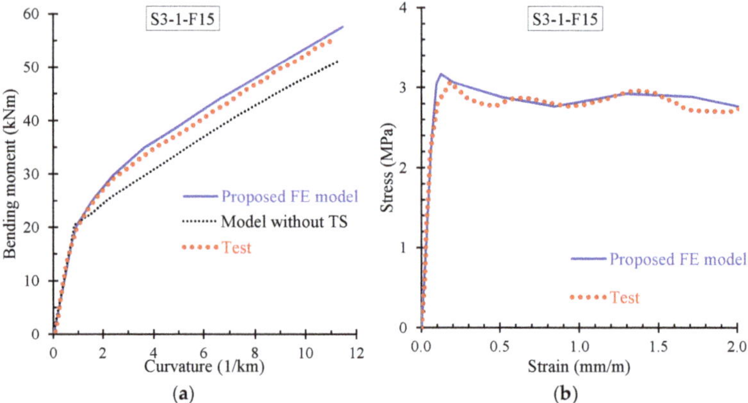

Figure 11. Comparisons between numerical curves and experimental results of beam S3-1-F15 in terms of: (**a**) bending moment versus curvature; and (**b**) residual stress versus strain.

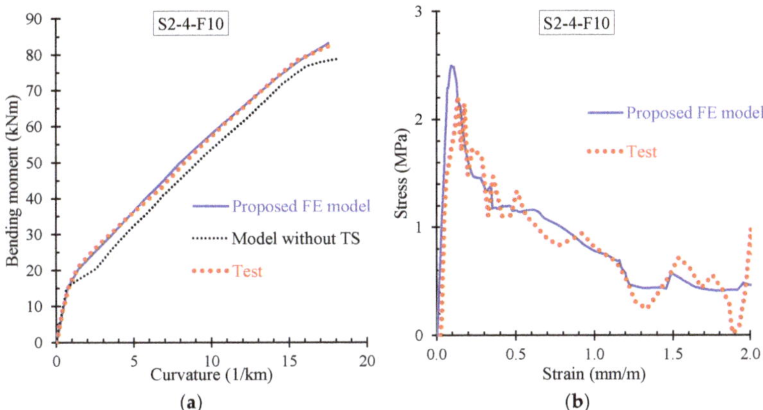

Figure 12. Comparisons between numerical curves and experimental results of beam S2-4-F10 in terms of: (**a**) bending moment versus curvature; and (**b**) residual stress versus strain.

It is noted that the point of the peak stress in the stress–strain diagrams of Figure 9b, Figure 10b, Figure 11b, and Figure 12b corresponds to the cracking point in the moment versus curvature diagram shown in Figure 9a, Figure 10a, Figure 11a, and Figure 12a, respectively. Since steel fibers mainly influence the post-cracking behavior of SFRC, the post-peak descending part of the residual stress versus strain curve depends on the volume fraction of the added steel fibers. Thus, SFRC mixtures with higher amount of fibers (for example, beam S3-1-F15 with $V_{SF} = 1.5\%$) exhibit higher post-cracking stress.

Further, the numerical predictions of the stress versus strain curves using the proposed nonlinear smeared crack model with tension softening and tension stiffening are in very good compliance with the experimentally measured values, as can clearly be observed for all examined beams in Figures 9b, 10b, 11b and 12b.

Numerical curves derived by the developed nonlinear FE model with and without tension softening and stiffening effect are also compared with the experimental ones in the diagrams of Figure 13 for the SFRC beams: (a) S3-2-7F; (b) B2-F03; (c) B3-F06; (d) S1-F05; (e) S1-F10; and (f) S1-F15 [66,70,71]. These comparisons reveal that in most of the examined cases the numerical predictions of the proposed FE model that considers the smeared crack analysis with tension softening and tension stiffening effect are in closer agreement with the test data than the numerical curves without this effect.

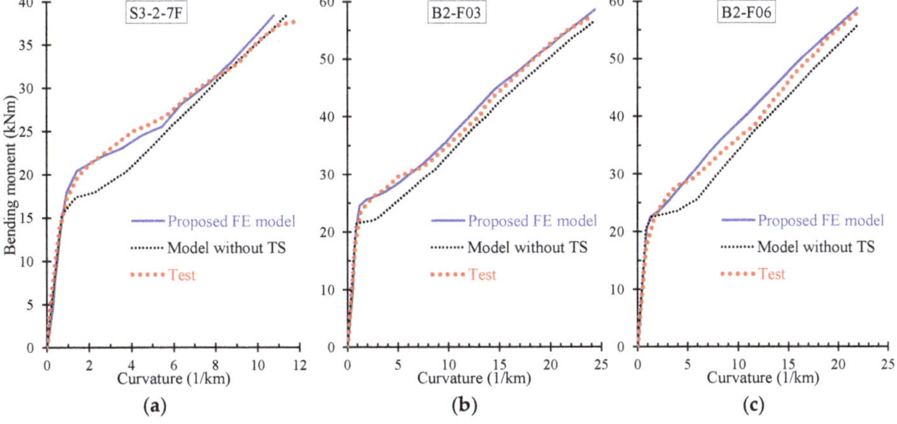

Figure 13. *Cont.*

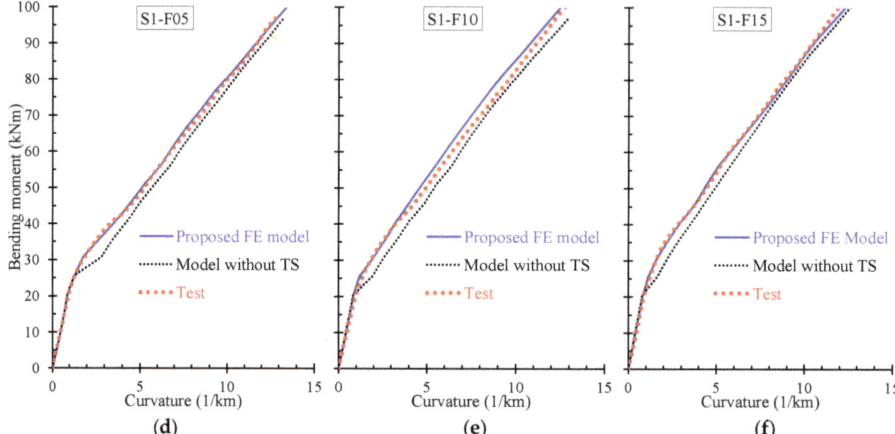

Figure 13. Comparisons between bending moment versus curvature numerical and experimental curves (model without TS: without tension softening and stiffening effect): (a) S3-2-7F; (b) B2-F03; (c) B3-F06; (d) S1-F05; (e) S1-F10; and (f) S1-F15.

The effectiveness of the proposed nonlinear FE model to predict accurately the flexural response of SFRC beams with different longitudinal reinforcement ratios is examined in the diagrams of Figure 14 in terms of bending moment versus curvature behavioral curves. The SFRC beams compared in the diagrams of Figure 14a–c has steel fiber volume fraction 0.5%, 1.0%, and 1.5%, respectively. Further, the ratio of the steel longitudinal reinforcement of the beams are as follows:

- Beams of series "S1" (S1-F05, S1-F10 and S1-F15): $\rho_l = 1.0\%$.
- Beams of series "S2" (S2-F05, S2-F10, S2-4-F10 and S2-F15): $\rho_l = 0.6\%$.
- Beams of series "S3" (S3-1-F05, S3-1-F10 and S3-1-F15): $\rho_l = 0.3\%$.

The comparisons shown in Figure 14 reveal that a good agreement between the numerical and the experimental bending moment versus curvature curves was achieved.

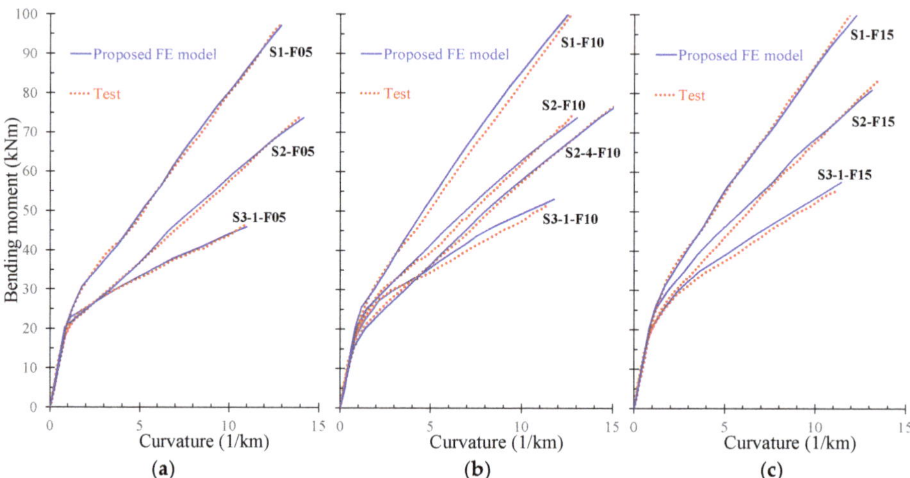

Figure 14. Numerical and test curves of SFRC with different steel fiber and longitudinal reinforcement ratios: (a) beams with $V_{SF} = 0.5\%$; (b) beams with $V_{SF} = 1.0\%$; and (c) beams with $V_{SF} = 1.5\%$.

Further, to establish the validity of the proposed nonlinear FE analysis, Table 4 presents the calculated values of the differences between the numerical predictions and the test results in terms of "calculation errors". Specifically, the discrepancy of the bending moment values between the numerical predictions and the tests for the same curvatures along the entire bending moment versus curvature diagrams of each examined beam was calculated to validate, statistically compare, and estimate the overall accuracy of the proposed model.

Table 4. Error calculation.

Ref.	Beam Name	V_{SF} (%)	MAE	SE	CV
[66]	S3-2-7F	0.29	3.8%	0.9%	23%
[67]	S3-1-F05	0.5	5.0%	1.1%	22%
	S3-1-F10	1.0	6.1%	0.9%	15%
	S3-1-F15	1.5	7.3%	1.4%	19%
[68]	S2-4-F10	1.0	4.5%	0.9%	19%
[69]	S2-F05	0.5	4.6%	0.7%	14%
	S2-F10	1.0	4.2%	0.7%	17%
	S2-F15	1.5	8.7%	1.1%	13%
[70]	B2-F03	0.3	5.2%	1.4%	27%
	B3-F06	0.6	5.7%	0.8%	14%
[71]	S1-F05	0.5	2.3%	0.5%	22%
	S1-F10	1.0	5.2%	1.1%	21%
	S1-F15	1.5	3.2%	1.2%	36%
	average:		5.1%	1.0%	20%

This discrepancy in terms of error percentage for each point was calculated according to the following known expression:

$$Error\ (\%) = \left| \frac{M_{FE\ model} - M_{exp}}{M_{exp}} \right| \times 100 \tag{39}$$

where $M_{FE\ model}$ and M_{exp} are the bending moment values derived from the proposed FE model and the experiments, respectively.

Table 4 presents the values of the mean absolute error (MAE), the standard error (SE), and coefficient of variation (CV). These values indicate that model predictions lead to an error below 9% for every SFRC beam examined and the average value of MAE of all beams is 5.1%.

4.2. Influence of Steel Fibers on the Flexural Performance

Figure 15 includes a diagram with the experimental and the numerical behavioral curves of beams S2-F05, S2-F10, and S2-F15 [69]. In the same figure, the schemes of the cracking pattern and the stress distribution of the beams derived by the proposed model at bending moment 50 kNm are also compared. For simplicity and comparison reasons, these numerical schemes show the plane y-z view of the 3D FE model of the beams and only beam S2-F15 is illustrated in both 3D and plane views. It is noted that SFRC beam S2-F15 with higher amount of steel fibers (V_{SF} = 1.5%) demonstrates higher strength and lower deformation than the other SFRC beams with 1.0% and 0.5% steel fiber volume fraction. Further, this SFRC beam (specimen S2-F15) exhibits increased number of cracks and decreased crack spacing and width with respect to the cracking performance of beam S2-F05 with lower amount of steel fibers (V_{SF} = 0.5%) due to the favorable contribution of the steel fibers. These observations are based on the numerical cracking pattern and stress distribution derived from the proposed FE analysis (Figure 15). The same observations have also been established by relevant tests of flexural SFRC beams [72–74].

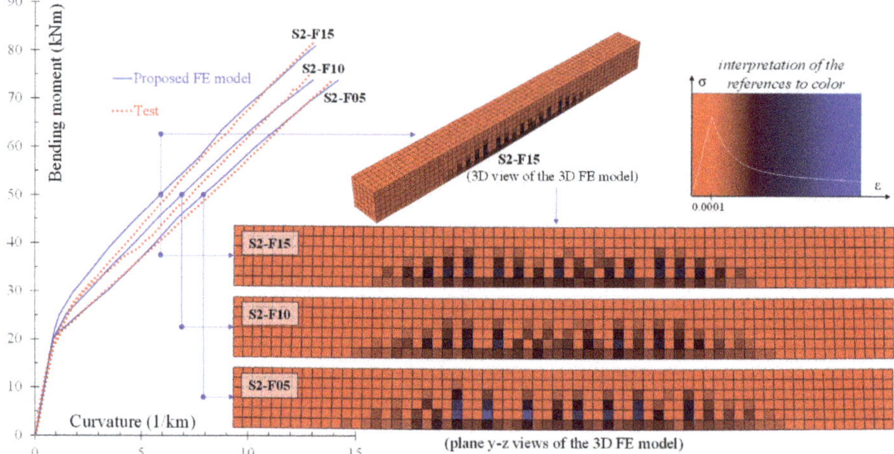

Figure 15. Influence of steel fibers on the bending moment versus curvature curves and the cracking pattern–stress distribution (plane y-z view of 3D model).

Further, the contribution of steel fibers on the increase of the residual strength and the overall post-cracking stress versus strain behavior of SFRC structural members with steel bars and stirrups is demonstrated in Figure 16. Figure 16a presents the residual stress versus strain response derived from the proposed model of beams S3-1-F05, S3-1-F10, and S3-1-F15 with V_{SF} = 0.5%, 1.0%, and 1.5% (beams of series "S3-1") and Figure 16b presents the corresponding numerical response of beams S2-F05, S2-F10, and S2-F15 with V_{SF} = 0.5%, 1.0%, and 1.5% (beams of series "S2"). In both examined cases, it is indicated that the increase of the volume fraction of the added steel fibers causes a gradual increase of the residual strength.

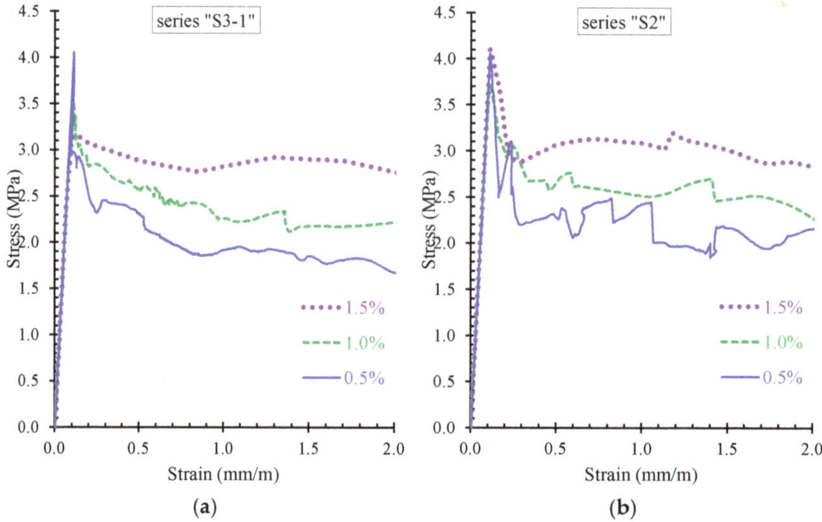

Figure 16. Influence of steel fibers on the residual strength of beams of series: (**a**) "S3-1"; and (**b**) "S2".

5. Conclusions

The influence of steel fibers on the short-term flexural behavior of realistic SFRC structural members reinforced with steel reinforcement was investigated by nonlinear 3D finite element (FE) analysis using software Abaqus. A database of 17 large-scale beam specimens collected from the literature was used as experimental baseline to perform the developed analysis that considers the tension stiffening effect and the nonlinearities of the materials by new and established constitutive relationships for steel fiber-reinforced concrete (SFRC) under compression and tension. The appropriate loading and constraint conditions were also considered in order for the performed FE analyses to be very close to experimental conditions. Based on the results of this study, the following conclusions can be drawn:

- The proposed FE model considers the contribution of the steel fibers according to their type, aspect ratio, and volume fraction to achieve a more realistic evaluation of the compressive and tensile behavior of SFRC utilizing well-established constitutive laws based on experimental data. Especially, for the simulation of the SFRC under tension a smeared crack approach is proposed that utilizes the fracture properties of the material and employs constitutive relationships of stress versus crack width with tension softening for the post-cracking SFRC tensile response instead of stress–strain laws. The post-cracking tensile behavior of the SFRC near the reinforcing bars is modeled by a tension stiffening model that considers the SFRC fracture properties and the reinforcement characteristics.

- The size of the fracture process zone of the SFRC under tension is attributed to the material, the type of loading, and the size of the specimen. An estimation of this length based on comparisons between numerical and experimental data is presented.

- The examined realistic SFRC structural members with increased values of the fiber factor, F, exhibit enhanced flexural performance in terms of initial stiffness, strength, deformation capacity, cracking behavior, and residual stress. The developed FE analysis predicts accurately the overall experimental flexural behavior and points out the contribution of the steel fibers on this improvement.

- Beams with higher amount of steel fibers (V_{SF} = 1.5%) demonstrated lower deformation than the other SFRC beams with 1.0% and 0.5% steel fiber volume fraction at the same level of the applied load. Further, the favorable contribution of the added steel fibers on the cracking performance of the examined SFRC beams is also highlighted in the numerical results of this study. Beams with 1.5% steel fiber volume fraction showed increased number of cracks and decreased crack spacing and width with respect to the corresponding beams with lower amount of fibers (V_{SF} = 0.5%).

- The favorable contribution of the added steel fibers on the post-cracking short-term behavior of realistic SFRC structural members with reinforcing bars and stirrups was revealed by the performed analyses. Numerical results of the examined SFRC beams show that the post-peak descending part of the residual stress versus strain curves depends on the volume fraction of the added steel fibers and SFRC mixtures with higher amounts of fibers exhibit higher post-cracking stress.

- Comparisons between experimental results and numerical predictions include typical bending moment versus curvature behavioral curves, deformation–cracking patterns, and residual stress versus strain diagrams. The proposed nonlinear smeared crack analysis with tension softening and tension stiffening effect was found to be reliable and capable of accurately simulating the flexural response of large-scale SFRC beams with steel reinforcement since the predicted behavioral curves are in good agreement with the corresponding experimental ones.

Author Contributions: All authors contributed extensively to this study, discussed the results and reviews, prepared the manuscript, and agreed to the amendments at all stages of the paper. V.K.K. developed the aspects of the numerical investigation and performed the FE analyses under the supervision of C.E.C. and C.G.K. All authors have read and agreed to the published version of the manuscript.

Funding: This research received no external funding.

Materials **2020**, *13*, 2698

Acknowledgments: Violetta K. Kytinou gratefully acknowledges the financial support received from Eugenides Foundation towards doctoral studies.

Conflicts of Interest: The authors declare no conflict of interest.

References

1. Cuenca, E.; Ferrara, L. Self-healing capacity of fiber reinforced cementitious composites. State of the art and perspectives. *KSCE J. Civ. Eng.* **2017**, *21*, 2777–2789. [CrossRef]
2. Guerini, V.; Conforti, A.; Plizzari, G.; Kawashima, S. Influence of steel and macro-synthetic fibers on concrete properties. *Fibers* **2018**, *6*, 47. [CrossRef]
3. Tsonos, A.G.; Stylianidis, K. Seismic retrofit of Beam-to-Column joints with high-strength fiber jackets. *Eur. Earthq. Eng.* **2002**, *16*, 56–72.
4. Chalioris, C.E.; Karayannis, C.G. Effectiveness of the use of steel fibers on the torsional behavior of flanged concrete beams. *Cem. Concr. Compos.* **2009**, *31*, 331–341. [CrossRef]
5. Tsonos, A.-D.G. Ultra-high-performance fiber reinforced concrete: An innovative solution for strengthening old R/C structures and for improving the FRP strengthening method. *WIT Trans. Eng. Sci.* **2009**, *64*, 273–284.
6. Vougioukas, E.; Papadatou, M. A model for the prediction of the tensile strength of fiber-reinforced concrete members, before and after cracking. *Fibers* **2017**, *5*, 27. [CrossRef]
7. Abambres, M.; Lantsoght, E.O.L. ANN-Based Shear Capacity of Steel Fiber-Reinforced Concrete Beams without Stirrups. *Fibers* **2019**, *7*, 88. [CrossRef]
8. Chalioris, C.E.; Sfiri, E.F. Shear performance of steel fibrous concrete beams. *Procedia Eng.* **2011**, *14*, 2064–2068. [CrossRef]
9. Ma, K.; Qi, T.; Liu, H.; Wang, H. Shear behavior of hybrid fiber reinforced concrete deep beams. *Materials* **2018**, *11*, 2023. [CrossRef]
10. Zhao, J.; Liang, J.; Chu, L.; Shen, F. Experimental study on shear behavior of steel fiber reinforced concrete beams with high-strength reinforcement. *Materials* **2018**, *11*, 1682. [CrossRef]
11. Torres, J.A.; Lantsoght, E.O.L. Influence of Fiber Content on Shear Capacity of Steel Fiber-Reinforced Concrete Beams. *Fibers* **2019**, *7*, 102. [CrossRef]
12. Lantsoght, E.O.L. Database of Shear Experiments on Steel Fiber Reinforced Concrete Beams without Stirrups. *Materials* **2019**, *12*, 917. [CrossRef] [PubMed]
13. Caggiano, A.; Gambarelli, S.; Martinelli, E.; Nisticò, N.; Pepe, M. Experimental characterization of the post-cracking response in hybrid steel/polypropylene fiber-reinforced concrete. *Constr. Build. Mater.* **2016**, *125*, 1035–1043. [CrossRef]
14. Smarzewski, P. Effect of curing period on properties of steel and polypropylene fiber reinforced ultra-high performance concrete. *Mater. Sci. Eng.* **2017**, *245*, 032059.
15. Mobasher, B.; Li, C.Y. Mechanical properties of hybrid cement-based composites. *ACI Mater. J.* **1996**, *93*, 284–292.
16. Kabir, M.R.; Alam, M.S.; Said, A.M.; Ayad, A. Performance of hybrid reinforced concrete beam column joint: A critical review. *Fibers* **2016**, *4*, 13. [CrossRef]
17. Bischoff, P.H. Tension stiffening and cracking of steel fiber-reinforced concrete. *ASCE J. Mater. Civ. Eng.* **2003**, *15*, 174–182. [CrossRef]
18. Barris, C.; Torres, L.; Baena, M.; Pilakoutas, K.; Guadagnini, M. Serviceability limit state of FRP RC beams. *Adv. Struct. Eng.* **2012**, *15*, 653–663. [CrossRef]
19. Torres, L.; Barris, C.; Kaklauskas, G.; Gribniak, V. Modelling of tension-stiffening in bending RC elements based on equivalent stiffness of the rebar. *Struct. Eng. Mech.* **2015**, *53*, 997–1016. [CrossRef]
20. Rimkus, A.; Gribniak, V. Experimental investigation of cracking and deformations of concrete ties reinforced with multiple bars. *Constr. Build. Mater.* **2017**, *148*, 49–61. [CrossRef]
21. Fields, K.; Bischoff, P.H. Tension stiffening and cracking of high-strength reinforced concrete tension members. *ACI Struct. J.* **2004**, *101*, 447–456.
22. Meskenas, A.; Kaklauskas, G.; Daniunas, A.; Bacinskas, D.; Jakubovskis, R.; Gribniak, V.; Gelazius, V. Determination of the stress-crack opening relationship of SFRC by an inverse analysis. *Mech. Compos. Mater.* **2014**, *49*, 685–690. [CrossRef]

23. Morelli, F.; Amico, C.; Salvatore, W.; Squeglia, N.; Stacul, S. Influence of tension stiffening on the flexural stiffness of reinforced concrete circular sections. *Materials* **2017**, *10*, 669. [CrossRef] [PubMed]
24. Gribniak, V.; Arnautov, A.K.; Norkus, A.; Kliukas, R.; Tamulenas, V.; Gudonis, E.; Sokolov, A.V. Steel fibers: Effective way to prevent failure of the concrete bonded with FRP sheets. *Adv. Mater. Sci. Eng.* **2016**. [CrossRef]
25. Gribniak, V.; Tamulenas, V.; Ng, P.-L.; Arnautov, A.K.; Gudonis, E.; Misiunaite, I. Mechanical behavior of steel fiber-reinforced concrete beams bonded with external carbon fiber sheets. *Materials* **2017**, *10*, 666. [CrossRef] [PubMed]
26. Gribniak, V.; Ng, P.-L.; Tamulenas, V.; Misiunaite, I.; Norkus, A.; Šapalas, A. Strengthening of fibre reinforced concrete elements: Synergy of the fibres and external sheet. *Sustainability* **2019**, *11*, 4456. [CrossRef]
27. Meda, A.; Minelli, F.; Plizzari, G.A. Flexural behaviour of RC beams in fibre reinforced concrete. *Compos. Part B: Eng.* **2012**, *43*, 2930–2937. [CrossRef]
28. Abrishami, H.H.; Mitchell, D. Influence of steel fibers on tension stiffening. *ACI Struct. J.* **1997**, *94*, 769–776. [CrossRef]
29. Fischer, G.; Li, V.C. Influence of matrix ductility on tension-stiffening behavior of steel reinforced engineered cementitious composites (ECC). *ACI Struct. J.* **2002**, *99*, 104–111.
30. Deluce, J.R.; Vecchio, F.J. Cracking behavior of steel fiber-reinforced concrete members containing conventional reinforcement. *ACI Struct. J.* **2013**, *110*, 481–490.
31. Meskenas, A.; Gelazius, V.; Kaklauskas, G.; Gribniak, V.; Rimkus, A. A new technique for constitutive modeling of SFRC. *Procedia Eng.* **2013**, *57*, 762–766. [CrossRef]
32. Gribniak, V.; Arnautov, A.K.; Norkus, A.; Tamulenas, V.; Gudonis, E.; Sokolov, A. Experimental investigation of the capacity of steel fibers to ensure the structural integrity of reinforced concrete specimens coated with CFRP sheets. *Mech. Compos. Mater.* **2016**, *52*, 401–410. [CrossRef]
33. Ng, P.; Gribniak, V.; Jakubovskis, R.; Rimkus, A. Tension stiffening approach for deformation assessment of flexural reinforced concrete members under compressive axial load. *Struct. Concr.* **2019**, *20*, 2056–2068. [CrossRef]
34. Dassault Systèmes Simulia System Information. In *Abaqus 2017 User's Manual*; Version 6.12.1; SIMULIA: Providence, RI, USA, 2017.
35. Gómez, J.; Torres, L.; Barris, C. Characterization and simulation of the bond response of NSM FRP reinforcement in concrete. *Materials* **2020**, *13*, 1770. [CrossRef]
36. Zhelyazov, T. Structural Materials: Identification of the constitutive models and assessment of the material response in structural elements strengthened with externally-bonded composite material. *Materials* **2020**, *13*, 1272. [CrossRef]
37. Lubliner, J.; Oliver, J.; Oller, S.; Oñate, E. A plastic-damage model for concrete. *Int. J. Solids Struct.* **1989**, *25*, 299–326. [CrossRef]
38. Genikomsou, A.S.; Polak, M.A. Finite element analysis of punching shear of concrete slabs using damaged plasticity model in ABAQUS. *Eng. Struct.* **2015**, *98*, 38–48. [CrossRef]
39. Hany, N.F.; Hantouche, E.G.; Harajli, M.H. Finite element modeling of FRP-confined concrete using modified concrete damaged plasticity. *Eng. Struct.* **2016**, *125*, 1–14. [CrossRef]
40. Chi, Y.; Yu, M.; Huang, L.; Xu, L. Finite element modeling of steel-polypropylene hybrid fiber reinforced concrete using modified concrete damaged plasticity. *Eng. Struct.* **2017**, *148*, 23–35. [CrossRef]
41. Karayannis, C.G. A numerical approach to steel-fibre reinforced concrete under torsion. *Struct. Eng. Rev.* **1995**, *7*, 83–91.
42. Chalioris, C.E. Steel fibrous RC beams subjected to cyclic deformations under predominant shear. *Eng. Struct.* **2013**, *49*, 104–118. [CrossRef]
43. Chalioris, C.E. Analytical approach for the evaluation of minimum fiber factor required for steel fibrous concrete beams under combined shear and flexure. *Constr. Build. Mater.* **2013**, *43*, 317–336. [CrossRef]
44. Lantsoght, E.O.L. How do steel fibers improve the shear capacity of reinforced concrete beams without stirrups? *Compos. Part B Eng.* **2019**, *175*, 107079. [CrossRef]
45. Marar, K.; Eren, O.; Yitmen, I. Compression specific toughness of normal strength steel fiber reinforced concrete (NSSFRC) and high strength steel fiber reinforced concrete (HSSFRC). *Mater. Res.* **2011**, *14*, 239–247. [CrossRef]

46. Aslani, F.; Nejadi, S.; Samali, N. Long-term flexural cracking control of reinforced self-compacting concrete one way slabs with and without fibres. *Comput. Concr.* **2014**, *14*, 419–444. [CrossRef]
47. Cagatay, I.; Dincer, R. Modeling of concrete containing steel fibers: Toughness and mechanical properties. *Comput. Concr.* **2011**, *8*, 357–369. [CrossRef]
48. Chalioris, C.E.; Liotoglou, F.A. Tests and simplified behavioral model for steel fibrous concrete under compression. In *Advances in Civil Engineering and Building Materials IV*; Chang, S.-Y., Al Bahar, S.K., Husain, A.-A.M., Zhao, J., Eds.; CRC Press/Balkema: Leiden, The Netherlands, 2015; pp. 195–199.
49. Lee, S.-C.; Oh, J.-H.; Cho, J.-Y. Compressive behavior of fiber-reinforced concrete with end-hooked steel fibers. *Materials* **2015**, *8*, 1442–1458. [CrossRef]
50. Zhao, M.; Zhang, B.; Shang, P.; Fu, Y.; Zhang, X.; Zhao, S. Complete stress–strain curves of self-compacting steel fiber reinforced expanded-shale lightweight concrete under uniaxial compression. *Materials* **2019**, *12*, 2979. [CrossRef]
51. Chalioris, C.E.; Panagiotopoulos, T.A. Flexural analysis of steel fibre-reinforced concrete members. *Comput. Concr.* **2018**, *22*, 11–25.
52. Choi, W.-C.; Jung, K.-Y.; Jang, S.-J.; Yun, H.-D. The influence of steel fiber tensile strengths and aspect ratios on the fracture properties of high-strength concrete. *Materials* **2019**, *12*, 2105. [CrossRef]
53. Wang, Z.L.; Wu, J.; Wang, J.G. Experimental and numerical analysis on effect of fiber aspect ratio on mechanical properties of SFRC. *Constr. Build. Mater.* **2010**, *24*, 559–565. [CrossRef]
54. Bezerra, A.; Maciel, P.S.; Corrêa, E.; Soares Junior, P.R.R.; Aguilar, M.T.P.; Cetlin, P.R. Effect of high temperature on the mechanical properties of steel fiber-reinforced concrete. *Fibers* **2019**, *7*, 100. [CrossRef]
55. Karayannis, C.G. Smeared crack analysis for plain concrete in torsion. *ASCE J. Struct. Eng.* **2000**, *126*, 638–645. [CrossRef]
56. Karayannis, C.G.; Chalioris, C.E. Experimental validation of smeared analysis for plain concrete in torsion. *ASCE J. Struct. Eng.* **2000**, *126*, 646–653. [CrossRef]
57. Rimkus, A.; Cervenka, V.; Gribniak, V.; Cervenka, J. Uncertainty of the smeared crack model applied to RC beams. *Eng. Fract. Mech.* **2020**, 107088. [CrossRef]
58. Karayannis, C.G. Nonlinear analysis and tests of steel-fiber concrete beams in torsion. *Struct. Eng. Mech.* **2000**, *9*, 323–338. [CrossRef]
59. Yoo, D.Y.; Yoon, Y.S.; Banthia, N. Predicting the post-cracking behavior of normal-and high-strength steel-fiber-reinforced concrete beams. *Constr. Build. Mater.* **2015**, *93*, 477–485. [CrossRef]
60. Michels, J.; Christen, R.; Waldmann, D. Experimental and numerical investigation on postcracking behavior of steel fiber reinforced concrete. *Eng. Fract. Mech.* **2013**, *98*, 326–349. [CrossRef]
61. Aydın, S. Effects of fiber strength on fracture characteristics of normal and high strength concrete. *Period. Polytech. Civ. Eng.* **2013**, *57*, 191–200. [CrossRef]
62. Şahin, Y.; Köksal, F. The influences of matrix and steel fibre tensile strengths on the fracture energy of high-strength concrete. *Constr. Build. Mater.* **2011**, *25*, 1801–1806. [CrossRef]
63. Kazemi, M.T.; Fazileh, F.; Ebrahiminezhad, M.A. Cohesive crack model and fracture energy of steel-fiber-reinforced-concrete notched cylindrical specimens. *ASCE J. Mater. Civ. Eng.* **2007**, *19*, 884–890. [CrossRef]
64. Barros, J.A.O.; Figueiras, J.A. Flexural behavior of SFRC: Testing and modeling. *ASCE J. Mater. Civ. Eng.* **1999**, *11*, 331–339. [CrossRef]
65. Bazant, Z.P.; Oh, B.H. Crack band theory for fracture of concrete. *Mater. Constr.* **1983**, *16*, 155–177. [CrossRef]
66. Kaklauskas, G.; Gribniak, V.; Bacinskas, D. Inverse technique for deformational analysis of concrete beams with ordinary reinforcement and steel fibers. *Procedia Eng.* **2011**, *14*, 1439–1446. [CrossRef]
67. Gribniak, V.; Kaklauskas, G.; Kwan, H.; Bacinskas, D.; Lbinas, D. Deriving stress-strain relationships for steel fiber concrete in tension from tests of beams with ordinary reinforcement. *Eng. Struct.* **2012**, *42*, 387–395. [CrossRef]
68. Gribniak, V.; Kaklauskas, G.; Torres, L.; Daniunas, A.; Timinskas, E.; Gudonis, E. Comparative analysis of deformations and tension-stiffening in concrete beams reinforced with GFRP or steel bars and fibers. *Compos. Part B Eng.* **2013**, *50*, 158–170. [CrossRef]
69. Kaklauskas, G.; Gribniak, V.; Meskenas, A.; Bacinskas, D.; Juozapaitis, A.; Sokolov, A.; Ulbinas, D. Experimental investigation of the deformation behavior of SFRC beams with an ordinary reinforcement. *Mech. Compos. Mater.* **2014**, *50*, 417–426. [CrossRef]

70. Meskenas, A.; Gribniak, V.; Kaklauskas, G.; Sokolov, A.; Gudonis, E.; Rimkus, A. Experimental investigation of cracking behaviour of concrete beams reinforced with steel fibres produced in Lithuania. *Balt. J. Road Bridge Eng.* **2017**, *12*, 82–87. [CrossRef]
71. Meskenas, A. Serviceability Analysis of Steel Fibre Reinforced Concrete Beams. Doctoral Dissertation, Vilnius Gediminas Technical University, Vilnius, Lithuania, 2018.
72. Banthia, N.; Soleimani, S.M. Flexural response of hybrid fiber-reinforced cementitious composites. *ACI Mater. J.* **2005**, *102*, 382–389.
73. Ahmed, S.F.U.; Maalej, M.; Paramasivam, P. Flexural responses of hybrid steel-polyethylene fiber reinforced cement composites containing high volume fly ash. *Constr. Build. Mater.* **2007**, *21*, 1088–1097. [CrossRef]
74. Chalioris, C.E.; Kosmidou, P.-M.K.; Karayannis, C.G. Cyclic response of steel fiber reinforced concrete slender beams: An experimental study. *Materials* **2019**, *12*, 1398. [CrossRef] [PubMed]

 *materials*

Article

Characterization and Simulation of the Bond Response of NSM FRP Reinforcement in Concrete

Javier Gómez *, Lluís Torres and Cristina Barris

AMADE, Polytechnic School, University of Girona, 17003 Girona, Spain; lluis.torres@udg.edu (L.T.); cristina.barris@udg.edu (C.B.)
* Correspondence: javier.gomez@udg.edu; Tel.: +34-972-418-817

Received: 6 February 2020; Accepted: 7 April 2020; Published: 9 April 2020

Abstract: The near-surface mounted (NSM) technique with fiber reinforced polymer (FRP) reinforcement as strengthening system for concrete structures has been broadly studied during the last years. The efficiency of the NSM FRP-to-concrete joint highly depends on the bond between both materials, which is characterized by a local bond–slip law. This paper studies the effect of the shape of the local bond–slip law and its parameters on the global response of the NSM FRP joint in terms of load capacity, effective bond length, slip, shear stress, and strain distribution along the bonded length, which are essential parameters on the strengthening design. A numerical procedure based on the finite difference method to solve the governing equations of the FRP-to-concrete joint is developed. Pull-out single shear specimens are tested in order to experimentally validate the numerical results. Finally, a parametric study is performed. The effect of the bond–shear strength slip at the bond strength, maximum slip, and friction branch on the parameters previously described is presented and discussed.

Keywords: CFRP; NSM; bond behavior; structural behavior; material characterization; numerical modeling

---

1. Introduction

The use of fiber reinforced polymer (FRP) materials for strengthening existing concrete structures has been widely studied during the last years because of the potential advantages of these materials compared to the conventional ones [1]. The most commonly used techniques for strengthening reinforced concrete (RC) structures with FRPs are the so-called externally bonded reinforcement (EBR) and near-surface mounted (NSM) reinforcement. While in the EBR methodology the FRP reinforcement is bonded on the exterior face of the structural element, in the NSM system the FRP is installed into groves cut in the concrete cover where it is bonded with the appropriate adhesive. The NSM technique has attracted the attention of researchers and industry in recent years due to several potential advantages compared to the EBR system, such as no need of specific surface preparation except grooving, better anchorage capacity, better protection in front of external agents like fire or vandalism, and no relevant change in the aesthetics of the structural element [2].

A number of research works have been carried out in order to obtain analytical solutions of the differential equations governing the bond performance of the FRP-to-concrete joint. This differential equation is based on the local bond–slip behavior, relating the bond and slip at every point along the bonded length. For the EBR technique, for instance, Yuan et al. proposed an analytical solution based on a local bilinear bond–slip law in [3] and a trilinear bond–slip one with an exponential descending branch in [4]. In a later work, Ali et al. [5] adapted Yuan's analytical solution to the NSM strengthening system. Even though these authors proposed closed form equations to calculate the global bond behavior, these are not always easy to implement and are only suitable for specific local bond–slip laws.

In recent years, experimental and numerical studies have provided solutions to the global bond behavior of the EBR strengthening technique [6–12], although less attention has been paid to the NSM strengthening systems [13,14]. Focusing specifically on the characterization of the local bond behavior of NSM FRP-to-concrete joints, several bond–slip laws have been proposed in the literature, with different shapes and stages. Some examples are the models proposed by Sena et al. [13], Borchert et al. [14], Zhang et al. [15], and Seracino et al. [16]. The effect of their differences on the structural bond response and design implications is not always straightforward, thus, development of numerical tools that allow obtaining and comparing the response for different models is of main interest.

The aim of this paper is to study the effect of the shape of the local bond–slip law on the global bond behavior of NSM FRP strengthening systems. For this purpose, a numerical procedure based on finite difference method is proposed to solve in a general way the differential equation that characterizes the bond behavior of the FRP-to-concrete joint, which can be applied independently of the specific local bond–slip law. The numerical results are then verified by comparison between numerical and experimental results. Finally, a parametric study to evaluate the effects of the parameters defining the local behavior of the NSM FRP strengthened bonded joint is carried out.

2. Bond Mechanisms in NSM FRP Strengthening Systems

In the NSM FRP strengthening system, the forces are transferred from the FRP material to the internal faces of the concrete groove through its perimeter in contact with the adhesive. This way, there are two main issues improving the transfer of stresses with respect to the EBR reinforcement: the higher ratio of the area of the perimeter in contact with the adhesive to the FRP area, and the transmission to the concrete material through a confined zone in the interior of the slot in the concrete cover.

The global bond stress-slip response for NSM strips subjected to a pull-out force is generally characterized by an initial, relatively stiff, linear behavior, followed by a nonlinear curve up to the maximum value of the bond stress (bond strength), after which the originated damage causes a softening branch. Moreover, in NSM strengthening systems, it has been observed the activation of a friction component for relatively large slips as an extension of the softening branch [1,14,17–19]. The presence of a friction branch has been also reported in the bond behavior of other strengthening materials as fiber reinforced cementitious matrix composites (FRCM) [20,21].

The bond mechanisms generated in the NSM FRP technique can lead to several types of failure modes, which are mainly influenced by the bonded length, FRP surface and shape, groove configuration, materials mechanical properties and adhesive properties [22,23]. In general, the failure modes can be grouped into three main categories: failure at the FRP-adhesive interface, failure at the epoxy–concrete interface, and adhesive cover splitting [1].

The global bond stress-slip performance of an NSM FRP strip bonded to a concrete block is the result of the local bond–slip behavior at every point along the bonded length, which can be considered the constitutive model characterizing the bond behavior of the NSM FRP element. Different models and shapes can be found in the literature for these local laws, and the assessment of their results is the main objective of this paper.

3. Assessment of the Bond–Slip Response of NSM FRP

3.1. Governing Equation and Global Response

In the single shear pull-out test (pull-push test), the load is transferred from the FRP to the concrete through the adhesive. Figure 1a shows the stress equilibrium of an infinitesimal element of FRP NSM concrete element of length dx. Figure 1b shows the stress equilibrium of the FRP-adhesive interface in an infinitesimal element of length dx.

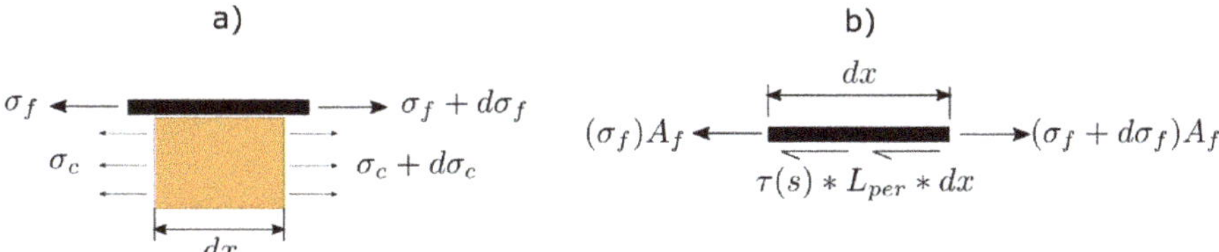

Figure 1. Stress equilibrium (**a**) in the FRP-concrete joint, and (**b**) in the FRP-adhesive interface.

From stress equilibrium between FRP and concrete, and the FRP-adhesive interface, Equation (1) and Equation (2) can be obtained.

$$A_f d\sigma_f + A_c d\sigma_c = 0,\tag{1}$$

$$\frac{d\sigma_f}{dx} - \frac{\tau(s) * L_{per}}{A_f} = 0,\tag{2}$$

where σ_f is the stress in the FRP, A_f is the FRP area, calculated as the product of the FRP thickness (t_f) by the FRP width (w_f), σ_c is the stress in the concrete, A_c is the area of concrete, $\tau(s)$ is the bond–slip law of the joint, and L_{per} is the is the intermediate perimeter in the adhesive.

The slip of the bonded joint can be defined as the relative displacement between the FRP and the concrete element.

$$s = u_f - u_c,\tag{3}$$

where s is the slip, u_f is the FRP displacement and u_c is the concrete displacement. Using the constitutive equations of the FRP and the concrete, Equation (4) and Equation (5) are obtained.

$$\frac{ds}{dx} = \varepsilon_f - \varepsilon_c = \frac{\sigma_f}{E_f} - \frac{\sigma_c}{E_c},\tag{4}$$

$$\frac{d^2 s}{dx^2} = \frac{1}{E_f}\frac{d\sigma_f}{dx} - \frac{1}{E_c}\frac{d\sigma_c}{dx},\tag{5}$$

where E_f is the elastic modulus of the FRP, ε_f is the strain in the FRP, E_c is the elastic modulus of the concrete, and ε_c is the strain in the concrete.

Finally, by substituting Equation (1) and Equation (2) into Equation (5), the differential equation governing the bond–slip behavior of a NSM FRP-to–concrete bonded joint is:

$$\frac{d^2 s(x)}{dx^2} - \tau(s) * L_{per}\left(\frac{1}{E_f * A_f} - \frac{1}{E_c * A_c}\right) = 0,\tag{6}$$

It is worth mentioning that usually the second term in the parenthesis corresponding to the concrete properties is much lower than the first and can be neglected (Ko et al. [24]).

Equation (6) can be solved analytically using the boundary conditions

$$\varepsilon_f = 0 \text{ at } x = 0,\tag{7}$$

$$s = s(L_b) \text{ at } x = L_b,\tag{8}$$

where ε_f is the strain in the FRP at the free end, L_b is the bonded length, and $s(L_b)$ is the applied slip at the loaded end.

This differential equation can be solved for specific shapes of the local bond-stress slip law [3]. In the rest of cases a numerical procedure is required.

The numerical model implemented in this work is based on the finite difference method and aims to solve the governing equation of the bonded joint for any type of local bond–slip law.

Following this methodology, the bonded length (L_b) is discretized into n uniform small increments of length $\Delta x = L_b/n$. The increments are delimited by $n+1$ points, which position is defined by $x_i = i \cdot \Delta x$ ($i = 0,1, \ldots n$) (Figure 2). The procedure is based on an incremental slip methodology that pretends to simulate a usual experimental test under monotonic increase of the loaded end slip (this way, possible snap-back effect after maximum load is not reproduced in the simulations, Zou et al. [21]). Starting from the loaded end and moving towards the free end, the procedure calculates, for every studied point, the slip and the load transmitted between two adjacent sections. For a certain value of slip at the loaded end, an iterative process is carried out in which $P(j,1)$ is a lower bound value of the initial load at the loaded end at iteration j. $P(j,1)$ is used to calculate the corresponding strain in the FRP (ε_{FRP}), the strain in the concrete (ε_c), and the bond–shear stress (τ) using the bond–slip law at the loaded point. Still, at iteration j, the load at every point i, $P(j,i)$, is calculated using Equation (9) along the FRP and towards the free end, by

$$P(j,i) = P(j,i-1) - \tau(j,i-1) * L_{per} * \Delta x, \tag{9}$$

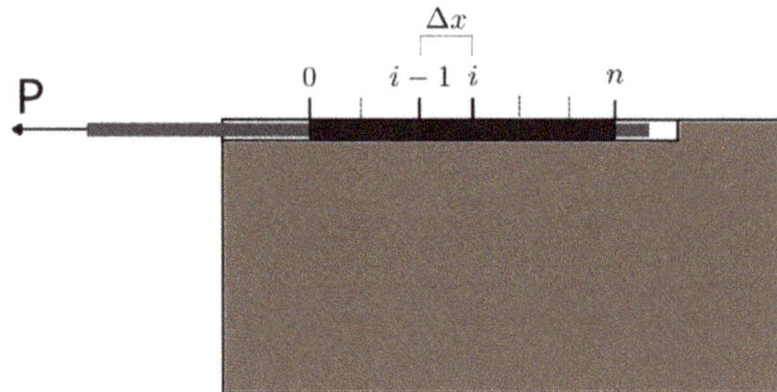

Figure 2. Scheme of the discretization along the FRP.

From the load profile obtained using Equation (9), the strain distribution in the concrete and the FRP can be obtained, and, subsequently, the slip profile along the FRP is calculated as

$$s(j,i) = s(j,i-1) - \left(\varepsilon_f - \varepsilon_c\right) * \Delta x, \tag{10}$$

Increments of P at the loaded end are applied until convergence is attained, which happens when the ε_f at the free end is less than a prescribed tolerance, set to be a value close to zero. Once convergence is achieved, the corresponding load is registered and the process is repeated for a new value of slip at the loaded end, defined as $s(j+1,1)$. At this point, the iterative procedure is again repeated. The procedure finishes when all the points of the load–slip curve are calculated. Figure 3 shows the flowchart of the numerical procedure.

3.2. Comparison of the Numerical Results with Existing Analytical Solutions

In order to verify the numerical procedure developed, a comparison with analytical solutions available for some specific laws is presented in this section. For the sake of simplicity, only the analytical solution for the bilinear bond–slip law proposed by Yuan et al. [3] is taken as a reference, considering a bond–shear strength (τ_{max}) of 15 MPa, a slip at the bond–shear strength (s_1) of 0.1 mm and a maximum slip (s_f) of 1.13 mm. These values have been chosen according to experimental results available in the literature. Four different situations are simulated by modifying the bonded length ($L_b = 200$ and 400 mm) and the FRP area ($A_f = 14$ and 30 mm^2). The results, presented in Figure 4, show good agreement between the analytical and the numerical solutions, and therefore the suitability of the developed procedure.

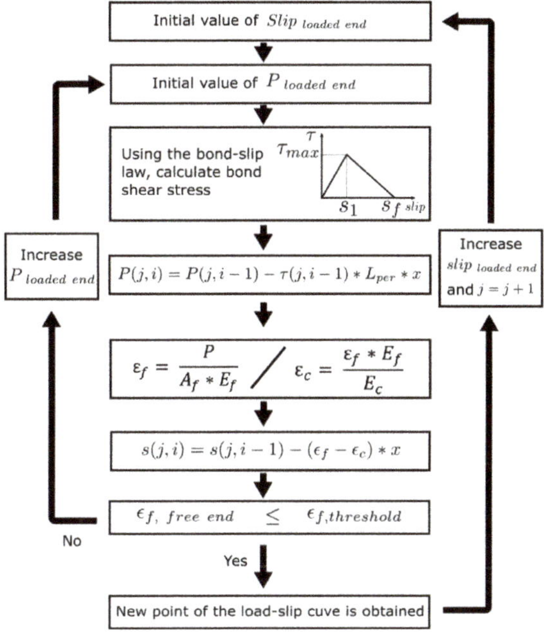

Figure 3. Flowchart of the numerical procedure.

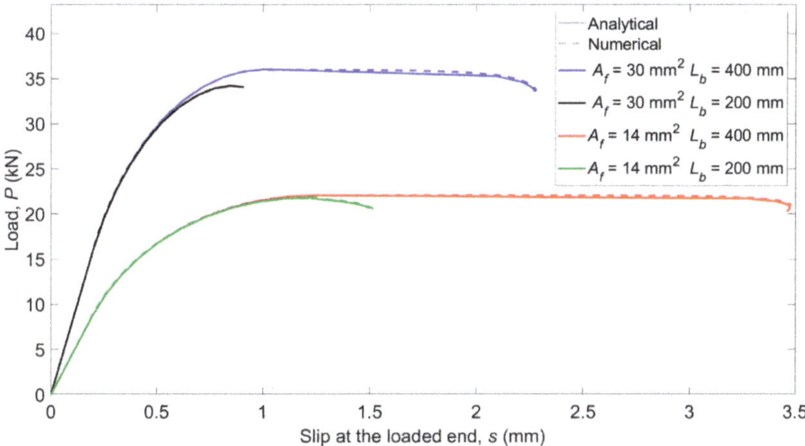

Figure 4. Comparison between the analytical and the numerical models.

4. Bond–Slip Behavior at a Local Level

In this work, six different models (Table 1) based on two of the most widely accepted constitutive laws for bond characterization of NSM FRP systems (the Bilinear law [16] and Borchert law [14]) are considered. The study focuses on the effect of the combination of their parameters on the structural global response of the bonded joint. From the bilinear law, four combinations are studied: i) a bilinear (BL) model; ii) a linear descending (LD) model, which does not consider the initial elastic branch; iii) a two stage bond–slip law with a non-linear ascending branch (TSANL) model, with the aim to study the effect of the shape of the ascending branch; and iv) a bilinear plus friction (BLF) model, with the purpose to obtain the effect of friction in a bilinear bond–slip law. On the other hand, from Borchert law (BO), one additional modification is proposed, consisting in suppressing the original bond–shear strength plateau (BONP).

Table 1. Equations of the bond–slip models used in the parametric study.

Bond–Slip Model	Equation
Linear Descending (LD)	$\tau(s) = \begin{cases} \tau_{max} * \dfrac{(s_f - s)}{s_f}, s > 0 \text{ and } s \le s_f \\[2mm] 0, s \ge s_f \end{cases}$
Bilinear (BL)	$\tau(s) = \begin{cases} \tau_{max} * \dfrac{s}{s_1}, s < s_1 \\[2mm] \tau_{max} * \dfrac{(s_f - s)}{(s_f - s_1)}, s > s_1 \text{ and } s \le s_f \\[2mm] 0, s \ge s_f \end{cases}$
Bilinear-friction (BLF)	$\tau(s) = \begin{cases} \tau_{max} * \dfrac{s}{s_1}, s < s_1 \\[2mm] \tau_{max} * \dfrac{(s_f - s)}{(s_f - s_1)}, s > s_1 \text{ and } s \le s_3 \\[2mm] \tau_f, s \ge s_3 \end{cases}$
Two-stage ascending non-linear (TSANL)	$\tau(s) = \begin{cases} \tau_{max} * \left(\dfrac{s}{s_1}\right)^{\alpha}, s < s_1 \\[2mm] \tau_{max} * \dfrac{(s_f - s)}{(s_f - s_1)}, s > s_1 \text{ and } s \le s_f \\[2mm] 0, s \ge s_f \end{cases}$
Borchert (BO)	$\tau(s) = \begin{cases} \tau_{max} * \left(\dfrac{s}{s_1}\right)^{\alpha}, s < s_1 \\[2mm] \tau_{max}, s \ge s_1 \text{ and } s \le s_2 \\[2mm] \tau_{max} - \dfrac{(\tau_{max} - \tau_f)}{(s_f - s_1)} * (s - s_1), s > s_2 \text{ and } s \le s_3 \\[2mm] \tau_f, s \ge s_3 \end{cases}$
Borchert no plateau ($s_1 = s_2$) (BONP)	$\tau(s) = \begin{cases} \tau_{max} * \left(\dfrac{s}{s_1}\right)^{\alpha}, s < s_1 \\[2mm] \tau_{max} - \dfrac{(\tau_{max} - \tau_f)}{(s_f - s_1)} * (s - s_1), s > s_1 \text{ and } s \le s_3 \\[2mm] \tau_f, s \ge s_3 \end{cases}$

Note: τ = bond–shear stress; s = slip; τ_{max} = bond–shear strength; s_1 = slip when the bond–shear strength is achieved; s_2 = final slip of the plateau; s_3 = initial slip of the friction plateau; s_f = maximum slip of the bonded joint; τ_f = friction bond–shear stress; α = experimental parameter that control the shape of the bond–slip law.

5. Global Bond–Slip Response for Different Local Bond–Slip Laws

The main objective of this section is to evaluate the influence of the local bond law on the global behavior of the FRP laminates bonded to concrete as NSM strengthening system, when subject to a pull-out force, using the numerical model described in Section 3, and the bond–slip laws introduced in Section 2.

5.1. Parameters Used

The values of the parameters that define the different local bond–slip laws, as well as the properties of materials and characteristics of specimens have been chosen according to previous experimental campaigns found in the literature [13–16,25,26], with the aim to simulate conditions as realistically as possible.

The elastic modulus of the laminate and concrete are 150 GPa and 33 GPa, respectively. The concrete block cross-section is 200 × 200 mm and the groove thickness and width are 5 mm and 15 mm, respectively. The bonded length (L_b) has been set as 200 mm, and the laminate section is 1.4 × 10 mm.

Regarding the bond–slip parameters, bond–shear strength has been set as 15 MPa, and the corresponding slip as 0.1 mm. In bond–slip models that have a friction branch, τ_f has been defined as the 35% of τ_{max}. The initial point of the plateau of BO model is $s_1 = 0.8 \cdot s_2$, and the experimental parameter α is 0.31 [14].

Another parameter that is crucial for the simulations is the fracture energy (G_f), typically defined as the area under the bond–slip law [6]. In those bond–slip laws with a friction branch, G_f is calculated according to Haskett et al. [27], as the area under the ascending and descending stages of the bond–slip law, without taking into account the friction stage, as illustrated in Figure 5. A value of 8.5 N/mm has been set for G_f in all the models [16,25,26].

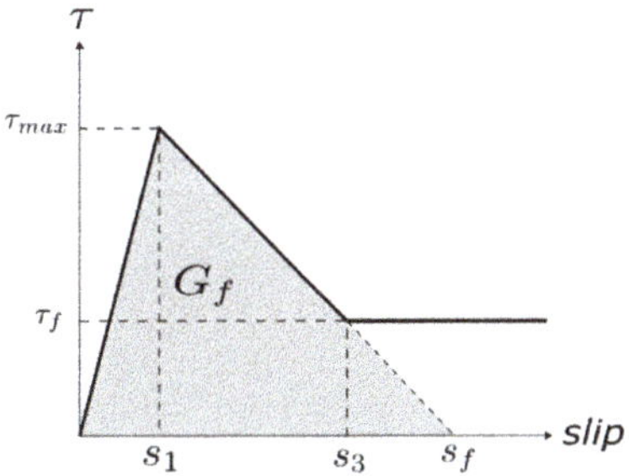

Figure 5. Fracture energy obtained from the bond–slip law.

The comparison between the studied bond–slip laws is shown in Figure 6. As indicated before, it can be observed that the bond–shear strength (τ_{max}) is 15 MPa, and the corresponding slip (s_1) is 0.1 mm, except for the LD model, which does not consider s_1.

5.2. Load–Slip Response

The comparison of the load–slip curves obtained from each bond–slip law is shown in Figure 7. Overall, two main tendencies can be observed. The first one (Figure 7a), followed by LD, BL and TSANL models, achieve a maximum load, which remains constant as the slip in the loaded end increases until failure. This plateau is obtained because the maximum activated length is attained when the slip in the loaded point is equal to s_f, meaning that this point does not carry load anymore (debonding). Therefore, the load capacity will remain constant and the slip will keep increasing until failure. In the second trend (Figure 7b), followed by BLF, BO, and BONP models, the load increases up to the maximum carrying capacity, where a sudden decrease takes place followed by a residual (friction) bond force. The difference between these two trends falls to the effect of the friction branch of the local bond–slip law after the loaded point reaches debonding.

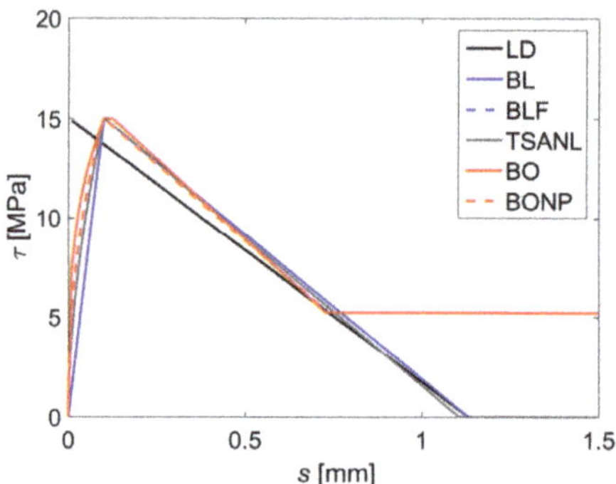

Figure 6. Comparison of the bond–slip models.

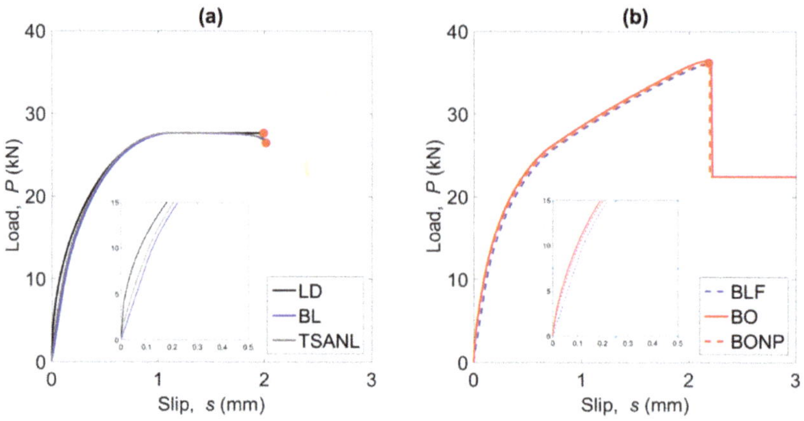

Figure 7. Comparison of the load–slip curves, (**a**) shows LD, BL and TSANL models, and (**b**) shows BLF, BO, and BONP models.

Focusing on the initial part of the load–slip curve (interior plot in Figure 7a) it can be seen that the LD model shows the highest stiffness, because the ascending branch of the bond–slip law has an infinite stiffness. Besides, it can be observed that bond–slip models with a non-linear ascending branch (TSANL, BO, and BONP) exhibit a stiffer branch of the load–slip curve than BL and BLF models caused by the fact that for a fixed value of s_1 the area under the ascending branch is greater in those cases with non-linear tendency.

5.3. FRP Strain, Bond–Shear Stress and Slip along the Bonded Length

Following the numerical procedure, the FRP strain, bond–shear stress and slip can also be obtained. Figure 8 shows the response along the bonded length for the different laws (loaded end at $x = 0$ mm, free end at $x = 200$ mm) evaluated at the situation of imminent failure (marked with a red dot in Figure 7). For 7.5 mm grooved specimens, the failure point was defined as the point before the load drops to 0, and for 10 mm grooved specimens, the failure was defined as the maximum load point.

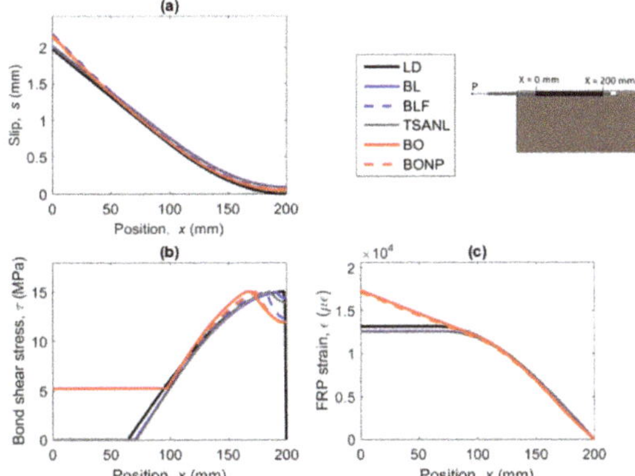

Figure 8. (**a**) FRP slip, (**b**) bond–shear stress, and (**c**) FRP strains along the bonded length for the different bond–slip models.

From Figure 8a it can be observed that the shape of the bond–slip law has a small effect on the slip profile along the FRP. It can also be seen that BLF, BO and BONP exhibit slightly higher values for the slip at the loaded end. Furthermore, the slip at the free end equals to 0 for the LD model, contrarily to the rest of models, which show a small value of slip.

An evident difference between the models with and without friction can be observed in the bond stress distribution along the bonded length (Figure 8b). Models that include friction still transmit bond–shear stress after the softening stage, whilst the other models exhibit a completely damaged zone, where no load is transmitted. Hence, for friction bond–slip models, the increase of the area under the bond–shear stress profile allows the joint to carry more load even though the zone near the loaded end is damaged.

Figure 8c shows the FRP strain distribution. In models without friction, the strain profile stabilizes when the maximum load is attained. On the other hand, strain values in the FRP do not stabilize for BLF, BO, and BONP models, because of the load keeps increasing after the loaded end surpasses the softening stage.

6. Experimental Program

In this section, an experimental program is presented and the results, in terms of load–slip curves are discussed. Then, a relatively simple methodology to obtain the bond–slip law from the experimental load–slip curve is presented. Finally, the theoretical results obtained from the numerical procedure are compared to experimental values.

6.1. Material Properties

The concrete used in the experimental campaign has been characterized according to the UNE 12390-3 [28] and the ASTM C469 / C469M-10 standard [29]. From the characterization tests on 150×300 mm cylindrical specimens, a compressive strength of 36.9 MPa and an elastic modulus of 46.8 GPa were obtained.

Carbon-FRP (CFRP) strips with 3×10 mm cross-section were used in the experimental campaign. Their mechanical properties were tested according to ISO 527-5 [30], obtaining an elastic modulus of 169.3 GPa and a tensile strength of 3205.9 MPa.

The bi-component epoxy resin used in this program was tested according to the ISO 527-2 standard [31] after 12 days of curing under 20°C and 55%RH, obtaining an elastic modulus of 10.7 GPa and a tensile strength of 27.9 MPa.

6.2. Experimental Details

6.2.1. Parameters of the Study

Four different NSM configurations were tested, combining two different groove thicknesses (7.5 mm and 10 mm) and two different bonded lengths (150 mm and 225 mm). Four specimens were tested per each configuration, giving a total of 16 specimens tested in a direct pull-out shear test.

The *fib* Bulletin 90 [1] limits the deepness of the groove in order to avoid epoxy cover splitting. ACI [32], in turn, suggests that the groove thickness should be, at least, 3 times the laminate thickness. In order to study the effect of the resin layer thickness, two groove thicknesses were defined in this study: 10 mm and 7.5 mm. The first one was set according to the ACI requirement, whilst the second groove thickness was chosen to satisfy the *fib* [1] condition that establishes that the minimum groove dimensions must be 1.5 or 2 times the laminate size.

Regarding the effect of the bonded length, Seracino et al. [33] used bonded lengths between 100 mm and 350 mm and concluded that the minimum L_b to achieve the maximum carrying capacity was 200 mm. Furthermore, Zhang et al. [34] tested NSM strengthened concrete elements with bonded lengths between 25 and 350 mm. From their study, it was observed that the experimental effective bonded lengths for the different specimens were between 150 and 175 mm.

The bonded lengths used in this experimental campaign were chosen to be above and below the values suggested in [33] and [34], in order to validate if the finite differences model is able to predict the response for short and long bonded lengths.

6.2.2. Test Setup

In order to study the effect of the bonded length and the groove width, four pull-out configurations were designed. Four specimens were tested for each configuration defined in Table 2.

Table 2. Characteristics of the specimens

Specimen ID	Bonded Length (mm)	Groove Thickness (mm)	Maximum Load (kN)	Failure Mode
NSM – 150 – 10	150	10	43.85	F-A
NSM – 225 – 10	225	10	54.74	F-A
NSM – 150 – 7.5	150	7.5	47.14	C-A
NSM – 225 – 7.5	225	7.5	57.71	C-A

Note: F-A = FRP-adhesive interface, C-A = concrete-adhesive interface.

The specimens were identified as NSM – L_b – t_g, where L_b stands for the bonded length and t_g for the groove thickness. The tests were carried out under a single shear test configuration, as shown in Figure 9. The specimen was rigidly fixed with a 60 mm wide plate to avoid translation in the vertical direction and also perpendicularly in the opposite side with a 50 mm plate to avoid rotation. At the top of the specimen a length of 50 mm was left unbonded to avoid stress concentrations in that zone [35]. The concrete block dimensions used on the short-bonded length specimens were 200 × 200 × 370 mm, and for the long-bonded length specimens 200 × 200 × 420 mm. A servo-hydraulic testing machine was used to apply the load. The test was performed under displacement control with a speed of 0.2 mm/min. One LVDT was placed in the loaded end to measure the relative displacement between the CFRP strip and the concrete.

6.3. Experimental Results

The results obtained from the experimental tests are shown in Figure 10, where the average load–slip curves for the four tested configurations are presented.

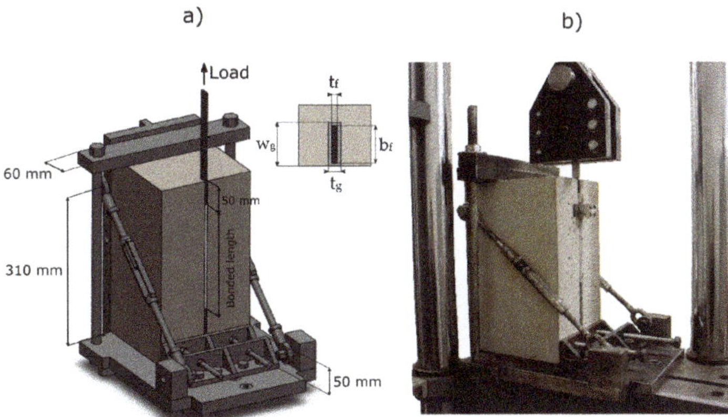

Figure 9. (**a**) scheme of the set-up for the pull-out single shear test, and (**b**) picture of the tests.

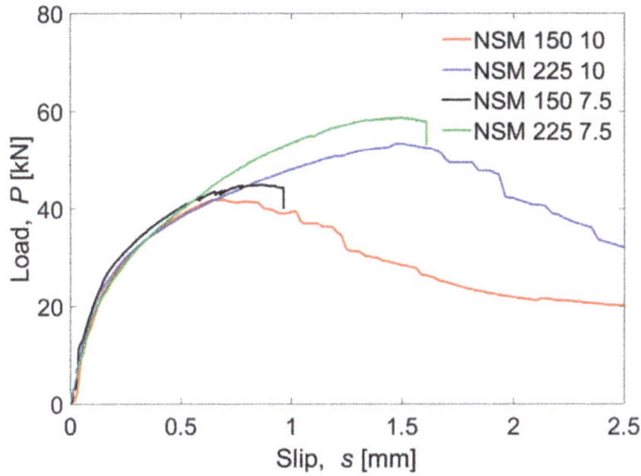

Figure 10. Experimental results from the single shear tests.

Failure loads for the narrowest groove specimens were slightly higher than for those with the widest grooves: the failure load of NSM-150-7.5 was 7.1% higher than NSM-150-10, and NSM-225-7.5 failure load was 10.0% higher than in the NSM-225-10 case. Besides, differences in the failure load can be observed: 10 mm grove thickness specimens failed in the FRP-adhesive interface while 7.5 mm grove thickness specimens failed in the resin–concrete interface.

As can be observed, in specimens with 10 mm groove thickness, after reaching the maximum load, the slip continued increasing although the load decreased, which could be interpreted as all the bonded length being activated and damaged, as well as friction effect taking place. On the other hand, failure of the specimens with 7.5 mm groove was sudden and instantaneous: after reaching the maximum load, the load capacity dropped abruptly with no friction effects. For this reason, in this study, two different bond–slip laws have been defined: a bilineal law for the case of 7.5 mm groove specimens and a bilineal plus friction law for the case of 10 mm groove specimens. Finally, it should be noticed that the slip at the end of the elastic stage and at the maximum load was similar both for specimens with 10 mm and 7.5 mm groove thickness.

Regarding the effect of the bonded length, it can be seen that the initial tendency of the load–slip curves was the same for both bonded lengths. On the other hand, as the bonded length increased, the maximum load carrying capacity increased as well. Increasing the bonded length 50% caused a load increase of 24.8% and a 22.4% for specimens with 10 mm and 7.5 mm groove thickness, respectively.

6.4. Bond–Slip Law Adjusted to the Experimental Results

From the experimental tests results, a bilinear bond–slip law could be obtained from the load–slip curves [36] for specimens with 7.5 mm groove and 10 mm groove. For the case of specimens with 7.5 mm groove thickness, the slip at the bond strength (s_1) was obtained at the end of the linear branch of the load–slip, whilst the maximum slip (s_f) could be estimated from the slip where the maximum load was achieved (Figure 11). Once s_1 and s_f were estimated, the value of the bond–shear strength (τ_{max}) was obtained from a least-squares approach that estimated the optimum value of τ_{max} to obtain the experimental load until P_{max}.

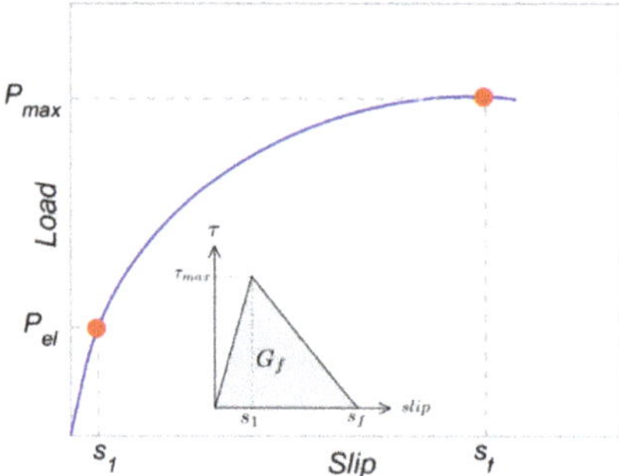

Figure 11. Bond–slip law parameters obtained from the load–slip curve for 7.5 mm grooved specimens.

For specimens with 10 mm groove thickness, which present a frictional stage after the maximum load is achieved, a trilinear bond–slip model with a friction branch has been adopted. In this model, s_1 and s_f are assumed equal to the values obtained for the 7.5 mm grooved specimens. Once the values of s_1 and s_f are defined, the methodology to calculate τ_{max} is the same as that defined for 7.5 mm grooved specimens. The friction bond–shear strength (τ_f) was defined as 35% of τ_{max}, following the reccomendations in [14].

Figure 12 shows the two adjusted bond–slip laws, for 7.5 mm and for 10 mm groove specimens, with their corresponding parameters.

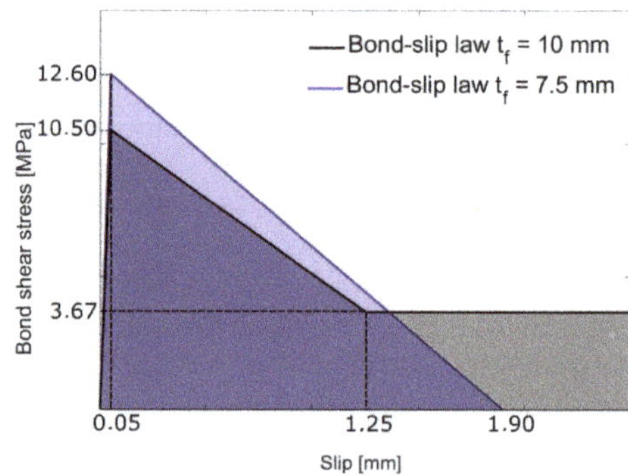

Figure 12. Experimental bond–slip laws for the NSM with 7.5 mm and 10 mm thickness grooves.

It can be observed that for the specimens with 7.5 mm groove thickness, τ_{max} is higher than for specimens with 10 mm groove thickness, implying higher fracture energy and justifies the higher maximum load experimentally obtained.

6.5. Comparison between Experimental and Numerical Results

By introducing the previously obtained bond–slip laws and the materials' mechanical properties in the finite differences model, the theoretical load–slip curves of the bonded joints are obtained. The comparison between the theoretical and experimental behavior is shown in Figure 13.

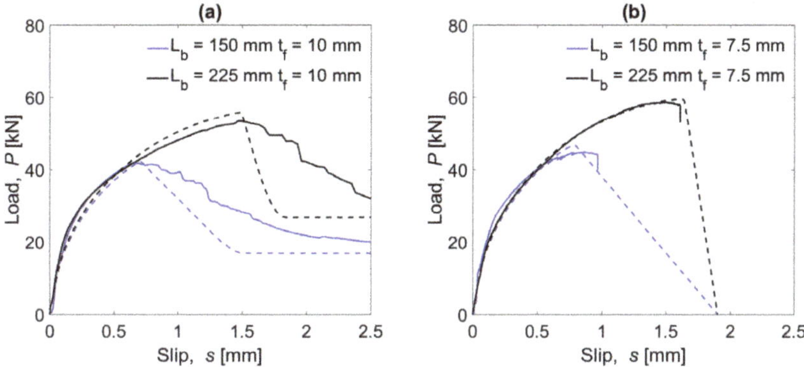

Figure 13. Comparison between the experimental (continuous lines) and the theoretical (dashed lines) values for the NSM with **(a)** 10 mm and **(b)** 7.5 mm groove thickness.

As expected, a close agreement between the predicted response and the experimental values is observed. Focusing on Figure 13a (10 mm groove thickness), it can be seen that the ascending stages of the load–slip curves and the maximum loads are correctly predicted, although the experimental results show a smoother decrease of the load once the maximum load is attained than the evolution predicted by the theoretical model. In the case of Figure 13b (7.5 mm groove thickness), the ascending branch and the maximum load are also correctly predicted.

7. Parametric Study

In the previous section, the bilinear and bilinear + friction bond–slip laws were implemented in the finite differences model. In this section, a parametric study is performed using the six different bond–slip models defined in Section 2 (Table 1) to better investigate the influence of other modifications of the bond–slip law.

The effect of four bond–slip law parameters on the maximum load (P_{max}) and on the effective bonded length (L_{eff}) are studied. The studied parameters are: i) the bond–shear strength (τ_{max}), ii) the slip at the bond–shear strength (s_1), iii) the maximum slip (s_f) and iv) the friction bond–shear strength (τ_f). L_{eff} is defined here as the bonded length needed to withstand the maximum stabilized load, therefore, it is only applicable to bond–slip models without friction branch. Computationally, the calculation of L_{eff} is measured from the first point that achieves s_f until the point that has less than 3% of the maximum FRP strain [5].

7.1. Effect of the Bond–Shear Strength (τ_{max})

Using the numerical model, the load–slip curve, and the slip, bond–shear stress and FRP strain profile along the bonded length are obtained. For the sake of brevity, only the response of the BL model is shown in Figure 14, however, the trends observed for the BL model can be extrapolated to the other bond–slip models. A range of τ_{max} between 10 MPa and 20 MPa has been considered taking $\tau_{max,0} = 10$ MPa as a reference for normalization of results and, being $s_f = 1.13$ mm, $s_1 = 0.1$ mm, $s_2 = 0.12$ mm, and $\tau_f = 5.25$ MPa.

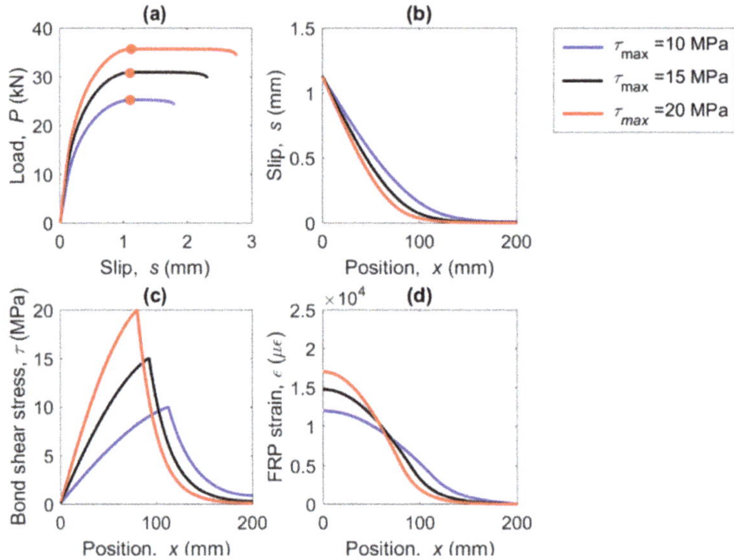

Figure 14. (**a**) Load–slip curve, (**b**) slip along the FRP, (**c**) bond–shear stress along the FRP, and (**d**) strains along the FRP for three values of bond–shear strength and a bilinear bond–slip law.

In Figure 14, the curves are represented for the situation where the slip at the loaded end arrives at s_f. It is clearly seen that P_{max} increases with τ_{max} (Figure 14a). This is because an increase of τ_{max} implies an increase of G_f, thus, a higher load is needed to damage the bonded joint. In Figure 14b, where the slip profile is shown, it can be observed that as the τ_{max} increases, the activated bonded length decreases. At $x = 200$ mm the slip in the free end when τ_{max} is 10 MPa, is equal to 0.0091 mm, but when τ_{max} is 20 MPa, the slip in the free end is negligible, meaning that the activated length decreases with the increase of τ_{max}.

In Figure 14c the bond stress profile is represented. In that case, it can be seen that as τ_{max} increases, the bond–shear profile becomes narrower because every segment along the bonded length is able to transfer a larger bond–shear force. Finally, as the bond–shear strength (τ_{max}) increases, the strain in the loaded end increases as well (Figure 14d). It is worth noticing that as the bond–shear strength capacity increases, the activated bonded length decreases.

In Figure 15, the increase of the maximum load is depicted in function of the increase of the τ_{max}, previously normalized with respect to $\tau_{max,0} = 10$ MPa. In general terms, a practically linear increase of the maximum load with the bond–shear strength can be observed for all the models. This is due to the fact that increasing τ_{max} without changing the value of s_f causes a linear increase of the fracture energy (G_f), leading to a situation where a higher load is needed to damage the bonded joint.

Moreover, it can be seen that LD, BL, and TSANL models show very similar results, meaning that these models are analogously affected by the increase of the bond–shear strength. These models show an increase of P_{max} of 41% when τ_{max} is 20 MPa. Finally, BLF, BO, and BONP, show lesser effect on the maximum load than the other models, because of the friction branch, which remains constant for all the cases. Although these models increase their G_f, the increase of the area under the bond–slip curve is smaller than in the non-friction cases. The increase of P_{max} in these models arrives to 30% when

τ_{max} is 20 MPa. Finally, the BLF model is slightly less affected than BO and BONP models, because of the non-linear ascending branch of Borchert models, which experiment a higher increase of G_f when τ_{max} increases.

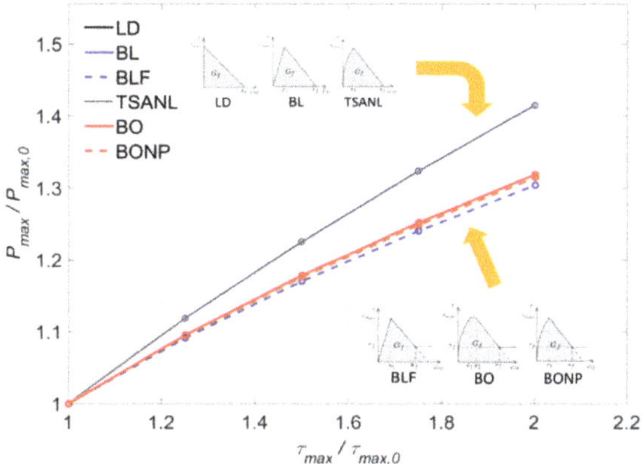

Figure 15. Variation of P_{max} versus τ_{max}.

The influence of τ_{max} on the effective length, L_{eff}, is shown in Figure 16 for models without friction. In general, a decrease of its value is observed as τ_{max} increases, as expected: if the bond–shear strength increases, every finite element of the bonded length can transfer a larger shear force, hence, less bonded length is needed to transfer the total load. It is worth noticing that the three models behave in a very similar way, indicating that assuming either the one stage model, the bilinear model or the model with a non-linear ascending stage, does not provide a difference on the effective bonded length for different bond strengths.

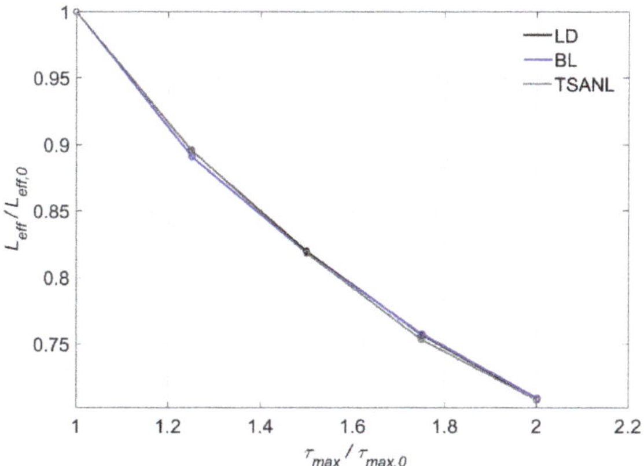

Figure 16. Variation of L_{eff} versus τ_{max}.

7.2. Effect of the Slip at the Bond–Shear Strength (s_1)

A range for s_1 between 0.1 mm and 0.5 mm has been considered, taking $s_{1,0} = 0.1$ mm for normalization of results, and keeping $\tau_{max} = 15$ MPa. The value of the G_f when $s_{1,0} = 0.1$ mm is equal to 8.5 N/mm. In order to meet the new value of s_1, G_f varies between 8.5 N/mm and 10.9 N/mm, depending on the model considered. In BO model, the difference between s_1 and s_2 has been kept constant for all cases. The linear descending (LD) model is not included because the slip at the bond–shear strength is 0, and if this point is shifted horizontally, it would become a bilinear (BL) model.

The results are plotted in Figure 17 in terms of increment of the maximum load with respect to the increment of s_1. As observed, in all models the load increment is considerably small (with a maximum variation of 8.9%). For the BL model, no effect on the maximum load is obtained for a substantial increment of s_1, because τ_{max} and s_f are constant, and therefore, the fracture energy (G_f) remains the same.

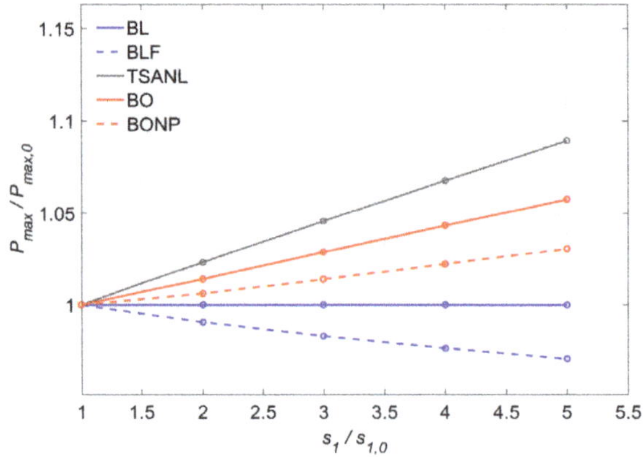

Figure 17. Variation of P_{max} versus s_1.

In the case of the BLF model, the maximum load decreases as the s_1 increases. Even though the fracture energy (G_f) remains constant for all the s_1 values, as can be seen in Table 3, the area under the total bond shear–slip curve, considering the three stages (elastic, softening and friction) sligthly decreases with the increment of s_1. It should be noticed, however, that the maximum load decrease only arrives up to 3% of the initial load.

Table 3. Variation of fracture energy for the BONP, BO and BLF models.

		BONP	**BO**	**BLF**
s_1 [mm]	$\dfrac{s_1}{s_{1,0}}$	$\left(\dfrac{G_f-G_{f,0}}{G_{f,0}}\right)*100$	$\left(\dfrac{G_f-G_{f,0}}{G_{f,0}}\right)*100$	$\left(\dfrac{G_f-G_{f,0}}{G_{f,0}}\right)*100$
0.10	1	0 %	0 %	0 %
0.20	2	4.64 %	6.85 %	0 %
0.30	3	9.29 %	13.71 %	0 %
0.40	4	13.94 %	20.56 %	0 %
0.50	5	18.59 %	27.42 %	0 %

It is also seen that bond–slip models that have a non-linear ascending branch (TSANL, BO, and BONP models) exhibit an some increase on the maximum load (P_{max}), caused by the increase of area under the elastic stage curve from the shifting of s_1. This effect can be observed in Table 3 through an increment of the fracture energy (G_f) with the increase of s_1. It should be noticed that the variation of the fracture energy for TSANL and BONP models is the same, therefore, in Table 3, only BONP model is showed. In BO and BONP cases, the increase of the fracture energy is mitigated by the decreasing of friction area between s_3 and s_f (Table 1): since the value of s_f is kept constant and s_1 increases, the value of s_3 shifts horizontally, causing a reduction of area between s_3 and s_f.

The TSANL model is more affected by the modification of s_1 than the other models, since it does not have a friction branch and the total area under the bond–slip law will not decrease with the shifting of s_1, and the increase of the fracture energy will not be mitigated. The increase of G_f in the TSANL model is the same as in the BONP model, since the definition of G_f has been considered as in Figure 5.

Figure 18 shows the variation of the effective bonded length with s_1 for TSANL and BL models. The effective bonded length (L_{eff}) increases as s_1 increases. Since higher slips are obtained in the elastic stage, a longer bonded length will be activated and L_{eff} will increase.

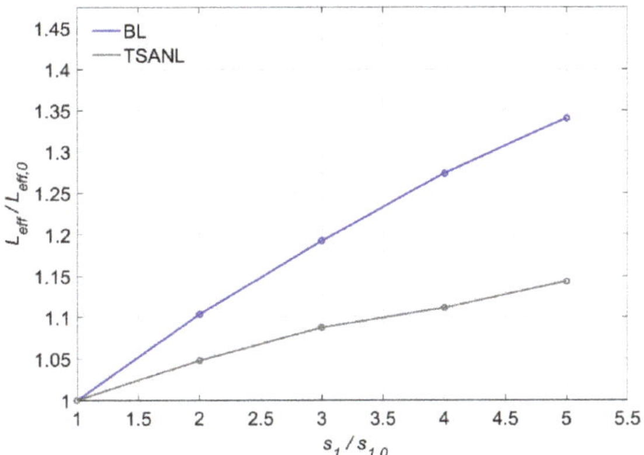

Figure 18. Variation of Leff versus s_1.

Moreover, the BL model is more sensitive to the variation of s_1 than the TSANL model, because the change in the non-linear ascending branch of the TSANL model slightly increases the area under the elastic branch, improving the resistance of the joint under in this stage. This way, the increase of slip caused by the s_1 shifting is mitigated by this increase of resistance.

7.3. Effect of the Maximum Slip (s_f)

Fixing the values of the bond–shear strength (τ_{max}) and the slip at the b ond shear strength (s_1) to 15 MPa and 0.1 mm, respectively, and setting the fracture energy (G_f) to the values obtained in Section 7.1 (ranging between 8.5 N/mm and 11.33 N/mm) for each model, the values of the initial maximum slip ($s_{f,0}$) are between 0.70 mm and 0.75 mm. Because the bond–slip models have different shapes, and therefore, different values of G_f, the maximum slip, which satisfies the conditions of τ_{max} and s_1, will be different for each bond–slip law, as seen in Figure 19.

Overall, in Figure 19 it can be seen that all the models show an increasing tendency of P_{max} with s_f, because as s_f increases, G_f increases as well. The LD model presents exactly the same curve as the BL, since in both cases G_f is the same. Moreover, as indicated in previous subsections, models without friction branch—such as LD, BL, and TSANL—are more sensitive to the increase of s_f because the shifting of the slip affects directly to G_f, P_{max} increases up to 41% when s_f is the highest value. It is worth noticing that the TSANL is slightly less sensitive because the area under the elastic branch is higher than in LD and BL models, causing that the increment of s_f affects proportionally less to G_f and P_{max}, around 13% when s_f is the maximum value. As for the models with friction branch—BLF, BO, and BONP—increase the P_{max} caused by the shifting of s_f is diminished by the effect of the friction branch in the total area as indicated in previous subsections.

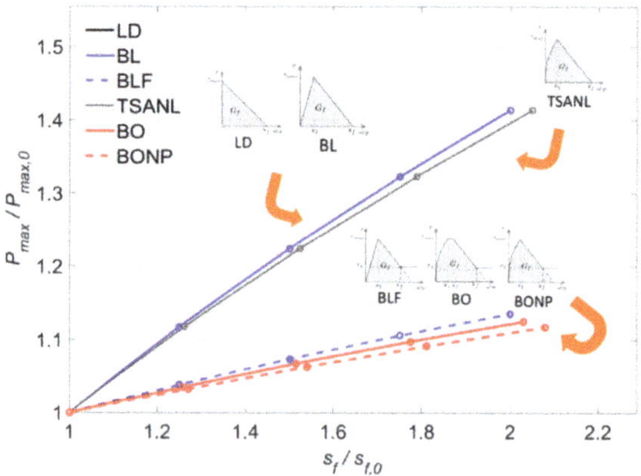

Figure 19. Variation of P_{max} versus s_f.

Figure 20 shows that L_{eff} increases with the increase of s_f for all models. A higher s_f allows the bonded joint to have higher slips while transferring load, and consequently, more bonded length is activated. LD model exhibits the highest influence of s_f and BL model is the less affected model.

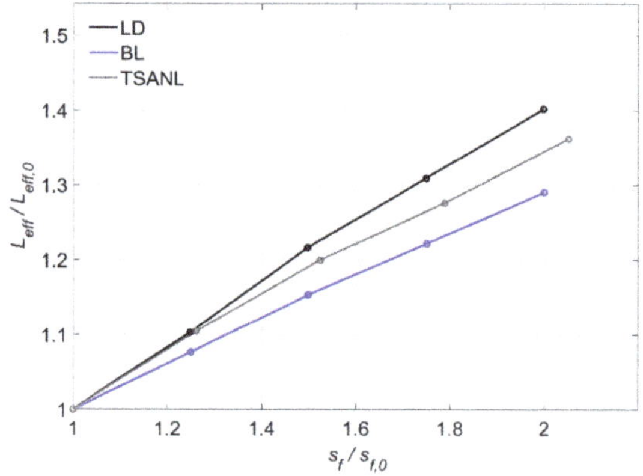

Figure 20. Variation of L_{eff} versus s_f.

7.4. Effect of the Friction Branch (τ_f)

Only BLF, BO, and BONP (with friction branch) models are included in this section. The bond–slip law parameters are set to $\tau_{max} = 15$ MPa, $s_1 = 0.1$ mm, and $G_f = 8.5$ N/mm. The bond–shear stress of the friction branch ranges between 10% and 40% of τ_{max}, therefore τ_f will vary between 1.5 MPa ($\tau_{f,0}$) and 6 MPa.

Figure 21 shows the variation of P_{max} with the increase of τ_f. It can be seen that all models present the same almost linear rising tendency, as expected. For example, for an increase of 2.5 times $\tau_{f,0}$ (i.e., $\tau_f = 3$ MPa), the increase of P_{max} is around 17%, while for 4 times $\tau_{f,0}$ ($\tau_f = 6$ MPa), P_{max} increases around 35%.

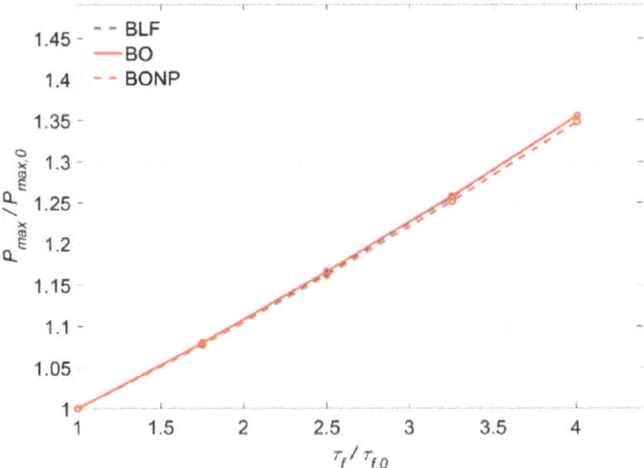

Figure 21. Variation of P_{max} versus τ_f.

8. Conclusions

A study on the effect of the local bond–slip law characterizing the material bond behavior and its parameters on the global structural bond response of NSM FRP strengthened RC elements has been presented. A numerical method has been developed in order to being able to solve the governing equations of the bonded joint for any type of bond–slip law.

From the comparison between the results of the different bond–slip laws, the following conclusions are obtained:

- In models not including a friction branch after softening, a maximum value of load is attained, which stabilizes for a certain value of the bonded length. In contrast, in models with friction component, the load continuously increases up to a certain slip beyond which only the friction component remains. Furthermore, models with non-linear ascending branch show a stiffener initial load–slip response.

- The non-linear ascending branch effect on the maximum load is practically negligible. Small differences of P_{max} are observed between BL and TSANL models, and BLF and BONP models, respectively.

- The shape of the bond–slip law has a small effect on the slip profile along the FRP. However, the bond stress and slip distribution at maximum state along the bonded length is strongly affected by the friction branch.

- From the comparison between numerical and experimental results, it can be concluded that:

- A close agreement between the finite differences model and the experimental results is obtained. The comparison between the load–slip curves obtained experimentally and numerically showed that the ascending part is correctly predicted until failure.

- A somewhat larger maximum load of around 7–10% was obtained for specimens with 7.5 mm groove thickness compared to those with 10 mm. As the groove thickness decreased, the maximum load of the bonded joint increased.

- Specimens with 10 mm groove thickness showed a behavior indicating the existence of a friction branch in their local bond law and the failure was in the FRP-adhesive interface. Conversely, the behavior of 7.5 mm groove thickness specimens was properly described using a bilinear function, while the failure was in the resin–concrete interface.

- From the parametric study carried out, the following conclusions can be drawn:

- As the bond–shear strength increases, the maximum load grows, and conversely, the effective bonded length decreases.

- The slip at the bond shear strength, s_1, has a small effect on the maximum load and an increasing effect on the effective bonded length.
- The maximum load and the effective bonded length increase with the maximum slip. Moreover, bond–slip laws without friction branch are much more sensitive to the shifting of the maximum slip.
- The presence of a friction stage in the local bond behavior causes an increase on the maximum load. As the bond–shear strength on the friction branch increases, the maximum load increases as well.

Author Contributions: Conceptualization, validation and formal analysis, J.G., C.B. and L.T.; methodology and investigation, J.G., C.B. and L.T.; writing—original draft preparation, J.G.; writing—review and editing, J.G., C.B. and L.T.; supervision, C.B. and L.T.; funding acquisition, C.B. and L.T. All authors have read and agreed to the published version of the manuscript.

Funding: This research was funded by the Spanish Government (MINECO), Project Ref. BIA2017-84975-C2-2-P.

Acknowledgments: The first author acknowledges the University of Girona for the IFUdG grant IFUDG2018/28. The authors also thank the support of SIKA Services AG for the supply of materials used in this study.

Conflicts of Interest: The authors declare no conflict of interest.

References

1. *fib* Bulletin 90. *Externally Applied FRP Reinforcement for Concrete Structures. International Federation for Structural Concrete*; fib Bulletin: Lausanne, Switzerland, 2019.
2. De Lorenzis, L.; Teng, J.G. Near-surface mounted FRP reinforcement: An emerging technique for strengthening structures. *Compos. Part B Eng.* **2007**, *38*, 119–143. [CrossRef]
3. Yuan, H.; Teng, J.G.; Seracino, R.; Wu, Z.S.; Yao, J. Full-range behavior of FRP-to-concrete bonded joints. *Eng. Struct.* **2004**, *26*, 553–565. [CrossRef]
4. Yuan, H.; Lu, X.; Hui, D.; Feo, L. Studies on FRP-concrete interface with hardening and softening bond–slip law. *Compos. Struct.* **2012**, *94*, 3781–3792. [CrossRef]
5. Ali, M.S.M.; Oehlers, D.J.; Griffith, M.C.; Seracino, R. Interfacial stress transfer of near surface-mounted FRP-to-concrete joints. *Eng. Struct.* **2008**, *30*, 1861–1868.
6. Chen, J.F.; Teng, J.G. Anchorage strength models for FRP and steel plates. *J. Struct. Eng.* **2001**, *127*, 784–791. [CrossRef]
7. D'Antino, T.; Pellegrino, C. Bond between FRP composites and concrete: Assessment of design procedures and analytical models. *Compos. Part B Eng.* **2014**, *60*, 440–456. [CrossRef]
8. Ferracuti, B.; Savoia, M.; Mazzotti, C. A numerical model for FRP-concrete delamination. *Compos. Part B Eng.* **2006**, *37*, 356–364. [CrossRef]
9. Lu, X.Z.; Teng, J.G.; Ye, L.P.; Jiang, J.J. Bond–slip models for FRP sheets/plates bonded to concrete. *Eng. Struct.* **2005**, *27*, 920–937. [CrossRef]
10. Biscaia, H.C.; Chastre, C.; Silva, M.A.G. Linear and nonlinear analysis of bond–slip models for interfaces between FRP composites and concrete. *Compos. Part B Eng.* **2013**, *45*, 1554–1568. [CrossRef]
11. Soares, S.; Sena-Cruz, J.; Cruz, J.R.; Fernandes, P. Influence of surface preparation method on the bond behavior of externally bonded CFRP reinforcements in concrete. *Materials (Basel)* **2019**, *12*, 414. [CrossRef]
12. Yin, Y.; Fan, Y. Influence of roughness on shear bonding performance of CFRP-concrete interface. *Materials (Basel).* **2018**, *11*, 1875. [CrossRef]
13. Sena-Cruz, J.; Barros, J.A.O. Bond between Near-surface mounted carbon-fiber-reinforced polymer laminate strips and concrete. *J. Compos. Constr.* **2004**, *8*, 519–527. [CrossRef]
14. Borchert, K.; Zilch, K. Bond behavior of NSM FRP strips in service. *Struct. Concr.* **2008**, *9*, 127–142. [CrossRef]
15. Zhang, S.S.; Teng, J.G.; Yu, T. Bond–slip model for interfaces between near-surface mounted CFRP strips and concrete. In Proceedings of the 11th International Symposium on Fiber Reinforced Polymers for Reinforced Concrete Structures (FRPRCS-11), Guimarães, Portugal, 26–28 June 2013; pp. 1–7.
16. Seracino, R.; Raizal Saifulnaz, M.R.; Oehlers, D.J. Generic debonding resistance of EB and NSM plate-to-concrete joints. *J. Compos. Constr.* **2007**, *11*, 62–70. [CrossRef]
17. Zilch, K.; Borchert, K. A general bond stress slip relationship for NSM FRP strips. In Proceedings of the 8th International Symposium on FRP Reinforcement for Concrete Structures, Patras, Greece, 16–18 July 2007.

18. Finckh, W.; Zilch, K. Strengthening and rehabilitation of reinforced concrete slabs with carbon-fiber reinforced polymers using a refined bond model. *Comput. Civ. Infrastruct. Eng.* **2012**, *27*, 333–346. [CrossRef]

19. Zilch, K.; Niedermeier, R.; Finckh, W. *Strengthening of Concrete Structures with Adhesively Bonded Reinforcement*; Ernst & Shon: Berlin, Germany, 2014.

20. D'Antino, T.; Colombi, P.; Carloni, C.; Sneed, L.H. Estimation of a matrix-fiber interface cohesive material law in FRCM-concrete joints. *Compos. Struct.* **2018**, *193*, 103–112. [CrossRef]

21. Zou, X.; Sneed, L.H.; D'Antino, T. Full-range behavior of fiber reinforced cementitious matrix (FRCM)-concrete joints using a trilinear bond–slip relationship. *Compos. Struct.* **2020**, *239*, 112024. [CrossRef]

22. Torres, L.; Sharaky, I.A.; Barris, C.; Baena, M. Experimental study of the influence of adhesive properties and bond length on the bond behavior of NSM FRP bars in concrete. *J. Civ. Eng. Manag.* **2016**, *22*, 808–817. [CrossRef]

23. Sharaky, I.A.; Torres, L.; Baena, M.; Miàs, C. An experimental study of different factors affecting the bond of NSM FRP bars in concrete. *Compos. Struct.* **2013**, *99*, 350–365. [CrossRef]

24. Ko, H.; Matthys, S.; Palmieri, A.; Sato, Y. Development of a simplified bond stress-slip model for bonded FRP-concrete interfaces. *Constr. Build. Mater.* **2014**, *68*, 142–157. [CrossRef]

25. Bilotta, A.; Ceroni, F.; Barros, J.A.O.; Costa, I.; Palmieri, A.; Szabó, Z.K.; Nigro, E.; Matthys, S.; Balazs, G.L.; Pecce, M. Bond of NSM FRP-strengthened concrete: Round robin test initiative. *J. Compos. Constr.* **2015**, *20*, 04015026. [CrossRef]

26. Ceroni, F.; Pecce, M.; Bilotta, A.; Nigro, E. Bond behavior of FRP NSM systems in concrete elements. *Compos. Part B Eng.* **2012**, *43*, 99–109. [CrossRef]

27. Haskett, M.; Oehlers, D.J.; Ali, M.S.M. Local and global bond characteristics of steel reinforcing bars. *Eng. Struct.* **2008**, *30*, 376–383. [CrossRef]

28. UNE 12390-3. *Testing Hardened Concrete. Part 3: Compressive Strength of Test Specimens*; AENOR: Madrid, Spain, 2003.

29. ASTM C469/C469M-10. *Standard Test Method for Static Modulus of Elasticity and Poisson's Ratio of Concrete in Compression*; ASTM International: West Conshohocken, PA, USA, 2010.

30. ISO 527-5:2009. *Determination of Tensile Properties—Part 5: Test Conditions for Unidirectional Fiber-Reinforced Plastic Composites*; ISO: Geneva, Switzerland, 2009.

31. ISO 527-2:2012. *Determination of Tensile Properties-Part 2: Test Conditions for Moulding and Extrusion Plastics. International Organisation for Standardization (ISO)*; ISO: Geneva, Switzerland, 2012.

32. ACI 440.2R-17. *Guide for the Design and Construction of Externally Bonded FRP Systems for Strengthening Existing Structures*; American Concrete Institute: Farmington Hills, MI, USA, 2017.

33. Seracino, R.; Jones, N.M.; Ali, M.S.; Page, M.W.; Oehlers, D.J. Bond strength of near-surface mounted FRP strip-to-concrete joints. *J. Compos. Constr.* **2007**, *11*, 401–409. [CrossRef]

34. Zhang, S.S.; Teng, J.G.; Yu, T. Bond strength model for CFRP strips near-surface mounted to concrete. *J. Compos. Constr.* **2014**, *18*, A4014003. [CrossRef]

35. Savoia, M.; Mazzotti, C.; Ferracuti, B. A new single-shear set-up for stable debonding of FRP-concrete joints. *Constr. Build. Mater.* **2009**, *23*, 1529–1537.

36. Emara, M.; Torres, L.; Baena, M.; Barris, C.; Cahís, X. Bond response of NSM CFRP strips in concrete under sustained loading and different temperature and humidity conditions. *Compos. Struct.* **2018**, *192*, 1–7. [CrossRef]

Article

Structural Materials: Identification of the Constitutive Models and Assessment of the Material Response in Structural Elements Strengthened with Externally-Bonded Composite Material

Todor Zhelyazov [1,2]

[1]   Structural Engineering and Composites Laboratory-SEL, Reykjavik University, Reykjavik IS-101, Iceland; elovar@yahoo.com
[2]   Department of Mechanics, Technical University of Sofia, Sofia 1000, Bulgaria

Received: 5 February 2020; Accepted: 4 March 2020; Published: 11 March 2020

Abstract: This article investigates the material behavior within multiple-component systems. Specifically, a structural concrete element strengthened to flexure with externally-bonded fiber-reinforced polymer (FRP) material is considered. Enhancements of mechanical performances of the composite structural element resulting from synergies in the framework of the multiple-component system are studied. The research work comprises the determination of the constitutive relations for the materials considered separately as well as the investigation of materials' response within a complex system such as the composite structural element. The definition of the material models involves a calibration of the model constants based on characterization tests. The constitutive relations are integrated into the finite element model to study the material behavior within the multiple-component system. Results obtained by finite element analysis are compared with experimental results from the literature. The finite element analysis provides valuable information about the evolution of some internal variables, such as mechanical damage accumulation. The material synergies find expression in the load-carrying capacity enhancement and the delay in the damage accumulation in concrete.

Keywords: FRP; concrete; damage; synergy; strengthening; finite element analysis

---

1. Introduction

This article investigates the structural response of materials considered either separately or as constituencies of a multiple-component composite system. The prediction of the behavior of a complex system is not always a straightforward task and requires the implementation of sophisticated techniques.

For the identification of multiple-component systems, one of the available approaches is the study of the mechanical wave propagation. The velocities of the compressional and shear waves depend on the elastic moduli of the continuum. Therefore, the estimation of these velocities is a basis of evaluation and testing of engineered materials (e.g., concrete) [1]. Specifically, the estimate of the ultrasound velocity provides information about material properties such as rigidity [2] stress state [3], and damage state [4].

For soft elastic materials, the study of wave propagation also has a number of practical applications. Among others, these are non-destructive defect detection [5,6] and acoustic tomography [7]. Some sources in the literature also report the investigation of inclusions-induced scattering of sound waves within the modeling of fiber-reinforced composites [8,9]. The mechanical properties of fiber-reinforced polymers (FRP) depend strongly on the inclusions, specifically on the volume fraction and spatial orientation of the fibers in the polymer matrix. The evaluation of the effective dynamic mass density of composites implies in-depth discussions [10–13]. Recently developed analytical frameworks [14,15], modeling the interaction and propagation of acoustic waves in soft, elastic material with scatterers (inclusions), allow for taking into account the form of the inclusions as well as their arrangement in the medium. Some literature sources report finite element analyses of the interaction between propagating sound waves and elastic structures embedded in the continuum [16,17]. Generally, the inclusions are either hard (steel, carbon) [18] or voids [19].

Given the complexity of the problem, reliable characterization procedures are necessary for the definition of the material models and the constitutive laws.

The behavior of the multiple-component system can be characterized by a remarkable synergy of the constituent materials. This hypothesis should be verified for different cases. On the one hand, the material synergies enhance the performances of an FRP-strengthened reinforced concrete beam in terms of increased load-carrying capacity and stiffness. One the other hand, negative effects due to material interaction might be also present in some cases. A well-known example is the so-called premature failure of FRP-strengthened structural elements [20–25]. Global failure might result from a local failure in concrete and might occur even if the FRP reinforcement still has significant reserves in strength. In this case, the prediction of the material response based on a characterization test of the composite material considered separately appears to be not adequate within the multiple-component system.

Adding external FRP reinforcement (i.e., increasing the number of bonded FRP plates) leads to an increase in the load-carrying capacity. The flexural performance enhancement of concrete/reinforced concrete beams strengthened with externally bonded FRP plates pioneered by [26], has been investigated in numerous research works (see among others [21,22,27–35]). Recently, researchers also focus on the investigation of optimal design of FRP strengthened beams [36] and on new solutions, such as near-surface mounted FRP reinforcement [37]. Another novelty is the adding of discreet steel fibers in concrete: the materials' synergies inhibit cracking and lead to further enhancement of the mechanical performances of the composite structural element [38]. In these techniques, material synergies enhance the mechanical performances of the composite structural elements. For a quasi-static loading, synergies result in a "delayed" global failure. They possibly modify the time rate of change of the damage accumulation. The verification of such a hypothesis requires quantification of degradation in concrete. The damage-based approach used in this study provides such an estimate.

The complex mechanical behavior of a composite structural element, specifically, a reinforced concrete beam strengthened to flexure with adhesively-bonded composite material, is the research object of this study. Assuming that the behavior of the composite structural element essentially depends on failure mechanisms in concrete, the investigation focuses on the strain-softening concrete response against a background defined by the other materials' (FRP, adhesive, steel) responses.

Finite element simulation can provide valuable insight into the evolution of some internal variables (e.g., the accumulated damage), which are difficult to measure experimentally. Material models, needed as an input for the finite element analysis, are generally defined based on experimental data. Model constants are identified in numerical procedures designed to search for the set of constants providing the best fit with experimental data. The following materials characterization tests are required to build a numerical model of an FRP-strengthened structural element: (a) compression tests on concrete cylindrical specimens; (b) tension by flexure tests on concrete prismatic specimens; (c) tension tests on steel specimens (for the internal steel reinforcement); (d) tension tests on specimens made of the adhesive; (e) tension tests on specimens of the composite material plate. The identification of the response of the composite material/concrete interface requires additional tests, such as the single-lap shear test [39–41].

Complex systems may include "unknown" parameters [42]. In the context of the present study, the term system, or a "multiple-component system," means a composite structural element made of components (i.e., materials) which exhibit various types of behavior but which complement each other to form the overall response of the structural element. From a certain point of view, the strengthened structural element can also be referred to as a "complex system." For example, the premature failure modes that prevent the full utilization of the externally bonded FRP reinforcement are experimentally identified, typically, for specific geometry and external reinforcement arrangement. Also, the "delay" in the damage accumulation in concrete (within the multiple-component system), leading to an enhancement of the mechanical performances of the composite structural element, needs more in-depth analysis.

A numerical model is built to predict the complex behavior of the composite structural element. The model assumes experimentally identified constitutive laws to simulate constituent materials. The model is verified through a comparison of numerical results with experimental data. For an illustrative example, the experimentally-obtained data on the evolution of the mid-span deflection is compared with predictions on the evolution of the same parameter obtained by finite element analysis. In this context, the finite element analysis sheds light on the interaction between different components of the system (i.e., the different materials). A constitutive relation defined based on damage mechanics fundamentals is chosen to model the behavior of concrete and to investigate the synergy effect improving structural performance. This approach provides an insight into the evolution of some internal (or non-observable) variables, such as the damage accumulated in the representative volume element. On the basis of the distribution of this variable obtained (for example) by finite element analysis, an accurate assessment of the resources remaining in the material subjected to a specified loading history can be assessed.

2. Materials' Behavior: Constitutive Relations

The behavior of materials is defined through the stress-strain relationship:

$$\sigma_i = \sum_{j=1}^{6} C_{ij}\varepsilon_j, \; i = 1\ldots6 \tag{1}$$

In Equation (1), σ_i are the stress components, C_{ij} is the stiffness matrix, and ε_j are the strain components. The stress-strain relationships, in their turn, are defined on the basis of empirical data obtained by identification tests on standard specimens. Therefore, to identify the response of an FRP-strengthened structural element, as a minimum, material characterization tests should be performed on specimens made of composite material, concrete, steel, and adhesive.

The formulation of the employed material models, as well as the identification tests used as a basis for their definition, are discussed in this section.

2.1. Composite Material

The FRP materials can be used as external reinforcement (sheets and plates externally bonded to the tensioned surfaces) or as internal reinforcement (reinforcing bars of filament in composite material analogous to the traditional steel reinforcement).

It is known from experiments that the mechanical response of the FRP materials is linear and elastic until failure. However, there might be some nuances in the modeling of the FRP material's behavior in the function of the material symmetries. For example, the unidirectional FRP sheets should be modeled by employing transverse isotropy. Thus, for a material in which the y–z plane is a plane of symmetry (x-, y- and z- are material directions: see Figure 1), the stiffness matrix is given by [43]:

$$C_{ij} = \begin{pmatrix} C_{11} & C_{12} & C_{12} & 0 & 0 & 0 \\ C_{12} & C_{22} & S_{23} & 0 & 0 & 0 \\ C_{12} & C_{23} & C_{22} & 0 & 0 & 0 \\ 0 & 0 & 0 & \frac{C_{22}-C_{23}}{2} & 0 & 0 \\ 0 & 0 & 0 & 0 & C_{55} & 0 \\ 0 & 0 & 0 & 0 & 0 & C_{55} \end{pmatrix} \tag{2}$$

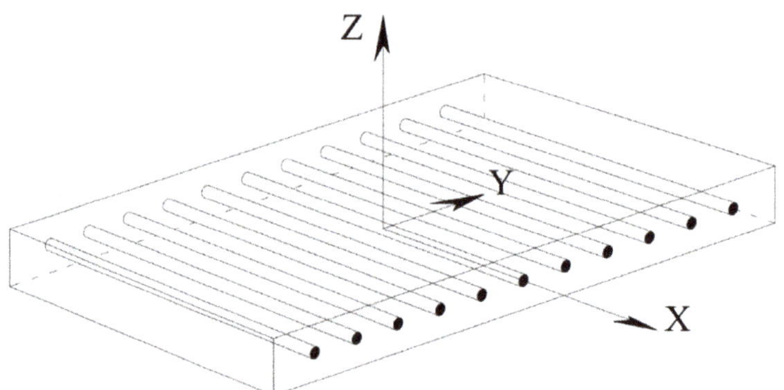

Figure 1. A model of unidirectional FRP plate.

Taking into account that the stiffness matrix is the inverse of the compliance matrix:

$$C_{ij} = S_{ij}^{-1} \tag{3}$$

the components of the stiffness matrix can be expressed in terms of the components of the compliance matrix as follows:

$$
\begin{aligned}
C_{11} &= \frac{S_{22}S_{33} - S_{23}^2}{S}, \quad C_{12} = \frac{S_{13}S_{23} - S_{12}S_{33}}{S}, \\
C_{22} &= \frac{S_{33}S_{11} - S_{13}^2}{S}, \quad C_{23} = \frac{S_{12}S_{13} - S_{23}S_{11}}{S}, \\
C_{55} &= \frac{1}{S_{55}}
\end{aligned}
\tag{4}
$$

where $S = S_{11}S_{22}S_{33} - S_{11}S_{23}^2 - S_{22}S_{13}^2 - S_{33}S_{12}^2 + 2S_{12}S_{23}S_{13}$. It should be noted that the elements of the compliance matrix are determined more directly than those of the stiffness matrix (see, for example, [44]).

The components of the compliance matrix are identified by performing tension tests on specimens made of composite material. Loading in different material directions is simulated by varying the angle between the applied load direction and the orientation of the fiber. Figure 2 shows a test specimen made of carbon fiber reinforced polymer (CFRP) and loaded in a direction parallel to the fiber orientation. The identified components of the elasticity tensor of the transversely isotropic profile are summarized in Table 1.

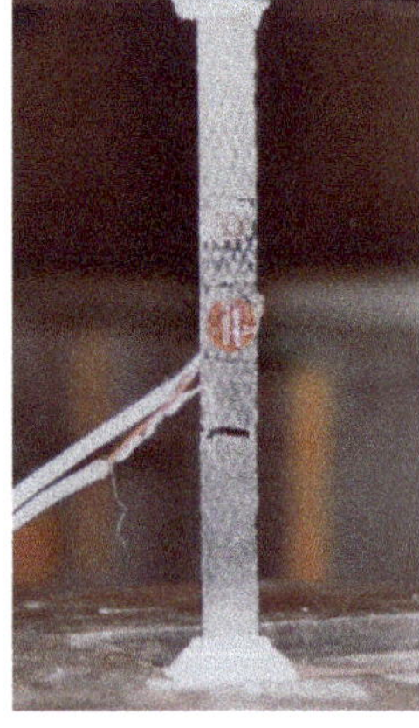

Figure 2. A test specimen of composite material: identification test [45].

Table 1. CFRP mechanical properties [45].

E_x	$E_y = E_z$	$\upsilon_{xy} = \upsilon_{xz}$	υ_{yz}	G_{xz}
(MPa)	(MPa)	-	-	(MPa)
120,000	27,800	0.35	0.18	4860

2.2. Concrete

Two approaches are possible to model the inelastic concrete response through the assessment of the accumulated damage. The elasticity tensor can be directly related to the current damage state, or the compliance matrix can be considered as a variable characterizing of the state of damage [46].

Within the first of the above-mentioned approaches, the inelastic mechanical response of concrete is modeled based on the stress-strain relationship of a linear, elastic material:

$$\sigma_{i,j} = \frac{v}{(1+v)(1-2v)} E \varepsilon_{kk} \delta_{ij} + \frac{1}{(1+v)} E \varepsilon_{ij} \tag{5}$$

by adding a damage variable [47,48]:

$$\sigma_{i,j} = \frac{v}{(1+v)(1-2v)} E(1-D) \varepsilon_{kk} \delta_{ij} + \frac{1}{(1+v)} E(1-D) \varepsilon_{ij}. \tag{6}$$

In Equations (5) and (6), E and v denote, respectively, Yong's modulus and the Poisson's ratio respectively, ε_{kk} is the trace of the strain tensor, and δ_{ij} is the unit tensor. The evolution of the damage variable in tension (D_t) and compression (D_c) is defined as follows [49]:

$$D_t = 1 - \frac{\varepsilon_{inf}(1 - A_t)}{\varepsilon_{eqv}} - \frac{A_t}{e^{B_t(\varepsilon_{eqv} - \varepsilon_{inf})}} , \tag{7}$$

$$D_C = 1 - \frac{\varepsilon_{inf}(1 - A_C)}{\varepsilon_{eqv}} - \frac{A_C}{e^{B_C(\varepsilon_{eqv} - \varepsilon_{inf})}} . \tag{8}$$

As can be seen, the evolution of either of the components of the damage variable is governed by the so-called equivalent strain:

$$\varepsilon_{eqv} = \sqrt{\sum \langle \varepsilon_i \rangle^2}. \tag{9}$$

The equivalent strain retains only the positive components of the principal strains ε_i by employing the Macaulay brackets:

$$\langle \varepsilon_i \rangle = \begin{cases} 0 & if \quad \varepsilon_i < 0 \\ \varepsilon_i & if \quad \varepsilon_i \geq 0 \end{cases} . \tag{10}$$

This feature of the model is in line with the phenomenological observation that cracking in concrete takes place in regions where tensile strains (stresses) are present. From a physical standpoint, the equivalent strain is related to the crack opening mode I [50]. In other words, cracking is generally related to tensile strains and stresses.

The behavior of concrete in compression was identified by testing standard cylindrical specimens (80 mm in radius and 320 mm in height) whereas the behavior of concrete in tension—by testing prismatic concrete specimens (400 mm in length and a cross-section of 100×100 mm; tension by flexure test).

The identification tests are used to characterize the concrete response not only by determining compression strength, tensile strength, and maximum strain, but also to identify the model constants in Equations (7) and (8). These constants are identified through the curve fitting procedure. To this end, finite element models designed to simulate identification tests, specifically compression tests on cylindrical concrete specimens and tension by flexure tests on prismatic concrete specimens (Figure 3a), are built. The model constants are calibrated with respect to the experimental data. The constants ε_{inf} (i.e., the damage threshold, see Figure 3b), A_c, and B_c are needed for the evaluation of the compression component of the damage variable (see Equation (8) and Figure 3c). The constants A_t and B_t are employed to define the evolution of the tensile component of the damage variable (Equation (7) and Figure 3d).

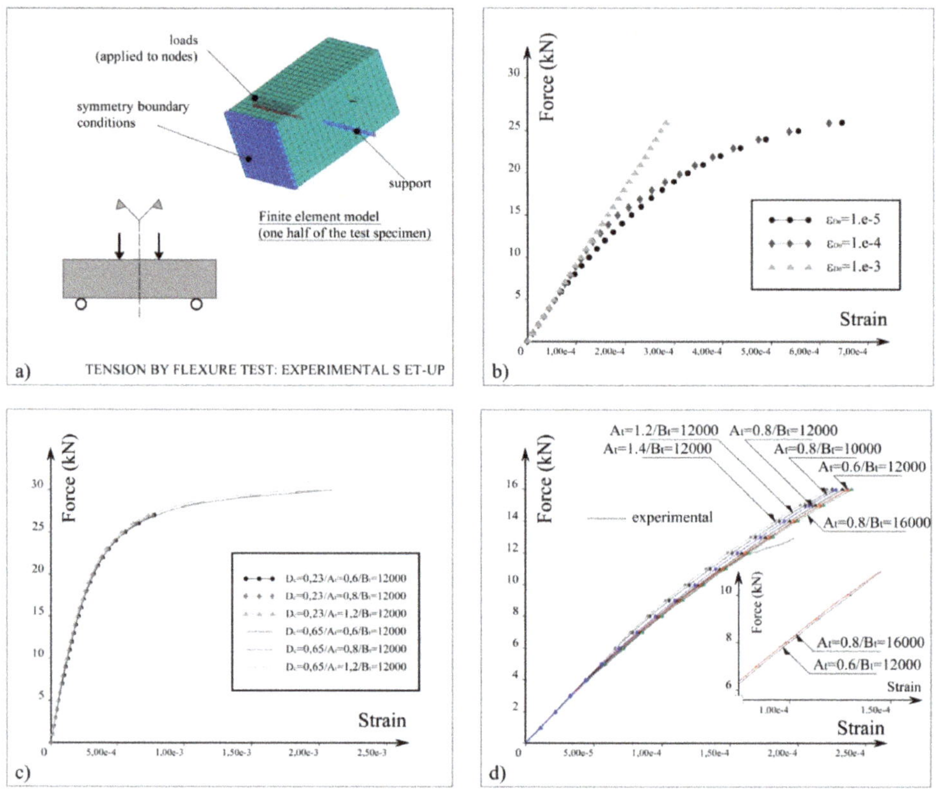

Figure 3. (**a**) Finite element model employed in the material characterization procedure and identification of the model constants; (**b**) dependency of the force-strain relationship on the damage threshold $\varepsilon_{D,0}$ (FE simulation of a compression test); for the simulation of the four-point bending tests, a value of the damage threshold is set to 5e-6.; (**c**) tuning of the material constants A_c and B_c (FE simulation of a compression test); (**d**) calibration of the constants A_t and B_t (tension by flexure test).

2.3. Steel

Steel response is characterized by an initial elastic phase as long as stress remains within the elastic domain. Upon reaching the yield stress, plastic flow is initiated. For common grades of steel reinforcement, it is followed by a hardening phase. A standard model based on the plasticity theory fundamentals is used [51,52]. In the large-displacements domain, a plastic behavior of steel reinforcement is expected. In this context, a true stress-strain curve is used, as a more representative measure of the state of the material, compared to the engineering stress-strain curve. The employed constitutive relation for steel uses an additive decomposition of the strain tensor into an elastic (ε_{el}) and plastic (ε_{pl}) part:

$$\varepsilon = \varepsilon_{el} + \varepsilon_{pl}. \tag{11}$$

The strain tensor increment can be obtained as a sum of elastic and plastic increments:

$$d\varepsilon = d\varepsilon_{el} + d\varepsilon_{pl}.$$ (12)

The yield criterion is a scalar function of the stress (σ) and a set of internal variables (ξ):

$$f(\sigma, \xi) = 0$$ (13)

The state variables at the mesoscale can be divided into observable (such as the total strain tensor) and internal (i.e., the elastic strain tensor, the plastic strain tensor, and the back strain tensor) variables [53].

Equation (13) defines the yield surface in the considered stress space. The evolution of the plastic strain is determined by the flow rule:

$$d\varepsilon^{pl} = d\lambda \frac{\partial Q}{\partial \sigma}$$ (14)

where Q is the plastic potential, and $d\lambda$ is the plastic strain increment. For kinematic hardening, the yield surface is defined by the following equation:

$$f(\sigma - \alpha, \xi) = 0$$ (15)

In Equation (15), α denotes the back stress tensor (that can be interpreted as a vector X_D pointing to the origin of the yield surface in the stress space; see Figure 4).

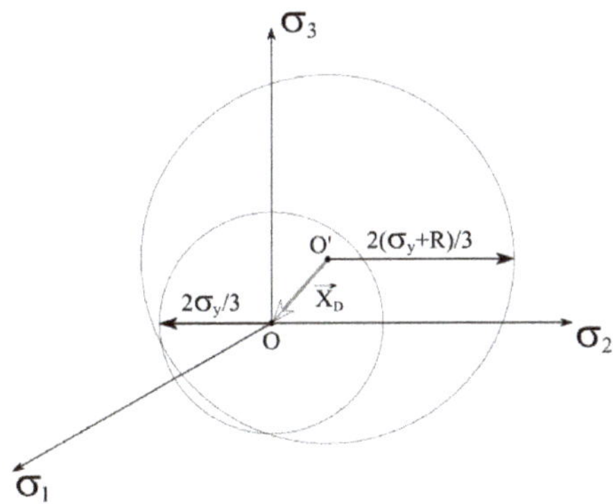

Figure 4. Geometrical interpretation of the back stress tensor.

3. Behavior of Materials as Components of a System

The material constitutive relations, formulated based on characterization tests, are implemented into the numerical model of a multiple-component system. In the present study, this is a reinforced concrete beam strengthened to flexure by adhesively bonded composite material. The behavior of the multiple-component system is also studied experimentally. The model reproduces the geometry and the reinforcement arrangement (both internal steel reinforcement and external CFRP reinforcement) of the test specimen. It is validated based on a comparison between numerical and experimental results on the evolution of the mid-span deflection.

3.1. Finite Element Model

Figure 5 shows the geometry of the modeled (and of the experimentally tested) specimens. The employed material models are listed in Table 2.

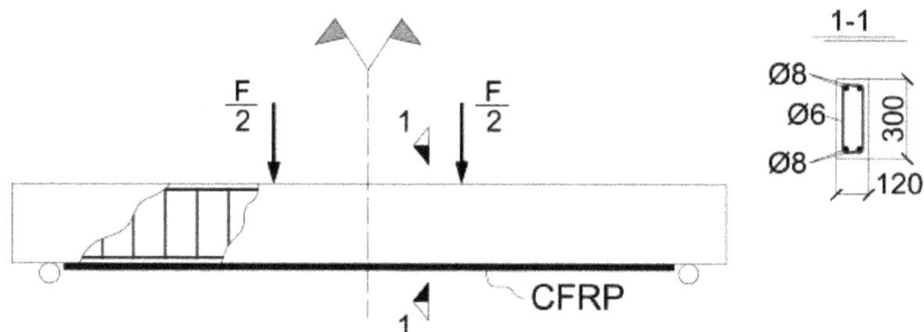

Figure 5. Geometry of the reinforced concrete beam strengthened with externally bonded CFRP and subjected to four-point bending test.

Table 2. Constitutive laws for constituent materials.

Material	Material Symmetries	Constitutive Law
Concrete	Isotropy	Elasticity coupled with damage
Steel	Isotropy	Bilinear kinematic hardening
CFRP	Transverse isotropy	Linear elasticity

Taking into account the existing symmetries of the experimentally tested specimens, a quarter-model space is considered. Appropriate symmetry boundary conditions are applied to nodes located in the symmetry planes. In the finite element (FE) model, boundary conditions are applied to lines (an idealized situation). Nodes attached to the line of the (idealized) contact between the roller support and the beam, are constrained to move in the global y-direction. One of these nodes is also constrained in global x-direction. Forces in the global y-direction are applied to nodes attached to lines defined with respect to the application point of the concentrated load (the z-coordinates of these nodes is equal to the z-coordinate of the applied load). The sum of the magnitudes of these forces equals the current value of the applied load.

A perfect bond is assumed for all the interfaces (steel/concrete and concrete/CFRP).

In the 3-D FE model, the concrete, the steel bars, steel stirrups, and the adhesive layer, are meshed with Solid 95 (3-D 20-node structural solid). Shell 186 (3-D 20-node layered structural solid) is used to model the external CFRP reinforcement. The simulation of damage accumulation should take into account the effect of localization (localization of damage). A possible option is to remesh damaged regions with a finer mesh. With the implementation of higher-order elements containing mid-side nodes, the need for mesh refinement during analysis can be avoided. On the other hand, higher-order finite elements are a favorable option if the solution implies a refinement of the initial finite element mesh in specified regions. Additionally, the use of finite elements with mid-side nodes is a prerequisite for more accurate analysis results because of the improved shape function.

With the hypothesis that the global response of the composite structural element is influenced mainly by the failure behavior of concrete, a finer mesh is generated for concrete. Additionally, the finite element size should comply with the requirements of the representative volume element size derived based on the energy balance in the framework of the continuum damage mechanics [53]. The sizes of the finite elements generated for CFRP reinforcement, the adhesive layer, and the steel reinforcement is defined so that a coherent mesh is obtained (i.e., smooth transitions between regions of different finite element sizes). Details on the generated finite element mesh, displayed in Figure 6, are summarized in Table 3.

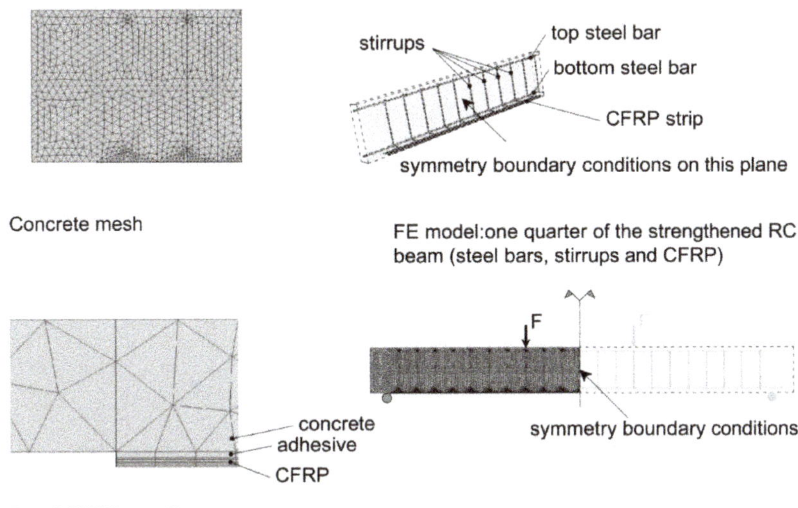

Figure 6. RC beam strengthened to flexure with externally bonded FRP material: finite element mesh.

Table 3. Size of the finite element mesh.

Component	Element Type	Element #	Node #
Concrete	SOLID 95	158,717	238,920
Steel	SOLID 95	4604	35,053
Adhesive layer	SOLID 95	1890	14,228
CFRP	SOLID 186	1890	14,868

The quasi-static loading is modeled by incrementally increasing the applied load. For each step of the applied load, a nonlinear static solution is performed. The damage-based constitutive relation for concrete is applied in the post-processing phase of each step. The damage variable is calculated based on the stress and strain distributions obtained in the finite element solution for the current value of the applied load. Before the next step of the solution, the material properties of the finite elements affected by damage are modified. Moreover, the finite elements in which the critical damage is reached are deactivated. In the next step, they don't contribute to the overall rigidity. Thus, with the simulation of the propagating cracks, the finite element model becomes eventually insufficiently constrained. It is no more possible to perform nonlinear static analysis, and the numerical routine stops.

Finite element simulations reproduce force-controlled tests. From an experimental standpoint, displacement-controlled tests generally provide more accurate and more reliable results. However, the focus here is on the accuracy of the results obtained by finite element simulation, assessed through a comparison with the experimental data.

3.2. Prediction of the Transient Mechanical Response of the Composite Structural Element: Validation

Figure 7 shows experimental results and results obtained by finite element analysis. The numerical models reproduce the geometry, the reinforcement arrangement, and the material properties of experimentally-tested beams [45]. From the experimental data available, for finite element simulations are selected the beams strengthened with three and five layers of CFRP. As shown in Figure 7, such a difference in the amount of the externally bonded reinforcement illustrates well the effects of synergy in the composite structural element.

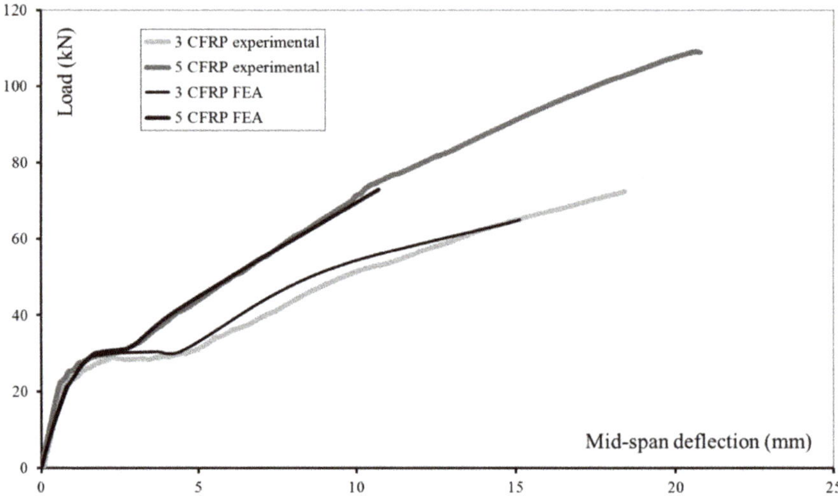

Figure 7. RC beam strengthened with three and five layers of externally bonded CFRP reinforcement: experimental results [45] and results obtained by finite element analysis (FEA).

By assuming a specified value for the thickness of the FRP fabric (for example, $t_{FRP} = 0.2$ mm), the cross-section of the FRP plate would be:

$$A_{FRP} = n \times t_{FRP} \times b_{FRP} \tag{16}$$

where n is the number of FRP layers and b_{FRP} stands for the thickness of the FRP plate. This cross-section can be further used in the design of the FRP-strengthened concrete element. In the present study, finite elements with the corresponding number of layers and material properties model the CFRP reinforcement.

The compliance between the numerical and experimental results is estimated as follows [54]:

$$f = \left(1 - \frac{\sqrt{\sum_{i=1}^{N}(F_{n,i} - F_{e,i})^2}}{\sqrt{\sum_{i=1}^{N}\left(F_{e,i} - F_m^{(e)}\right)^2}}\right) \times 100 \tag{17}$$

In Equation (17), "F_n" and "F_e" denote the vectors containing numerical and experimental results respectively, N is the number of components of the compared vectors, and $F_m^{(e)}$ is the mean of the components of the vector F_e:

$$F_m^{(n)} = \frac{\sum_{i=1}^{N} F_{e,i}}{N}. \tag{18}$$

A higher value of f corresponds to a better fit with the experimental data. The comparison between experimental and numerical results shows that for the beam strengthened with three layers of CFRP, the value of f is 75%, and for the beam, strengthened with five layers of CFRP—93%.

The good agreement between experimental data and FE results (Figure 7) shows that the model can be employed in the analysis of problems involving the interaction of multiple material models, depending on a variety of factors. In a simplified analysis, for example, based on the effective moments of inertia of the composite element's cross-section, three phases can be identified in the evolution of the mid-span deflection [55]. In the first phase, deflection develops until flexural cracks initiate in the tensioned zone of the cross-section. Al materials behave linearly within the elastic domain. The second phase (initiated with cracking in concrete) continues until the yielding stress in the internal flexural steel reinforcement is reached. The damage accumulation in concrete provokes nonlinearities in the global response of the composite structural element: the smooth transition between phase 1 and phase 2 depicted in Figure 7. Damage accumulation results in crack initiation and propagation in concrete. Cracks (upon intersection) activate the flexural steel reinforcement and the CFRP reinforcement. Moreover, in the cracked regions, in the tensile zone, only steel and CFRP contribute to the overall bending resistance. Between cracks, synergy provided by the concrete may be taken into account. With the progressive cracking in concrete and the failure of the tensile steel reinforcement in the cracked regions, only the CFRP reinforcement provides bending resistance in the tensile zone. The compression zone was not considered in the analysis of the other two phases. Taking into account the highly asymmetric (tension-compression) concrete response, at the beginning of phase three, concrete in the compression zone still has significant reserves in load-carrying capacity. The overall behavior of the structural element is governed by the composite material - linear and elastic until failure.

The analysis based on the effective moments of inertia is discussed here with the only purpose to provide an intuitive classification of the different phases in the response of a composite structural element subjected to a quasi-static loading.

The analysis of both experimental and numerical results (Figure 7) leads to the conclusion that the performances of the FRP-strengthened structure element are enhanced in terms of load-carrying capacity (delayed damage accumulation with the increase of the amount of the external FRP reinforcement). The failure load increase demonstrates the enhancement of the load-carrying capacity. The delay in the damage accumulation is associated with the increase in the load corresponding to the end of the initial elastic phase. The synergy of materials can explain these observations.

As can be seen in Figure 7, the model of the multiple-component system provides accurate results in good agreement with experimental data for most of the loading history. The transitions between the initial elastic phase and the subsequent phase of increasing displacements, without a significant increase in the applied load, as well as the initiation of the hardening phase in the overall time history, are well captured. However, in the numerical solution, the load-carrying capacity of the CFRP-strengthened RC beam is underestimated. Indeed, $F_{max,n} = 0.9 \times F_{max,exp}$ where $F_{max,n}$ is the failure load predicted by the numerical simulation and $F_{max,exp}$ is the failure load obtained in the experiment. In the numerical analyses, the failure load and the maximum displacement are less than those obtained in the experimental study. This underestimation is referred to as a "premature numerical failure." It can be attributed to excessive distortions in the initially generated finite element mesh. One possible solution to this problem could be the generation of new finite element mesh based on the current position of nodes, choosing one of the steps preceding the step at which the routine stops. The development of such an algorithm is the subject of ongoing research. Possibly, after testing, it will be possibly integrated into the analyses of structures' responses within the large displacement domain. On the other hand, a modification of the solution options (number of steps and substeps) might increase the accuracy of the predictive results in the last phase of the loading history. A detailed and in-depth explanation of the premature, numerical failure also requires the extension of the database available. A conclusion made based on a large number of comparisons between numerical and experimental results would be more relevant.

The strategy chosen to simulate the strain-softening concrete response, see Equations (6)–(8), involves the calculation of damage variable for each increment of the driving force parameter and for each finite element used to model concrete. Technically, the calculated damage variables are stored in predefined arrays. This algorithm makes possible the simulation of the initiation and propagation of the damage in an initially undamaged, homogeneous, and isotropic medium. In Figure 8, the propagation of the damaged zone in concrete is visualized. For three consecutive values of the applied load ($F_1 < F_2 < F_3$), finite elements of a particular degree of mechanical damage are not displayed. Damage is initiated in the vicinity of the mid-span of the strengthened beam and propagates towards the supports.

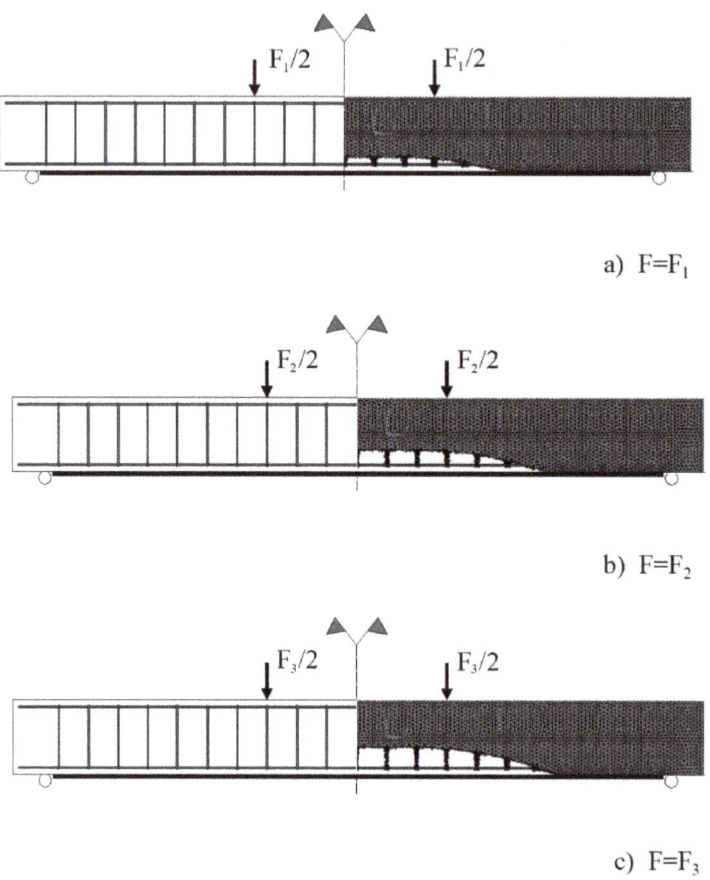

Figure 8. Finite element simulation of the damage propagation in concrete.

With the increase of the applied load, mechanical damage accumulates, and the damaged zone propagates in concrete. When the damage variable, calculated for a given finite element, reaches a predefined critical value, the procedure deactivates this finite element in the next step of the incremental solution. As a result of the application of this algorithm, zones of zero rigidity nucleate and propagate in concrete. In the finite element simulation, the behavior of the flexural steel reinforcement and the externally bonded CFRP reinforcement intersecting a new-formed zone of zero rigidity (i.e., a crack) is analogous to their responses within an experimental study. Strains in CFRP and steel, within the regions of the intersection with cracks, are higher to resist the tensile force.

The finite element analysis sheds light on the debonding failure mechanism. Analysis of previous (mainly experimental) research leads to a classification of the debonding failure modes into two types [56]. Debonding of the external FRP reinforcement can be initiated at or near one of the plate ends and then propagates away from the plate end. Alternatively, debonding failure mode initiates at an intermediate flexural or flexural-shear crack and then propagates towards the plate end. Based on the visualization of the damaged zone propagation obtained by finite element analysis, it can be concluded that the failure mode triggered in the modeled beam is "intermediate crack induced interfacial debonding." The interface failure depends on a variety of factors, such as the roughness of the surfaces in contact [57], exposure to aggressive environments [58], etc. With the hypothesis of a perfect interface (what excludes the above mentioned two failure modes), debonding results from a failure in concrete, in the vicinity of the adhesive joint.

4. Conclusions

The mechanical behavior of structural materials considered as elements of a multiple-component system has been investigated. Material models have been defined based on empirical data obtained in material characterization tests.

A finite element model has been built to shed light on the unknowns related to the composite action: synergies in the materials' response and material failure behavior within the structural element. The numerical model has been validated through a comparison with experimental data. The comparisons showed that the numerical model predicts with an adequate accuracy the experimentally-obtained evolution of the observable variable. The verification results demonstrate the capability of the numerical model to reproduce accurately the strain-softening concrete response taking into account the behavior of the other materials (FRP, adhesive, steel). The discussion focuses on concrete behavior since it is estimated that the global failure of the studied structural element results from a local failure in concrete.

A specific feature of the proposed approach is the possibility to track the evolution of the damage accumulation in concrete, which is directly related to the initiation and propagation of cracks on the macroscopic scale. Based on this monitoring, the load-carrying capacity and stiffness of the composite structural element remaining after a specified period of exploitation can be rationally estimated.

In the reported study, the enhancement of the mechanical performances of the composite structure demonstrates the synergy of the constituent materials. The external adhesively bonded CFRP reinforcement is a factor that enhances the load-carrying capacity of the strengthened beams. Also, it results in the delay in the failure mechanisms taking place in concrete.

Funding: This research received no external funding.

Acknowledgments: The experimental program discussed in the present study was carried out at the University of Reims Champagne—France, within the Ph.D. research work of the author, supported by the French Government. The early stages of the numerical investigation were carried out at the University of Reims Champagne and the University of Lille, LML—France. All support to the preparation and the realization of the experimental program, as well as the enlightening discussions, are gratefully acknowledged.

Conflicts of Interest: The authors declare no conflict of interest.

References

1. Planès, T.; Larose, E. A review of ultrasonic Coda Wave Interferometry in concrete. *Cem. Concr. Res.* **2013**, *53*, 248–255. [CrossRef]
2. Anderson, D.A.; Seals, R.K. Pulse velocity as a predictor of 28 and 90 day strength. *ACI J. Proc.* **1981**, *78*, 116–122.
3. Lillamand, I.; Chaix, J.-F.; Ploix, M.-A.; Garnier, V. Acoustoelastic effect in concrete material under uni-axial compressive loading. *NDT E Int.* **2010**, *43*, 655–660. [CrossRef]
4. Berthaud, Y. Damage measurement in concrete via an ultrasonic technique. Part I experiment. *Cem. Concr. Res.* **1991**, *21*, 73–82. [CrossRef]
5. Ruffino, E.; Delsanto, P. Scattering of ultrasonic waves by void inclusions. *J. Acoust. Soc. Am.* **2000**, *108*, 1941. [CrossRef] [PubMed]
6. Lombard, O.; Barrière, C.; Leroy, V. Nonlinear multiple scattering of acoustic waves by a layer of bubbles. *Europhys. Lett.* **2015**, *112*, 24002–24007. [CrossRef]
7. Craster, R.V.; Guenneau, S. *Acoustic Metamaterials: Negative Refraction, Imaging, Lensing and Cloaking*; Springer Science & Business Media: Berlin/Heidelberg, Germany, 2012; Volume 166.
8. Varadan, V.K.; Ma, Y.; Varadan, V.V. Multiple scattering of compressional and shear waves by fiber-reinforced composite materials. *J. Acoust. Soc. Am.* **1986**, *80*, 333–339. [CrossRef]

9. Varadan, V.K.; Varadan, V.V.; Pao, Y.-H. Multiple scattering of elastic waves by cylinders of arbitrary cross section. I. SH waves. *J. Acoust. Soc. Am.* **1978**, *63*, 1310–1319. [CrossRef]

10. Berryman, J.G. Long-wavelength propagation in composite elastic media I. Spherical inclusions. *J. Acoust. Soc. Am.* **1980**, *68*, 1809–1819. [CrossRef]

11. Berryman, J.G. Long-wavelength propagation in composite elastic media II. Ellipsoidal inclusions. *J. Acoust. Soc. Am.* **1980**, *68*, 1820–1831. [CrossRef]

12. Mei, J.; Liu, Z.; Wen, W.; Sheng, P. Effective mass density of fluid-solid composites. *Phys. Rev. Lett.* **2006**, *96*, 024301. [CrossRef] [PubMed]

13. Mei, J.; Liu, Z.; Wen, W.; Sheng, P. Effective dynamic mass density of composites. *Phys. Rev. B* **2007**, *76*, 134205. [CrossRef]

14. Skvortsov, A.; MacGillivary, I.; Sharma, G.S.; Kessissoglou, N. Sound scattering by a lattice of resonant inclusions in a soft medium. *Phys. Rev. E* **2019**, *99*, 063006. [CrossRef] [PubMed]

15. Sharma, G.S.; Skvortsov, A.; MacGillivary, I.; Kessissoglou, N. On superscattering of sound waves by a lattice of disk-shaped cavities in a soft material. *Appl. Phys. Lett.* **2020**, *116*, 041602. [CrossRef]

16. Hladky-Hennion, A.C.; Decarpigny, J.N. Analysis of the scattering of a plane acoustic wave by a doubly periodic structure using the finite element method: Application to Alberich anechoic coatings. *J. Acoust. Soc. Am.* **1991**, *90*, 3356–3367. [CrossRef]

17. Hennion, A.C.; Bossut, R.; Audoly, C.; Decarpigny, J.N. Analysis of the scattering of a plane acoustic wave by a periodic elastic structure using the finite element method: Application to compliant tube gratings. *J. Acoust. Soc. Am.* **1990**, *87*, 1861–1870. [CrossRef]

18. Sharma, G.S.; Skvortsov, A.; MacGillivray, I.; Kessissoglou, N. Acoustic performance of periodic steel cylinders embedded in a viscoelastic medium. *J. Sound Vib.* **2019**, *443*, 652–665. [CrossRef]

19. Sharma, G.S.; Skvortsov, A.; MacGillivray, I.; Kessissoglou, N. Sound absorption by rubber coatings with periodic voids and hard inclusions. *Appl. Acoustics* **2019**, *143*, 200–210. [CrossRef]

20. Teng, J.G.; Chen, J.F.; Smith, S.T.; Lam, L. *FRP-Strengthened RC Structures*; Wiley: Hoboken, NJ, USA, 2002.

21. Ritchie, P.A.; Thomas, D.A.; Lu, L.W.; Connely, G.M. External reinforcement of concrete beams using fibre reinforced plastics. *ACI Struct. J.* **1991**, *88*, 490–500.

22. Saadatmanesh, H.; Ehsani, M.R. RC beams strengthened with GFRP plates, I: Experimental study. *J. Struct. Eng. ASCE* **1991**, *117*, 3417–3433. [CrossRef]

23. Takeda, K.; Mitsui, Y.; Murakami, K.; Sakai, H.; Nakamura, M. Flexural behaviour of reinforced concrete beams strengthened with carbon fibre sheets. *Compos. Part A* **1996**, *27*, 981–987. [CrossRef]

24. Garden, H.N.; Hollaway, L.C.; Thorne, A.M. A preliminary evaluation of carbon fibre reinforced polymer plates for strengthening reinforced concrete members. *Proc. Inst. Civ. Eng. Struct. Build.* **1997**, *123*, 127–142. [CrossRef]

25. Ross, C.A.; Jerome, D.M.; Tedesco, J.W.; Hughes, M.L. Strengthening of reinforced concrete beams with externally bonded composite laminates. *ACI Struct. J.* **1999**, *96*, 212–220.

26. Meier, U.; Kaiser, H. Strengthening of structures with CFRP laminates. In *Advanced Composites Materials in Civil Engineering Structures*; ASCE: New York, NY, USA, 1991.

27. Hollaway, L.C.; Leeming, M.B. (Eds.) *Strengthening of Reinforced Concrete Structures*; Woodhead Publishing: Cambridge, UK; CRC: Boca Raton, FL, USA, 1999.

28. Chajes, M.J.; Thomson, T.A.; Januszka, T.F.; Finch, W.W. Flexural strengthening of concrete beams using externally bonded composite materials. *Constr. Build. Mat. Oxford UK* **1994**, *8*, 191–201. [CrossRef]

29. Sharif, A.; Al-Sulaimani, G.J.; Basunbul, I.A.; Baluch, M.H.; Ghaleb, B.N. Strengthening of initially loaded reinforced concrete beams using frp plates. *ACI Struct. J.* **1994**, *91*, 160–168.

30. Wei, A.; Saadatmanesh, H.; Ehsani, M.R. RC beams strengthened with FRP plates. II: Analysis and parametric study. *J. Struct. Eng. ASCE* **1991**, *117*, 3434–3455.

31. Triantafillou, T.C.; Plevris, N. Strengthening of RC beams with epoxy-bonded fibre-composite materials. *Mater. Struct.* **1992**, *25*, 201–211. [CrossRef]

32. Roberts, T.M. Approximate analysis of shear and normal stress concentrations in the adhesive layer of plated RC beams. *Struct. Eng.* **1989**, *67*, 229–233.

33. Roberts, T.M.; Haji-Kazemi, H. Theoretical study of the behaviour of reinforced concrete beams strengthened by externally bonded steel plates. *Proc. Inst. Civ. Eng.* **1989**, *87*, 39–55. [CrossRef]

34. Van Gemert, D.; Maesschalck, R. Structural repair of a reinforced concrete plate by epoxy bonded external reinforcement. *Int. J. Cem. Compos. Lightweight Concr.* **1983**, *5*, 247–255. [CrossRef]
35. Jones, R.; Swamy, R.N.; Bloxham, J.; Bouderbalah, A. Composite behaviour of concrete beams with epoxy bonded external reinforcement. *Int. J. Cem. Compos. Lightweight Concr. Lond.* **1980**, *2*, 91–107.
36. Gribniak, V.; Misiunaite, I.; Rimkus, A.; Sokolov, A.; Šapalas, A. Deformations of FRP–concrete composite beam: Experiment and numerical analysis. *Appl. Sci.* **2019**, *9*, 5164. [CrossRef]
37. Sharaky, I.A.; Baena, M.; Barrisa, C.; Sallam, H.E.M.; Torres, L. Effect of axial stiffness of NSM FRP reinforcement and concrete cover confinement on flexural behaviour of strengthened RC beams: Experimental and numerical study. *Eng. Struct.* **2018**, *173*, 987–1001. [CrossRef]
38. Gribniak, V.; Ng, P.-L.; Tamulenas, V.; Misiūnaitė, I.; Norkus, A.; Šapalas, A. Strengthening of fibre reinforced concrete elements:: Synergy of the fibres and external sheet. *Sustainability* **2019**, *11*, 4456. [CrossRef]
39. Rossetti, V.A.; Galeota, D.; Giammatteo, M.M. Local bond stress-slip relationships of glass fibre reinforced plastic bars embedded in concrete. *Mater. Struct.* **1995**, *28*, 340–344. [CrossRef]
40. Chajes, M.J.; Finch, W.W.; Januszka, T.F.; Thomson, T.A. Bond and Force Transfer of Composite-Material Plates Bonded to Concrete. *ACI Struct. J.* **1996**, *93*, 209–217.
41. Ferracuti, B.; Savoia, M.; Mazzotti, C. Interface law for FRP–concrete delamination. *Compos. Struct.* **2007**, *80*, 523–531. [CrossRef]
42. Thurner, S.; Hanel, R.; Klimek, P. *Introduction to the Theory of Complex Systems*; Oxford University Press: Oxford, UK, 2018.
43. Kaw, A.K. *Mechanics of Composite Materials*; Taylor & Francis: Boca Raton, FL, USA, 2006; ISBN: 1420058290, 9781420058291.
44. Jones, R.M. *Mechanics of Composite Materials*; Taylor and Francis: Philadelphia, PA, USA, 1999; ISBN: 156032712X, 9781560327127.
45. Zhelyazov, T. Strengthening of Reinforced Concrete Structures by Adhesively Bonded Composite Materials, Behavior of Structures Subjected to Bending. Ph.D. Thesis, University of Reims, Reims, France, 2008.
46. Ortiz, M. A constitutive theory for the inelastic behavior of concrete. *Mech. Mater.* **1985**, *4*, 67–93. [CrossRef]
47. Lemaître, J.; Dasmorat, R. *Engineering Damage Mechanics: Ductile, Creep, Fatigue and Brittle Failures*; Springer: Berlin/Heidelberg, Germany, 2004.
48. Marigo, J.J. Damage Law Formulation for an Elastic Material. *Proc. Acad. Sci. Paris French* **1981**, *292*, 1309–1312.
49. Mazars, J.J. Damage mechanics application to nonlinear response and failure behavior of structural concrete. Master's Thesis, Paris 6 University, Paris, France, 1984.
50. Prisco, M.; Mazars, J. Crush-Crack: A non-local damage model for concrete. *Mech. Cohesive-Frict. Mater.* **1996**, *1*, 321–347. [CrossRef]
51. Hill, R. *The Mathematical Theory of Plasticity*; Oxford University Press: New York, NY, USA, 1983.
52. Rice, J.R. Continuum mechanics and thermodynamics of plasticity in relation to microscale deformation mechanisms. In *Constitutive Equations in Plasticity*; Argon, A., Ed.; MIT Press: Cambridge, MA, USA, 1975; pp. 23–79.
53. Lemaitre, J. *A Course on Damage Mechanics*; Springer: Berlin/Heidelberg, Germany, 1996.
54. Eem, S.-H.; Jung, H.-J.; Koo, J.-H. Modeling of Magneto-Rheological Elastomers for Harmonic Shear Deformation. *IEEE Trans. Magn.* **2012**, *48*, 3080–3083. [CrossRef]
55. Daugevičius, M.; Valivonis, J.; Skuturna, T. Prediction of Deflection of Reinforced Concrete Beams Strengthened with Fiber Reinforced Polymer. *Materials* **2019**, *12*, 1367. [CrossRef] [PubMed]
56. Smith, S.T.; Teng, J.G. FRP-strengthened RC beams. I: Review of debonding strength models. *Eng. Struct.* **2002**, *24*, 385–395. [CrossRef]
57. Yin, Y.; Fan, Y. Influence of Roughness on Shear Bonding Performance of CFRP-Concrete Interface. *Materials* **2018**, *11*, 1875. [CrossRef]
58. Liang, H.; Li, S.; Lu, Y.; Yang, T. Reliability Analysis of Bond Behaviour of CFRP–Concrete Interface under Wet–Dry Cycles. *Materials* **2018**, *11*, 741. [CrossRef]

 *materials*

Article

Effect of SiC Reinforcement and Its Variation on the Mechanical Characteristics of AZ91 Composites

Anil Kumar [1,2], Santosh Kumar [2], Nilay Krishna Mukhopadhyay [3], Anshul Yadav [4] and Jerzy Winczek [5,*]

[1] Department of Mechanical Engineering, Kamla Nehru Institute of Technology, Sultanpur 228118, India; anilk@knit.ac.in

[2] Department of Mechanical Engineering, Indian Institute of Technology (BHU), Varanasi 221005, India; santosh.kumar.mec@itbhu.ac.in

[3] Department of Metallurgical Engineering, Indian Institute of Technology (BHU), Varanasi 221005, India; mukho.met@iitbhu.ac.in

[4] Membrane Science and Separation Technology Division, CSIR-Central Salt and Marine Chemicals Research Institute, Bhavnagar 364002, India; anshuly@csmcri.res.in

[5] Faculty of Mechanical Engineering and Computer Science, Częstochowa University of Technology, 42-201 Częstochowa, Poland

* Correspondence: winczek@imipkm.pcz.czest.pl

Received: 25 September 2020; Accepted: 29 October 2020; Published: 31 October 2020

Abstract: In this study, the processing of SiC particulate-strengthened magnesium alloy metal matrix composites via vacuum supported inert atmosphere stir casting process is presented. The effects of small variations in the SiC particulate (average size 20 μm) reinforcement in magnesium alloy AZ91 were examined. It was found that with the addition of SiC particulate reinforcement, the hardness improved considerably, while the ultimate tensile and yield strength improved slightly. The density and *porosity* of the magnesium alloy-based composites increased with the increase in the wt.% of SiC particulates. The tensile and compressive fracture study of the fabricated composites was also performed. The tensile fractures were shown to be mixed-mode fractures (i.e., ductile and cleavage). The fractured surface also disclosed tiny dimples, micro-crack, and cleavage fractures which increases with increasing reinforcement. For the compression fracture, the surface microstructural studies of AZ91 displayed major shear failure and demonstrated the greater shear bands when compared to AZ91/SiC composites, which instead revealed rough fracture surfaces with mixed-mode brittle and shear features.

Keywords: metal matrix composites; SiC; AZ91; mechanical characterization; magnesium alloy

1. Introduction

Magnesium and its composites can supplant steel, aluminium, and plastic-based components due to being ultra-lightweight and having good thermal conductivity. Initially, there was a restriction on the use of magnesium alloys due to the high cost. However, as the cost of magnesium alloys is gradually decreasing, the interest in magnesium alloys has increased.

Magnesium alloy composites are reinforced by various ceramics particles, metals, and carbon nanotubes. Ceramic particles such as SiC, Al_2O_3, TiC, MgO, and graphite are mostly preferred reinforcements [1–4]. The synthesis of the metal matrix composites (MMCs) are fabricated using different technologies such as stir casting [5], powder metallurgy, gas infiltration [6], squeeze casting, spray deposits [7], injection moulding [8] and in situ techniques [9]. There are some other methods of processing magnesium alloys, such as powder metallurgy [10] and friction stir processing [11]. Stir casting is a cost-effective process for the fabrication of magnesium alloy-based

MMCs. The manufacturing of Mg alloys and its composites is an immense challenge for scientists and engineers because of their very high affinity towards environmental oxygen. Different casting techniques have been established for Mg alloys and their composites [12]. Stir casting is an efficient casting procedure for the fabrication of particulate-reinforced MMC because of its flexibility, simplicity and lower manufacturing cost [13]. The primary issues identified in stir casting are agglomeration, floating or settling of reinforced particles, and chemical reactions between the reinforcement and matrix alloys. These issues can be avoided by mixing particles in a semi-solid state. Vacuum and inert atmospheres are also generated in the die holder area to prevent oxidation of cast composites. Cast iron was used to make a die in two parts for easy removal of the cast product. The precise method of stir casting process was discussed in our previous work [14].

A new reinforcement method was developed by Wu et al. [15] in which MMCs infiltrate each other within the three-dimensional reinforcements. They concluded that magnesium matrix composite intertwined by stainless steel reinforcement exhibit the better mechanical behaviour. The microstructures of in situ reinforcements Al-Ti-B-C-Ce, Al-Ti-C, and Al-Ti-B, were examined by Tian et al. [16] using a combination of the scanning electron microscope (SEM), X-ray diffraction (XRD) spectroscopy, and TEM. The tensile strength at room temperature (RT) and at 350 °C was enhanced by 19.0%, and 18.4%, respectively. Titanium particulate-reinforced magnesium alloy composites demonstrated better ductility than the ceramic particles. The issue of wettability of pure titanium in pure magnesium was investigated by Kondoh et al. [17]. They concluded that magnesium composites reinforced with 3 wt.% Ti particles showed considerably improved tensile strength and yield strength with good elongation. In situ composites have been developed by adding ceric ammonium nitrate (CAN) into magnesium melt at temperatures of 665 °C and 875 °C. Mechanical behaviour of developed MMCs were evaluated through compression, hardness, and scratch tests [18]. The observations reveal the formation of MgO, CeO_2, and $CeMg_{12}$ phases in various sizes and shapes. Furthermore, these particles have been attributed enhanced mechanical properties. AZ91 reinforced with four distinct concentrations (0, 0.3, 0.6, 1) wt.% of WS_2 microparticles have been manufactured using the stir casting method. The XRD and SEM of the equal channel angular pressing (ECAP) deformed samples were examined. A significant improvement was observed in yield strength (YS), ultimate tensile strength (UTS) and elongation [19]. The optimum values of mechanical properties at different % of WS_2 and passes were determined. The unified effects of the number of ECAP passes and variation in content of SiC particulate were considered by Huang and Ali [20]. It was found that even distribution of matrix grain size and SiC particle segregation depend on the number of ECAP passes. The experimental results revealed that the elastic modulus increased on increasing the reinforcement from 2% to 5%.

Mechanical testing results are essential for comparing the quality, alloy growth, and decrease in non-ferrous materials. Many researchers have fabricated nanocomposites reinforced with Al_2O_3 nanoparticles by the ultrasonic-assisted semi-solid stirring method into a cylinder component. Jiang et al. [21] investigated microstructure, mechanical, and wear behaviour of the rheoformed composite parts. The ultimate tensile strength, yield strength, and compression strength are increased with content of TiC in AZ91. Extruded magnesium alloy and composite may exhibit better mechanical properties in comparison with un-extruded alloy and composites like copper and aluminium [22]. Magnesium alloy shows enhanced solidification behaviour over other cast metal, such as aluminium and copper alloys [23]. Casting is the leading synthesis method for the Mg alloy parts which represents 98% of the structural applications of the magnesium [24].

In this study, an indigenous stir casting setup is used to examine the effect of variation (2%, 5%, 8% and 11%) of SiC particulate. The incorporation of an inert atmosphere is also used to provide defect-free cast product. The selection of the size of reinforcement (size and percentage), along with the processing method used in this study has not been attempted by earlier researchers. This research contributes to a new dimension in processing and testing of magnesium alloy composites, which will have wider applications in automobile and aerospace industries. The variation in the size and amount of reinforcement plays a significant role in controlling the mechanical characteristics of the composite. This processing method has an enormous impact on the mechanical behaviour of the composites [25–28]. Physical tests, such as density and *porosity* measurements, have also been done. The surface morphology of the fracture surface in tensile and compression testing have been examined using FE-SEM.

2. Materials and Methods

2.1. Selection of Material

The commercial magnesium alloy AZ91 was used as matrix materials in this investigation. The elemental analysis of the alloy was done through optical emission spectroscopy. The main alloying elements of the base alloy are given in Table 1.

Table 1. The elemental composition of AZ91.

Mg	Al	Zn	Mn	Si	Other Elements
89.47	8.87	1.02	0.18	0.09	rest

SiC particulates (average size of 20 μm) were used as reinforcement. Various proportions (2%, 5%, 8% and 11%) of the reinforcement were added to the magnesium alloy AZ91. The basis for the selection of reinforcement is the bonding and wettability between the reinforcement and matrix materials. The morphology of the SiC particulates reinforcement with help scanning electron microscope is given in Figure 1, and Figure 2 shows the EDAX of SiC particulates.

Figure 1. Morphology of SiC particulate.

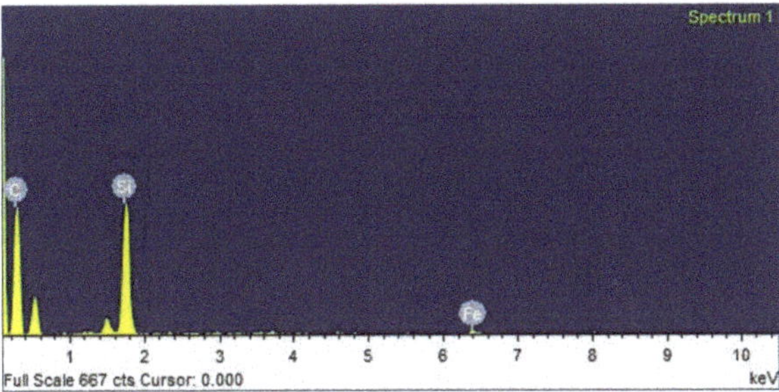

Figure 2. EDAX of SiC particulate.

2.2. Stir Casting Process

The stir casting method is appropriate for Mg alloy and its composites. The addition and mixing of particles in the matrix were performed using different stirrers. Liquid magnesium has a considerable tendency to be oxidized, and it burns unless proper attention is devoted to ensuring its surface against oxidation. Protection of the liquid Mg with the aid of appropriate flux was used before the application of gaseous shielding. The safest means of shielding liquid magnesium alloy is to make a vacuum followed by impingement of argon gas. In the present study, the Mg alloy-based composite was developed by the stir casting procedure using a vacuum and inert gas. A sound casting product was obtained by the above-described stir casting process. The diameter and length of the cast sample were 3.9 cm and 20.0 cm, respectively.

2.3. Density Measurement

Density (ρ) measurements of magnesium alloy (AZ91) and its composites were performed on the polished sample. The density of the AZ91 alloy and its composites was experimentally determined using the Archimedes principle [29]. Filtered water was selected as an immersion fluid. Three samples were arbitrarily selected and were thoroughly weighed both in the air and when fully immersed in distilled water [30]. All weights were measured with an electronic balance (accuracy of 0.001 g).

The following equation was used to calculate experimental density:

$$\rho_e = \frac{W_a \rho_w}{W_a - W_w} \tag{1}$$

where ρ_e, W_a, ρ_w, W_w are the observed density (g/cm^3), weight of the specimen in air, density of the water and the weight of the specimen in water, respectively.

The theoretical densities of each of the composites were calculated using the rule of mixtures (assuming that there is no Mg/Al/Zn-SiC interfacial reaction) equation, as shown below:

$$\rho_{th} = V_r \rho_r + (1 - V_r) \rho_m \tag{2}$$

where ρ_{th}, V_r, ρ_r and ρ_m are theoretical density, volume fraction of reinforcement, density of reinforcement and matrix, respectively.

The *porosity* was calculated using the following equation

$$porosity = \frac{\rho_{th} - \rho_e}{\rho_{th}} \times 100 \tag{3}$$

For theoretical computation of the value of density of the composites, density values of 1.81 g/cm^3 for the matrix material and 3.216 g/cm^3 for the SiC particles were used.

2.4. X-ray Diffraction (XRD) Studies

The XRD analysis was carried out on the polished specimens [31] of the monolithic magnesium alloy AZ91 and its composites using MiniFlex 300/600 Regaku tabletop XRD diffractometer (Tokyo, Japan) to determine the possible phases available. The samples were exposed to Cu K-α radiation (λ = 1.5406 Å) at a scanning speed of 10°/min and step width of 0.020 deg. The scan range was 10–90° with a continuous scan mode. The Bragg's angle, the value of the interplanar spacing (d) and intensity were later matched with the corresponding standard data from Mg, Mg$_{17}$Al$_{12}$, SiC and other phases.

2.5. Mechanical Characterization

2.5.1. Vickers Microhardness

The microhardness (Vickers) test was performed for the entire composite to estimate the homogeneity and uniform distribution of reinforced particles. The hardness values were recorded with different wt.% of SiC reinforcement. The hardness measurements were conducted on LECO's micro-Vickers hardness machine (St. Joseph, MI, USA) across the polished surface of composites by applying a load of 9.8 N. The pyramidal diamond indenter with a facing angle of 136° was used for indentation tests.

2.5.2. Tensile Test

The tensile testing of particulate-reinforced metal matrix composites was performed to estimate the mechanical properties of vacuum-assisted stir die-cast composites. The tensile samples were prepared from composites with different percentages of SiC particles, as well as commercial magnesium alloy AZ91. The gauge length and diameter of the tensile test sample was 20 mm and 8 mm, respectively, as per the ASTM E8/E8M- 16a standard [32]. The tensile tests were carried out at RT on the Instron-4208 under an initial strain rate of 0.005 s.

2.5.3. Compressive Test

Uniaxial compressive tests at RT were carried out on cylindrical monolithic sample according to ASTM standard E9 [33]. The sample length and diameter were 12 mm and 8 mm, respectively, to make the aspect ratio (l/d) of 1.5. Aimil makes semi-automatic universal testing machine was used to performed compression tests.

2.5.4. Factograph of Tensile and Compressive Tests

Fractured surface studies were conducted on the tensile and compressive fractured specimen's surfaces of monolithic magnesium alloy and its composites to provide various possible fracture mechanisms operating insight into the sample during tensile and compressive loading. These studies were performed using a ZEISS FE-SEM (Jena, Germany) at different magnifications.

3. Results and Discussion

The results of the density measurements of monolithic AZ91 and its composites are presented in Table 2. The measured densities of the composites are remarkably close to the theoretical densities. Thus, near-dense and *porosity*-free composites can be consistently produced using the vacuum-assisted stir casting methodology adopted in the current study. However, as the amount of reinforcement increases, the *porosity* is also found to increase slightly. The density values of the composites are in general higher than those of the monolithic alloys.

Table 2. Density and *porosity* measurements of AZ91 alloy and its composite (AZ91/SiC).

Materials	Theoretical Density (g/cm^3)	Experimental Density (g/cm^3)	Porosity (%)
AZ91	1.81	1.80 ± 0.02	0.53 ± 0.02
AZ91 + 2% SiC	1.84	1.81 ± 0.01	1.32 ± 0.02
AZ91 + 5% SiC	1.88	1.85 ± 0.02	1.51 ± 0.01
AZ91 + 8% SiC	1.92	1.88 ± 0.01	1.92 ± 0.02
AZ91 + 11% SiC	1.96	1.92 ± 0.01	2.11 ± 0.01

Furthermore, the density values are found to increase with increasing content of SiC particulate, due to the presence of the comparatively higher density SiC particulate in the magnesium alloy [34]. The *porosity* also exhibits a gradual increase as the weight percentage of the reinforcement in the composite increases. The increase in *porosity* in composites is due to the increase in micro-voids in the vicinity of massive SiC particles. Similar observations have also been reported by other authors [35–39] with respect to magnesium alloy metal matrix composites reinforced with SiC particulates and fabricated through stir casting techniques.

Figure 3 shows the X-ray diffraction analysis results corresponding to magnesium alloy AZ91 and its composites with 2, 5, 8, and 11 wt.% reinforcement with SiC particulates. The lattice spacings (d) obtained are matched with that of magnesium, SiCp, and $Mg_{17}Al_{12}$ phases based on the data available in JCPDS. This reveals the presence of the magnesium matrix, $Mg_{17}Al_{12}$, the intermetallic phases, and the reinforcement SiCp. All the peaks are identified according to the intensity plot of the XRD data. The results confirm the absence of MgO, the formation of which is prevented due to the incorporation of the vacuum and the inert argon gas atmosphere maintained during the melting and casting.

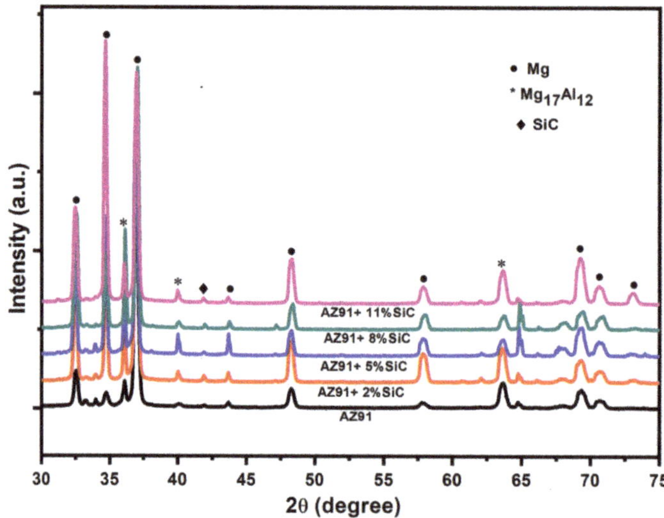

Figure 3. X-ray diffraction pattern of magnesium alloy and its composites containing different amounts (wt.%) of SiC (2%, 3%, 8% and 11%).

A rectangular cross-section sample (15 mm × 30 mm) of the composite samples was used to measure the microhardness. The microhardness data were examined at ten different locations over the cross-section (at the load of 9.81 N) of the AZ91 alloy and SiC particulate-reinforced composites. There is a slight variation in hardness values at different points across the cross-section of the AZ91 alloy and its composites (Figure 4). This establishes the uniform distribution of SiC particles within the magnesium alloy matrix. The variation in hardness in AZ91 alloy is due to the presence of the hard-intermetallic phase in magnesium alloy. This variation is because of the presence of different sizes of reinforcement, as well as slight accumulation of particles in some places. Table 3 shows the average variation in microhardness values of the magnesium alloy (AZ91) and its composites reinforced with SiC particulate with weight percentage variation (2%, 5%, 8%, and 11%). From Table 3, it can be observed that there is an increase in the hardness when the content of SiC reinforcement is enhanced from 2 to 11 wt.%. The Vickers hardness value increases by 28% on the addition of 2% of SiC particulate, and it further increases by up to 80% on addition of 11% of SiC particulate. The observed hardness values of composites with different percentages of SiC (2%, 5%, 8% and 11%) particulate are found to be higher than that of magnesium alloy because of the finer grain size and interactive influence of the presence of SiC particulate, which restricts the localized matrix deformation during the indentation of the composites.

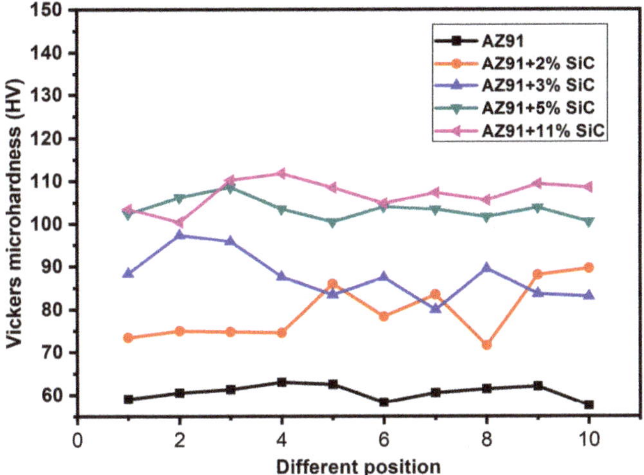

Figure 4. Microhardness data of magnesium alloy and its composites with different (2%, 3%, 8% and 11%) weight percentages at different locations.

Table 3. Vickers microhardness of magnesium alloy and its composite.

Material	Hardness (HV)
AZ91	60 ± 1.8
AZ91 + 2% SiC	77 ± 2.4
AZ91 + 5% SiC	85 ± 3.5
AZ91 + 8% SiC	102 ± 2.5
AZ91 + 11% SiC	108 ± 3

The presence of hard SiC reinforcement will increase the load-bearing capability and also restrict the deformation of the matrix by constraining the dislocation movement [40]. The hardness value of the magnesium alloy composite rises with an increase in the wt.% of SiC particulate reinforcement.

The ultimate tensile strength (UTS) and yield strength (YS) of the AZ91 and its composites with different weight percentages of SiC (2%, 5%, 8%, and 11%) are given in Table 4. Figure 5 shows the engineering stress–strain curve of magnesium alloy AZ91 and its composites under tensile loading. The graphical bar chart representations of the yield strength and ultimate tensile strength are presented in Figure 6. The ultimate tensile strength (UTS) of AZ91 is found to be 188 MPa. The ultimate tensile strength of the 2% SiC particulate-reinforced composite is lower than that of the magnesium alloy, and it increases when increasing the weight percentage of reinforcements. This result is in agreement with the investigation reported in literature [23]. However, the ultimate tensile strength of 11% SiC particulate-reinforced composites is higher than that of AZ91.

Table 4. The ultimate tensile strength and yield strength.

Materials	UTS (MPa)	YS (MPa)	Elongation (%)
AZ91	188 ± 4.5	121 ± 2.8	5 ± 1.1
AZ91 + 2% SiC	114 ± 3.4	73 ± 1.6	10 ± 1.3
AZ91 + 5% SiC	138 ± 2.6	64 ± 1.2	8 ± 0.9
AZ91 + 8% SiC	141 ± 2.7	97 ± 1.4	12 ± 1.2
AZ91 + 11% SiC	196 ± 3.8	126 ± 2.0	7 ± 1.4

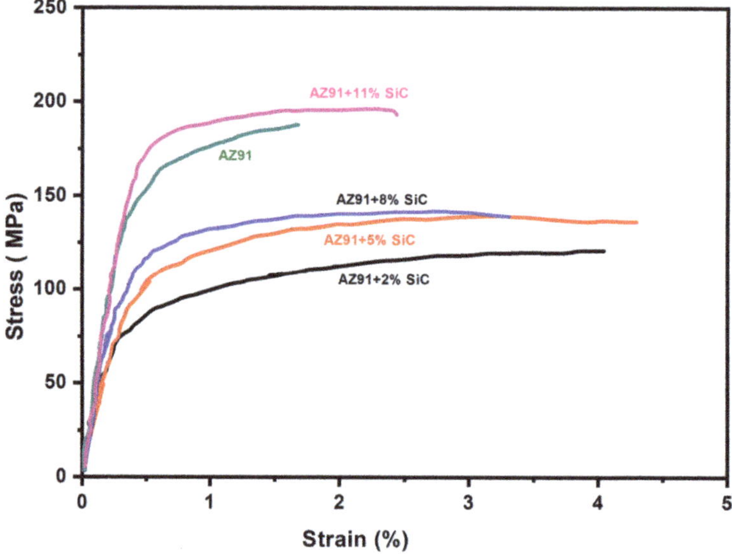

Figure 5. Engineering stress–strain curve of AZ91 and its composites reinforced with SiC particulates.

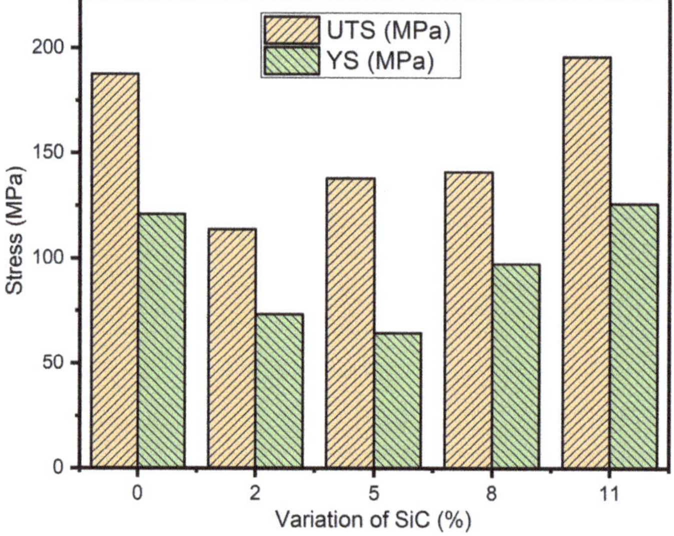

Figure 6. Variation of ultimate tensile strength and yield strength with the variation of SiC particulates.

The difference of ultimate tensile strength in AZ91 alloys and SiC particulate-reinforced composite is because of its processing method, which is stir casting in the present study. For the as-cast AZ91 composite, the ultimate tensile strength is generally lower than that of the as-cast AZ91 [23] due to the addition of secondary hard particles and the presence of *porosity*, reducing tensile strength in the as-cast state. Further secondary processing such as extrusion, forging and rolling operation may have enhanced the properties due to the formation of the strong bond between reinforcement, as well as the absence of micro-voids and *porosity*, which are present in the as-cast composites.

The yield strength (YS) of the SiC reinforced composites is observed to improve with an increase in weight percentage of SiC particles in the magnesium alloy (AZ91) composites as depicted in Table 4 and Figure 6. The different strengthening mechanisms may contribute to increased yield strength of the metal matrix composite.

The compression tests of AZ91 alloy and its composites were performed, and the results are presented in Table 5. The ultimate compressive strength (UCS) of the composite materials reinforced with SiC particulates is higher than that of the unreinforced AZ91 magnesium alloy, as depicted in Figure 7. The UCS of the composites increases further as the weight percentage of SiC particulate increases in the base alloy, AZ91. The UCS of the composites reinforced with different wt.% of SiC increases because of the addition of hard particles in a softer matrix, which increases load-bearing capacity. Significant improvements in ultimate compressive strengths are observed in the SiC particulate-reinforced composites in comparison with the monolithic alloy. The enhancement in ultimate compressive strength may be attributed to the partial closing of the small microscopic cracks and voids during compressive loading. The comparable observation is also reported for SiC_p reinforced AZ92 magnesium alloy composites [41]. The compressive strength is increased by 3% on the addition of 2% of SiC particulates, and it is further increased up to 10% on acquisition of 11% of SiC particulate. The compressive strength of AZ91 alloy is sufficiently high, and there is no appreciable increment in compressive strength on the addition of SiC reinforcements.

Table 5. Ultimate compressive strengths of AZ91 alloy and its composites.

Materials	Ultimate Compressive Strength (MPa)
AZ91	308 ± 5.3
AZ91 + 2% SiC	316 ± 2.8
AZ91 + 5% SiC	323 ± 3.5
AZ91 + 8% SiC	331 ± 6.0
AZ91 + 11% SiC	340 ± 2.5

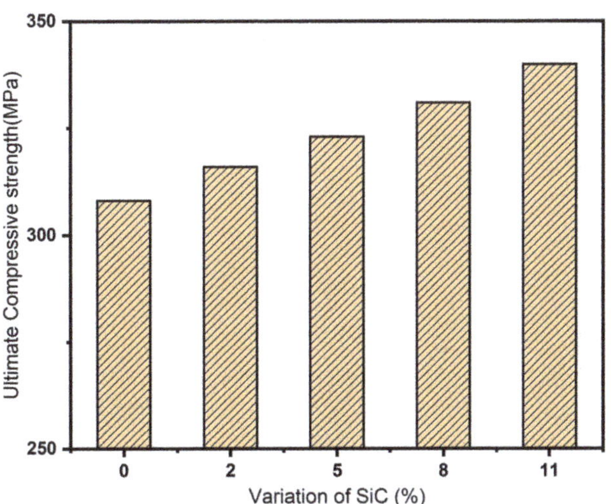

Figure 7. Representation of ultimate compressive strength of alloy and composites with different percentages of SiC.

The fracture surfaces of the magnesium alloy AZ91 and as-cast composites reinforced with different wt.% (2%, 5%, 8% and 11%) of SiC particulates were rough, with a normal height variation of 2 mm. Figure 8 shows SEM images of the fracture surfaces of the as-cast magnesium alloy AZ91 at different magnifications. Figure 9 shows SEM images of the fracture surfaces of the as-cast magnesium alloy AZ91-based composites reinforced with different weight percentages of SiC particulates. The main features of the fracture exterior were close to concentrated SiC particulates and agglomeration of SiC particles. The de-bonding of the particle–matrix interface, cracks on the particle, and inter-granular cracks in the matrix were also present. From the fracture surface investigation, it can be seen that fracture occurred via two modes in the case of the composite. In the matrix phase, ductile fracture of the matrix predominantly takes place, in which nucleation of voids plays an important role.

Brittle fracture occurs when the concentration of SiC particles is high. The composite tends to exhibit a more brittle fracture morphology than the matrix alloy. The cracking of the composite is initiated by de-bonding on the interface of the matrix and SiC reinforcement. In the case of the matrix, the fracture takes place by cleavage mode, combined with the ductile feature. However, the composites exhibit mixed-mode fracture (i.e., ductile and cleavage), as well as particle de-bonding. The matrix particle boundary would be mostly controlled by mechanical bonding; hence, when the maximum load is attained, de-bonding of SiC particulates occurs, rather than particle fracturing. The fracture surface characteristics of the composite also shows that the tiny size dimple, micro-crack, and cleavage fracture increases with increase in reinforcement.

In the compression test of AZ91 magnesium alloy and its composites, the fracture occurred nearly 45-degree angle concerning compression test axis and showed dominant shear failure.

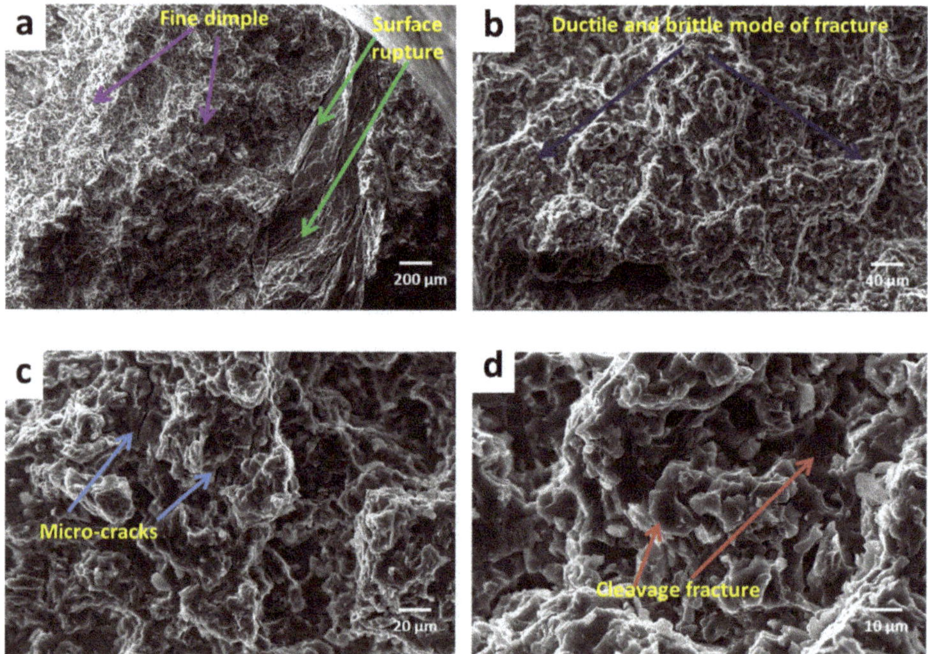

Figure 8. SEM tensile fractograph of cast magnesium alloy AZ91 at (**a**) 100×, (**b**) 500×, (**c**) 1000× and (**d**) 2000×.

Figure 10 shows the fractography of the compressive fracture surfaces of AZ91 magnesium alloy-based composites. Figure 11 shows the SEM micrograph of the fracture surface of AZ91 at various magnifications, i.e., 41×, 100×, 200× and 500×. Figure 12 shows the SEM micrograph of the magnesium alloy AZ91-based composites reinforced with a different weight percentage of SiC particulates. The fracture surfaces of AZ91 exhibit major shear failure and show a greater number shear bands when compared to AZ91/SiC composites, which instead show uneven fracture surfaces with mixed-mode, shear and brittle characteristics. Short and long shear bonds are seen in the fractured surfaces, along with the shear twinning mode of plastic deformation. The rough, brittle and shear modes are common fracture modes in magnesium alloys and its composites.

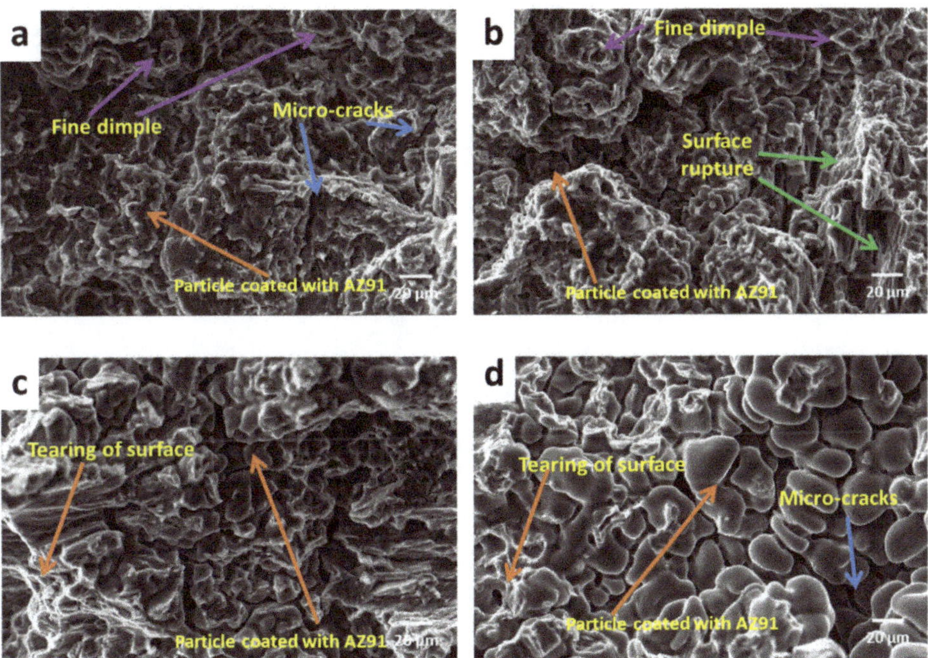

Figure 9. SEM tensile fractograph of cast magnesium alloy AZ91-based composites: (**a**) 2% SiC, (**b**) 5% SiC, (**c**) 8% SiC, (**d**) 11% SiC.

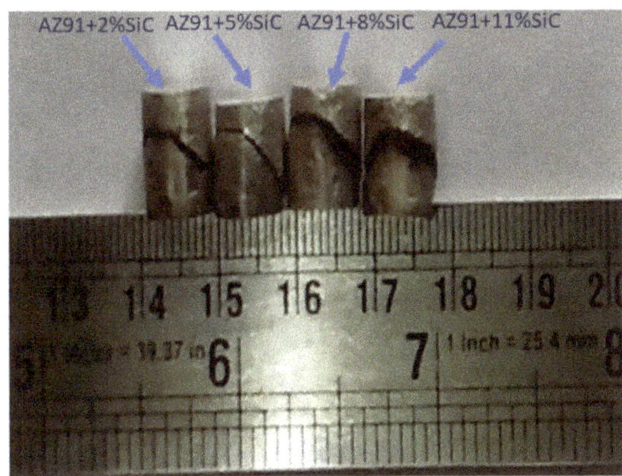

Figure 10. Fractured samples in a compression test.

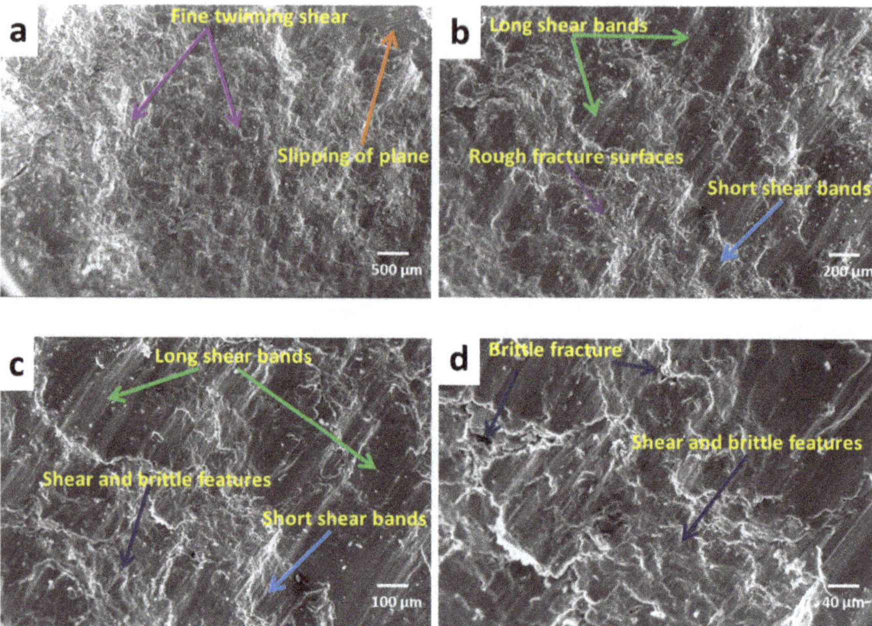

Figure 11. SEM compressive facto-graph of AZ91 at different magnifications: (**a**) 41×, (**b**) 100×, (**c**) 200×, (**d**) 500×.

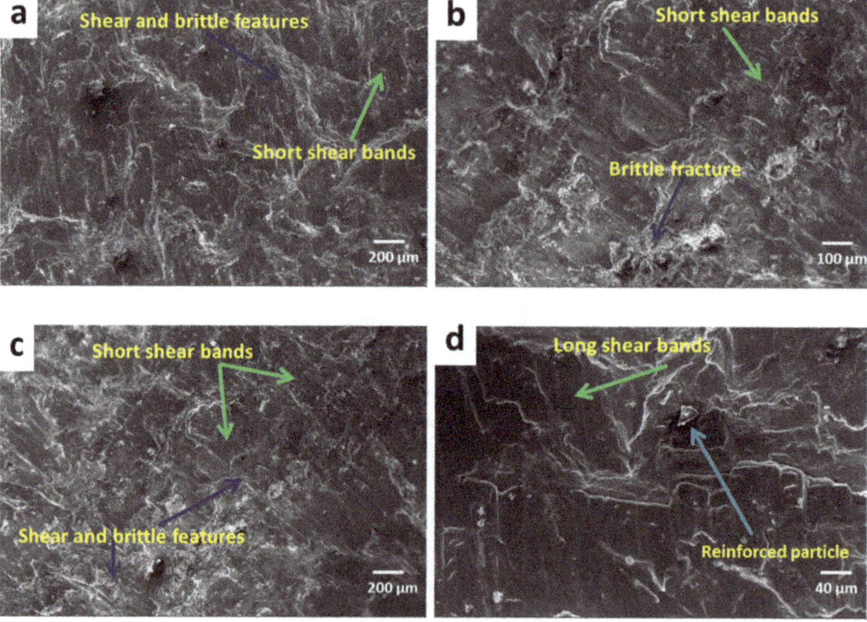

Figure 12. SEM compressive fractograph of AZ91-based composites reinforced with (**a**) 2% SiC, (**b**) 5% SiC, (**c**) 8% SiC, (**d**) 11% SiC.

4. Conclusions

A technique for preparing defect-free magnesium alloy by casting with improved physical and mechanical properties is presented in this study. The product developed using this casting technique can be used in different engineering applications, such as in aerospace, defence and automobile applications. The following conclusions were drawn based on experiments performed in this study:

- An increase in density was noticed owing to the increase in the amount of high-density reinforcement in the matrix, and an increase in *porosity* was observed because of the increase in micro-voids in the vicinity of the reinforcements.
- The XRD peaks confirmed the presence of Mg, $Mg_{17}Al_{12}$ and SiC Phases, indicating that there is no intermetallic bonding between the reinforcement and matrix materials under the recorded processing temperature.
- The Vickers microhardness value was increased by 28% with the addition of 2% SiC particulate, and it was further increased by up to 80% with the addition of 11% SiC particulate. This is because of the increase in load bearing capacity when adding hard particles to soft magnesium alloy.
- The tensile strength was found to initially decrease with the addition of reinforcements (up to 2%), and then increase with the increase in the percentage of SiC particulates (5% to 11%) in magnesium alloys. On increasing the content of the reinforcement, grain refinement occurs, and this led to an increase in ductility.
- The compressive strength increased by 3% on addition of 2% SiC particulate, and it further increased by up to 10% on addition of 11% SiC particulate. The increase in compressive strength can be attributed to the grain refinement, which occurs when increasing the percentage of SiC particulate.
- The composite materials exhibited mixed-mode fracture (i.e., ductile and cleavage). The fracture surface morphology of the composites materials also revealed that the tiny dimples, micro-cracks, and cleavage fractures increased with an increase in reinforcement. This is attributed to the fact that the ductility increases with increasing weight percentage of SiC particulate.
- The compression fractured surface of AZ91 exhibited dominant shear failure and showed more shear bands in comparison to the AZ91/SiC composites, which showed rough fractured surfaces with a mixed mode of shear and brittle features. Strain localization and slightly plastic deformation were observed in composite materials. This is because of the fact that the alloy exhibited more brittleness than the reinforce composites.

Finally, on the basis of the above facts, it can be concluded that the addition of SiC particulates in magnesium alloy enhanced the physical and mechanical properties. Therefore, SiC-reinforced magnesium alloy can be developed for applications in the aerospace, defence and automobile sectors.

Author Contributions: Conceptualization, A.K., S.K., N.K.M. and A.Y.; methodology, A.K., S.K., N.K.M., A.Y. and J.W.; formal analysis, A.K., S.K., N.K.M., A.Y. and J.W.; resources, S.K. and N.K.M.; data curation, A.K., A.Y. and J.W.; writing—original draft preparation A.K., A.Y. and J.W.; writing—review and editing, A.K., S.K., N.K.M., A.Y. and J.W.; supervision, S.K. and N.K.M.; project administration, A.K., S.K., N.K.M., A.Y. and J.W. All authors have read and agreed to the published version of the manuscript.

Funding: This research received no external funding.

Conflicts of Interest: The authors declare no conflict of interest.

References

1. Lu, L.; Thong, K.K.; Gupta, M. Mg-based composite reinforced by Mg_2Si. *Compos. Sci. Technol.* **2003**, *63*, 627–632. [CrossRef]
2. Zhang, X.; Zhang, Q.; Hu, H. Tensile behaviour and microstructure of magnesium AM60-based hybrid composite containing Al_2O_3 fibres and particles. *Mater. Sci. Eng. A* **2014**, *607*, 269–276. [CrossRef]

3. Lu, D.; Jiang, Y.; Zhou, R. Wear performance of nano-Al_2O_3 particles and CNTs reinforced magnesium matrix composites by friction stir processing. *Wear* **2013**, *305*, 286–290. [CrossRef]

4. Paramsothy, M.; Tan, X.H.; Chan, J.; Kwok, R.; Gupta, M. Al_2O_3 nanoparticle addition to concentrated magnesium alloy AZ81: Enhanced ductility. *J. Alloys Compd.* **2012**, *545*, 12–18. [CrossRef]

5. Gui, M.; Han, J.; Li, P. Fabrication and Characterization of Cast Magnesium Matrix Composites by Vacuum Stir Casting Process. *J. Mater. Eng. Perform.* **2003**, *12*, 128–134. [CrossRef]

6. Tun, K.S.; Gupta, M. Improving mechanical properties of magnesium using nano-yttria reinforcement and microwave assisted powder metallurgy method. *Compos. Sci. Technol.* **2007**, *67*, 2657–2664. [CrossRef]

7. Ye, H.Z.; Liu, X.Y. Review of recent studies in magnesium matrix composites. *J. Mater. Sci.* **2004**, *39*, 6153–6171. [CrossRef]

8. Loh, N.H.; Tor, S.B.; Khor, K.A. Production of metal matrix composite part by powder injection molding. *J. Mater. Process. Technol.* **2001**, *108*, 398–407. [CrossRef]

9. Wang, H.Y.; Jiang, Q.C.; Li, X.L.; Wang, J.G. In situ synthesis of TiC/Mg composites in molten magnesium. *Scr. Mater.* **2003**, *48*, 1349–1354. [CrossRef]

10. Sharma, N.; Singh, G.; Sharma, P.; Singla, A. Development of Mg-Alloy by Powder Metallurgy Method and Its Characterization. *Powder Metall. Met. Ceram.* **2019**, *58*, 163–169. [CrossRef]

11. Singh, K.; Singh, G.; Singh, H. Review on friction stir welding of magnesium alloys. *J. Magnes. Alloys* **2018**, *6*, 399–416. [CrossRef]

12. Luo, A.A. Magnesium casting technology for structural applications. *J. Magnes. Alloys* **2013**, *1*, 2–22. [CrossRef]

13. Hashim, J.; Looney, L.; Hashmi, M.S.J. Metal matrix composites: Production by the stir casting method. *J. Mater. Process. Technol.* **1999**, *92*, 1–7. [CrossRef]

14. Kumar, A.; Kumar, S.; Mukhopadhyay, N.K. Introduction to magnesium alloy processing technology and development of low-cost stir casting process for magnesium alloy and its composites. *J. Magnes. Alloys* **2018**, *6*, 245–254. [CrossRef]

15. Wu, S.; Wang, S.; Wen, D.; Wang, G.; Wang, Y. Microstructure and mechanical properties of magnesium matrix composites interpenetrated by different reinforcement. *Appl. Sci.* **2018**, *8*, 2012. [CrossRef]

16. Tian, L.; Guo, Y.; Li, J.; Xia, F.; Liang, M.; Duan, H.; Wang, P.; Wang, J. Microstructures of three in-situ reinforcements and the effect on the tensile strengths of an Al-Si-Cu-Mg-Ni alloy. *Appl. Sci.* **2018**, *8*, 1523. [CrossRef]

17. Kondoh, K.; Kawakami, M.; Umeda, J.; Imai, H. Magnesium matrix composites reinforced with titanium particles. *Mater. Sci. Forum* **2009**, *618*, 371–375. [CrossRef]

18. Singh, H.; Kumar, D.; Singh, H. Development of magnesium-based hybrid metal matrix composite through in situ micro, nano reinforcements. *J. Compos. Mater.* **2020**, 002199832094643. [CrossRef]

19. Abbas, A.; Huang, S.J. Investigation of severe plastic deformation effects on microstructure and mechanical properties of WS2/AZ91 magnesium metal matrix composites. *Mater. Sci. Eng. A* **2020**, *780*, 139211. [CrossRef]

20. Huang, S.J.; Ali, A.N. Experimental investigations of effects of SiC contents and severe plastic deformation on the microstructure and mechanical properties of SiCp/AZ61 magnesium metal matrix composites. *J. Mater. Process. Technol.* **2019**, *272*, 28–39. [CrossRef]

21. Jiang, Q.C.; Wang, H.Y.; Ma, B.X.; Wang, Y.; Zhao, F. Fabrication of B 4 C particulate reinforced magnesium matrix composite by powder metallurgy. *J. Alloys Compd.* **2005**, *386*, 177–181. [CrossRef]

22. Steglich, D.; Tian, X.; Bohlen, J.; Kuwabara, T. Mechanical Testing of Thin Sheet Magnesium Alloys in Biaxial Tension and Uniaxial Compression. *Exp. Mech.* **2014**, *54*, 1247–1258. [CrossRef]

23. Luo, A. Processing, microstructure, and mechanical behavior of cast magnesium metal matrix composites. *Metall. Mater. Trans. A* **1995**, *26*, 2445–2455. [CrossRef]

24. Avedesian, M.M.; Baker, H. *ASM Specialty Handbook: Magnesium and Magnesium Alloys*; ASM International Materials Park: Russell Township, OH, USA, 1999; ISBN 978-0-87170-657-7.

25. Lloyd, D.J. Particle reinforced aluminium and magnesium matrix composites. *Int. Mater. Rev.* **1994**, *39*, 1–23. [CrossRef]

26. Chawla, N.; Chawla, K.K. *Metal Matrix Composites*; Springer: New York, NY, USA, 2013; ISBN 9781461495482. [CrossRef]

27. Friedrich, H.E.; Mordike, B.L. *Magnesium Technology: Metallurgy, Design Data, Applications*; Springer: Berlin, Germany, 2006. [CrossRef]

28. Gupta, M.; Lai, M.O.; Saravanaranganathan, D. Synthesis, microstructure and properties characterization of disintegrated melt deposited Mg/SiC composites. *J. Mater. Sci.* **2000**, *35*, 2155–2165. [CrossRef]

29. Mohazzab, P. Archimedes' Principle Revisited. *J. Appl. Math. Phys.* **2017**, *5*, 836–843. [CrossRef]

30. Hassan, S.F. Effect of primary processing techniques on the microstructure and mechanical properties of nano-Y_2O_3 reinforced magnesium nanocomposites. *Mater. Sci. Eng. A* **2011**, *528*, 5484–5490. [CrossRef]

31. Jiang, F.; Hirata, K.; Masumura, T.; Tsuchiyama, T.; Takaki, S. Effect of the Surface Layer Strained by Mechanical Grinding on X-ray Diffraction Analysis. *ISIJ Int.* **2018**, *58*, 376–378. [CrossRef]

32. ASTM E8/E8M-16a. *Standard Test Methods for Tension Testing of Metallic Materials*; ASTM International: West Conshohocken, PL, USA, 2016. [CrossRef]

33. ASTM E 09. *Standard Test Methods of Compression Testing of Metallic Materials at Room Temperature*; ASTM International: West Conshohocken, PL, USA, 2019. [CrossRef]

34. Gertsberg, G.; Aghion, E.; Kaya, A.A.; Eliezer, D. Advanced production process and properties of Die Cast magnesium composites based on AZ91D and SiC. *J. Mater. Eng. Perform.* **2009**, *18*, 886–892. [CrossRef]

35. Poddar, P.; Srivastava, V.C.; De, P.K.; Sahoo, K.L. Processing and mechanical properties of SiC reinforced cast magnesium matrix composites by stir casting process. *Mater. Sci. Eng. A* **2007**, *460*, 357–364. [CrossRef]

36. Hassan, S.F.; Gupta, M. Development of high performance magnesium nano-composites using nano-Al_2O_3 as reinforcement. *Mater. Sci. Eng. A* **2005**, *392*, 163–168. [CrossRef]

37. Hassan, S.F.; Gupta, M. Effect of length scale of Al_2O_3 particulates on microstructural and tensile properties of elemental Mg. *Mater. Sci. Eng. A* **2006**, *425*, 22–27. [CrossRef]

38. Nguyen, Q.B.; Gupta, M. Increasing significantly the failure strain and work of fracture of solidification processed AZ31B using nano-Al_2O_3 particulates. *J. Alloys Compd.* **2008**, *459*, 244–250. [CrossRef]

39. Liu-Qing, Y.; Yong-lin, K.; Fan, Z.; Jun, X.U. Microstructure and mechanical properties of rheo-diecasting AZ91D Mg alloy. *Trans. Nonferrous Met. Soc. China* **2010**, *20*, s862–s867. [CrossRef]

40. Aravindan, S.; Rao, P.V.; Ponappa, K. Evaluation of physical and mechanical properties of AZ91D/SiC composites by two step stir casting process. *J. Magnes. Alloys* **2015**, *3*, 52–62. [CrossRef]

41. Viswanath, A.; Dieringa, H.; Kumar, K.K.; Pillai, U.T.S.; Pai, B.C. Investigation on mechanical properties and creep behavior of stir cast AZ91-SiCp composites. *J. Magnes. Alloys* **2015**, *3*, 16–22. [CrossRef]

Publisher's Note: MDPI stays neutral with regard to jurisdictional claims in published maps and institutional affiliations.

Article

Fabrication and Mechanical Characterization of Dry Three-Dimensional Warp Interlock Para-Aramid Woven Fabrics: Experimental Methods toward Applications in Composite Reinforcement and Soft Body Armor

Mulat Alubel Abtew [1,2,3,4], **Francois Boussu** [2,3], **Pascal Bruniaux** [2,3] **and Han Liu** [1,*]

[1] College of Textile and Clothing Engineering, Soochow University, 178 G.J.D. Road, Suzhou 215021, China; mulat-alubel.abtew@ensait.fr

[2] Faculty of Science and Technology, University of Lille, Nord de France, 59000 Lille, France; Francois.boussu@ensait.fr (F.B.); pascal.bruniaux@ensait.fr (P.B.)

[3] GEMTEX Laboratory, ENSAIT, 2 Allée Louise et Victor Champier, 59056 Roubaix, France

[4] Faculty of Textiles—Leather Engineering and Industrial Management, Gheorghe Asachi' Technical University of Iasi, Dimitrie Mangeron Bd. 53, 700050 Iasi, Romania

* Correspondence: hanl@suda.edu.cn

Received: 23 July 2020; Accepted: 19 September 2020; Published: 23 September 2020

Abstract: Recently, three-dimensional (3D) warp interlock fabric has been involved in composite reinforcement and soft ballistic material due to its great moldability, improved impact energy-absorbing capacity, and good intra-ply resistance to delamination behaviors. However, understanding the effects of different parameters of the fabric on its mechanical behavior is necessary before the final application. The fabric architecture and its internal yarn composition are among the common influencing parameters. The current research aims to explore the effects of the warp yarn interchange ratio in the 3D warp interlock para-aramid architecture on its mechanical behavior. Thus, four 3D warp interlock variants with different warp (binding and stuffer) yarn ratios but similar architecture and structural characteristics were engineered and manufactured. Tensile and flexural rigidity mechanical tests were carried out at macro- and meso-scale according to standard EN ISO 13 934-1 and nonwoven bending length (WSP 90.5(05)), respectively. Based on the results, the warp yarn interchange ratio in the structure revealed strong influences on the tensile properties of the fabric at both the yarn and final fabric stages. Moreover, the bending stiffness of the different structures showed significant variation in both the warp and weft directions. Thus, the interchange rations of stuffer and binding warp yarn inside the 3D warp interlock fabric were found to be very key in optimizing the mechanical performance of the fabric for final applications.

Keywords: 3D warp interlock fabric; warp yarn interchange ratio; mechanical test; mechanical characterization; fiber-reinforced composite; soft body armor; para-aramid fiber

---

1. Introduction

Textile materials nowadays are widely used in various technical applications, including composite reinforcement for aerospace, transport, military, and other applications. The structure of the fabric involved in technical applications should have higher mechanical performance compared to the structure involved in conventional applications such as clothing and home furnishings. Previously, two-dimensional (2D) woven (in the form of plain, twill, and satin) and unidirectional (UD) fabric structures were mainly used and discussed regarding their mechanical performance under different loading conditions [1,2]. Apart from 2D and UD fabrics, three-dimensional (3D) woven fabric have also become promising for use in composite materials, such as fibrous reinforcement and protective solutions, as well as other technical applications [3,4]. Its involvement in the above application is mainly because it has an improved capacity to absorb energy by higher intra-ply resistance to delamination, saves on cost, and has high production rates [5–7]. Moreover, such fabric also provides better shaping ability for 3D shape solutions by linking different yarns through different weave styles at the required thickness [8–11]. However, its mechanical properties are much different from those of conventional 2D woven and unidirectional fabric due to its complex fabric structure. This has attracted the attention of various researchers to understand the mechanical behavior before recommending particular applications. Based on this, various researchers have been intensively investigating not only the final mechanical behavior under static or dynamic loading at the dry and composite reinforcement stages, but also the influence of various factors, including fabric type, fabric architecture, fiber type, yarn density, and weave parameters [12–14].

The fabric weave type and structure are among the most important factors that greatly affect the mechanical properties of dry 3D warp interlock fabrics [15–20]. In addition, detailed experimental observations of 3D woven composites have indicated that the geometry of the fabric has a dominant role in determining the mechanical properties and associated failure mechanisms [21]. Another interesting feature of 3D warp interlock fabric is that it not only provides room for managing different weft layer parameters inside the fabric architecture, but also allows relatively accurate control of the yarn's evolution in the structure [22]. A research work experimentally studied the effect of the fabric structure on the tensile behavior of 3D woven composite using four 3D woven multilayer fabric structures as reinforcement [16]. According to the investigation, the fabric structure showed a great effect not only on the tensile strength but also on the dimensional stability of the composites. Similarly, another study also supported the effect of fabric structure on the overall mechanical properties of 3D textile composites [17]. The effect of fabric type on the mechanical behavior of reinforced composites using unidirectional (UD), two-dimensional (2D), three-dimensional (3D) orthogonal, 3D angle-interlock, and 3D warp interlock multilayers as reinforcement was also comprehensively studied. The results showed that composites made with 3D woven fabric had considerably better impact resistance, knife penetration resistance, and dynamic mechanical analysis (DMA) behavior as compared with their UD and 2D counterparts [18].

Apart from the type of weave architecture and raw material type (fiber and yarn), warp yarn crimp values in the woven structure and their combinations (hybridization) also affect the mechanical behavior of both dry 3D warp interlock fabrics and composites [23–26]. A study on the tensile properties of 3D woven reinforced composite made of Kevlar fiber revealed higher tensile strength as compared to nylon-reinforced composite. Moreover, a hybridized composite made of both Kevlar–glass and Kevlar–glass–nylon fibers showed great improvement in tensile strength as compared to a monolithic composite made of glass fiber [24]. The effect of fiber type and hybridization on the mechanical properties of three-dimensional angle-interlock composite fabrics has also been investigated. Bandaru et al. [25] investigated the effect of fabric hybridization on the mechanical behavior. For this, two types of fabrics, one made of Kevlar and basalt yarns individually and another made of a combination of Kevlar and basalt yarns, were developed and tested with the quasi-static tensile test. Based on the results, hybridization improved the tensile behavior of the three-dimensional angle-interlock composite fabric. The weaving process and its parameters, including the positioning of stuffer, fiber volume fractions, and linking warp yarns in the 3D orthogonal and warp interlock woven fabric structure, were also shown to have an influence on the mechanical properties of the fabrics [27–29].

In addition, yarn and weave densities also play an essential role in the mechanical properties of 3D woven fabrics [30,31]. To validate this, an experimental study was conducted on 3D layer-to-layer glass/epoxy woven composite structures made with different pick densities. In general, the study revealed that reducing the waviness and misalignment of load-carrying fibers improved the mechanical properties of the material [30]. Nasrun et al. [32] investigated the effects of weft density on the mechanical tensile strength behavior of 3D angle interlock woven fabric. Four fabrics based on different weft densities (12, 16, 20, and 24 picks per cm) using polyester plied yarn with 100 Tex were produced and tested based on ASTM standards. According to the results, the tensile strength of 3D angle interlock woven fabric could be improved with increased weft density. The effect of Z-yarn on the mechanical properties of 3D weave architecture was also studied and discussed based on the fabric geometry and 3D finite element simulations [33]. An interlocking pattern is another factor that affects the final mechanical performance of three-dimensional orthogonal layer-to-layer interlock composites. Three types of orthogonal layer-to-layer interlock fabrics—warp, weft, and bi-directional interlock composite—were tested for their mechanical performance [34]. Due to their lower crimps, the warp and weft interlock composites showed better tensile behavior as compared to the bi-directional interlock composite. The off-axis angles among the different yarns while developing the 3D woven fabrics also showed an influence on the mechanical properties of the materials [35,36].

Apart from the influencing parameters described above, the current study investigates and discusses the effect of the warp yarn interchange ratio on the mechanical behavior of 3D warp interlock fabrics. We designed, manufactured, and experimentally investigated the mechanical behaviors of 3D warp interlock para-aramid fabrics under quasi-static conditions. For this investigation, four 3D warp interlock para-aramid fabric architectures considering the binding–stuffer warp yarn interchange ratio were designed and fabricated. All designed orthogonal layer-to-layer (O-L) fabric architectures were manufactured considering the same warp and weft yarn densities. The test program includes uniaxial yarn tensile tests, flexural rigidity tests, and a uniaxial fabric tensile test. Characterization of high-performance 3D warp interlock para-aramid fabrics provides an understanding for different technical applications, including ballistics, for better performance.

2. Materials and Methods

2.1. Materials

The arrangement, movement, and deformational behavior of different warp (stuffer and binding) and weft yarns can affect the overall properties of fabric. According to Reference [37], stuffer and binding warp yarns can define the in-plane and through-thickness properties respectively, of the final fabric, whereas weft yarn also helps to define the number of fabric layers as well as determine the transverse properties of the fabric. Figure 1 shows the general interlacing structure of the different yarns in 3D woven fabric. To understand the effects of the warp yarn interchange ratio on the fabric's mechanical behaviors, four 3D warp interlock fabric architectures based on different binding and stuffer warp yarn interchange ratios were designed and manufactured.

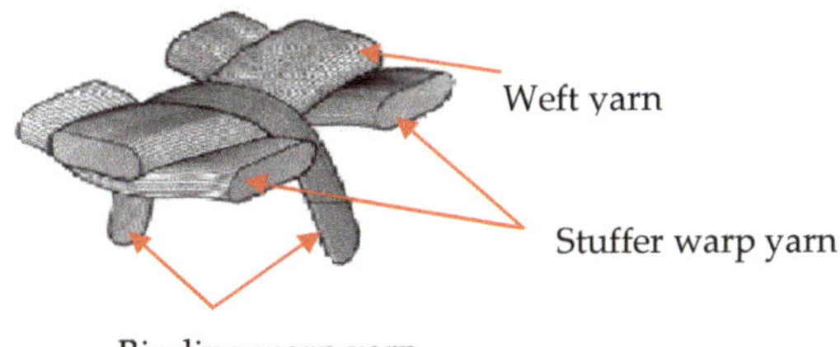

Figure 1. General schematic diagram of deformations of binding, stuffer, and weft yarns in three-dimensional (3D) woven fabric [38].

All of the 3D warp interlock fabric architectures were manufactured using high-performance 930dtex p-aramid fibers (Twaron® f1000), delivered by Teijin Aramid (Wuppertal-Elberfeld, Germany), a subsidiary of the Teijin Group, the Netherlands in ENSAIT-GEMTEX Laboratory, Roubaix. Such fiber is used to produce the most recommended 2D woven fabrics (Twaron CT-709) by the Teijin Company for the development of body armor due to its good ballistic performance and high level of molding capability. The 3D warp interlock fabric architectures were designed with different binding and stuffer warp yarn interchange proportions in the fabric repeat unit: variant A: 100% binding, 0% stuffer, variant B: 66.7% binding, 33.3% stuffer, variant C: 50% binding, 50% stuffer, and variant D: 33.3% binding, 66.7% stuffer, as shown in Figure 2.

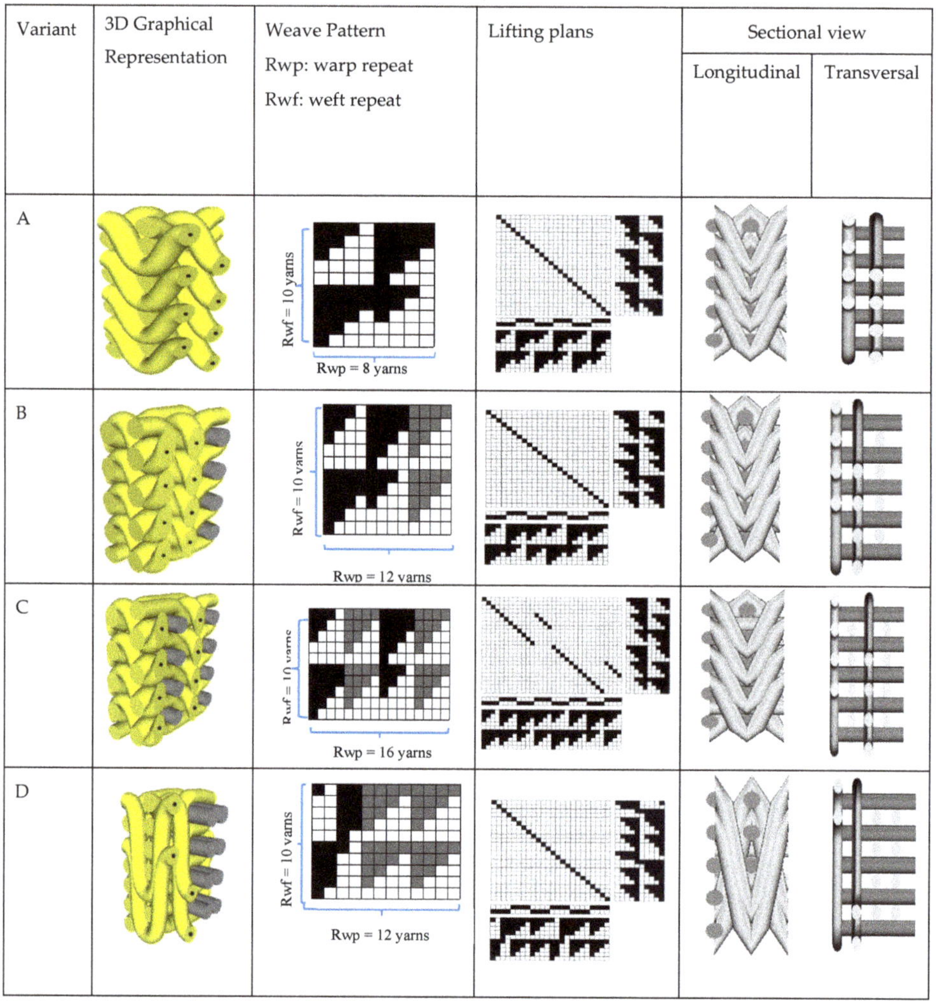

Variant	3D Graphical Representation	Weave Pattern Rwp: warp repeat Rwf: weft repeat	Lifting plans	Sectional view	
				Longitudinal	Transversal
A		Rwf = 10 yarns Rwp = 8 yarns			
B		Rwf = 10 yarns Rwp = 12 yarns			
C		Rwf = 10 yarns Rwp = 16 yarns			
D		Rwf = 10 yarns Rwp = 12 yarns			

Figure 2. Three-dimensional (3D) graphic representation and weave pattern of repeat unit for 3D warp interlock fabric with different interchanging ratios between binding and stuffer warp yarn.

Unlike the warp yarn composition in the structure and the fabric thickness, all developed 3D warp interlock fabrics were designed as orthogonal layer-to-layer (O-L) and had the same fiber type, number of weft layers, and yarn density. High-performance fibers with 2.35 mN/Tex of tenacity, 225 N strength at break, and 3.45% of elongation at break were used for all the fabrics. Each multilayer 3D warp interlock fabric also involved five weft layers. Also, 48 ends/cm/panel and 50 picks/cm/panel yarn densities were used in the warp and weft directions respectively, when manufacturing the fabrics. The theoretical fabric weight was computed as 970 g/m^2 for all fabric structures. The average fabric thickness was measured as 1.42, 1.44, 1.52, and 1.63 mm for variants A, B, C, and D, respectively. Such thickness variation among fabrics arises due to the interlacing behavior of each weft layer by the different binding warp yarns.

Commercially available software, TexGen® and WiseTex®, were used to develop the 3D geometric and weave peg plans of the fabrics, respectively. The fabricated multilayer fabrics were then woven by using a modified semi-automatic ARM dobby loom. The ARM dobby loom (with a 50 cm width) was designed with an adapted warp beam creel to properly accommodate up to 24 warp beams and can produce all kinds of 3D warp interlock fabrics. It also has a yarn guiding system to separate, accommodate, and guide warp yarn ends before weaving head sections, as shown in Figure 3.

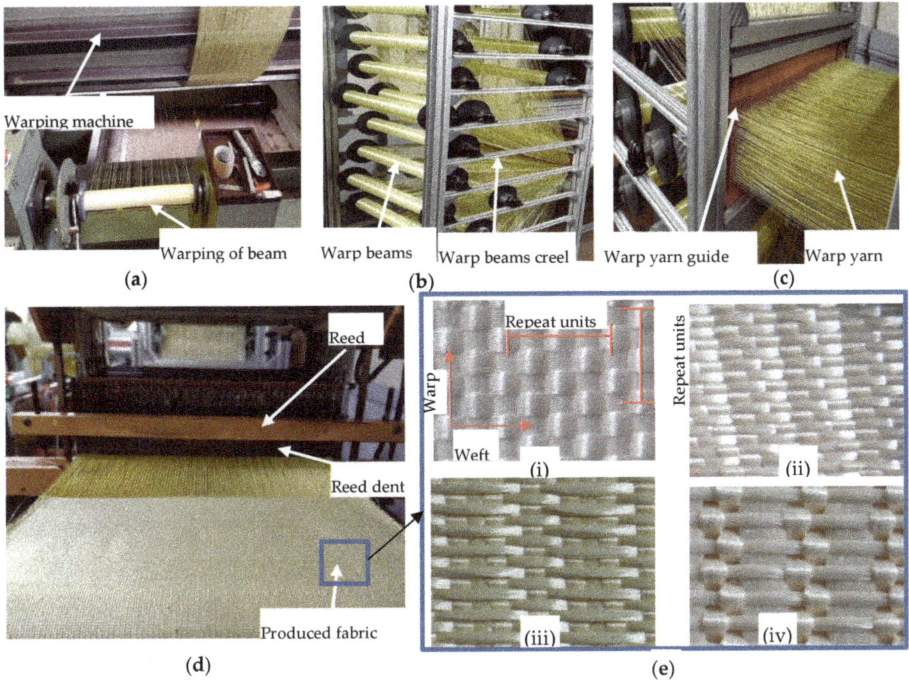

Figure 3. Manufacturing process of 3D warp interlock fabric using semi-automatic loom: (**a**) warping process and winding of warp ends to warp beam, (**b**) warp beam arranged in adapted warp beam creel, (**c**) warp yarn guiding system, (**d**) weaving machine head and weaving process, and (**e**) top surface view of produced 3D warp interlock fabrics ((**i**) variant A, (**ii**) variant B, (**iii**) variant C, (**iv**) variant D).

2.2. Experimental Testing Methods

In this section, different testing methodologies and procedures for experimental testing, measuring, and characterizing the produced 3D warp interlock fabrics will be explained. Uniaxial tensile tests on different types of yarn in the fabric (Manufactured by Instron, Norwood, MA, USA and having two clamps initially 200 mm apart and a speed of 100 mm/min at room temperature), flexural rigidity tests (using Cantilever customized in GEMTEX Laboratory, Roubaix, France), and uniaxial fabric tensile tests (Manufactured by Instron, Norwood, MA, USA) were performed to characterize and understand the effects of the binding:stuffer yarn interchange ratio on the mechanical properties of 3D warp high-performance interlock p-aramid fabrics. Moreover, various fabric properties including actual thickness (mm) and density (g/m^2) were computed. The average actual fabric thickness and weight of the 3D warp interlock fabrics were precisely measured according to NF EN ISO 5084 [39] and NF EN 12127 [40] standards, respectively. The top and cross-sectional views of the produced 3D warp interlock fabrics were examined using a portable optical microscope (dnt professional liquid crystal display (LCD) digital microscope portable camera equipped with Universal Serial Bus (USB)/Thin Film Transistor (TFT) 5 MPix zoom 20 to 500× dnt DigiMicro Lab 5.0). Figure 3e shows the top views of the 3D warp interlock p-aramid fabrics.

2.2.1. Yarn and Fabric Uniaxial Tensile Testing Setup and Procedure

For textile materials that are intended for use in various applications that demand high mechanical behavior (e.g., composite reinforcement and soft ballistic material), we need a better understanding of their behavior at different levels. Thus, investigating the mechanical behavior of the material at the yarn level greatly helps in determining the yarn mechanics in applications at both dry and composite stages. For example, yarn density, modulus of elasticity and deformation, yarn strain and stress on impact, yarn-to-yarn friction, yarn tenacity, etc., are some of the critical properties that affect the material's ballistic performance [41–43]. In this section, 10 samples of p-aramid yarns from each 3D warp interlock fabric structure, considering the warp (stuffer and binding) and weft (different weft layers) directions, were carefully drawn without damaging the fibers/filaments to investigate and understand the mechanical behavior.

The uniaxial tensile test helps in investigating parameters including the stress–strain relationship, elastic modulus (E), maximum stress (σmax), and maximum strain (εmax) values of the yarn in both the machine (warp) and cross (weft) directions. In our investigation, the uniaxial tensile tests were conducted using an Instron 8516 universal testing machine manufactured by Instron (Norwood, MA, USA) with a 5 KN load cell at a velocity of 50 mm/min, as shown in Figure 4a. To guarantee repeatability of the results, the entire yarn test was performed with 10 replicas for each sample in both the warp and weft directions (Figure 4b). Each yarn sample was prepared with a total length of 250 mm and firmly fixed at both ends at 200 mm distance using two adapted steel clamps to avoid any slippage while testing. The fabric sample tensile behavior was also performed on the Instron 5900 testing machine with a 250 KN load cell according to EN ISO 13 934-1 standard [44], as shown in Figure 4d. A rectangular 300 × 50 mm sample that was adhesively bonded to 50 mm using resin at both ends was prepared to avoid slippage between the sample and the clamp jaw. Before every test, the two separate jaws (movable top clamp and fixed lower jaw) of the testing machine were set at a distance of 200 mm between them. The sample was firmly mounted between the upper and lower clamps to coincide with the resin bonded size to avoid any load displacement reading errors due to slippage between the sample and the clamp. Three samples for each fabric type were tested at 100 mm/min in the weft and warp directions to ensure repeatability of the investigation (Figure 4b). For each test, forces vs. deformation values with time duration were automatically recorded for all samples. Moreover, the tensile testing machine was also checked after each test so as to achieve accurate results by avoiding slippage between the sample and the clamp. The average tensile properties of the samples in the weft and warp directions were then analyzed using the extracted data during the test. To determine the effect of the warp yarn ratio on the breaking strength and elongation of 3D warp interlock fabrics, the measurement results obtained from the tests were analyzed and evaluated.

2.2.2. Flexural Rigidity Test of Fabrics

The effects of stuffer and binding warp yarn interchange ratio on the flexural rigidity properties of the 3D warp interlock fabrics were investigated. A bending test under mass was performed using a fabric stiffness testing apparatus following the Standard Test Methods for Nonwoven Bending Length (WSP 90.5(05)) under the principle of the cantilever. For this test, 300×50 mm^2 rectangular strips of samples for the warp and weft directions were prepared. Before testing, the samples were kept in standard atmosphere conditions (relative humidity (RH) 65% ± 2% and temperature 20 ± 2 °C) to bring them to moisture equilibrium as directed by ERT 60.2-99 and ISO 554. Figure 5 shows pictorial and schematic illustrations of the fabric stiffness testing apparatus while testing the flexural rigidity properties of a sample. The apparatus was designed with bending curvature according to ISO 4604 with a fixed angle (41.5°) and its platform was set up to accommodate proper samples while testing. Five samples in both directions (weft and warp) were examined for each fabric type, and the bending length was computed based on the average values. All samples were weighed using a digital balance with approximately ±0.001 g precision based on TS 251 before every test.

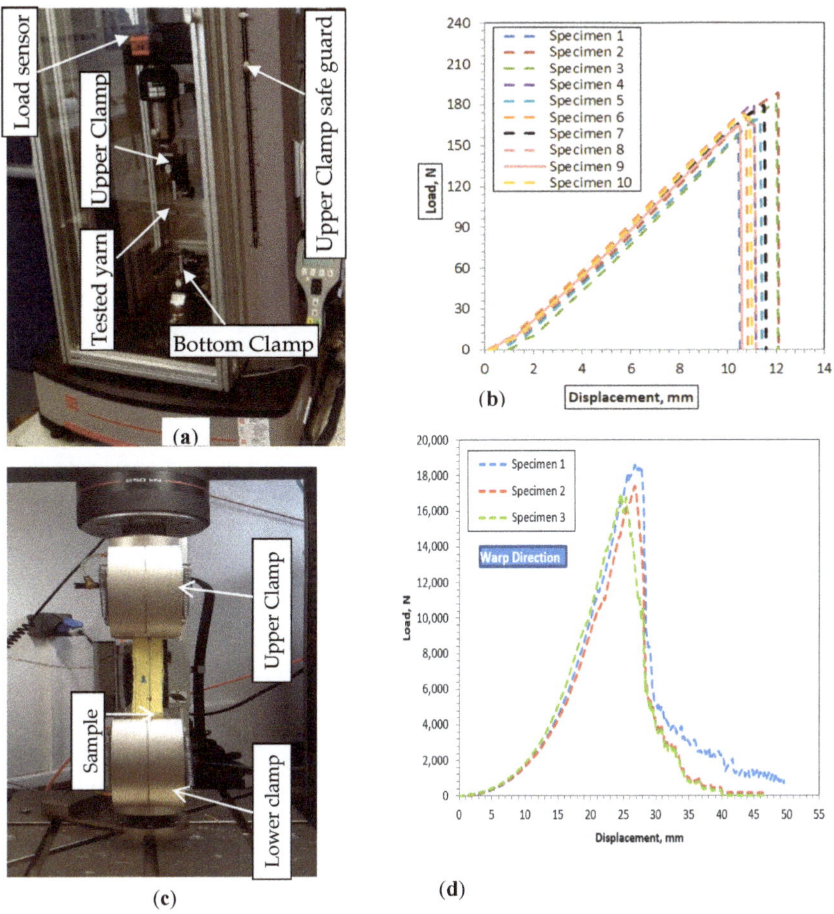

Figure 4. Uniaxial tensile testing: (**a**) yarn tensile testing device setup, (**b**) examples of uniaxial warp-binding tensile test results for sample fabric D in the machine direction (MD), (**c**) fabric tensile testing device setup, and (**d**) examples of uniaxial fabric tensile test results of samples in the machine (warp) direction for variant A.

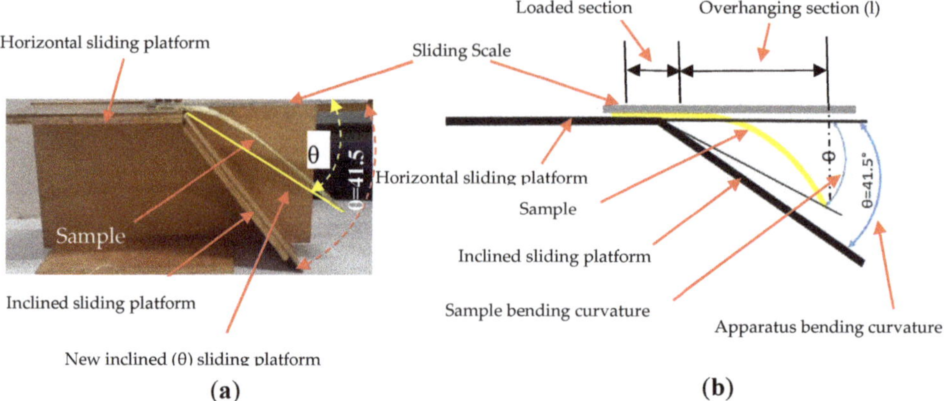

Figure 5. (**a**) Photograph and (**b**) schematic of flexural rigidity test of 3D warp interlock fabrics in stiffness testing apparatus.

During testing, the sample was properly positioned on the horizontal platform, on which one edge was fixed and the other was free to hang on the platform. The sample was allowed to slide on the horizontal sliding platform by pushing gently at a regular rate using a regular sliding scale. In our case, the investigation to compute flexural rigidity was performed in two ways. Some tests were performed until the sample was overhanging by its weight and the sample edges at the front touched the inclined sliding platform (41.5°). In other tests, if the overhanging sample did not touch the inclined platform, the overhanging sample (l) and the sample bending curvature (θ) were measured for further computation. For both cases, the flexural rigidity of the sample was then computed considering overhanging length (l), bending length, and the sample's areal weight and bending curvature. The flexural bending rigidity (N.m) was computed based on Equation (1):

$$G = \frac{1}{\frac{\tan\theta}{\cos\frac{\theta}{2}}} \times \frac{\rho \times l^3}{8} \tag{1}$$

where G is fabric flexural bending rigidity, ρ is fabric sample weight per unit area (mass per unit area × gravitational acceleration), l is overhanging length, and θ is bending curvature.

In general, the average bending length of the samples can be calculated using Equation (2):

$$C = \frac{l}{2} \tag{2}$$

where C is the sample bending length and l is the sample overhanging length after the bending test.

When the front edges of the sample touched the inclined sliding platform (41.5°) of the apparatus, the fabric flexural bending rigidity could be calculated by Equation (3):

$$G = \frac{1}{\frac{\tan\theta}{\cos\frac{\theta}{2}}} \times \frac{\rho l^3}{8}, \text{ for } \theta = 41.5°, \frac{1}{\frac{\tan\theta}{\cos\frac{\theta}{2}}} = 1 \tag{3}$$

Here, $G = 1 \times \frac{\rho l^3}{8}$, since l/2 is the bending length and ρ is mass per unit area multiplied by acceleration due to gravity, and the fabric flexural bending rigidity can be simplified as follows (Equation (4)):

$$G = W \times g \times c^3 \tag{4}$$

where G is the flexural bending rigidity of the sample (N m), W is the sample unit areal weight (g/m^2), C is the average bending length of the sample (mm), and g is gravitational acceleration (m/s^2).

3. Results and Discussion

In this section, the mechanical behavior of the developed 3D warp interlock fabrics considering different binding and stuffer warp yarn compositions is discussed, and the yarns in the 3D fabric structure are investigated and discussed. The fabric and its corresponding yarn tensile behavior were evaluated based on the stress–strain response curve and the average maximum tensile stress (MPa) and strain (%) values at the fracture point. The rigidity behavior of the fabrics was also assessed based on flexural bending rigidity (N.m) and fabric bending length. Finally, the effects of warp yarn composition on the waviness of the yarn in the 3D warp interlock fabrics in both directions were explained in terms of crimp percentage.

3.1. Yarn Uniaxial Tensile Property

A yarn's tensile property can be defined as the maximum applied force/load that is required to break the yarn. Understanding this property is very important not only because it is a key parameter for the fabrication of yarn, but it directly influences the strength of the developed fabrics.

3.1.1. Stuffer and Binding Warp Yarn Testing

The 3D warp interlock fabric structure not only has weft and warp yarns in the plane direction, but also another warp yarn type through the thickness direction [37]. This reinforces the fabric in three directions, and each group of yarns provides a specific function to the structure. The composition of one or more types of warp yarn mainly depends on the final application. The 3D warp interlock fabric can be represented with different weft layers of fabric interlaced through the thickness direction by using binding warp yarns. Thus, the fabric could present different mechanical behaviors due to different interlacement and binding positions of the yarns. Figure 6a,b shows the average stress–strain curves for warp-binding and warp-stuffer yarn for the 3D warp interlock fabrics. Based on the stress–strain curves, it is also possible to analyze the mechanical properties of each yarn in the fabric. The tensile stress–strain curve shows similar trends for the binding and stuffer yarn in the fabrics. As observed in Figure 6a, except for variant C, the higher the proportion of binding warp yarn in the fabric, the lower the tensile modulus (E) in the warp direction. The average tensile modulus (E) of binding warp yarn for variant D was found to be higher compared to the other samples. Variant C showed the lowest tensile modulus (E), followed by B and D.

This is because the binding warp yarn in variant C occupies a lesser interlacement depth along with higher stuffer yarn to bind the wet layers. As shown in Figure 6b, a very similar tensile modulus (E) of the stuffer yarn was observed for all fabrics. However, the tensile modulus of the stuffer yarn was found to be a little bit higher for samples with a smaller proportion of stuffer yarn, and vice versa. Variant B shows a higher tensile modulus, whereas variant D recorded the lowest.

Table 1 shows the tensile stress and strain at failure for binding and stuffer warp yarns of the developed variants.

Table 1. Tensile stress and strain at failure for binding and stuffer warp yarns.

Sample	Binding Warp Direction		Stuffer Warp Direction	
	Tensile Stress, MPa	Strain at Failure, %	Tensile Stress, MPa	Strain at Failure, %
Variant A	191.11 ± 6.9	5.80 ± 0.9	-	-
Variant B	212.04 ± 7.6	5.56 ± 1.1	222.10 ± 10.1	5.22 ± 0.2
Variant C	188.18 ± 7.9	6.01 ± 0.76	212.73 ± 9.6	5.13 ± 0.11
Variant D	227.9 ± 10.9	5.73 ± 0.6	237.60 ± 12.1	5.76 ± 0.46

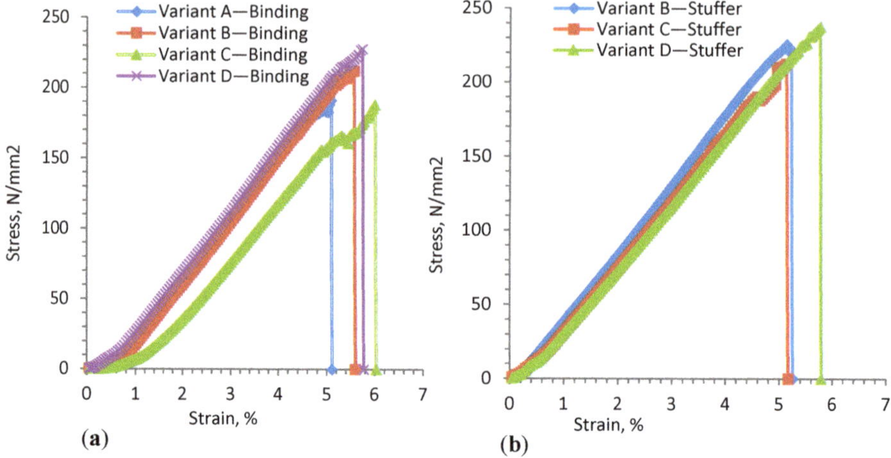

Figure 6. Typical stress–strain curves of (**a**) warp-binding yarn and (**b**) warp-stuffer yarn for different 3D warp interlock fabric architecture.

Figure 7a,b also shows the average maximum stress at failure (σmax) and tensile failure strain (εmax) for the respective warp yarns. The maximum stress is higher in the case of fill yarns; however, the failure strain is quite similar in both directions. The average maximum stress and tensile strain at failure for binding warp yarn of variant A showed lower values than other fabric types. The maximum tensile failure strain of binding warp yarn was obtained for variant C, followed by variant D, whereas the maximum failure stress of binding warp yarn was obtained for variant D. The elongation failure of binding warp yarn could be affected by the occurrence of filament degradation due to friction between the yarns while linking the weft layers in the weaving process more than stuffer warp yarn. However, unlike the binding yarn, the maximum stress and tensile strain at failure of stuffer yarn show a similar trend in all samples. Both stress and strain at failure were higher for variant D, followed by variants B and C.

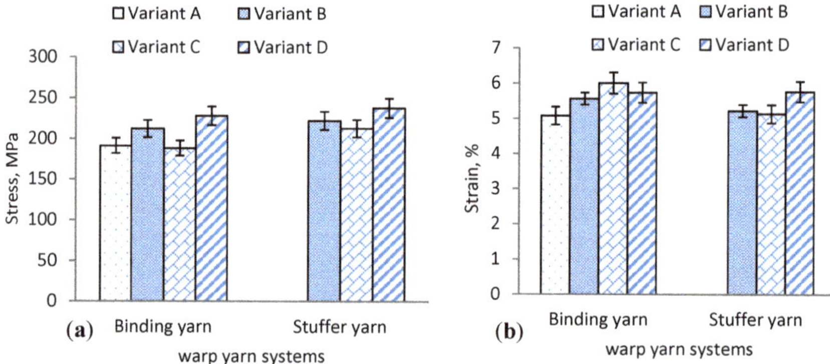

Figure 7. (**a**) Average maximum tensile stress and (**b**) tensile failure strain of tested sample for binding and stuffer warp yarns for different fabrics.

3.1.2. Weft Layer Yarn Testing

The tensile behaviors of weft yarn in the 3D warp interlock fabrics can be influenced by various parameters. In Figure 8a–d, the tensile stress–strain behaviors of weft layer yarns of 3D warp interlock fabrics are shown to generally have more or less a similar trend. However, both the binding and stuffer warp yarn interchange ratio and the location of weft yarn in the weft layer can influence the tensile properties of the corresponding weft yarn. The fabric with the highest or lowest proportion of binding warp yarn exhibited no effect on the final weft yarn tensile properties. The tensile stress and strain at failure for weft layer yarns of the developed variants are summarized in Table 2.

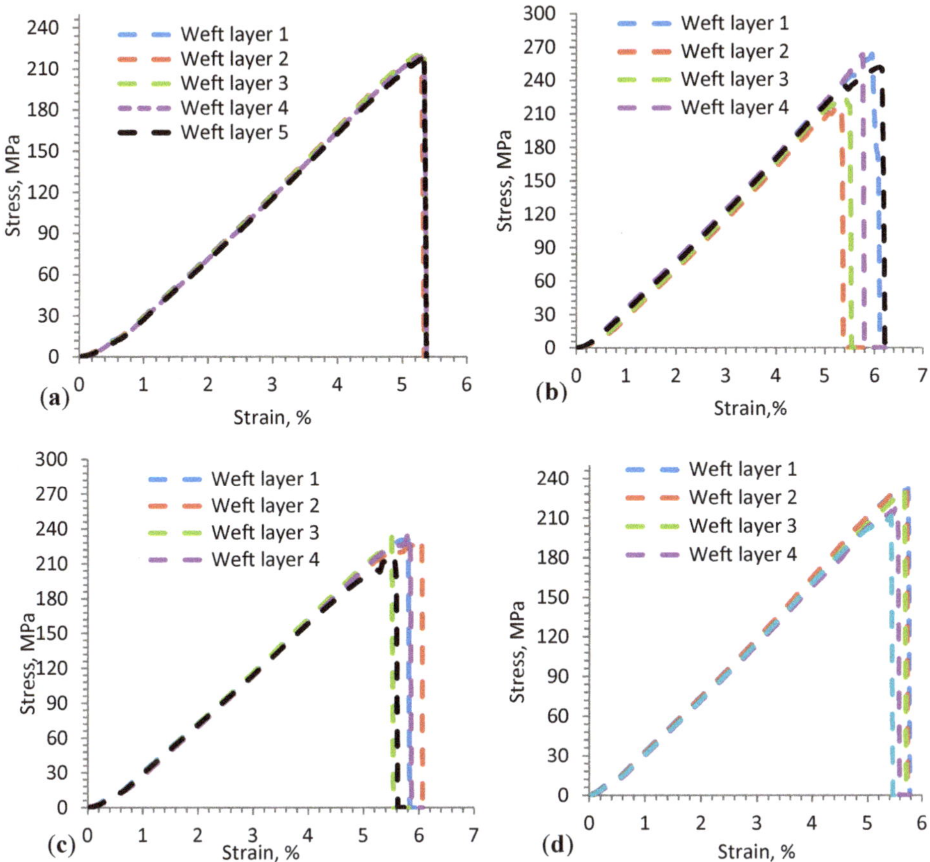

Figure 8. Typical stress–strain curves of layers 1–5 of weft yarn for (**a**) variant A, (**b**) variant B, (**c**) variant C, and (**d**) variant D 3D warp interlock fabric architecture.

Table 2. Tensile stress and strain at failure for weft layer yarns in developed variants.

Weft Layer	Variant A		Variant B		Variant C		Variant D	
	Tensile Stress, MPa	Strain at Failure, %	Tensile Stress, MPa	Strain at Failure, %	Tensile Stress, MPa	Strain at Failure, %	Tensile Stress, MPa	Strain at Failure, %
layer 1	216.5 ± 10.9	5.3 ± 0.6	263.5 ± 11.6	5.9 ± 0.5	231.5 ± 11.9	5.8 ± 0.6	233.1 ± 6.9	5.6 ± 0.7
layer 2	217.1 ± 9.9	5.3 ± 0.7	217.0 ± 9.2	5.3 ± 0.9	231.8 ± 10.2	6.0 ± 0.9	231.9 ± 8.9	5.5 ± 0.8
layer 3	224.8 ± 11.2	5.3 ± 0.6	225.9 ± 10.1	5.4 ± 0.9	233.3 ± 12.4	5.5 ± 0.7	231.7 ± 11.1	5.7 ± 1.1
layer 4	211.1 ± 8.9	5.4 ± 0.68	263.0 ± 10.59	5.8 ± 0.6	237.6 ± 10.4	5.8 ± 1.0	218.2 ± 10.9	5.5 ± 1.0
layer 5	214.2 ± 7.2	5.5 ± 0.8	251.6 ± 9.8	6.1 ± 0.9	217.6 ± 8.9	5.5 ± 0.8	211.7 ± 6.9	5.4 ± 0.7

For example, variants A and D show approximately similar tensile strength (E) of weft yarn at failure within the respective layers as compared to variants C and B. Moreover, even though the positions of weft yarn in the fabric layer affects its tensile properties, clear trends were not observed for the fabric variants.

Moreover, based on the stress–strain curves, the mechanical properties of the 3D warp interlock fabric samples were characterized based on their unique tensile behavior of stress (MPa) and corresponding strain (%) at the breaking point. Figure 9a,b shows the average maximum stress (MPa) and strain (%) values at the fracture point for the samples in the warp and weft directions. These values were found to be more or less similar in variants A and D compared to variants B and C. Weft yarn in layer 1 shows maximum load and tensile strain (%) in variant B as compared to the majority of weft layers, whereas weft yarn in layer 5 shows lower maximum tensile strain (%) than other weft layers.

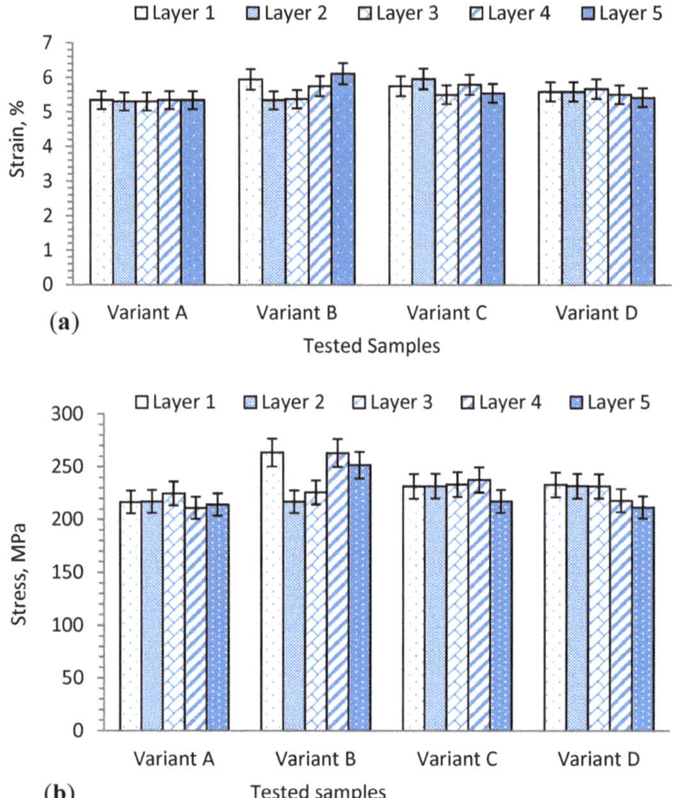

Figure 9. (**a**) Average maximum tensile failure strain and (**b**) maximum tensile failure stress of tested sample for weft yarns in different layers.

3.2. Fabric Uniaxial Tensile Property

Uniaxial tensile stress vs. tensile strain of the 3D warp interlock orthogonal layer-to-layer fabrics made with different warp yarn ratios was experimentally examined in the warp and weft directions. In this section, these test results will be discussed. The stress (MPa) vs. strain (%) curves of the three tensile tests for 3D warp interlock fabric samples are described in Figure 10a,b. The test was performed until the sample reached the maximum deformation state and failed. In general, the stress–strain curve of the samples indicates a similar progression, where the tensile stress values become higher as the strain value increases. The stress–strain curve comprises three main parts: the first section shows a nonlinear curve (crimp area), the second section shows higher stress vs. strain values (elongation area), and the last section shows declining stress–strain values. In the first section, the crimping area at the beginning of the curve shows almost negligible values in the applied loading direction and deformation in the displacement direction at the beginning, and then gradually increases. This is mainly due to the straightening of yarn in the fabric before real deformation occurs along the load direction. When it is closely observed, the crimp area (marked with a black circle) was found to be similar for all samples in the weft direction.

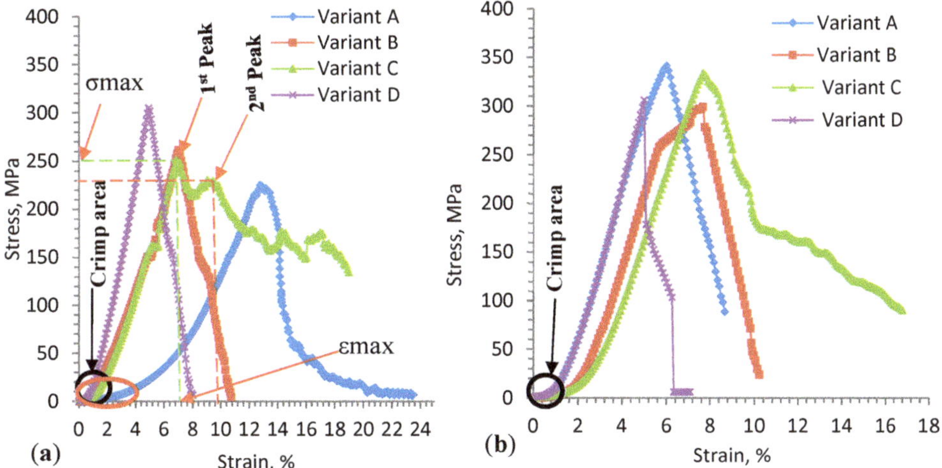

Figure 10. Typical stress–strain curves of 3D warp interlock fabric architectures in (**a**) warp (machine) direction, and (**b**) weft (cross) direction.

However, due to the different warp yarn (binding and stuffer yarn) composition in the various fabrics, all samples showed different crimp areas in the warp direction (black and red). For example, variant A shows more straightening area (marked with a red circle) as compared to the other samples in the warp direction, whereas variant D shows a higher crimp area, followed by variants B and C. The maximum strain absorbed in the crimp area for weft and warp directions was approximately 1.13% and 1.72%, respectively.

The second section of the stress–strain curve (Figure 10a,b) shows different linear progression with the rapid growth of tensile stress and strain until the sample reached its breaking point in warp and weft directions. The tensile stress vs. strain curve shows more or less a similar trend for all samples in the weft direction due to the balanced weft yarn composition in the fabric. Even though the curve is linearly progressive, the different warp yarn system leads to varying tensile stress and strain relationships in the warp direction. Most obviously, due to the balanced proportions of stuffer and binding warp yarns in the fabric, the typical stress–strain curve of variant C displays two main peak points in the warp direction (Figure 10a). The dominant first and second failure peaks were due to the failure of stuffer and binding warp yarns, respectively. Such condition arises mainly from the actual length difference between stuffer and binding warp yarns in the fabric while loading. On the contrary, different peak points cannot be seen for the other samples with either the same or one dominant warp yarn (stuffer or binding) system in the warp direction. For instance, variant A (100% binding warp yarn) shows a gentler progressive slope and single tensile failure peak with lower tensile modulus (E) in the warp direction, as shown in Figure 10a, whereas variant D (66.7% stuffer warp yarn) demonstrates a steeper linear progressive curve with a single peak and higher tensile modulus (E). The higher tensile modulus value of variant D is due to its higher stiffness behavior, which comes from the higher proportion of stuffer warp yarn in the fabric in the warp direction, whereas the lower value for variant A is due to the higher waviness of the binding warp yarn in the fabric, which needs more load to deform the sample. Variants B and C lie in between and show similar linear progressive slope with approximately equal tensile modulus (E) until the breaking peak point in the warp direction (Figure 10a).

Unlike the warp direction, the stress vs. strain of fabrics in the weft direction show almost linear and elastic progressive curves with approximately similar trends (Figure 10b). This might be because they have the same weft yarn composition and density among all fabric in the weft direction. However, a pressing effect of the warp yarn on the weft yarn during the weaving process could affect the slope of the curve and tensile strength values. Variants A and C show approximately similar tensile stress (E), with higher values of 341.40 ± 10 MPa and 333.82 ± 12 MPa respectively, as compared to variants B (297.31 ± 9.8 MPa) and D (306.10 ± 8.2 MPa). The tensile stress and strain at failure for the variants are summarized in Table 3. Here, the value of each sample is the average of three samples tested. This is due to the presence of higher weft yarn undulations in variants A and D, which can absorb more exerting forces in the loading direction. Variant B also recorded better tensile modulus (E) as compared to variant C. Even though the 3D warp interlock fabrics have the same structure and areal density, they show different maximum tensile strain (εmax) and stress (σmax) at failure. Figure 11 shows the fabric samples' tensile strain and strength at failure in the warp and weft direction.

Table 3. Tensile stress and strain at failure for variants.

Sample	Warp Direction		Weft Direction	
	Tensile Stress, MPa	Strain at Failure, %	Tensile Stress, MPa	Strain at Failure, %
Variant A	224.80 ± 11.2	12.88 ± 1.1	341.40 ± 10	5.88 ± 0.68
Variant B	261.34 ± 12.3	6.97 ± 0.67	297.31 ± 9.8	7.38 ± 0.9
Variant C	250.84 ± 9.8	8.97 ± 0.7	333.82 ± 12	7.55 ± 0.69
Variant D	304.94 ± 10.9	4.89 ± 0.64	306.10 ± 8.2	4.88 ± 0.76

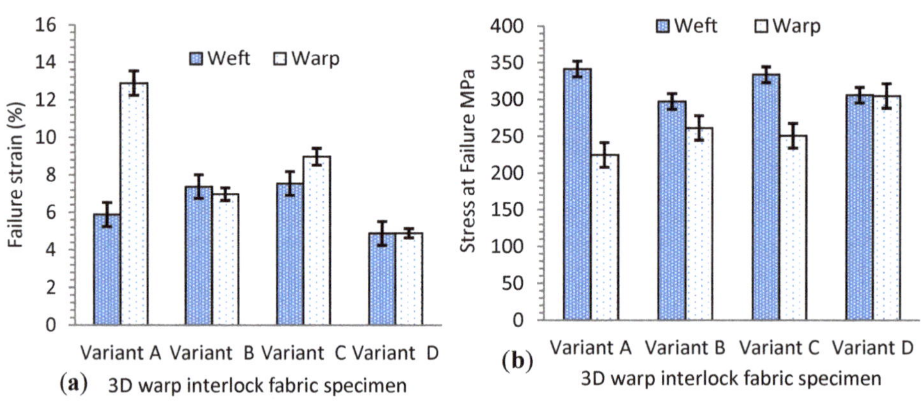

Figure 11. (a) Average maximum tensile strain (εmax), and (b) tensile stress (σmax) at failure for 3D warp interlock fabric samples in warp and weft directions.

Considering the weft direction, the fabric samples did not show a significant difference in maximum tensile strain (εmax) and stress (σmax) at failure, as shown in Figure 11a,b. This is because they had the same weft yarn composition in their 3D architecture. The slight difference arises from the stressing effects of warp yarn while interlacing to form the fabric. For example, variants A and B failed at tensile strain values around 5.88% ± 0.68% and 7.38% ± 0.9% when the stress was applied along the weft direction, whereas variants C and D failed at strain values of 7.55% ± 0.69% and 4.88% ± 0.76%. However, due to the different warp yarn proportions in the fabric architecture, the maximum tensile strain (εmax) and stress (σmax) at failure were found to be different in the warp direction in the 3D warp interlock fabric samples. For example, variant A (100% binder warp yarn) showed higher tensile strain at failure in the warp direction (12.88% ± 1.1%) than the weft direction (5.88% ± 0.68%) due to the higher undulating property of binder warp yarn in the fabric architecture. On the contrary, variant D (66.7% stuffer, 33.3% binding warp yarn) showed nearly equal tensile strain at failure in both the warp and weft direction (4.89% ± 0.64% and 4.88% ± 0.76%). This is because both directions have approximately equal undulation of warp and weft yarns in the fabric structure. Similarly, the tensile stress at failure for the fabric samples was also affected by the warp yarn composition in the fabric structure. For example, the tensile responses of variant D were found to be equal in the warp and weft directions due to the approximately similar undulation behavior of yarn in both directions, whereas variants A, B, and C showed different tensile stress at failure in the warp and weft directions due to their various binding warp yarn composition (Figure 11b).

In addition, tensile fabric damage to the samples after each test was also observed and analyzed. Figure 12 shows pictures after testing 3D warp interlock fabric samples at tensile failure state in the warp and weft direction.

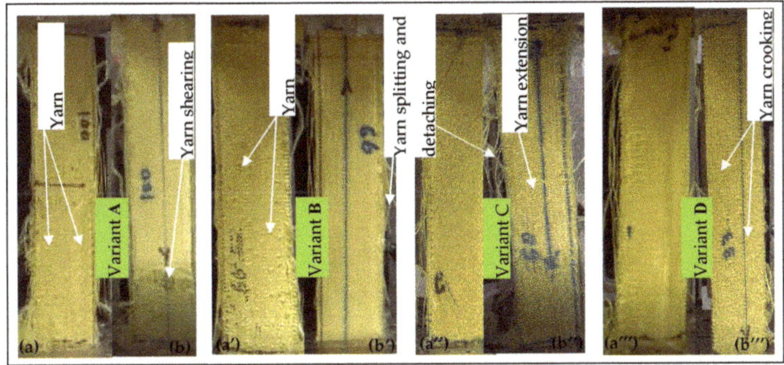

Figure 12. Photographs of 3D warp interlock fabrics at tensile failure in the (**a,a′,a″,a‴**) weft and (**b,b′,b″,b‴**) warp direction.

The damage mechanism of the fabrics was observed at the tensile deformational failure state based on the macroscopic level. The tensile fabric damage of all samples in the weft direction showed a similar trend, mainly from crooking of weft yarns, as shown in Figure 12a,a′,a″,a‴. The tensile failure mechanism in the warp direction was different due to the different warp yarn system in each fabric.

For example, variant A exhibited shearing of yarns throughout the sample width near the gripping end until failure. Variants B and C showed similar tensile failure with yarn straightening and extension in the warp direction. Variant D showed similar tensile damage to the other samples in the weft direction. This is because variant D comprised more stuffer than binding warp yarn in the warp direction, which brings similar yarn orientation in the fabric to weft yarn. Regardless of yarn ratio in the fabric structure and the testing direction (warp or weft), all samples showed some yarn splitting from the edges followed by separation along the fabric length. Moreover, higher weft yarn undulation (waviness) due to the involvement of more binding warp yarn in the fabric enhances the tensile failure stress in the respective direction. Such yarn crimp during interlacing in the weaving process could bring extra inter-yarn friction between the yarns and enhance the tensile stress at failure. For instance, the tensile stress of variant A (341.40 ± 10 MPa) and variant D (304.94 ± 10.9 MPa) was found to be higher compared to other fabric samples in the weft and warp direction. Variant B (261.34 ± 12.3 MPa, 297.31 ± 9.8 MPa) and variant C (250.84 ± 9.8 MPa, 333.82 ± 12 MPa) showed different tensile stress at failure in the warp and weft direction due to their higher proportion of binding warp yarn. Based on these observations, the warp yarn composition in the warp directions can influence 3D warp interlock fabric's maximum tensile strength and strain in the warp direction, as well as in the weft direction. For example, variant D, made with more stuffer warp yarn, showed approximately the same maximum tensile stress at failure in the weft and warp directions. Variant A, with a higher proportion of binding warp yarn (100%), showed a great difference in maximum tensile stress and strain in the warp and weft directions.

3.3. Fabric Flexural Rigidity Behavior

Apart from tensile and other mechanical properties, it is also important to investigate flexibility properties, which determine drape comfort and handling of textile materials in various applications. As explained earlier, various factors can affect fabric's bending behaviors, such as material type, fabric structure, areal density, fabric size, etc. In this section, the influence of warp yarn composition in 3D warp interlock p-aramid fabric on its bending behaviors is discussed.

Except for variant A in the warp direction, due to their higher flexural rigidity properties, the other samples were examined for their bending curvature angle (θ) at specific bending length (C). Based on the obtained bending angle and bending length (Figure 5), it is possible to compute the flexural rigidity of each sample using Equation (1). Figure 13a shows the average flexural rigidity values of 3D warp interlock fabrics with warp yarn type composition in the warp and weft directions. It can be seen that fabric with higher stuffer warp yarn composition shows the highest flexural rigidity as compared to fabric with low or no stuffer warp yarn in the warp direction. For example, variant D, made with higher stuffer warp yarn composition, had the highest specific flexural rigidity, followed by variants C and B, whereas variant A, which comprises only binding warp yarn, shows the lowest flexural rigidity. Here, two things are observed and outlined. First, a higher interchange ratio of stuffer yarn greatly affects the bending stiffness of fabric in the longitudinal direction.

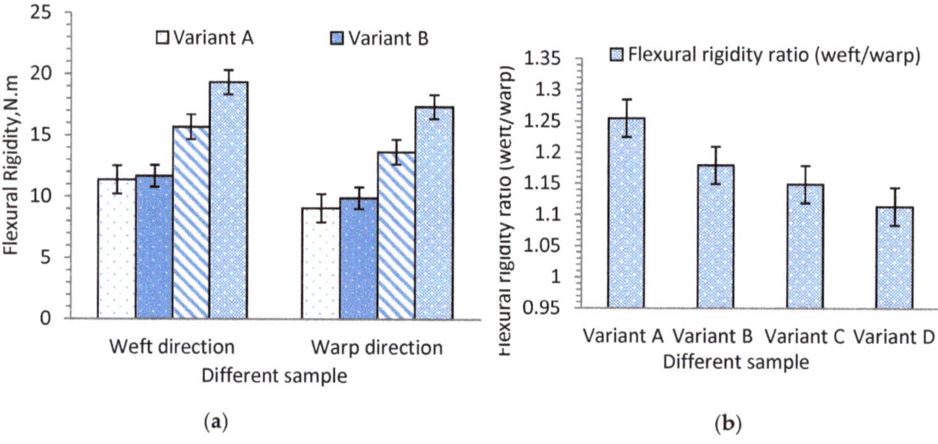

(a)   (b)

Figure 13. (**a**) Flexural rigidity value of 3D warp interlock fabric samples in weft and warp directions and (**b**) flexural rigidity ratio (weft/warp).

For example, it is clearly shown that the fabric sample with 66.7% stuffer warp yarn is about twice as stiff as the samples with no stuffer yarn in the warp direction. It is expected that this trend will be significant and will be reduced for fabrics with less stuffer yarn in their warp composition. The flexural rigidity in the weft direction is also slightly influenced by the warp yarn composition. Even though the weft density and composition were the same for all fabric structures, those in the weft direction were found to be slightly higher than those in the warp direction. This is because weft yarns are generally straighter than warp yarns, which makes a significant difference in bending stiffness. Figure 13b shows the flexural rigidity ratio of each fabric in the weft and warp directions. For example, the flexural rigidity of variant D in the weft direction was found to be higher than that in the warp direction. It can be seen that further reducing the proportion of stuffer yarn in the warp direction helps to reduce the stiffness and results in much lower stiffness in the weft direction. For example, variant A, with no stuffer yarn, shows much less stiffness in the warp direction than the weft direction. Without considering other mechanical properties, increasing the proportion of stuffer yarn in the warp yarn composition of 3D warp interlock fabric is an effective way to obtain higher bending stiffness properties in the warp and weft directions. It is of great interest to compare the two 3D warp interlock fabrics, one with no stuffer yarn and the other with 66.7% stuffer warp yarn. Ideally, the bending stiffness for both should be the same in the weft direction because they both have the same type of yarn, yarn density, and yarn arrangements. However, the difference is due to the orthogonal characteristics of the warp yarn composition, which influences the arrangement of the weft yarn. Thus, fabrics with more stuffer warp yarn have less stress on the weft yarn, which sustains the bending stiffness ability.

4. Conclusions

The main aim of the research was to explore and understand the effects of the warp yarn interchange ratio inside the 3D warp interlock p-aramid architectures on its mechanical behavior. For this, three-dimensional (3D) warp interlock p-aramid fabrics with different warp yarn ratios (variant A: 100% binding, 0% stuffer, variant B: 66.7% binding, 33.3% stuffer, variant C: 50% binding, 50% stuffer, and variant D: 33.3% binding, 66.7% stuffer) were fabricated and characterized under quasi-static conditions. Based on the tensile test results, fabric with more stuffer yarn (variant D) revealed higher tensile stress (E; 304.9 MPa) than fabric with less stuffer yarn (variant A, 224.8 MPa, variant B, 261.3 MPa, and variant C, 250.8 MPa). In addition, variant C, with a balanced warp yarn ratio, exhibited two tensile failure points due to the length difference between the two warp yarns. On the contrary, the stress–strain curve in the weft direction showed a linear and progressive trend because the samples had the same weft yarn ratios. Unlike weft direction, variants A (341.4 MPa) and C (333.82 MPa) showed approximately similar tensile stress, with higher values as compared to variants B (297.2 MPa) and C (306.1 MPa) in the warp direction. This is due to the loading effect of warp yarn on weft yarn during the weaving process which showed an influence on tensile strength. Besides, the warp yarn ratio in each variant also affected the maximum tensile strain (εmax) and stress (σmax) at failure in the warp direction. Fabric with a higher binding warp yarn ratio showed higher tensile strain at failure due to the higher crimp properties of the binding warp yarn in the fabric architecture. Variant A showed higher tensile failure strain in the warp (12.88%) than the weft (5.88%) direction, whereas variant D showed approximately similar tensile failure strain in both directions (4.88%) due to the similar undulation of warp and weft yarns. Unlike the warp direction, variants A and B failed at a maximum tensile strain (εmax) of around 5.88% and 7.8% and variables C and D failed at 7.55% and 4.8% respectively, in the weft direction. In addition, variant D (with more stuffer warp yarn) had the highest flexural rigidity (17.36 N.m) compared to fabric with little or no stuffer warp yarn (variant C, 13.66 N.m, variant B, 9.88 N.m, and variant A, 9.05 N.m) in the warp direction. Besides, the flexural rigidity in the weft direction was influenced by the warp yarn ratio and was higher than the respective warp direction due to the smaller waviness properties compared to warp yarns. For example, the fabric with 66.7% stuffer and 33% binding warp yarn (variant D) was about twice as stiff as the ones with no stuffer warp yarn (variant A) in the warp direction.

Author Contributions: Conceptualization, M.A.A.; methodology, M.A.A. and F.B.; software, M.A.A. and F.B.; validation, M.A.A., F.B., and P.B.; formal analysis, M.A.A.; investigation, M.A.A.; resources, F.B. and P.B.; data curation, M.A.A., F.B., P.B., and H.L.; writing—original draft preparation, M.A.A.; writing—review and editing, M.A.A., F.B., P.B., and H.L.; visualization, M.A.A., F.B., P.B., and H.L.; supervision, F.B. and P.B.; project administration, P.B. and F.B.; funding acquisition, P.B., F.B., and H.L. All authors have read and agreed to the published version of the manuscript.

Funding: This research was funded by the European Erasmus Mundus Program as part of the Sustainable Management and Design for Textiles (SMDTex) Project, project number SMDTex-2016-6, and the APC was funded by National Natural Science Foundation of China (Grant Number: 61906129), China Postdoctoral Science Foundation (Grant Number: 2019M661929), Jiangsu Province Postdoctoral Science Foundation (Grant Number 2019Z285), and the Scientific Research Project of Fiber Materials and Products for Emergency Protection and Public Safety supported by the National Advanced Functional Fiber Innovation Center (Grant Number: 2020-fx010036) and Third-Phase Priority Academic Program Development of Jiangsu Higher Education Institutions [PAPD].

Materials **2020**, *13*, 4233

Acknowledgments: The authors would like to thank Lucas Putigny, Nicola Dumont, and Fredrick Veyet at ENSAIT-GEMTEX Lab for their technical help during the manufacturing of the fabric structure.

Conflicts of Interest: The authors declare no conflict of interest. The funders had no role in the design of the study; in the collection, analyses, or interpretation of data; in the writing of the manuscript, or in the decision to publish the results.

References

1. Moure, M.; Feito, N.; Aranda-Ruiz, J.; Loya, J.; Rodriguez-Millán, M. On the characterization and modelling of high-performance para-aramid fabrics. *Compos. Struct.* **2019**, *212*, 326–337. [CrossRef]
2. Abtew, M.A.; Boussu, F.; Bruniaux, P.; Loghin, C.; Cristian, I.; Chen, Y.; Wang, L. Forming characteristics and surface damages of stitched multi-layered para-aramid fabrics with various stitching parameters for soft body armour design. *Compos. Part A Appl. Sci. Manuf.* **2018**, *109*, 517–537. [CrossRef]
3. Abtew, M.A.; Boussu, F.; Bruniaux, P.; Loghin, C.; Cristian, I.; Chen, Y.; Wang, L. Ballistic impact performance and surface failure mechanisms of two-dimensional and three-dimensional woven p-aramid multi-layer fabrics for lightweight women ballistic vest applications. *J. Ind. Text.* **2019**, 1–3. [CrossRef]
4. Abtew, M.A.; Boussu, F.; Bruniaux, P.; Loghin, C.; Cristian, I. Engineering of 3D warp interlock p-aramid fabric structure and its energy absorption capabilities against ballistic impact for body armour applications. *Compos. Struct.* **2019**, *225*, 111179. [CrossRef]
5. Bilisik, K.; Karaduman, N.S.; Bilisik, N.E.; Bilisik, H.E. Three-dimensional fully interlaced woven preforms for composites. *Text. Res. J.* **2013**, *83*, 2060–2084. [CrossRef]
6. Khokar, N. 3D-Weaving: Theory and Practice. *J. Text. Inst.* **2001**, *92*, 193–207. [CrossRef]
7. Abtew, M.A.; Boussu, F.; Bruniaux, P.; Loghin, C.; Cristian, I. Enhancing the Ballistic Performances of 3D Warp Interlock Fabric Through Internal Structure as New Material for Seamless Female Soft Body Armor Development. *Appl. Sci.* **2020**, *10*, 4873. [CrossRef]
8. Abtew, M.A.; Bruniaux, P.; Boussu, F.; Loghin, C.; Cristian, I.; Chen, Y.; Wang, L. A systematic pattern generation system for manufacturing customized seamless multi-layer female soft body armour through dome-formation (moulding) techniques using 3D warp interlock fabrics. *J. Manuf. Syst.* **2018**, *49*, 61–74. [CrossRef]
9. Chen, X.; Lo, W.-Y.; Tayyar, A.; Day, R. Mouldability of Angle-Interlock Woven Fabrics for Technical Applications. *Text. Res. J.* **2002**, *72*, 195–200. [CrossRef]
10. Boussu, F.; Legrand, X.; Nauman, S.; Binetruy, C. Mouldability of angle interlock fabrics. In Proceedings of the FPCM-9, the 9th International Conference on Flow Processes in Composite Materials, Montréal, QC, Canada, 8–10 July 2008.
11. Dufour, C.; Boussu, F.; Wang, P.; Soulat, D.; Ghys, P. Forming behaviour of 3D warp interlock fabric to produce tubular cross composite part. In Proceedings of the 14th AUTEX World Textile Conference, Bursa, Turkey, 26–28 May 2014.
12. Corbin, A.-C.; Boussu, F.; Ferreira, M.; Soulat, D. Influence of 3D warp interlock fabrics parameters made with flax rovings on their final mechanical behaviour. *J. Ind. Text.* **2018**, *49*, 1123–1144. [CrossRef]
13. Chen, X.; Spola, M.; Paya, J.G.; Sellabona1, P.M. Experimental Studies on the Structure and Mechanical Properties of Multi-layer and Angle-interlock Woven Structures. *J. Text. Inst.* **1999**, *90*, 91–99. [CrossRef]
14. Abtew, M.A.; Boussu, F.; Bruniaux, P.; Loghin, C.; Cristian, I.; Chen, Y.; Wang, L. Influences of fabric density on mechanical and moulding behaviours of 3D warp interlock para-aramid fabrics for soft body armour application. *Compos. Struct.* **2018**, *204*, 402–418. [CrossRef]
15. Chen, X.; Zanini, I. An Experimental Investigation into the Structure and Mechanical Properties of 3D Woven Orthogonal Structures. *J. Text. Inst.* **1997**, *88*, 449–464. [CrossRef]
16. Gu, H.; Zhili, Z. Tensile behavior of 3D woven composites by using different fabric structures. *Mater. Des.* **2002**, *23*, 671–674. [CrossRef]
17. Wang, Y.; Zhao, D. Effect of Fabric Structures on the Mechanical Properties of 3-D Textile Composites. *J. Ind. Text.* **2006**, *35*, 239–256. [CrossRef]
18. Behera, B.; Dash, B. Mechanical behavior of 3D woven composites. *Mater. Des.* **2015**, *67*, 261–271. [CrossRef]
19. Dai, S.; Cunningham, P.; Marshall, S.; Silva, C. Influence of fibre architecture on the tensile, compressive and flexural behaviour of 3D woven composites. *Compos. Part A Appl. Sci. Manuf.* **2015**, *69*, 195–207. [CrossRef]

20. Umer, R.; Alhussein, H.; Zhou, J.; Cantwell, W. The mechanical properties of 3D woven composites. *J. Compos. Mater.* **2016**, *51*, 1703–1716. [CrossRef]

21. Sheng, S.Z.; Van Hoa, S. Modeling of 3D Angle Interlock Woven Fabric Composites. *J. Thermoplast. Compos. Mater.* **2003**, *16*, 45–58. [CrossRef]

22. Hu, J. *3D Fibrous Assemblies, Properties Applications and Modelling of Three Dimensional Textile Structure*; Woodhead Publishing in Textiles: Cambridge, UK, 2008.

23. Corbin, A.-C.; Kececi, A.; Boussu, F.; Ferreira, M.; Soulat, D. Engineering Design and Mechanical Property Characterisation of 3D Warp Interlock Woven Fabrics. *Appl. Compos. Mater.* **2018**, *25*, 811–822. [CrossRef]

24. Isa, M.; Ahmed, A.; Aderemi, B.; Taib, R.; Mohammed-Dabo, I. Effect of fiber type and combinations on the mechanical, physical and thermal stability properties of polyester hybrid composites. *Compos. Part B Eng.* **2013**, *52*, 217–223. [CrossRef]

25. Bandaru, A.K.; Patel, S.; Sachan, Y.; Ahmad, S.; Alagirusamy, R.; Bhatnagar, N. Mechanical characterization of 3D angle-interlock Kevlar/basalt reinforced polypropylene composites. *Polym. Test.* **2016**, *55*, 238–246. [CrossRef]

26. Bandaru, A.K.; Sachan, Y.; Ahmad, S.; Alagirusamy, R.; Bhatnagar, N. On the mechanical response of 2D plain woven and 3D angle-interlock fabrics. *Compos. Part B Eng.* **2017**, *118*, 135–148. [CrossRef]

27. Boussu, F.; Picard, S.; Soulat, D. Interesting mechanical properties of 3D warp interlock fabrics. In *Narrow and Smart Textiles*; Springer International Publishing: Cham, Switzerland, 2018; pp. 21–31.

28. Dash, A.K.; Behera, B.K. Role of stuffer layers and fibre volume fractions on the mechanical properties of 3D woven fabrics for structural composites applications. *J. Text. Inst.* **2018**, *110*, 614–624. [CrossRef]

29. Abtew, M.A.; Boussu, F.; Bruniaux, P.; Loghin, C.; Cristian, I.; Chen, Y.; Wang, L. Yarn degradation during weaving process and its effect on the mechanical behaviours of 3D warp interlock p-aramid fabric for industrial applications. *J. Ind. Text.* **2020**, 1–24. [CrossRef]

30. Dahale, M.; Neale, G.; Lupicini, R.; Cascone, L.; McGarrigle, C.; Kelly, J.; Archer, E.; Harkin-Jones, E.; McIlhagger, A. Effect of weave parameters on the mechanical properties of 3D woven glass composites. *Compos. Struct.* **2019**, *223*, 110947. [CrossRef]

31. Chou, S.; Chen, H.-C.; Chen, H.-E. Effect of weave structure on mechanical fracture behavior of three-dimensional carbon fiber fabric reinforced epoxy resin composites. *Compos. Sci. Technol.* **1992**, *45*, 23–35. [CrossRef]

32. Nasrun, F.M.Z.; Yahya, M.F.; Ghani, S.A.; Ahmad, M.R. Effect of weft density and yarn crimps towards tensile strength of 3D angle interlock woven fabric. *AIP Conf. Proc.* **2016**, *1774*, 020003.

33. Rao, M.; Sankar, B.; Subhash, G. Effect of Z-yarns on the stiffness and strength of three-dimensional woven composites. *Compos. Part B Eng.* **2009**, *40*, 540–551. [CrossRef]

34. Umair, M.; Hamdani, S.T.A.; Asghar, M.A.; Hussain, T.; Karahan, M.; Nawab, Y.; Ali, M. Study of influence of interlocking patterns on the mechanical performance of 3D multilayer woven composites. *J. Reinf. Plast. Compos.* **2018**, *37*, 429–440. [CrossRef]

35. Labanieh, A.R.; Liu, Y.; Vasiukov, D.; Soulat, D.; Panier, S. Influence of off-axis in-plane yarns on the mechanical properties of 3D composites. *Compos. Part A Appl. Sci. Manuf.* **2017**, *98*, 45–57. [CrossRef]

36. Zhang, D.; Sun, M.; Liu, X.; Xiao, X.; Qian, K. Off-axis bending behaviors and failure characterization of 3D woven composites. *Compos. Struct.* **2019**, *208*, 45–55. [CrossRef]

37. Boussu, F.; Cristian, I.; Nauman, S. General definition of 3D warp interlock fabric architecture. *Compos. Part B Eng.* **2015**, *81*, 171–188. [CrossRef]

38. Cox, B.; Dadkhah, M.; Morris, W.; Flintoff, J. Failure mechanisms of 3D woven composites in tension, compression, and bending. *Acta Met. Mater.* **1994**, *42*, 3967–3984. [CrossRef]

39. AFNOR.EN NF ISO 5084. Textiles–Determination of Thickness of Textiles and Textile Products; ISO TC 38: 1996. pp. 1–5. Available online: https://www.iso.org/standard/23348.html (accessed on 18 September 2020).

40. AFNOR.EN N 12127. Textiles–Fabrics-Determination of Mass Per Unit Area Using Small Samples. 1998, pp. 1–10. Available online: https://www.en-standard.eu/bs-en-12127-1998-textiles-fabrics-determination-of-mass-per-unit-area-using-small-samples/ (accessed on 18 September 2020).

41. Roylance, D. Ballistics of Transversely Impacted Fibers. *Text. Res. J.* **1977**, *47*, 679–684. [CrossRef]

42. Chu, T.-L.; Ha-Minh, C.; Imad, A. A numerical investigation of the influence of yarn mechanical and physical properties on the ballistic impact behavior of a Kevlar KM2® woven fabric. *Compos. Part B Eng.* **2016**, *95*, 144–154. [CrossRef]

43. Cheeseman, B.A.; Bogetti, T.A. Ballistic impact into fabric and compliant composite laminates. *Compos. Struct.* **2003**, *61*, 161–173. [CrossRef]
44. AFNOR. NF EN ISO 13934-1. Tensile Properties of Fabrics, Part 1: Determination of Maximum Force and Elongation at Maximum Force Using the Strip Method. 2013, p. 10. Available online: https://www.iso.org/standard/60676.html (accessed on 18 September 2020).

Article

Characteristics of Heat Resistant Aluminum Alloy Composite Core Conductor Used in overhead Power Transmission Lines

Kun Qiao [1,*], **Anping Zhu** [2], **Baoming Wang** [3], **Chengrui Di** [1], **Junwei Yu** [2,3] **and Bo Zhu** [2,3,*]

1   School of Mechanical, Electrical & Information Engineering, Shandong University, Weihai 264209, China; dicr0918@sdu.edu.cn
2   Carbon Fiber Engineering Research Center, Shandong University, Jinan 250061, China; zhuanping2019@163.com
3   School of Materials Science and Engineering, Shandong University, Jinan 250061, China; 201820371@mail.sdu.edu.cn (B.W.); junweiyu@mail.sdu.edu.cn (J.Y.)
*   Correspondence: qiaokun@sdu.edu.cn (K.Q.); zhubo@sdu.edu.cn (B.Z.)

Received: 7 February 2020; Accepted: 27 March 2020; Published: 31 March 2020

Abstract: The heat resistant aluminum alloy wire composite material core conductor (ACCC/HW) which was used in overhead transmission lines is developed and studied in this work. The composite material core is carbon fiber/glass cloth reinforced modified epoxy resin composite. Tensile stress tests and stress-strain tests of both composite core and conductor are taken at 25 °C and 160 °C. Sag test, creep test and current carrying capacity test of composite conductor are taken. The stress of composite conductor are 425.2 MPa and 366.9 MPa at 25 °C and 160 °C, respectively. The sag of conductor of 50 m length are 95 mm, 367 mm, and 371 mm at 25 °C, 110 °C, and 160 °C, respectively. The creep strain are 271 mm/km, 522 mm/km, and 867 mm/km after 10 years under the tension of 15% RTS (Rated Tensile Strength), 25% RTS and 35% RTS at 25 °C, and 628 mm/km under 25% RTS at 160 °C, according to the test result and calculation. The carrying capacity of composite conductor is basically equivalent to ACSR (Aluminum Conductor Steel Reinforced). ACCC/HW is suitable in overhead transmission lines, and it has been used in 50 kV power grid, according to the results.

Keywords: aluminum alloys; composite materials; epoxy resins; power cables; transmission lines

---

1. Introduction

In recent years, the electrical energy demand has significantly increased around the world [1]. Only in China, the total annual electricity consumption has grown from 4.2×10^{12} kWh to 6.3×10^{12} kWh between 2010 and 2017. However, the existing overhead power grids reached the transmission limit. There are two solutions to this problem, which are building new transmission lines or adopting large-capacity high-temperature low-sag (HTLS) conductors. At the same time, the most commonly used conductor ACSR, which has been studied and applied for more than one-hundred years [2–6], cannot meet the requirement. Accordingly, when considering land resources and economic benefits, it is necessary to develop new-style overhead conductors, which will increase the transmission capacity and not change the current power transmission distribution.

Several new kinds of conductor with composite materials have been developed to solve the above problem, such as ACCR (Aluminum Conductor ceramic fiber reinforced aluminum matrix Composite core Reinforced) [7,8] and ACCC (Aluminum Conductor Composite Core) [9–12]. Table 1 lists the structures and basic mechanical properties of ACSR, ACCR, and ACCC.

These conductors are composed of inner supporting cores and outer wires. The difference of the materials that were used in support cores is the key distinction between ACSR, ACCR, and ACCC.

By comparison, ACCC with higher specific strength, lighter density, and higher tensile strength is more suitable for the HTLS overhead conductor. Because carbon fiber reinforced composite materials have the characteristics of light weight, high strength, corrosion resistance, and mass production. Additionally, carbon fiber reinforced composite materials are used in the fields of aerospace, sporting goods, rail transportation, pressure vessels, and so on [13–15].

Table 1. Structures and basic mechanical properties of aluminum conductor steel reinforced (ACSR), (Aluminum Conductor ceramic fiber reinforced aluminum matrix Composite core Reinforced (ACCR), and Aluminum Conductor Composite Core (ACCC) [16].

Conductor Type	ACSR	ACCR	ACCC
Support core	Steel core	Ceramic fiber reinforced aluminum matrix core	Carbon fiber reinforced resin matrix core
Conductive wire	Aluminum	Aluminum alloy	Soft aluminum
Support core density (g/cm^3)	7.8	3.3	1.8
Support core tensile stress (MPa)	1300	1275	2400
Support core modulus (GPa)	200	216	130

ACCC reportedly exhibits high-strength, low-sag, large current carrying capacity, and high temperature resistance [11,12], which is composed of carbon/glass hybrid fiber reinforced epoxy resin composite and T-type soft aluminum wires, abbreviated to ACCC/TW (trapezoidal) [10]. The properties of ACCC/TW and its composite core have been studied for several years, such as mechanical properties [10], galvanic corrosion properties [11], excessive bending effect [12], and fatigue failure mechanism [17]. When compared to circular cross section heat resistant aluminum wires, the wires of this ACCC/TW conductor are made up of a softer aluminum alloy, and their trapezoidal shapes allows for them to be twisted more tightly. However, their stranding process is more difficult and more expensive.

The cable composed of composite core and heat resistant aluminum alloy wires, abbreviated to ACCC/HW, might have better mechanical properties, heat resistance, and large current carrying capacity when compared with ACSR, because ACSR employs steel core and aluminum alloy wires, with the similar size, ACCC/HW can take more aluminum alloy wires. However, the performance of ACCC/HW has not been publicly reported.

The characteristics of ACCC/HW are tested and studied in this paper, such as tensile strength, strain-stress behavior, sag, creep deformation, and current carrying capacity at room temperature or at high temperature. At the same time, this work compares the properties of ACCC/HW and ACSR.

2. Materials and Methods

2.1. Structure and Material

Figure 1 shows the conductor configuration of ACCC/HW in this work, and Figure 2 shows ACSR. ACCC/HW is composed of carbon fiber/glass fiber cloth reinforced modified epoxy resin composite core and two external layers of heat-resistant aluminum alloy wires. The difference between ACCC/HW and ACSR is the supporting core material. Table 2 lists the structural parameters of ACCC/HW and ACSR of the similar specification.

Presently, the publicly reported ACCC overhead conductor is ACCC/TW, as shown in Figure 3. When compared with ACCC/TW, there are two features of ACCC/HW. Firstly, the conductive wire is heat-resistant aluminum alloy wire with circular section, and the wire used in ACCC/TW is soft aluminum wire with non-circular section. Secondly, the outer layer material of carbon fiber composite core used in ACCC/HW is bi-directional glass fiber cloth reinforced composite, and the outer material of ACCC/TW core is the unidirectional glass fiber reinforced composite.

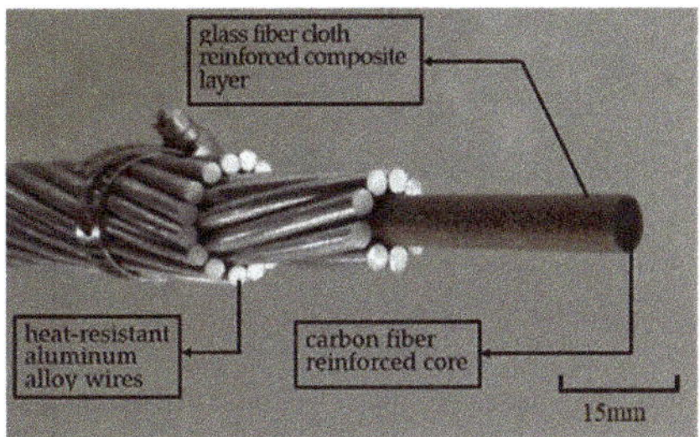

Figure 1. Conductor configuration of heat resistant aluminum alloy wire composite material core conductor (ACCC/HW).

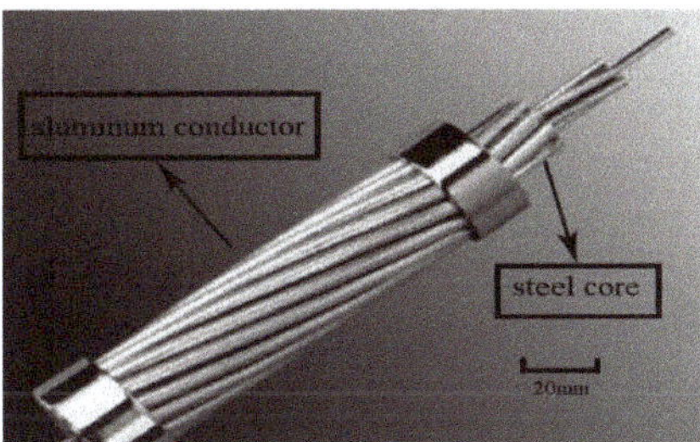

Figure 2. Conductor configuration of ACSR.

Table 2. Main structural parameters of ACCC/HW and ACSR of the similar specification.

	Properties	ACSR(LION)	ACCC/HW
Aluminum wire	Number	30	25
	Diameter (mm)	3.18	3.68 (inner layer) 3.43 (outer layer)
	Sectional area (mm^2)	238.3	243.0
Supported core	Material of supported core	Steel	Carbon fiber reinforced composite
	Number	7	1
	Diameter (mm)	9.54	7.46
	Sectional area (mm^2)	55.6	43.7
The conductor	Mass per unit length (kg/km)	1093.4	742.9
	Diameter (mm)	22.3	21.74
	Sectional area (mm^2)	293.9	286.7

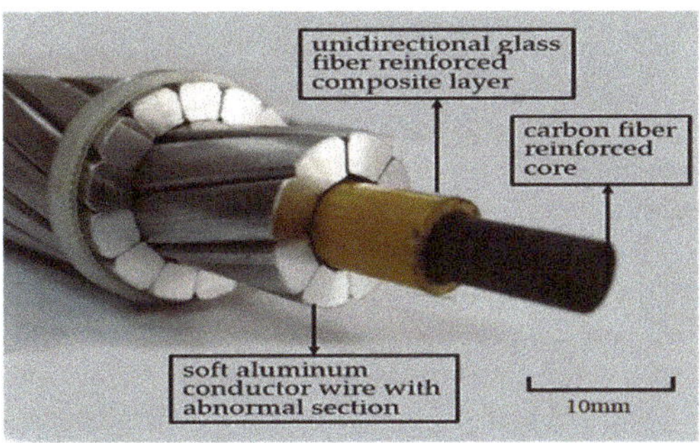

Figure 3. Conductor configuration of carbon/glass hybrid fiber reinforced epoxy resin composite and T-type soft aluminum wires (ACCC/TW).

The aluminum wire with non-circular section in ACCC/TW has certain radian, so the outer aluminum wires of ACCC/TW are stranded more closely. When comparing ACCC/HW with the same diameter, ACCC/TW contains more conductive aluminum area and it can conduct more electric quantity. However, the manufacturing process of soft aluminum wire with radian and non-circular section are more difficult, and the cost is relatively higher, especially in the case of ACCC/TW with small cross section. Therefore, with comprehensive consideration, ACCC/HW might be more suitable for the ACCC used in conductors of small size.

In this paper, the reinforcements of composite core are unidirectional carbon fiber and outer bi-directional glass fiber cloth. When compared with unidirectional reinforcements, the outermost bi-directional glass fabric can realize hoop enhancement and improve the anti-splitting performance and the anti-friction performance of the composite core.

The pultrusion process prepared the carbon fiber/glass fiber cloth reinforced epoxy composite core used in ACCC/HW. The carbon fiber is 12 K commercial type manufactured. The glass fiber cloth is woven by non-alkali high-strength glass fiber. The epoxy resin system is composed of modified multi-functional group epoxy resin, modified anhydride curing agent, and other assistants. The aluminum wire is a conventional heat-resistant aluminum alloy wire. The strength of aluminum wire after stranding is 166 MPa (internal layer) and 171 MPa (outer layer), respectively, and the direct current resistivity at 20 °C is 0.028147 $\Omega \cdot mm^2/m$.

2.2. Characterization

The ACCC/HW studied in this paper is only used in China now, so the test methods in this work are according to Chinese standard GB/T32502-2016 [18] "Overhead electrical stranded conductors composite core supported/reinforced". Additionally, Figure 1 shows the ACCC/HW studied in this work is based on the physical model.

The basic mechanical property tests of both the composite core and ACCC/HW were taken in this study. The basic mechanical properties concluded tensile strength at 25 °C and 160 °C, and tensile-strain behavior at 25 °C. Deformation resistant property analysis and current-carrying capacity test of ACCC/HW were also taken. The deformation resistant properties included the sag at heating up temperature and the creep performances at 25 °C and 160 °C.

2.2.1. Tensile Stress Test

It is necessary to use special fittings, which should not only be closely connected with the composite core and the aluminum wires to effectively transfer load because of the difference between ACCC/HW and ACSR in structure, but also avoid the damage of carbon fiber composite core in the installation and use process. ACSR mainly adopts the compression connection mode, and the external pressure tightly combines the aluminum wire and the steel core because of the plastic deformation of metal materials. However, if ACCC/HW is connected in this way, the structure of the carbon fiber composite core will be destroyed, and the mechanical properties will be deteriorated. Therefore, special fittings were used, which were composed of inner cone ring, outer cone ring, steel anchor, inner liner, and outer liner, as well as current inducing plater, as shown in Figure 4. Inner cone ring is at the outside of the composite core, and outer cone ring is at the outside of inner cone ring, to protect composite core from destruction by extrusion. Inner liner and outer liner contact with heat-resistant aluminum alloy wires.

The tensile test and the stress-strain test were carried out on 500 kN tension test machine, and the strength was the average of three test results. In room temperature tensile test at 25 °C, the effective lengths of ACCC/HW and the composite core were 10 m and 1 m, respectively. In the high temperature tensile test at 160 °C, the effective lengths of ACCC/HW and the composite core were 12 m and 1 m, respectively. After ACCC/HW was heated to 160 °C through the 50 kV high current heating system and preserved for three hours, the tensile test of ACCC/HW was carried out. After the composite core heated to 160 °C through special heating furnace and preserved for 401 hours, the test of composite core was carried out.

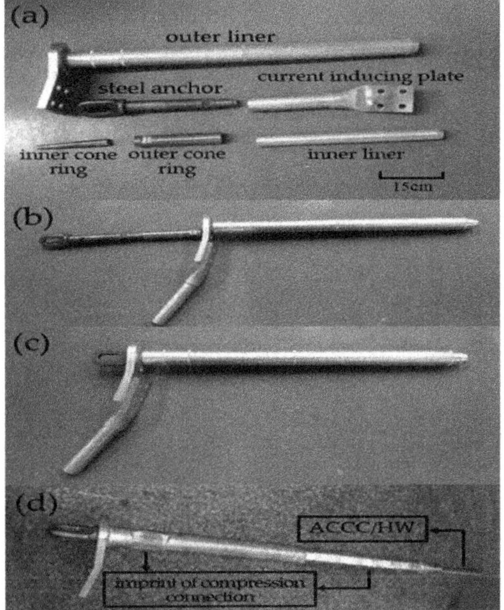

Figure 4. Special fittings used in this work. (**a**) detail sketch; (**b**) assembly process; (**c**) assembly process; (**d**) ready for use.

2.2.2. Stress-Strain Test

ACCC/HW was applied load with four loading- load keeping-unloading cycles to characterize the stress-strain performance. The effective length of the ACCC/HW was 12 m, and the length of extensometer was 2 m. Firstly, 2% RTS (Rated Tensile Strength) was exerted to straighten the conductor, and the RTS is 100kN. Figure 5 shows the loading procedure. The stress-strain test of the composite core was loaded in only one cycle, and the effective length of the composite core was 5 m.

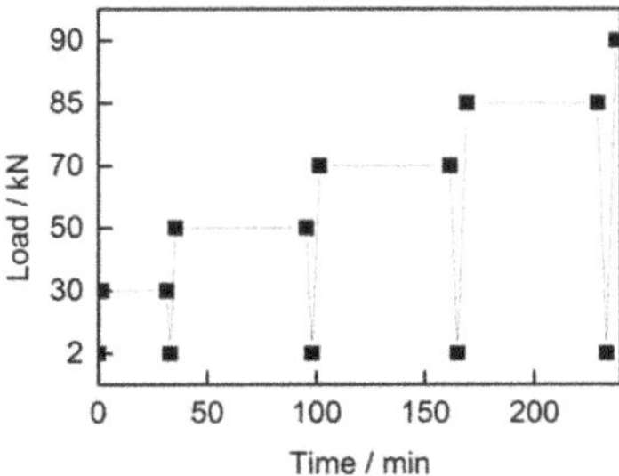

Figure 5. Stress-strain loading procedure of ACCC/HW.

2.2.3. Sag Test

In the sag test, ACCC/HW was heated through AC low voltage large current heating system. The temperature was measured through six pairs of thermocouples and the average temperature was taken. The test temperature was from 20 °C to 200 °C and the initial tension was 25% RTS. The sample test span was 50 m and the length of the heating span was 60 m. The sag of ACCC/HW at different temperatures was tested.

2.2.4. Creep Test

The creep test of ACCC/HW was taken on the numerically controlled creep test machine according to IEC61395-1998 [19]. The effective length of ACCC/HW and the extensometer were 14m and 2000 ± 1 mm. The elongation of ACCC/HW under the tensions of 15% RTS, 25% RTS, and 35% RTS for different time were tested at 25 °C. The total test time was 1000 h. The high temperature creep test was carried out at 160 ± 3 °C and the tension was 25% RTS.

2.2.5. Current-Carrying Capacity

The current-carrying capacity was tested according to IEC 1597-1995 [20]. Under the conditions of 25% RTS tension, the heating length of 13 m, no breeze, no sunshine, and natural convection, the current-carrying capacity of ACCC/HW was tested. Afterwards, the current-carrying capacity in different environments was calculated according to the calculation parameters that were recommended by China Electric Power Industry (wind speed was 0.5 m/s, sunshine intensity was 1000 W/m, conductor surface absorption coefficient was 0.9, conductor radiation coefficient was 0.9, ambient temperature was 20~45 °C, and conductor working temperature was 70~160 °C).

3. Results and Discussion

3.1. Basic Mechanical Property

The fracture forces of ACCC/HW at 25 °C and at 160 °C are 121.9 kN (the average of 124.5 kN, 119.0 kN, and 122.1 kN) and 105.2 kN (the average of 105.9 kN, 104.6 kN, and 105.2 kN), and the stresses are 425.2 MPa and 366.9 MPa. The fracture forces of the composite core at 25 °C and at 160 °C are 113.5 kN (the average of 122.1 kN, 109.8 kN, and 108.8 kN) and 106.1 kN (the average of 106.1 kN, 104.6 kN, and 107.6 kN), and the stresses are 2597.3 MPa and 2427.9 MPa. Table 3 shows the stress-strain test results of ACCC/HW, and Figure 6 shows the curve. The elasticity modulus of the composite core is 137.7 GPa, and Figure 7 shows the stress-strain curve.

Table 3. Stress-Strain Test Results of ACCC/HW.

Tension and Preservation Time	Load (kN)	ACCC/HW	
		Stress (MPa)	Strain (%)
30%RTS 30 min	30	105.0	0.14
50%RTS 60 min	50	175.0	0.35
70%RTS 60 min	70	245.0	0.62
85%RTS 60 min	85	297.5	0.83
90%RTS 0 min	90	305.0	0.84

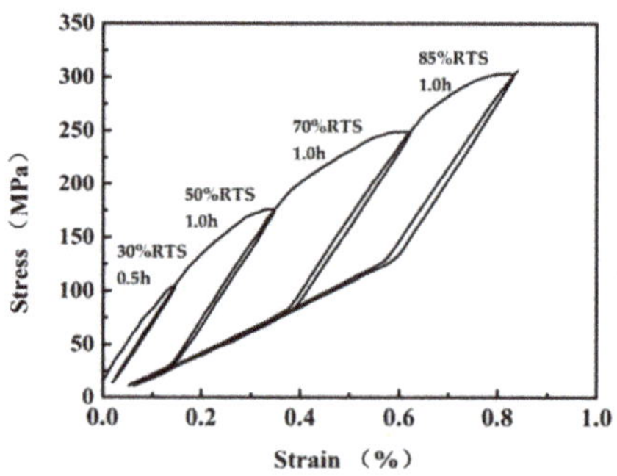

Figure 6. Stress-strain curve of ACCC/HW.

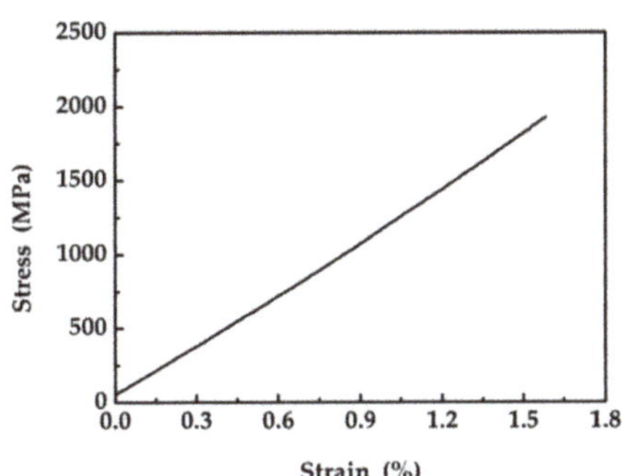

Figure 7. Stress-strain curve of composite core.

As shown in Figure 6, after loading to 30% RTS in the first cycle, the conductor is basically in the stage of elastic deformation. When the stress reaches 70 MPa, the slope of the stress-strain curve becomes smaller, but the change of slope is relatively small. During the second cyclic loading process to 50% RTS, the strain increases with the increase of stress at the beginning. Before the stress reaches 100MPa, the stress-strain loading curve basically coincided with the first unloading curve. After the stress reaches 100 MPa, the strain continuously increases with the increase of stress, but the increase speed obviously decreases. When the load reaches 50% RTS and the stress is 175 MPa, the load is no longer increased, and then kept for 60 min. In this case, the stress no longer increases, but the strain continuously increases. In the unloading process, the slope of the stress-strain curve is relatively large, and basically parallel to that before loading to 100 MPa in the second loading process. When the stress is unloaded to 30 MPa, the slope of the curve begins to decrease. During the third cyclic loading, the stress-strain curve changes for three times in the loading stage. At the initial stage of loading, the curve coincided with that in the last stage of the second unloading. When the stress reaches 3 0MPa, the slope of the curve suddenly increases and is then basically parallel to the curve in the initial stage of the second unloading process. When the stress reaches the peak point of the second curve at about 175 MPa, the slope of the curve decreases again, and the curve continuously increases slowly until the load reaches 70% RTS. Subsequently, the stress loading is stop, and the load is kept for 60 min. Similar to the second loading process, the stress no longer increases, while the strain continuously increases. The change of the curve in the third unloading process is basically the same as that in the second unloading process. At the beginning, the curve slope is large. After the stress reaches 83 MPa, the slope of the curve decreases until the stress is unloaded to the initial load. The curve change trend in the fourth process of loading and unloading is basically the same as that of the third time, with the difference in that the turning points of the loading curves are 83 MPa and 248 MPa, and the turning points of the unloading curves are 305 MPa and 129 MPa, respectively. The composite core is in an elastic deformation stage and the strain is below 1.1%, according to Figure 7.

Aluminum is metal material with elastic deformation and plastic deformation. Therefore, during the whole loading process of ACCC/HW, the aluminum strand has experienced two stages of elastic deformation and plastic deformation. In the first loading process, the curve changes according to the linear law, which indicated that the conductor is at the stage of elastic deformation. When the stress reaches 70 MPa, the curve is out of line, because plastic deformation occurs after the stress of the aluminum strand reaches the elastic limit. The 30% RTS is kept for 30 min. and then unloaded. It can be found that the unloading curve does not coincide with the loading curve, which results from the permanent plastic deformation of aluminum. The second loading process is basically the same as the first loading process. However, the elastic limit of aluminum is increased to 100 MPa because of the strain hardening of metal materials. In the process of unloading, the load of the aluminum strand is at the stage of plastic deformation, and the load is mainly supported by the composite core, so the slope of the unloading curve is larger. When the stress is unloaded to a certain stage, the aluminum strand is recovered to the elastic deformation stage and continues to bear load. The slope of the curve becomes smaller. Therefore, the inflection point occurred at 30 MPa of the stress-strain unloading curve. The third and fourth loading processes are basically the same as that of the second time.

The final modulus of ACCC/HW is only 71 GPa, which is slightly lower than the modulus of ACSR. However, Figure 6 obviously shows that the final non-recoverable plastic deformation of ACCC/HW is only 0.5‰, which is much smaller than the non-recoverable deformation of ACSR. Therefore, the safety of ACCC/HW is substantially improved.

3.2. Resistance to Deformation of ACCC/HW

Figure 8 shows the sag test results of ACCC/HW and the sag curve of ACSR [21] with similar structure.

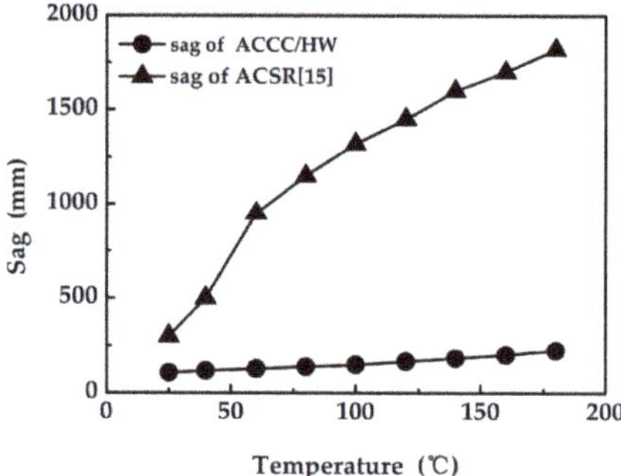

Figure 8. The sag curve of ACCC and ACSR with similar structure [21].

The sags of ACCC/HW at 25 °C, 110 °C, and 160 °C are 95 mm, 367 mm, and 371 mm, and the knee-point temperature is 110 °C, according to Figure 8. When the temperature is over 110 °C, the increase in sag is very small. The sags of ACSR with similar structure at 25 °C, 70 °C, 110 °C, and 160 °C are 231 mm, 1040 mm, 1420 mm and 1750 mm, respectively. The knee-point temperature of ACSR is 70 °C. Below the knee-point temperature, the sags of ACCC/HW and ACSR increases with the rise of temperature, and the sag of ACSR increases more rapidly. Above the knee-point temperature, the sag of the ACSR continuously increases, but the increase speed is smaller than that below the knee-point temperature. Additionally, above the knee-point temperature, the sag of ACCC/HW increases very slowly, which can be basically ignored. Based on the analysis, the sag of ACCC/HW is much smaller than that of ACSR, and it has higher safety.

The essential reason of sag is the thermal expansion property of material, and the thermal expansion is mainly determined by the thermal expansion coefficient of material. Below, the knee-point temperature point, the load is mainly borne by the core and aluminum conductor, and both the core material and aluminum conductor determine the sag. Above the transition temperature point, the load is mainly borne by the core, and the core material determines the sag. The thermal expansion coefficient of composite core, steel core, and aluminum conductor are 1.6×10^{-6}, 11.6×10^{-6}, and 23.6×10^{-6}. Accordingly, the sag of ACCC/HW is smaller than that of ACSR below 110 °C, and much smaller than that of ACSR above 110 °C, only about 20% of that of ACSR.

There are many ways to estimate the sag of ACSR [22], and several calculation formulas are given, such as (1) [22].

$$D \cong \frac{\omega S^2}{8H} \tag{1}$$

where D is the sag, ω is the resultant unit weight of the conductor, S is the span length, and H is the horizontal tension. In this sag test, the span length of ACCC/HW is 50 m, the total tension is 25 kN, and the weight is 742.9 kg/km. These test data do not match (1). Therefore, more sag test of the new type conductor should be taken to discover the law of sag phenomenon.

The sag characteristics of ACCC/HW that are discussed above are in short-term force. However, ACCC/HW in use is under a long-term tension state, and the deformation situation becomes very complicated. The deformation of the overhead conductor includes non-permanent deformation and permanent deformation. Non-permanent deformation includes elastic deformation and thermal deformation, the characterization physical quantities of which are the elasticity modulus E and the thermal expansion coefficient discussed above. Permanent deformation includes residual deformation and creep deformation [22]. The influence time of the residual deformation is very short when compared to the creep deformation. Therefore, the permanent deformation is mainly based on creep deformation in the long-term use.

When the temperature and tension are determined, the creep formula is formally consistent with the empirical creep formula of ACSR wire, as shown in (2) [23],

$$\varepsilon = at^{\mu} \tag{2}$$

where, ε is the creep elongation in t hours, mm/km; a is the coefficient; t is the creep time, h; and, μ is the coefficient determined in the test.

The creep test results of the conductors are conducted with power fitting to derive the ACCC/HW creep equation. The creep equations at 25 °C and at 160 °C that are derived from the test results are (2) to (5).

$$\varepsilon_{(10-10000)} = 31.52t^{0.189}\,(15\%\ RTS) \tag{3}$$

$$\varepsilon_{(10-10000)} = 56.08t^{0.196}\,(25\%\ RTS) \tag{4}$$

$$\varepsilon_{(10-10000)} = 99.74t^{0.195}\,(35\%\ RTS) \tag{5}$$

$$\varepsilon_{(10-10000)} = 60.20t^{0.206}\left(25\%\ RTS, 160\,^{\circ}C\right) \tag{6}$$

The creep of the conductor is affected by the metallurgical creep and the strain due to the geometrical settlement [23]. The geometrical settlements of the conductors in the creep test are the same, so the material creep mainly causes the differences of test results. The creep of the conductor is the permanent deformation that is caused by slip and dislocation in the metal crystal of aluminum due to the long-time force. Even if the load is no longer growing in the test, the creep deformation will increase with time. However, the speed of creep deformation decreases with time.

According to (3) to (6), the creep deformation of ACCC/HW increases at the power function with time. Coefficient a is the creep deformation of the conductor in one hour, and it is determined by the material, temperature, and stress. Exponent μ represents the growth rate of the creep deformation with time. It is affected by the material, structure, tension of ACCC/HW, as well as temperature. Coefficient a are 31.52, 56.08, and 99.74 at 15% RTS, 25% RTS, and 35% RTS, respectively. The initial creep deformation of the conductor increases with the increase of tension at 25 °C. Exponent μ are 0.189, 0.196, and 0.195 at 15% RTS, 25% RTS, and 35% RTS, respectively. The growth rate of creep deformation increases with tension growing below 25% RTS, and it remains largely unchanged between 25% RTS and 35% RTS.

Under the tension of 25% RTS, the coefficient a at 25 °C and 160 °C are little different, which are 56.08 and 60.20. Exponent μ are 0.195 and 0.206, respectively. The creep speed of the conductor increases faster at 160 °C, but the difference from that at 25 °C is still quite small. This is because 160 °C is higher than the knee-point temperature, and the strength is mainly borne by the composite core.

According to the equations, it is derived that the creep deformation amounts of ACCC/HW under the tension of 15% RTS, 25% RTS, and 35% RTS in 10 years (87600 h) are 271 mm/km, 522 mm/km, and 867 mm/km. At 160 °C, the creep deformation under the tension of 25% RTS in 10 years is 628 mm/km, which is smaller than 0.6‰ of the length of ACCC/HW. The test result is much smaller than the creep deformation of aluminum or steel wires, because the creep of the composite core is very small. When the aluminum strand is deformed to a certain degree, the load is transferred to the carbon fiber composite core. In this case, the tensile strength of the aluminum strand is small and the creep is small, which leads to the small creep deformation of ACCC/HW.

3.3. Carrying Capacity

Carrying capacity is one of the main parameters for evaluating the power transmission capacity of conductors, and it is related to the conductor structure, surface state, resistance, operating temperature, environment temperature, sunshine intensity, and wind speed, etc. Table 4 lists the test results of the carrying capacity of ACCC/HW under the conditions of no wind, no sunshine, and natural convection, and Table 5 shows the calculation results.

Table 4. Carrying capacity test results under the conditions of no wind, no sunshine, and natural convection.

Conductor Temperature/°C	Environment Temperature/°C	Carrying Capacity/A
61.0	23.0	329
80.5	23.0	468
100.1	23.1	565
120.8	23.4	631
140.2	23.8	698
160.7	23.9	767

Table 5. Carrying capacity results calculated according to the parameters that are recommended by China's electric power industry.

Conductor Temperature (°C)	Environment Temperature (°C)					
	20	25	30	35	40	45
70	643	599	552	500	442	376
80	713	675	635	592	546	495
90	779	746	710	673	634	591
100	828	798	767	734	700	663
110	878	850	822	792	761	729
120	923	897	871	844	816	787
130	965	941	917	892	867	840
140	1004	982	960	937	913	889
150	1041	1021	1000	978	956	934
160	1077	1057	1038	1018	997	976

Under the same conductor operating temperature, the carrying capacity decreases with the increase of the environment temperature, as shown in Table 5. At the same environment temperature, the carrying capacity increases with the increase of the conductor operating temperature. At the environment temperature of 40 °C, under the same calculation parameters, the calculated carrying capacity of 240/55 type ACSR with the similar structural parameters are 445A at 70 °C, 554A at 80 °C and 641A at 90 °C. Compared with the data in Table 5, under the conditions of operating temperature below 90 °C, the same environment temperature and the same operating temperature, the carrying capacity of ACSR is 1% greater than that of ACCC/HW. It can be considered that the carrying capacities of these two kinds of conductors are basically the same. The aluminum strands mainly determine the current carrying capacity. The sectional area of aluminum strands of ACCC/HW and ACSR are basically the same, so the current carrying capacities are also essentially the same. However, ACCC/HW does not have the magnetic hysteresis loss and thermal effect caused by the steel core.

In addition, a higher carrying capacity of the conductor can lead to serious heating situation and higher operating temperature. At a high temperature over 100 °C, the strength of ACSR reduces, the sag increases, and the security is greatly decreased. Moreover, the service life is also affected.

Therefore, ACSR is not suitable for use in a high temperature environment. Many countries have raised the requirements for the maximum allowable operating temperature of ACSR [24]. The maximum allowable temperatures are 50 °C in Sweden, 70 °C in China, 80 °C in Germany, Holland and Switzerland, and 90 °C in the United States. The maximum allowable temperature limits the use of ACSR, as well as the carrying capacity of ACSR. In contrast, ACCC/HW is not subject to this limit. According to the test results, ACCC/HW has higher strength and smaller sag at 160 °C. Therefore, ACCC/HW is suitable for operating at high temperature. China has stipulated that the two long-term maximum allowable temperatures for ACCC are 120 °C and 160 °C. At the environment temperature of 40 °C, the carrying capacities of ACCC/HT are 816A at the operating temperature of 120 °C and 997A at the operating temperature of 160 °C. The high carrying capacity of ACCC/HW can be brought into play.

Therefore, there is little difference between ACSR and ACCC/HW in terms of carrying capacity at the temperature below 90 °C. However, the maximum allowable temperature of ACSR is 70 °C, while ACCC/HW can be used in a high temperature environment. ACCC/HW has greater carrying capacity at high temperature, which is of great significance in improving the transmission efficiency of the whole transmission grid.

4. Conclusions

ACCC/HW has been in a trial process in 50 kV line in China for more than 10 years. The stresses of ACCC/HW at 25 °C and 160 °C are 425.2 MPa and 366.9 MPa. The sag of ACCC/HW with length of 50 m are 95 mm at 25 °C, 367mm at 110 °C, and 371 mm at 160 °C, respectively. According to the formula that was obtained from the experimental results, the creep deformation amounts of ACCC/HW under the tension of 15% RTS, 25% RTS, and 35% RTS at 25 °C after 10 years are 271 mm/km, 522 mm/km, and 867 mm/km. The creep deformation under 25% RTS at 160 °C after 10 years is 628 mm/km. The carrying capacity of ACCC/HW is basically equivalent to traditional aluminum conductor steel reinforced (ACSR), and higher than ACSR at a high temperature. ACCC/HW shows excellent mechanical behavior and good current capacity, especially in high temperature operation. Therefore, ACCC/HW is suitable to use in the overhead conductor line, and it has great application potential. ACCC/HW behave the above good performance, because carbon fiber reinforced composite core is high-strength, low-sag, and high-heat resistant, and ACCC/HW takes advantage of carbon fiber reinforced composite. However, carbon fiber reinforced composite is more brittle when compared with metal, and more caution should be taken in ACCC/HW construction and maintenance, such as employing larger diameter reels and using special fittings. Moreover, more tests and studies of ACCC/HW should be taken to estimate the application performance. At the same time, future tests and studies should focus more on the overall toughness assessment and aging performances to evaluate and improve the safety and applicability.

Author Contributions: Conceptualization, K.Q. and B.Z.; methodology, C.D.; validation, K.Q., B.Z. and A.Z.; formal analysis, J.Y. and A.Z.; investigation, B.W.; resources, K.Q.; data curation, B.W.; writing—original draft preparation, K.Q. and C.D.; writing—review and editing, K.Q.; visualization, J.Y.; supervision, B.Z.; project administration B.Z. All authors have read and agreed to the published version of the manuscript.

Funding: This research was funded by National Natural Science Foundation of China (Grant No. 51473088) and National Key Research and Development Plan of China (Project No. 2016YFC0301402).

Acknowledgments: The authors gratefully acknowledge the supports of carbon fiber reinforced composite samples test provided by Weihai Institute of Metrology in China.

References

1. Jones, R.V.; Lomas, K.L. Determinants of high electrical energy demand in UK homes: Socio-economic and dwelling characteristics. *Energy Build.* **2015**, *101*, 24–34. [CrossRef]

2. Chen, G.H.; Wang, X.; Wang, J.Q.; Liu, J.J.; Zhang, T.; Tang, W.M. Damage investigation of the aged aluminium cable steel reinforced (ACSR) conductors in a high-voltage transmission line. *Eng. Failure Anal.* **2012**, *19*, 13–21. [CrossRef]

3. Levesque, F.; Goudreau, S.; Langlois, S.; Legeron, F. Experimental study of dynamic bending stiffness of ACSR overhead conductors. *IEEE Trans. Power Deliv.* **2015**, *30*, 2252–2259. [CrossRef]

4. Ma, X.C.; Gao, L.; Zhang, J.X.; Zhang, L.C. Fretting wear behaviors of aluminum cable steel reinforced (ACSR) conductors in high-voltage transmission line. *Metals* **2017**, *7*, 373. [CrossRef]

5. Lalonde, S.; Guilbault, R.; Langlois, S. Numerical analysis of ACSR conductor–clamp systems undergoing wind-induced cyclic loads. *IEEE Trans. Power Deliv.* **2018**, *33*, 1518–1526. [CrossRef]

6. Wang, J.J.; Chan, J.K.; Graziano, J.A. The lifetime estimate for ACSR single-stage splice connector operating at higher temperatures. *IEEE Trans. Power Deliv.* **2011**, *26*, 1317–1325. [CrossRef]

7. Phillips, J.; Ryan, M.; Glass, D. Thinking outside the Box: Tennessee Valley Authority Uses the ACCR Conductor on a 500-kV Transmission Line. In Proceedings of the Electrical Transmission and Substation Structures Conference, Branson, MO, USA, 27 September–1 October 2015.

8. *Aluminum Conductor Composite Reinforced Technical Notebook (477 kcmil family) Conductor & Accessory Testing*; 3M Corp.: St. Paul, MN, USA, 2003.

9. Waters, D.H.; Hoffman, J.; Hakansson, E.; Kumosa, M. Low-velocity impact to transmission line conductors. *Int. J. Impact Eng.* **2017**, *106*, 64–72. [CrossRef]

10. Alawar, A.; Bosze, E.J.; Nutt, S.R. A composite core conductor for low sag at high temperatures. *IEEE Trans. Power Deliv.* **2005**, *20*, 2193–2199. [CrossRef]

11. Hakansson, E.; Predecki, P.; Kunosa, M.S. Galvanic corrosion of high temperature low sag aluminum conductor composite core and conventional aluminum conductor steel reinforced overhead high voltage conductors. *IEEE Trans. Reliab.* **2015**, *64*, 928–934. [CrossRef]

12. Burks, B.; Middleton, J.; Armentrout, D.; Kumosa, M. Effect of excessive bending on residual tensile strength of hybrid composite rods. *Compos. Sci. Technol.* **2010**, *70*, 1490–1496. [CrossRef]

13. Di, C.R.; Yu, J.W.; Wang, B.M.; Lau, A.K.T.; Zhu, B.; Qiao, K. Study of hybrid nanoparticles modified epoxy resin used in filament winding composite. *Materials* **2019**, *12*, 3853. [CrossRef] [PubMed]

14. Yu, H.L.; Zhao, H.; Shi, F.Y. Bending performance and reinforcement of rocker panel components with unidirectional carbon fiber composite. *Materials* **2019**, *12*, 3164. [CrossRef] [PubMed]

15. Wang, Y.R.; Liu, W.S.; Qiu, Y.P.; Wei, Y. A one-component, fast-cure, and economical epoxy resin system suitable for liquid molding of automotive composite parts. *Materials* **2019**, *11*, 685. [CrossRef] [PubMed]

16. Chen, Y.; Song, F.R.; Zhou, G.H.; Yi, L. *Technology and Application of Carbon Fiber Composite Conductor*; China Electric Power: Beijing, China, 2015; p. 4.

17. Burks, B.; Middleton, J.; Kumosa, M. Micromechanics modeling of fatigue failure mechanisms in a hybrid polymer matrix composite. *Compos. Sci. Technol.* **2012**, *72*, 1863–1869. [CrossRef]

18. *GB/T32502-2016; Overhead Electrical Stranded Conductors Composite Core Supported/Reinforced*; The Standards Press of China (SPC): Beijing, China, 2016.

19. *IEC 61395-1998; Overhead Electrical Conductors—Creep Test Procedures for Stranded Conductors*; British Standards Institution: London, UK, 1998.

20. *IEC 1597-1995; Overhead Electrical Conductors—Calculation Methods for Stranded Bare Conductors*; British Standards Institution: London, UK, 1995.

21. Dong, G.L.; Gong, J.G.; Yu, H.Y.; Yu, C.S. *Design, Construction, Operation and Maintenance of Aluminium Composite Core*; China Electric Power: Beijing, China, 2009; p. 36.

22. Dong, X.Y. Analytic Method to Calculate and Characterize the Sag and Tension of Overhead Lines. *IEEE Trans. Power Deliv.* **2016**, *31*, 2064–2071. [CrossRef]

23. Albizu, I.; Mazon, A.J.; Zamora, I. Flexible strain-tension calculation method for gap-type overhead conductors. *IEEE Trans. Power Deliv.* **2009**, *24*, 1529–1537. [CrossRef]

24. Nuchprayoon, S.; Chaichana, A. Cost evaluation of current uprating of overhead transmission lines using ACSR and HTLS conductors. In Proceedings of the EEEIC/I&CPS Europe, Milan, Italy, 6–9 June 2017.

 *materials*

Article

High Temperature Sensing and Detection for Cementitious Materials Using Manganese Violet Pigment

Rajagopalan Sam Rajadurai and Jong-Han Lee *

Department of Civil Engineering, Inha University, Incheon 22212, Korea; samraj9593@gmail.com
* Correspondence: jh.lee@inha.ac.kr; Tel.: +82-32-860-7564

Received: 29 January 2020; Accepted: 21 February 2020; Published: 22 February 2020

Abstract: In recent years, advanced materials have attracted considerable interest in the field of temperature detection and sensing. This study examined the thermochromic properties of inorganic manganese violet (MV) with increasing temperature. According to the thermochromic test, the material was found to have reversible and irreversible color change properties. The MV pigment was then applied to cementitious material at ratios of 1%, 3%, and 5%. The mixed cement samples with MV pigment were heated in a furnace, and digital images were captured at each temperature interval to evaluate the changes in the color information on the surface of the specimen. The mixed samples exhibited an irreversible thermochromic change from dark violet to grayish green above 400 °C. At the critical temperature of 440 °C, the *RGB* values increased by approximately 22%–55%, 28%–68%, and 7%–25%, depending on the content of MV pigment. In Lab space, the *L* value increased by approximately 23%–60% at 440 °C. The *a* value completely changed from positive to negative, and the *b* value changed from negative to positive. All the values differed according to the content of MV pigment at room temperature but approached similar ranges at the critical temperature, irrespective of the amount of MV pigment. To assess the changes in their microstructure and composition, scanning electron microscopy and energy dispersive X-ray spectroscopy were performed on the samples exposed to temperatures ranging from room temperature to 450 °C.

Keywords: irreversible thermochromic; cement composite; manganese violet; temperature indication; heat monitoring

---

1. Introduction

Cementitious materials are some of the most essential and reliable construction materials used in a range of applications, from building, bridges, and underground structures to special structures, such as nuclear power plants, chimneys, and reactors. The cement-based structures are normally exposed to room temperature but can sometimes be exposed to high temperature environments. When cementitious materials and structures are exposed to high temperatures, the material and mechanical properties can deteriorate, such as decreases in compressive and tensile strength, elastic modulus, cracking, and spalling [1,2]. In addition, complex physical and chemical changes occur when exposed to elevated temperatures [3,4]. Primarily, the physically combined water begins to evaporate, and the C-S-H (Calcium-Silicate-Hydrate) starts to dehydrate at approximately 100 °C [5,6]. The C-S-H decomposes gradually between around 100 °C to 300 °C [6]. From approximately 500 °C, the portlandite begins to disintegrate, and the calcite begins to decompose [3,6]. These phenomena lead to a deterioration of the material, such as a decay of strength and elastic modulus, as well as cracking and spalling of concrete.

After the cement-based structures are damaged by high temperatures, most of the structures can work perfectly after proper repair and maintenance instead of rebuilding the structures. To ensure that a structure can be repaired or replaced, an appropriate evaluation method is needed to define the damaged part of the structure due to the high temperature. At the same time, the evaluation should be fast and easy after an event. Typically, core strength and rebound hammer tests are performed to define the strength decay of the high-temperature damaged structures. Furthermore, the ultrasonic pulse velocity and the impact echo method can be used to detect the degree of damage. Similarly, some techniques, such as infrared thermal imaging, drilling resistance, petrographic examination, thermoluminescence, and the Windsor probe test, also detect damage to concrete [7–9]. On the other hand, these non-destructive techniques provide an indirect indication of damage to the structure and have limited application to the wide and large area of a real structure in practice.

The evaluation of high-temperature damaged structures also uses color changes. Normal cementitious materials containing siliceous aggregate tend to change color to pink or red at 300–600 °C, to whitish gray at 600–900 °C, and to buff at 900–1000 °C [10–14]. Moreover, the strength of concrete begins to decrease from 300 °C and drops by approximately 50% to 60% at approximately 500 °C [15,16]. Therefore, a color change to pink or red can indicate a deterioration of the material and mechanical properties. However, the degree of color change is difficult to define and recognize, particularly by the naked eye. Moreover, the color changes are less prominent for calcareous and igneous aggregates. Therefore, some studies based on color changes roughly estimated the temperature of the cementitious materials and were lack in the application to define the temperature information.

Color is the most immediate visible aspect in terms of exposure to high temperatures. To define the temperature information more precisely, a color change pigment was essentially mixed with cementitious materials. Some materials have a property to change color in response to the exterior environment, such as temperature and light [17]. The materials, which were recently recognized to exhibit a distinct color change in response to change in temperature, are found in different forms, such as metal oxides, polymers, pigments, and leuco dyes. Reversible thermochromic materials are used in the applications of smart windows, ultra-thin films, temperature sensors, cool colored, and protective coatings [18–26]. The irreversible thermochromic materials experience a permanent color change at certain temperatures and are widely used in applications, such as aeronautical components, furnaces, damage warnings, and thermal flow sensors [27–29]. Previous studies focused on metal structures but not on cementitious materials and structures widely used in the construction field.

In this study, manganese violet pigment was identified in the furnace test to undergo an efficient color change from dark violet to grayish yellow with temperatures above 400 °C. The manganese violet pigment, which is composed of manganese dioxide, ammonium dihydrogen phosphate, and phosphoric acid, presents a dark violet color due to the presence of phosphate and ammonia [30,31]. The color change to grayish yellow results from the evaporation of water and the liberation of ammonia from the pigment particles. The manganese violet pigment was then applied to white cement at a ratio of 1%, 3%, and 5% of the total mass. The white cement is similar to ordinary Portland cement except for the color. The white color was attributed to the decrease in the amount of chromium, manganese, iron, copper, vanadium, nickel, and titanium in the ordinary Portland cement. The white cement confers an advantage for the excellent exposure of color when mixed, and the material properties are unaffected due to a decrease in the above components [32]. The pigment mixed samples were heated in a furnace from room temperature to 450 °C. The mixed samples underwent a typical thermochromic change to a grayish green color. The thermochromic change was observed visually by the naked eye and was evaluated using digital images captured at each temperature interval.

Digital images are typically recorded in *RGB* color space, which is widely used in cameras and monitors by the percentage of red, green, and blue constituents. The *RGB* values of each component are in the range of zero to 255 [33]. The disadvantage of *RGB* color space is dependent on each device, which has its own different color sensor characteristics and produces different *RGB* output responses. Unlike the *RGB*, *Lab* color space can be used as a device-independent model when representing color. The color information in both *RGB* and *Lab* color spaces obtained from the digital images were analyzed, and the color information with temperatures was quantified. From the above techniques, a reliable method for temperature detection can be obtained by integrating the color change data and with those of the microstructural and the compositional changes. The microstructure changes caused by the high temperatures were studied by scanning electron microscopy (SEM). In addition, the changes in their composition due to the high temperature were examined using energy dispersive X-ray spectroscopy (EDX). Therefore, this study analyzed the color information visually, the changes in the *RGB* and *Lab* color spaces obtained from the digital images, and the microstructure and compositional changes with the application of high temperatures.

2. Experimental Program

2.1. Specimen Preparation and Test Procedure

For the application of manganese violet (MV) pigment [34] to the sensing and detection of high temperature in cementitious materials, this study conducted the thermochromic tests of MV pigment and mixed samples. Table 1 lists the typical characteristics of MV pigment. The MV pigment sample was first prepared and placed in a porcelain crucible for the thermochromic test. The pure sample was heated from room temperature to 450 °C in an electrical furnace. The color of the sample was examined at 100 °C, 200 °C, 300 °C, and 350 °C. Due to the fact that the sample was expected to change its color above 350 °C, the change in the color was examined from 350 °C to 450 °C at 10 °C intervals. The pigment sample was maintained at each target temperature for approximately 10 minutes to induce a uniform temperature distribution in the sample. Subsequently, the color change on the surface of the pigment was recorded using a digital camera. In addition, cementitious mixed samples were manufactured with white cement and MV pigment. White cement is typically similar to ordinary Portland cement, but has a high degree of whiteness due to a decrease in the amount of chromium, manganese, iron, copper, vanadium, nickel, and titanium. White cement also has distinct color recognition and decreases the strength reduction when mixed with color pigments. Therefore, white Portland cement is commonly used for colored concrete and mortar combined with pigments and precast concrete members required for high early strength.

Table 1. Typical characteristics of MV pigment.

Parameter	Characteristics
Color	Violet
Chemical characterization	Manganese III ammonium pyrophosphate
Density	$2.7 \sim 2.9 \text{ kg/m}^3$
Bulk density	0.60 g/cm^3
Average particle size	$2.30 \text{ } \mu m$
pH value	$2.5 \sim 4.7$
Thermal decomposition	$>400 \text{ °C}$

The ACI-212 (American concrete institute) [35] requires that the addition of pigment should not exceed 10% of the weight of the cementitious material and recommends less than 6% to have little or no effect on the mechanical properties of fresh and hardened concrete, which would also be economical. Therefore, this study defined the mixed ratios of the MV pigment as 1%, 3%, and 5% of the weight of white cement. The pigment was added and thoroughly mixed under a dry condition. After mixing with cement material, thermochromic tests were carried out in an electrical furnace to examine the color change of the mixed samples with increasing temperature from room temperature to 450 °C.

After the thermochromic heating tests were performed on the MV pigment and the mixed samples, the samples were collected before and after heating to the target temperatures. The fragments of the collected samples were positioned on the sample holder to perform different investigations. SEM examined the microstructural changes, and EDX was performed to analyze the compositional changes of the samples exposed to high temperatures.

2.2. Color Measurement and Analysis

To investigate the thermal color alteration of the samples, a SONY ILCE-TR digital camera was installed parallel to the samples at a distance of approximately 30 cm, and pictures of the surfaces of the heated samples were taken at each temperature interval. The images taken in the normal background and stored in JPEG format were processed to quantify the color information in both *RGB* and Lab color spaces.

The *RGB* color space is widely used for image processing and graphics resulting from cameras and monitors. In the *RGB* color space, every color is represented by the spectral components of *R* (red), *G* (green), and *B* (blue). However, the color intensities in the *RGB* space depend strongly on device sensors and characteristics, which could produce different *RGB* color output responses. On the other hand, *Lab* color space shows a representative color independent of the devices and includes all the colors in the spectrum, as well as colors outside human perception. Therefore, the *Lab* color space with high accuracy and portability is largely used in practical applications, such as printing, automobile, textiles, and plastics. In the Lab color space, the *L* value is a measure of lightness ranging from 0 to 100, which represents complete darkness or opaqueness to complete transparence, respectively. *a* and *b* range from −128 to 128. Positive and negative *a* indicate the redness and greenness, respectively, and positive and negative *b* is *a* measure of yellowness and blueness, respectively.

3. Thermochromic Characteristics of MV Pigment

3.1. Thermochromic Color Change of MV Pigment with High Temperatures

As the MV pigment samples were exposed to different high temperatures, the color changes were observed with increasing temperature from room temperature. Figure 1 presents the color changes of the samples at each temperature. The experimental results showed that the MV pigment experiences a systematic color change with increasing temperatures, as expressed in the following chemical equation:

$$2MnNH_4P_2O_7 \underset{(dark\ violet)}{} \overset{RT\ -\ 370\ ^\circ C}{\longleftrightarrow} \underset{(blue)}{Mn_2P_4O_{13}(NH)_2} \overset{410\ ^\circ C}{\longrightarrow} \underset{(grayish\ yellow)}{Mn_2P_4O_{12}} \tag{1}$$

The dehydration process, which means the removal of water from pigment particles, occurred in between room temperature and 370 °C. Due to water driven out of the pigment particles, the color of the pigment changed from violet to reversible blue. When the pigment was exposed to the atmosphere again, it reacted with the ambient moisture to restore its original violet color. Immediately after 370 °C, the violet color started to diminish, and the white color appeared partially on the surface of the pigment particles. However, the color change was inadequate to be observed precisely until 400 °C. When the temperature reached 410 °C, the color of the pigment completely changed from violet to grayish yellow. The stable color was maintained even after heating to higher temperatures. This was an irreversible reaction associated with both the evaporation of water and the liberation of ammonia from the pigment particles at temperatures higher than 400 °C. That is, when cooled to room temperature, the MV pigment did not recover its original color.

In addition, the color change was evaluated using *RGB* and *Lab* color spaces. Figure 2 shows the change in the *RGB* values, averaged from the images at each temperature, with increasing temperature. The *RGB* values decreased slightly from (68, 34, 96) to (50, 31, 79) from room temperature to 200 °C, respectively, and the *RGB* values were almost constant with increasing temperature until 370 °C. After that, all the *RGB* values began to increase gradually toward the whiteness color intensity up to 400 °C. The *RGB* values showed a sudden hike to (126, 128, 114) at 410 °C, which is the transition point of a color change from violet to white. Above 410 °C, the RGB values were stable as an irreversible white color. As a result, the *RGB* values at 410 °C were approximately 1.85, 3.76, and 1.19 times those at room temperature, respectively.

Figure 3 shows the change in the color values in the *Lab* space. Similar to the *RGB* color values, the *Lab* values were relatively constant from room temperature to 370 °C. At higher temperatures, the *L* value began to gradually increase toward lightness; *b* value also gradually increased, while the *a* value decreased. At 410 °C, the *Lab* color values showed a rapid change to around (52, −3, 6), respectively. The color change means a change from darkness to lightness. That is, the lightness index of the *L* value increased by approximately 248% from 21 at room temperature. The *a* value decreased from positive (28 at room temperature) to negative (−3 at 410 °C), which indicates a greenness intensity, whereas the *b* value increased from negative (−30 at room temperature) to positive (6 at 410 °C), which indicates yellowness. These three *Lab* color values at 410 °C were combined to produce a grayish yellow color. Above 410 °C, little change in the *Lab* values was observed.

Figure 1. Color changes in the MV pigment with temperatures (Scale: 1000 × 1000 pixels).

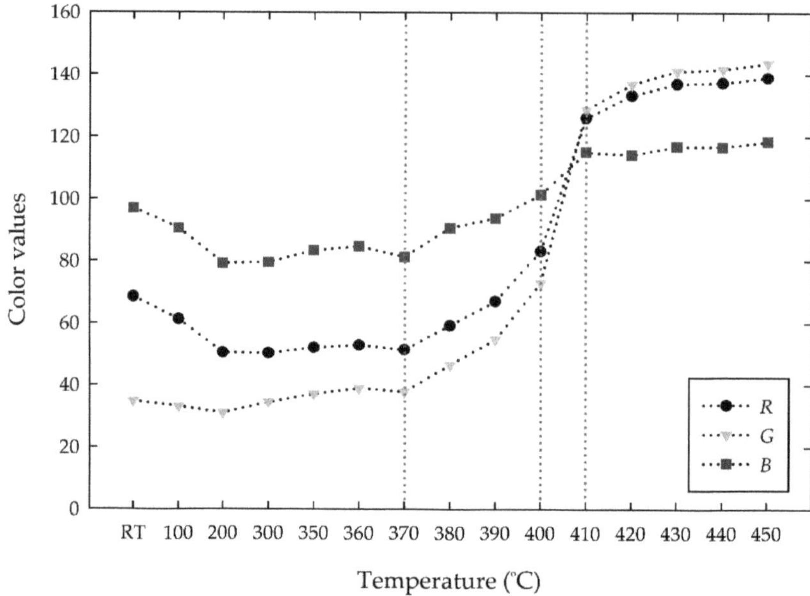

Figure 2. Changes in the *RGB* values of the MV pigment with increasing temperature.

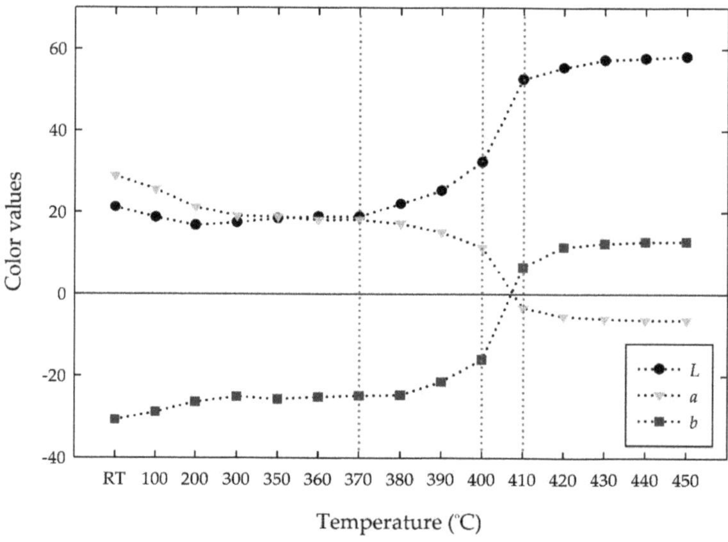

Figure 3. Changes in the *Lab* values of the MV pigment with increasing temperature.

3.2. SEM Micrograph and EDX Analysis of the MV Pigment

For further characterization, SEM (S-4300 FE-SEM, Hitachi, Tokyo, Japan) was performed on the MV pigment at room temperature and at 410 °C to follow the microstructural changes because of the heat treatment. Figure 4 presents SEM images of the morphology and structure of the MV pigment. From the figure, the MV pigment at room temperature revealed a cluster of rod-shaped, spherical, and hexagonal crystal structures. As the temperature was increased to 410 °C, the rod-shaped and hexagonal crystal characteristics of the MV pigment tended to change through the loss of water and ammonia. The dehydration process at high temperatures disintegrated their edges and provided a cluster of ring-shaped void structures.

(a)   (b)

Figure 4. SEM images of the MV pigment: (**a**) Room temperature; (**b**) 410 °C.

EDX analysis (EMAX-7000, HORIBA EDX system, Kyoto, Japan) was carried out to assess the changes in the mineralogical composition of the samples caused by the increase in temperature, as displayed in Figure 5. In the unaltered sample at room temperature, the EDX pattern exhibited the presence of more intense peaks of C, O, P, and Mn with the mass contents of 11.36%, 48.91%, 16.30%, and 14.56%, respectively, and less intense Al, Si, K, and Fe peaks. When exposed to 410 °C, the mass of C, O, and Mn decreased by 39%, 5%, and 1.35%, respectively, and P increased by 4.6%, in which the mass contents of the C, O, P, and Mn changes to 6.89%, 46.37%, 17.05%, and 14.37%, respectively.

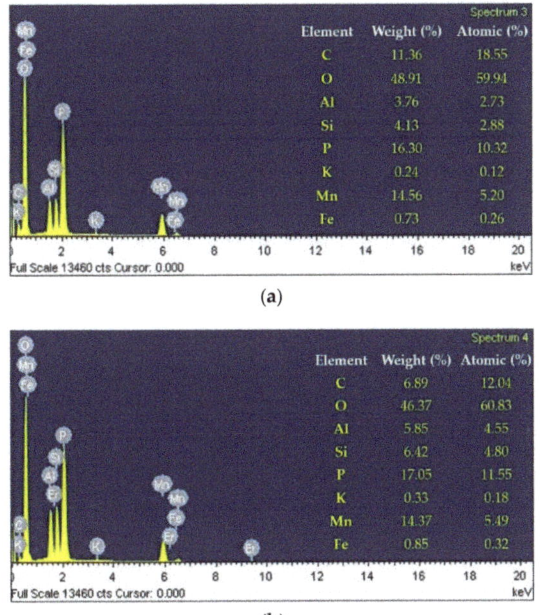

(a)

(b)

Figure 5. EDX analysis of the MV pigment: (**a**) Room temperature; (**b**) 410 °C.

4. Application of MV Pigment with Cementitious Material

4.1. Thermochromic Change of White Cement Mixed Samples with MV Pigment

White Portland cement is widely used in cement composites for the good exposure of color with pigments. Primarily, white cement remains a stable silver or pale white under high temperature conditions. Therefore, this study manufactured the white cement composite samples with the MV pigment. The MV pigment was separately added in 1%, 3%, and 5% of the weight of white cement. The mixed samples were heated individually in the furnace from room temperature to 450 °C. The photographs were recorded to examine the color of the samples at 100 °C, 200 °C, 300 °C, 350 °C, and 370 °C, and after that, every 10 °C up to 450 °C. Figure 6 shows the original and fully changed colors of the mixed samples at room temperature and 440 °C, respectively.

Figure 6. Color changes in the white cement and MV mixed samples exposed to different temperatures (Scale: 1000 × 1000 pixels).

With the addition of white cement, the color of the pigment changed from violet to blue at room temperature. When heated, the mixed samples exhibited similar thermochromic changes as the pure MV pigment. Between room temperature to 410 °C, the mixed samples underwent a process of water removal from the pigment, but the color change was insufficient to be observed with the naked eye. Subsequently, the surface color of the mixed samples tended to fade away. When the temperature reached 440 °C, the mixed samples with 1%, 3%, and 5% of the MV pigment completely changed to a grayish green color. The color change was attributed to both the evaporation of water and the decomposition of ammonia from the pigment particles. The mixed samples did not recover their original color upon cooling to room temperature, indicating an irreversible color change. The color change occurred at 410 °C in the pure pigment but at 440 °C for the mixed samples. This might be due to the reduced temperature intrusion between the layers due to the mixing of white cement.

4.2. SEM Micrograph and EDX Analysis of White Cement Mixed Samples

Figure 7 shows SEM images of the white cement to characterize the changes in the morphology associated with high temperatures. At room temperature, the particles of the white cement have a structural diversity with globular shapes, including hexagonal, spherical, and some irregular tiny particles. When heated to 440 °C, a slight change was found due to the breakage of particles, and the white cement was relatively stable. Figure 8 presents SEM micrographs of the mixed sample at room temperature and at 440 °C, respectively. From the figure, the mixed sample at room temperature exhibits triangular, hexagonal, and spherical particles with a small cluster of rod-shaped structures due to the presence of MV pigment. At 440 °C, the hexagonal and rod-shaped cluster structures disappeared due to the dehydration of MV particles, which changed the mixed sample to irregular bundled structures.

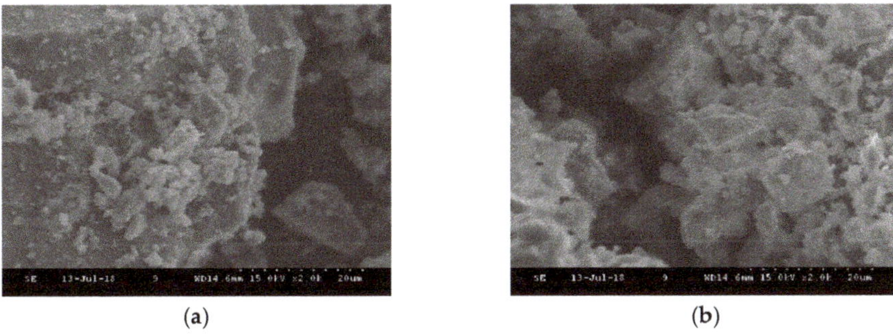

(a) (b)

Figure 7. SEM images of the white cement: (**a**) Room temperature; (**b**) 440 °C.

(a) (b)

Figure 8. SEM images of the mixed samples: (**a**) Room temperature; (**b**) 440 °C.

Table 2 presents the results of EDX analysis for the white cement and the mixed samples. In the white cement, the samples are predominantly composed of C, O, and Ca, with the mass contents of 14.9%, 47.35%, and 25.09%, respectively. At 440 °C, the C, O, and Ca contents were 15.35%, 48.03%, and 26.85%, which correspond to an increase of approximately 3%, 1.5%, and 7%, respectively.

The mixed sample consists of high peaks of C, O, and Ca with the mass components of 17.18%, 40.43%, and 19.19%, respectively, and low peaks of Mg, Al, Si, P, S, Mn, Sn, Sb, I, and Fe. The mixed sample with MV pigment included new elements of P, Mn, Sn, Sb, and I, which consisted of 2.05%, 1.78%, 1.24%, 8.03%, and 2.21%, respectively. When the temperature reached 440 °C, the C increased by 41.4%, and the O and Ca contents decreased by 3.3 and 11%, respectively. The mass of the new elements of P, Mn, Sn, Sb, and I by the addition of MV at 440 °C changed to 1.54%, 1.72%, 0.89%, 6.67%, and 2.07%, which correspond to a 24%, 3%, 28%, 16%, and 6% decrease, respectively.

Table 2. EDX analysis of the white cement mixed samples—weight (%).

Element	White Cement (%)		WC + MV Sample (%)	
	Room Temperature	440 °C	Room Temperature	440 °C
C	14.90	15.35	17.18	24.29
O	47.35	48.03	40.43	39.09
Mg	2.15	2.50	2.53	3.98
Al	1.19	0.65	1.47	0.54
Si	4.22	2.52	2.84	1.09
P	-	-	2.05	1.54
S	1.06	0.99	1.06	0.51
K	0.39	-	-	-
Ca	25.09	26.85	19.19	17.05
Mn	-	-	1.78	1.72
Sn	-	-	1.24	0.89
Sb	-	-	8.03	6.67
I	-	-	2.21	2.07
Ti	-	0.25	-	-
Fe	2.07	2.88	-	0.57
Ge	0.47	-	-	-
Zr	0.91	-	-	-

4.3. Thermochromic Analysis in the RGB Color Spaces

The color changes of the white cement mixed samples were evaluated using *RGB* and *Lab* color spaces. Figure 9 shows the changes in *RGB* values with increasing temperature. The white cement showed stable and high *RGB* values from room temperature to 450 °C, as shown in Figure 9a. At room temperature, the *RGB* color intensities of white cement were (165, 179, 186), which represents a grayish blue color. Heating of the white cement barely changed the color and thus displayed stable values with the mean *RGB* values of (164, 177, 183) from room temperature to 450 °C. Figure 9b–d show the changes in the *RGB* values of the mixed samples with increasing temperature. The addition of 1% pigment to white cement decreased the *RGB* values at room temperature from approximately (165, 179, 186) to (127, 125, 134), which corresponds to a 22% to 30% (R: 22%, G: 30%, B: 27%) decrease. In the mixed samples with 3 % and 5% pigments, the *RGB* values decreased by approximately 30% to 42% (R: 30%, G: 42%, B: 32%) and 33% to 47% (R: 34%, G: 47%, B: 33%), respectively. The decrease in the *RGB* values in the mixed samples indicates the disappearance of the whiteness intensity and the development of a mild violet color with the addition of 1%, 3%, and 5% pigments. With increasing temperature from room temperature to 410 °C, the mixed samples with 1%, 3%, and 5% pigments were almost stable with mean *RGB* values of approximately (127, 131, 133), (108, 108, 120), and (99, 97, 116), respectively, which are very close to those at room temperature. At higher temperatures, the *RGB* increased gradually toward the whiteness intensity, and the color completely changed at 440 °C. The *RGB* values of the mixed samples with 1%, 3%, and 5% MV at 440 °C were similarly (156, 168, 144), (145, 155, 133), and (157, 164, 146), respectively. Considering the reference *RGB* values of gray and green colors, the color obtained in the mixed samples at 440 °C can be defined as a grayish green color. Above 440 °C, the *RGB* values were stable and irreversible.

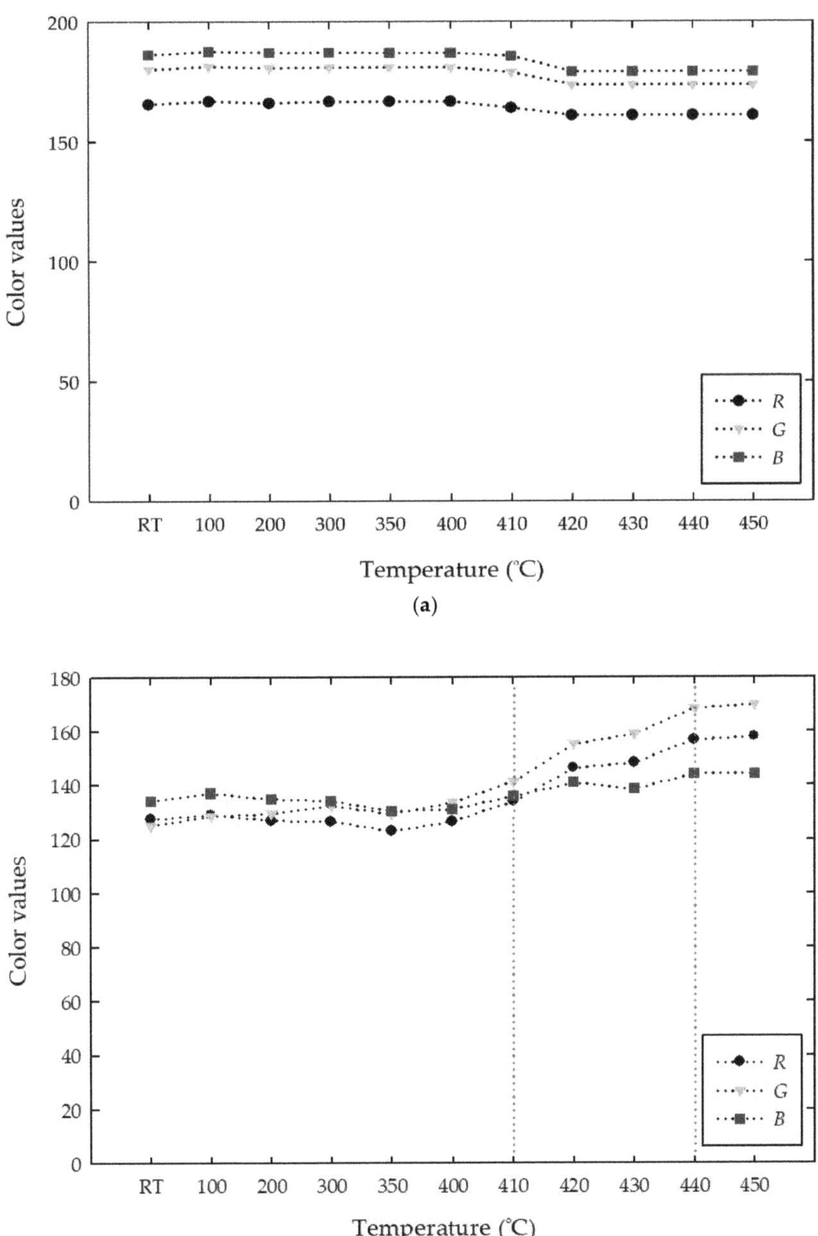

(a)

(b)

Figure 9. *Cont.*

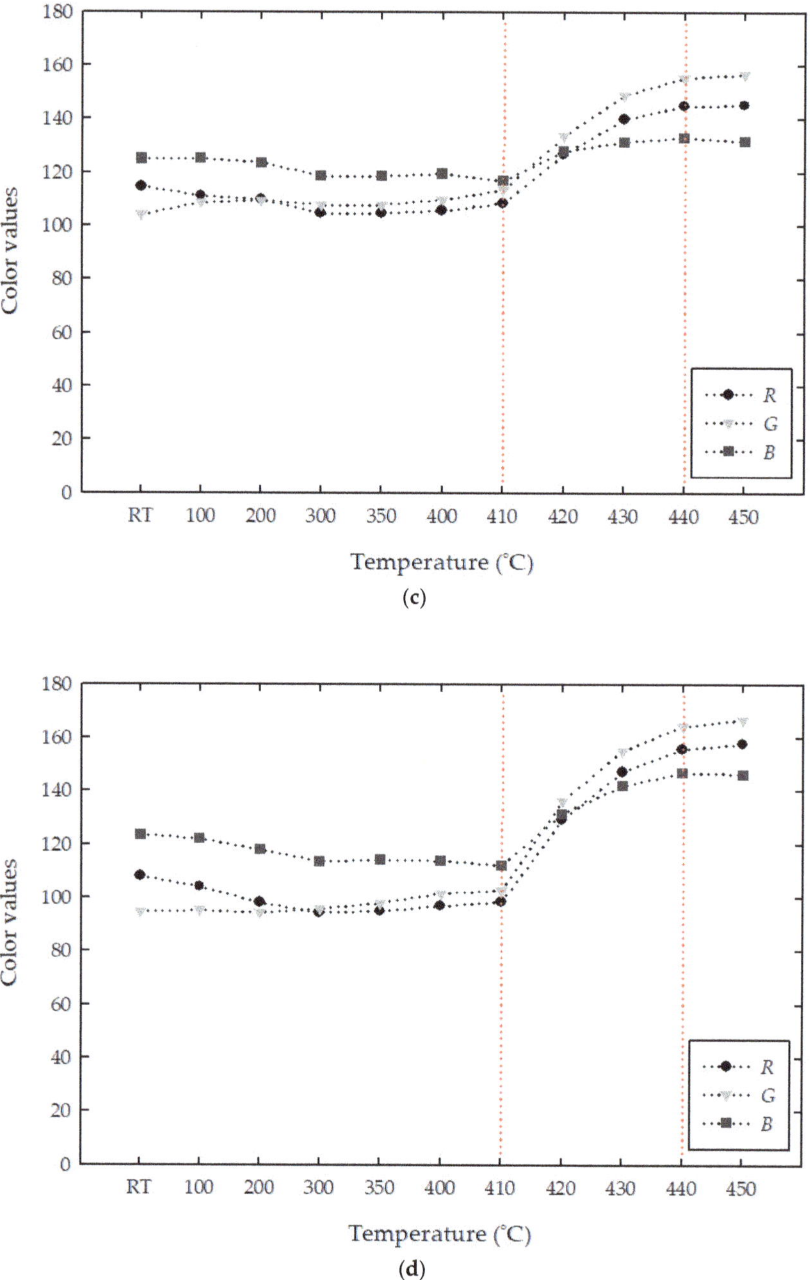

Figure 9. Changes in the *RGB* values of the white cement (WC) and MV mixed samples with increasing temperature: (**a**) WC; (**b**) WC + 1% MV; (**c**) WC + 3% MV; and (**d**) WC + 5% MV.

With increasing pigment content, the mean *RGB* values of the mixed samples from room temperature to 410 °C, in which the color of the samples was stable, were compared with those at a critical temperature of 440 °C, as shown in Figure 10. Compared to *R* values in between room temperature and 410 °C, those at 440 °C increased by approximately 22%, 33%, and 55% in the WC + 1% MV, WC + 3% MV, and WC +5 % MV specimens, respectively, as shown in Figure 10a. At a critical temperature of 440 °C, which completely changed to a grayish green color, the mixed samples with 1%, 3%, and 5% pigment showed R values in the range of 144 to 155. This indicates that the R value decreased due to the inclusion of pigment at room temperature, but increased to a similar range at a critical temperature of 440 °C. Figure 10b shows the change in G values for the white cement mixed samples. With increasing MV pigment, the G values decreased at room temperature. At 440 °C, the G values of the white cement with 1%, 3%, and 5% of MV pigment increased by 28%, 42%, and 68%, respectively, and showed a similar range of 155 to 168, irrespective of the content of MV pigment. Figure 10c shows the change in the B value for the white cement mixed samples. The B intensities, which were also stable between room temperature and 410 °C, decreased from approximately 183 to 116 with increasing content of MV pigment from zero to 5%. The B value at the critical temperature of 440 °C increased to 144, 133, and 146, which corresponds to a 7%, 10%, and 25% increase for the white cement samples with 1%, 3%, and 5%, respectively. The addition of pigments decreased the B values, but the amount of the increase is relatively small compared to that of the R and G values.

In the RGB color intensities, the incorporation of MV pigment particles in 1%, 3%, and 5% decreased the RGB intensities, which were stable between room temperature and 410 °C. The color intensities began to increase from 410 °C and completely changed to grayish green at 440 °C, at which point little difference of the RGB values was found among the white cement mixed samples. That is, with increasing pigment content, the RGB intensities decreased at room temperature but obtained relatively constant values at 440 °C. The RGB values and color intensities after 440 °C exhibited a slight difference, irrespective of the amount of MV pigment added to the white cement. Therefore, the magnitude of the change in the RGB values at 440 °C is also discussed. The changing magnitude of the RGB values mainly depends on the pigment to white cement ratio, as shown in Figure 11. With increasing addition of MV pigment from 1% to 5%, the magnitude of the change in the RGB almost linearly increased. According to the linear trend analysis, the R and G values showed a 10.9 and 13.0 increase at 440 °C per 1% addition of MV pigment to white cement, respectively. The approximately 6.5 increase in the B value per 1% MV pigment was relatively small. This suggests that the R and G values, which show a higher change at 440 °C than the B value, would be an index to instantly identify and evaluate cementitious materials subjected to a critical high temperature.

The color changes of the white cement mixed samples are related to the increasing number of RGB values. Therefore, the total color changes in the RGB color intensities are defined as the Euclidean distance $(\overline{d})$ using the following equation:

$$\overline{d} \;=\; \sqrt{(\Delta R)^2 + (\Delta G)^2 + (\Delta B)^2} \tag{2}$$

where ΔR, ΔG, and ΔB represent the differences of the R, G, and B values, respectively, 440 °C and those between room temperature to 410 °C. These also show a linear increase with increasing MV pigment ratio, as shown in Figure 12. The increasing rate of $\overline{d}$ is approximately 14.7 per 1% addition of MV pigment to white cement. As a result, the total color changes of white cement mixed with MV pigment can be effectively determined using the increases in RGB color values and $\overline{d}$.

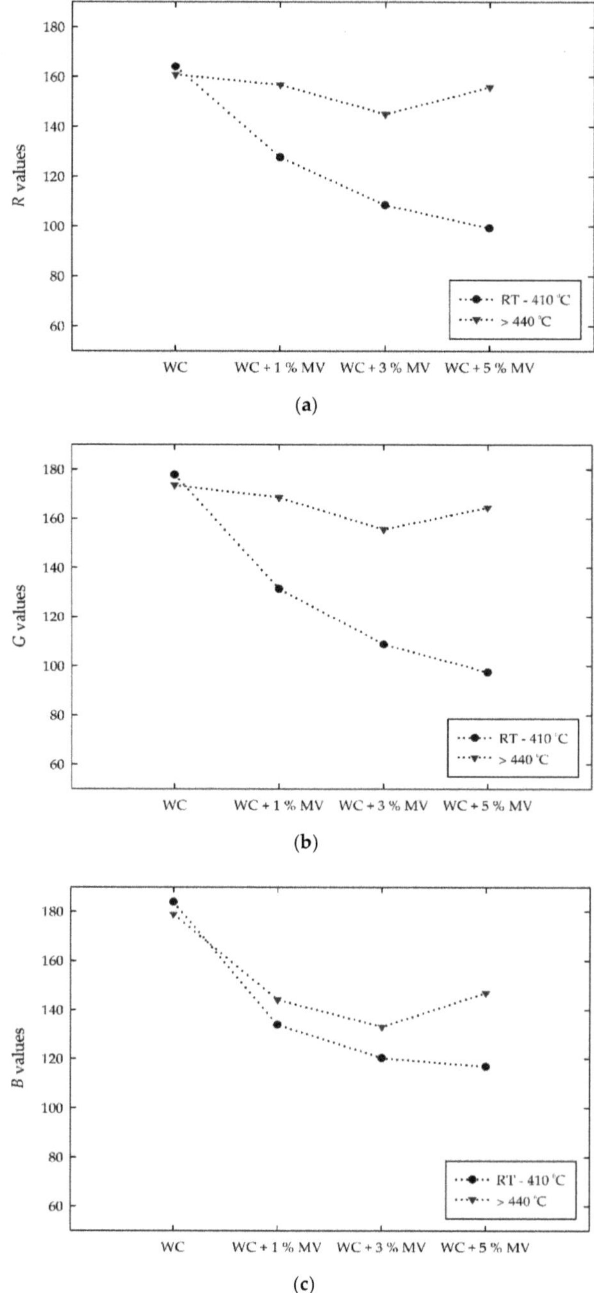

Figure 10. Comparisons of the *RGB* values for the white cement (WC) mixed samples with increasing temperatures: (**a**) *R* values; (**b**) *G* values; and (**c**) *B* values.

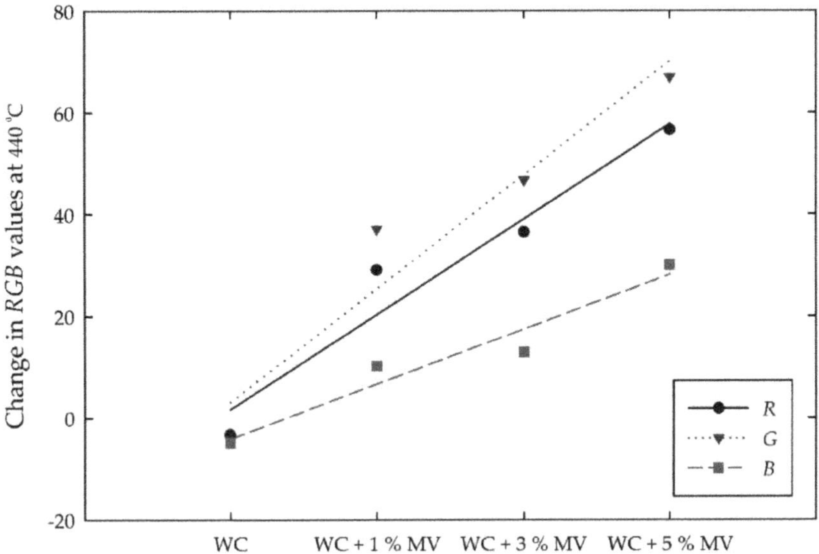

Figure 11. Changes in the *RGB* values for the white cement (WC) mixed samples with increasing MV pigment contents at 440 °C.

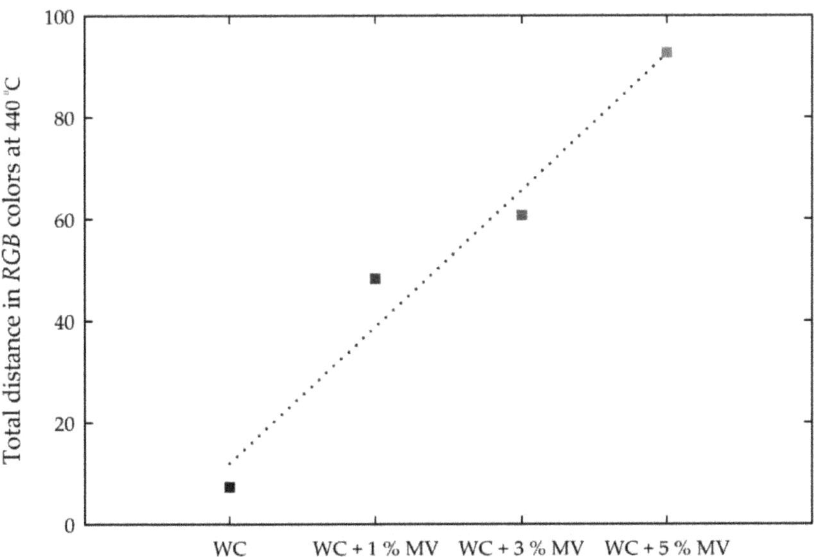

Figure 12. Total distance in the *RGB* intensities for the white cement (WC) mixed samples with increasing MV pigment content at 440 °C.

4.4. Thermochromic Analysis in the Lab Color Spaces

Figure 13 shows the changes in the *Lab* values of the white cement mixed samples with increasing temperature. The white cement exhibited a constant high L value of approximately 72, and the a and b values present negative constant values of approximately −3 and −4, respectively, with increasing temperature from room temperature to 450 °C, as shown in Figure 13a. This suggests that the white cement is stationary with increasing temperature. The addition of MV pigment decreased the L and b values, whereas it increased the a value at room temperature, as shown in Figure 13b–d. The mixed samples with 1%, 3%, and 5% of the MV pigment showed a decrease in L value from 72 to 52, 45, and 40, corresponding to a 28%, 36%, and 43% decrease, respectively. The a value, which was −3 at room temperature for white cement, changed to positive values (redness degrees) of 2, 8, and 10 for the mixed samples with 1%, 3%, and 5% of MV pigment, respectively. In contrast, the b intensity, which was negative for the white cement, showed a further decrease to a negative value (blueness degree). The change in the *Lab* values influences the development of color on the mixed samples. That is, the WC + 1% MV sample with the Lab of (52, 2, −4) represents a dark grayish blue color, and WC + 3% MV and WC + 5% MV samples with the *Lab* of (45, 8, −10) and (40, 10, −13), respectively, illustrate dark grayish violet color at room temperature. As the temperature increased from room temperature to 410 °C, the L value was relatively constant, the a value gradually decreased from positive to negative, and the b value increased from negative to positive. However, the changes in the *Lab* values were too low to observe the color change of the mixed samples until 410 °C.

After 410 °C, the color of the samples started to show a grayish green color with increasing L value and changing a from positive to negative and b from negative to positive values. The increase in L value increases the lightness intensity, and the decrease in a and increase in b indicate the greenness and yellowness intensities, respectively. At 440 °C, the mixed samples completely changed to a grayish green color and showed Lab values of (64, −7, 9) for 1% MV, (62, −7, 10) for 3% MV, and (64, −6, 9) for 5% MV samples. As discussed in the *RGB* color spaces, the *Lab* values were also similar at the critical temperature of 440 °C, irrespective of the amount of MV pigment added to the white cement. Above the critical temperature, the mixed samples exhibited stable and invariable *Lab* values.

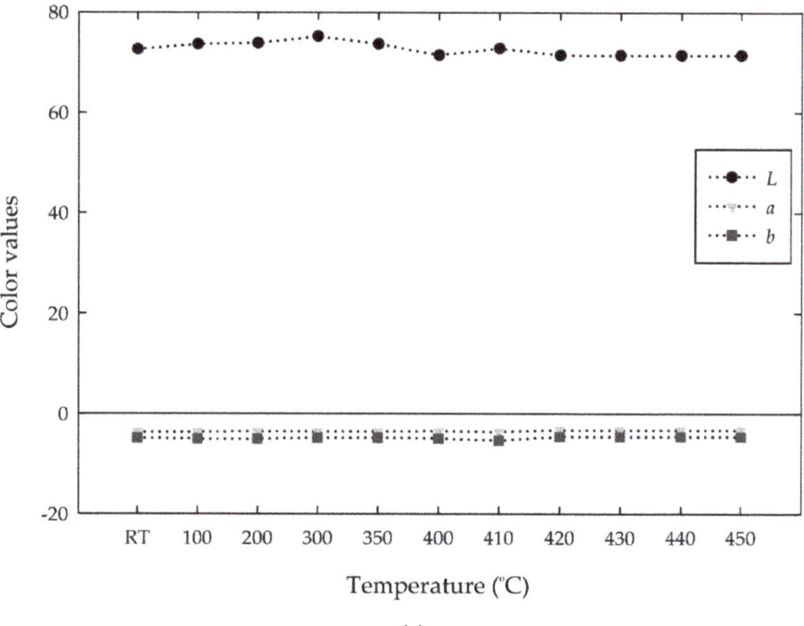

(a)

Figure 13. *Cont.*

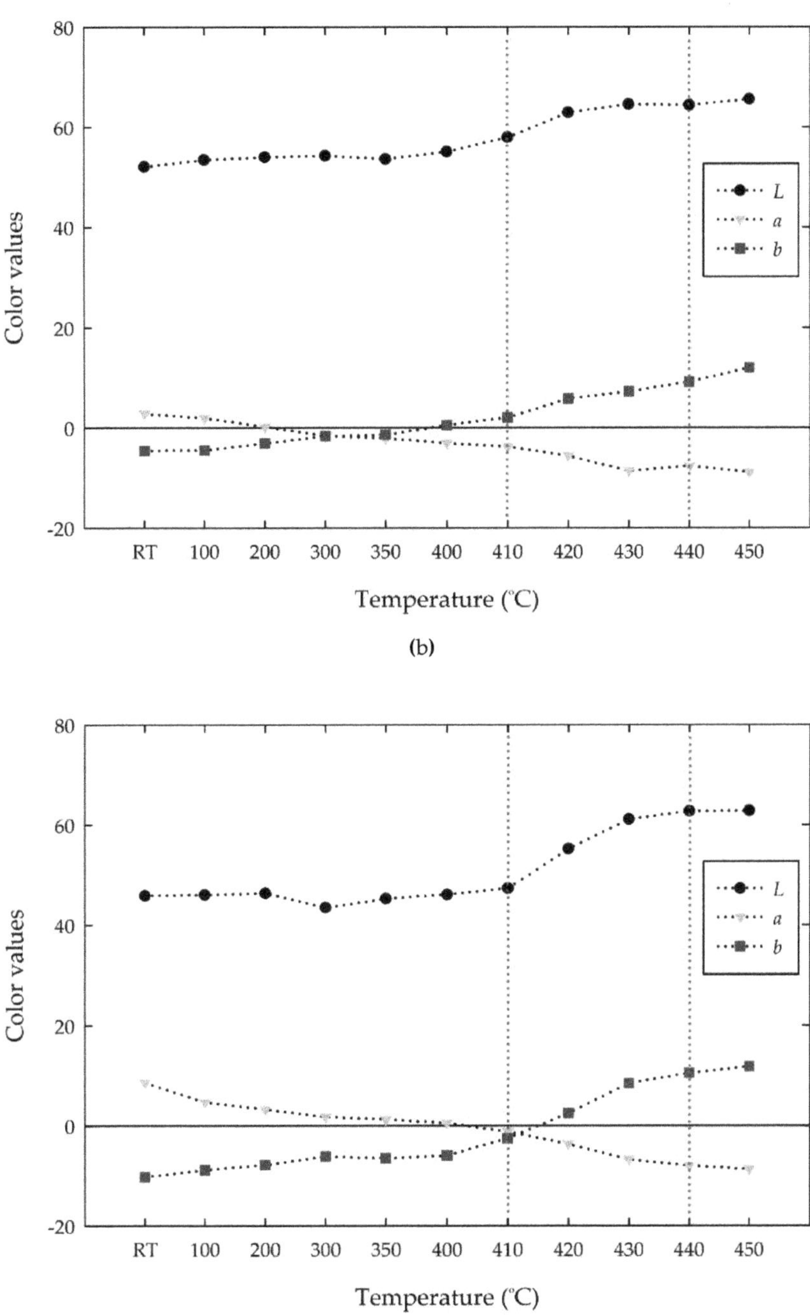

(b)

(c)

Figure 13. *Cont.*

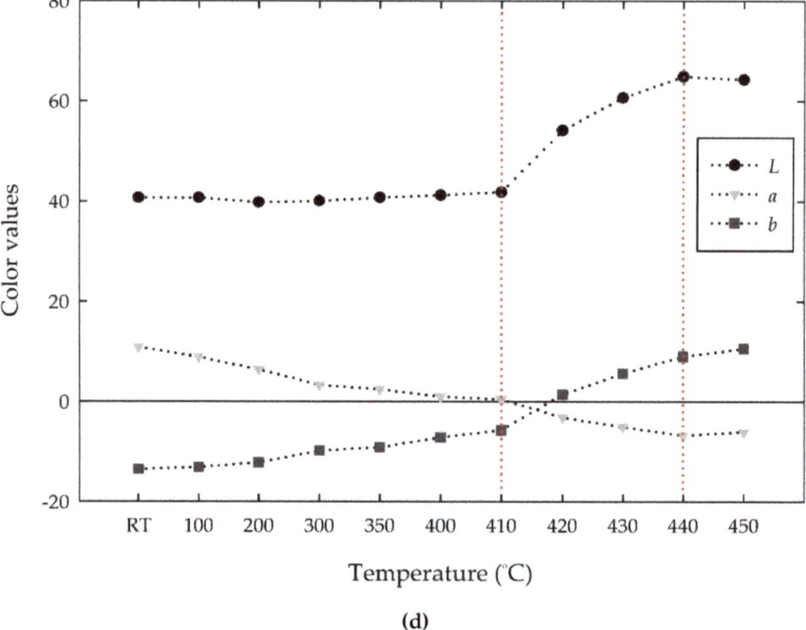

(d)

Figure 13. Changes in the *Lab* values of the white cement (WC) and MV mixed samples with increasing temperature: (**a**) WC; (**b**) WC + 1% MV; (**c**) WC + 3% MV; and (**d**) WC + 5% MV.

Figure 14 compares the *Lab* values of the mixed samples averaged from room temperature to 410 °C with those at a critical temperature of 440 °C. As shown in Figure 14a, the L value (or, the lightness intensity) decreased when the pigment concentration was increased from zero to 5%. However, the L value at 440 °C showed a similar level, ranging from 62 to 64, which indicates a grayish color. Figure 14b shows the change in a value. Compared to those between room temperature and 410 °C, the a values at the critical temperature of 440 °C decreased from positive to negative, resulting in a very similar range of −7 to −6 for all three mixed samples. In contrast, the b value increased from negative to positive, ranging from 9 to 10, which illustrates the increase in yellowness intensity, when the mixed samples reached the critical temperature of 440 °C, as shown in Figure 14c.

Based on the mean *Lab* values between room temperature and 410 °C, in which no major change was obtained in all the white cement mixed samples, Figure 15 shows the magnitude of the change in the *Lab* values at 440 °C. The L and b values changed to positive values, and the a values changed to negative values. That is, with increasing MV pigment from zero to 5%, the L and b values increased almost linearly from −1.9 to 23.8 and from 0.3 to 20.6, respectively, whereas the a value decreased from 0.4 to −11.1. A linear trend analysis showed that the magnitude of the L and b values at 440 °C increased by approximately 4.7 and 3.6, respectively, and the a value, the change of which was relatively small, decreased by approximately 2.1 per 1% increase in MV pigment.

In addition, the total changes in the *Lab* intensities at the critical temperature of 440 °C were calculated, as shown in Figure 16. Similar to Equation (2), the total change was defined as the Euclidean distance ($\bar{d}$), using the difference between the mean *Lab* values from room temperature to 410 °C and those at 440 °C. As the content of the MV pigment added to white cement increased from zero to 5%, $\bar{d}$ tended to increase almost linearly at a rate of approximately 5.8 per 1% addition of MV pigment to white cement. Along with *RGB* intensities, the *Lab* changes can be used to determine the color change at a critical temperature with the white cement-mixed samples containing MV pigment.

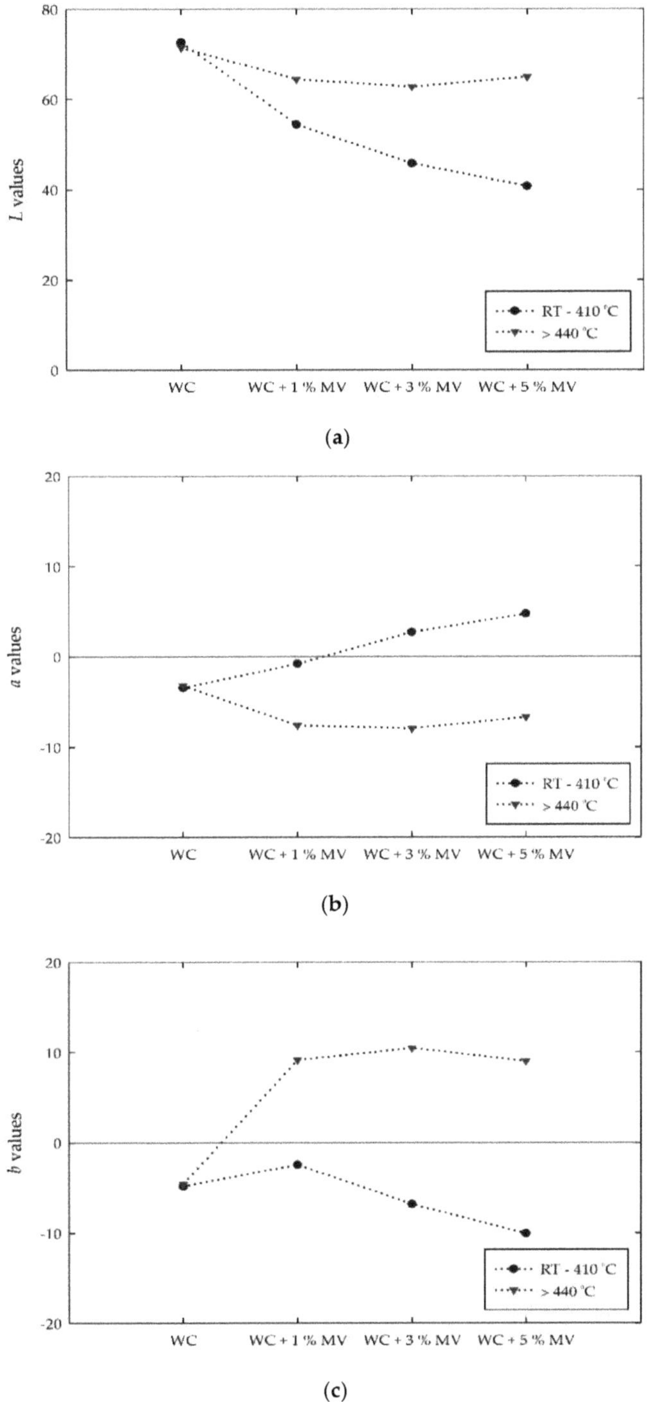

Figure 14. Comparisons of the *Lab* values for the white cement (WC) mixed samples with increasing temperature: (**a**) *L* values; (**b**) *a* values; and (**c**) *b* values.

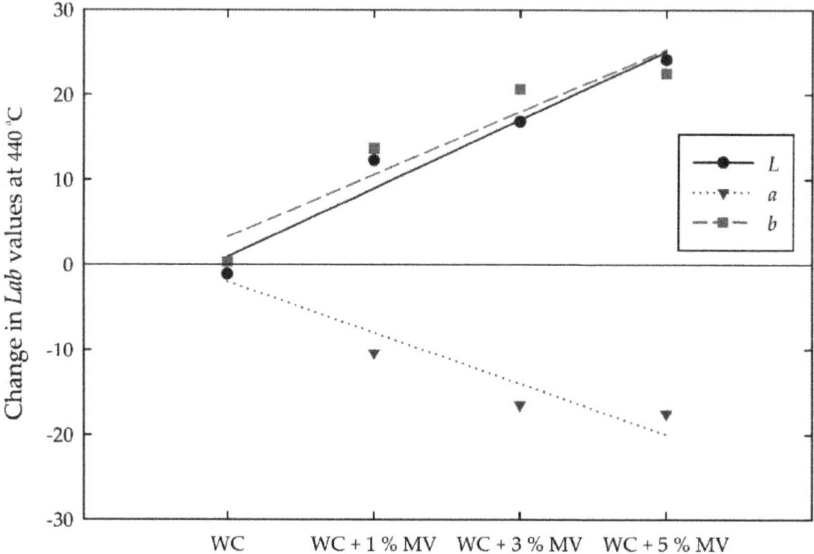

Figure 15. Changes in the *Lab* values for the white cement (WC) mixed samples with increasing MV pigment content at 440 °C.

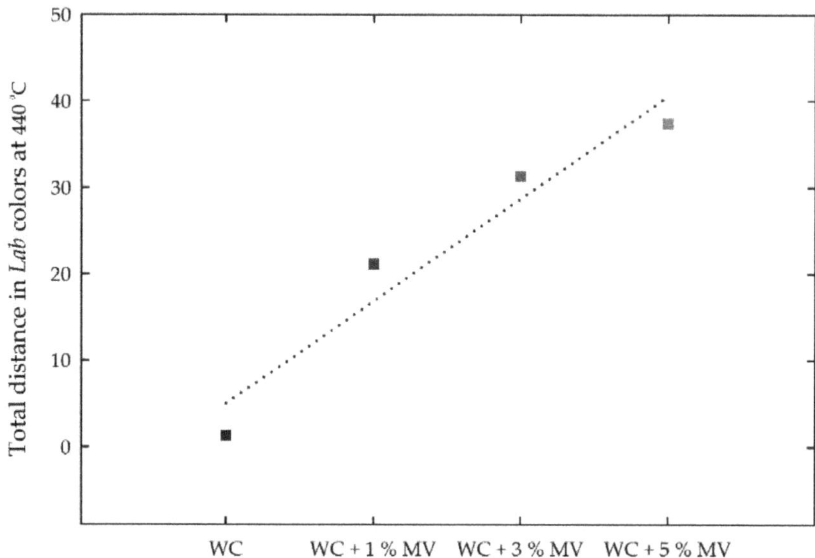

Figure 16. Total distance in the *Lab* intensities for the white cement (WC) mixed samples with increasing MV pigment content at 440 °C

5. Conclusions

This study proposed an irreversible thermochromic cementitious material using MV pigment. According to the thermochromic tests, the MV pigment underwent a reversible change from violet to blue at temperatures lower than 400 °C, but the color change was barely discernable by the naked eye.

When the temperature reached 410 °C, the MV pigment completely changed from violet to grayish yellow, which is associated with the evaporation of water and the liberation of ammonia from the pigment particles. The changed color was maintained at temperatures higher than 410 °C and when the temperature was returned to room temperature.

In the analysis of MV pigment in the *RGB* and *Lab* color spaces, the values were relatively constant from room temperature to 370 °C. At higher temperatures, the *RGB* values increased gradually toward the whiteness color intensity until 400 °C. The *L* and *b* values also gradually increased after 370 °C, while the *a* value decreased. At 410 °C, when the color completely changed from dark violet to grayish yellow, the *RGB* and *Lab* provided a sudden change and remained constant after the critical temperature of 410 °C. In particular, the *a* value changed from positive to negative, and the *b* value changed from negative to positive.

The MV pigment was then mixed with white cement at ratios of 1%, 3%, and 5% of the mass of the white cement. The color of the mixed samples was dark grayish blue and violet, depending on the content of MV pigment. The colors at room temperature were retained until 410 °C. Hence, the *RGB* and *Lab* intensities were relatively stable and constant. The color started to change at temperatures higher than 410 °C and completely turned to grayish green at 440 °C. In accordance with the change in color, the *RGB* increased toward the whiteness intensity. Compare to those before the thermochromic change occurred, the *RGB* values at 440 °C increased by approximately 22%–55%, 28%–68%, and 7%–25%, respectively. The *RGB* values almost linearly increased with the increasing content of MV pigment.

Similarly, the *L* value increased by approximately 23%–60% at 440 °C. The *a* value changed from positive to negative, and the *b* value changed from negative to positive. The changes in the Lab values also increased almost linearly with increasing MV pigment content. According to the linear trend analyses, the magnitude of the *Lab* values at 440 °C changed by approximately 4.7, 3.6, and −2.1, respectively, and the *RGB* values increased by approximately 10.9, 13.0, and 6.5, respectively, per 1% increase in MV pigment. In addition, the total changes in the *RGB* and *Lab* color intensities at the critical temperature of 440 °C, defined as the Euclidean distance, showed the increasing rates of approximately 14.7 and 5.8 per 1% addition of MV pigment to white cement.

However, the level of the *RGB* and *Lab* values and the color intensity of the mixed samples at 440 °C were in very similar ranges, irrespective of the amount of MV pigment added to white cement. Above a critical temperature of 440 °C and back to room temperature, the mixed samples exhibited stable and irreversible *RGB* and *Lab* values. Therefore, with a direct visual inspection of the distinct color change, the changes in the *RGB* and *Lab* values can provide an instant and promising index to identify the magnitude and dispersion of critical temperatures in cementitious materials and structures with MV pigment.

SEM images showed that the rod-shaped and hexagonal crystal structures of the MV pigment at room temperature were modified to a cluster of void structures at 410 °C. The EDX analysis results of the MV pigment, comprised of 11.36%, 48.91%, 16.30%, and 14.56% of C, O, P and Mn, respectively, changed to 6.89%, 46.37%, 17.05%, and 14.37%, after exposure to 410 °C, respectively. In the SEM images, the white cement, composed of hexagonal and spherical shapes with some irregular tiny particles, was relatively stable with slight changes due to the breakage of particles at 440 °C. The mixed samples consisted of triangular, hexagonal, and spherical particles with a small cluster of rod-shaped structures at room temperature. At 440 °C, the hexagonal and rod-shaped cluster structures disappeared due to the dehydration of MV particles, which changed into irregular bundled structures. In the EDX analysis, the white cement, which is mainly composed of C, O, and Ca with the mass contents of 14.9%, 47.35%, and 25.09% at room temperature, increased by 3%, 1.5%, and 7%, respectively, at 440 °C. The mixed sample exhibited C, O, and Ca with the mass components of 17.18%, 40.43%, and 19.19%, respectively, and P, Mn, Sn, Sb, and I with the mass of 2.05%, 1.78%, 1.24%, 8.03%, and 2.21%, respectively, due to the addition of MV pigment. At 440 °C, the mass of C, O, and Ca changed to 24.29%, 39.09%, and 17.05%, respectively, and that of P, Mn, Sn, Sb, and I decreased by 24%, 3%, 28%, 16%, and 6%, respectively.

Author Contributions: J.-H.L. and R.S.R. planned and designed the experiments; R.S.R. performed the experiments. J.-H.L. analyzed the data; J.-H.L. and R.S.R. wrote and revised the paper. All authors have read and agreed to the published version of the manuscript.

Acknowledgments: This research was supported by a grant from the Mid-career Research Program (NRF-2019R1A2C1006494) through the National Research Foundation (NRF) Korea.

Conflicts of Interest: The authors declare no conflict of interest.

References

1. Kim, K.Y.; Yun, T.S.; Park, K.P. Evaluation of pore structures and cracking in cement paste exposed to elevated temperatures by X-ray computed tomography. *Cem. Concr. Res.* **2013**, *50*, 34–40. [CrossRef]
2. Lee, J.; Choi, K.; Hong, K. Color and Material Property Changes in Concrete Exposed to High Temperatures. *J. Asian Arch. Build. Eng.* **2009**, *8*, 175–782. [CrossRef]
3. Zhang, Q.; Ye, G. Dehydration kinetics of Portland cement paste at high temperature. *J. Therm. Anal. Calorim.* **2012**, *110*, 153–158. [CrossRef]
4. Vejmelková, E.; Koňáková, D.; Scheinherrová, L.; Doleželová, M.; Keppert, M.; Cerny, R. High temperature durability of fiber reinforced high alumina cement composites. *Constr. Build. Mater.* **2018**, *162*, 881–891. [CrossRef]
5. Sabeur, H.; Platret, G.; Vincent, J. Composition and microstructural changes in an aged cement pastes upon two heating–cooling regimes, as studied by thermal analysis and X-ray diffraction. *J. Therm. Anal. Calorim.* **2016**, *126*, 1023–1043. [CrossRef]
6. Collier, N.C.; Collier, N. Transition and decomposition temperatures of cement phases-a collection of thermal analysis data. *Ceram. Silik.* **2016**, *60*, 338–343. [CrossRef]
7. Osumi, A.; Enomoto, M.; Ito, Y. Basic study of an estimation method for fire damage within concrete sample using high-intensity ultrasonic waves and optical equipment. *Jpn. J. Appl. Phys.* **2014**, *53*, 7. [CrossRef]
8. Shariati, M.; Ramli-Sulong, N.H.; Mohammad Mehdi Arabnejad, K.H.; Shafigh, P.; Sinaei, H. Assessing the strength of reinforced concrete structures through ultrasonic pulse velocity and schmidt rebound hammer tests. *Sci. Res. Essays* **2011**, *6*, 213–220.
9. Colombo, M.; Felicetti, R. New NDT techniques for the assessment of fire-damaged concrete structures. *Fire Saf. J.* **2007**, *42*, 461–472. [CrossRef]
10. Annerel, E.; Taerwe, L. Revealing the temperature history in concrete after fire exposure by microscopic analysis. *Cem. Concr. Res.* **2009**, *39*, 1239–1249. [CrossRef]
11. Short, N.; Purkiss, J.; Guise, S. Assessment of fire damaged concrete using colour image analysis. *Constr. Build. Mater.* **2001**, *15*, 9–15. [CrossRef]
12. Hager, I. Colour Change in Heated Concrete. *Fire Technol.* **2013**, *50*, 945–958. [CrossRef]
13. Annerel, E.V.R.; Taerwe, L. Assessment Techniques for the Evaluation of Concrete Structures After Fire. *J. Struct. Fire Eng.* **2013**, *4*, 123–130.
14. Felicetti, R. Combined while-drilling techniques for the assessment of deteriorated concrete cover. In Proceedings of the 7th International Symposium on Nondestructive Testing in Civil Engineering, Nantes, France, 30 June–3 July 2009.
15. Hager, I. Behaviour of cement concrete at high temperature. *Bull. Pol. Acad. Sci. Tech. Sci.* **2013**, *61*, 145–154. [CrossRef]
16. Lee, J. The effect of high temperature on color and residual compressive strength of concrete. In Proceedings of the 7th International Conference on Fracture Mechanics of Concrete and Concrete Structures. High Performance, Fiber Reinforced Concrete, Special Loadings and Structural Applications, Jeju, South Korea, 23–28 May 2010; pp. 1772–1775.
17. Ferrara, M.; Bengisu, M. Materials that Change Color. In *Springer Briefs in Applied Sciences and Technology*; Springer Science and Business Media LLC: Heidelberg, Germany, 2013; pp. 9–60.
18. Zhang, D.; Sun, H.-J.; Wang, M.-H.; Miao, L.-H.; Liu, H.-Z.; Zhang, Y.-Z.; Bian, J. VO_2 Thermochromic Films on Quartz Glass Substrate Grown by RF-Plasma-Assisted Oxide Molecular Beam Epitaxy. *Mater.* **2017**, *10*, 314. [CrossRef] [PubMed]

19. Karlessi, T.; Santamouris, M.; Synnefa, A.; Assimakopoulos, D.; Didaskalopoulos, P.; Apostolakis, K. Development and testing of PCM doped cool colored coatings to mitigate urban heat island and cool buildings. *Build. Environ.* **2011**, *46*, 570–576. [CrossRef]

20. Ma, Y.; Zhu, B.; Wu, K. Preparation and solar reflectance spectra of chameleon-type building coatings. *Sol. Energy* **2001**, *70*, 417–422. [CrossRef]

21. Yan, D.; Lu, J.; Ma, J.; Wei, M.; Evans, D.G.; Duan, X. Reversibly Thermochromic, Fluorescent Ultrathin Films with a Supramolecular Architecture. *Angew. Chem. Int. Ed.* **2010**, *50*, 720–723. [CrossRef]

22. Kamalisarvestani, M.; Saidur, R.; Mekhilef, S.; Javadi, F. Performance, materials and coating technologies of thermochromic thin films on smart windows. *Renew. Sustain. Energy Rev.* **2013**, *26*, 353–364. [CrossRef]

23. Parnklang, T.; Boonyanuwat, K.; Mora, P.; Ekgasit, S.; Rimdusit, S. Form-stable benzoxazine-urethane alloys for thermally reversible light scattering materials. *Express Polym. Lett.* **2019**, *13*, 65–83. [CrossRef]

24. Yu, J.-H.; Nam, S.-H.; Lee, J.W.; Boo, J.-H. Enhanced Visible Transmittance of Thermochromic VO_2 Thin Films by SiO_2 Passivation Layer and Their Optical Characterization. *Materials* **2016**, *9*, 556. [CrossRef]

25. Seeboth, A.; Ruhmann, R.; Mühling, O. Thermotropic and Thermochromic Polymer Based Materials for Adaptive Solar Control. *Materials* **2010**, *3*, 5143–5168. [CrossRef] [PubMed]

26. Srirodpai, O.; Wootthikanokkhan, J.; Nawalertpanya, S.; Yuwawech, K.; Meeyoo, V. Preparation, Characterization and Thermo-Chromic Properties of EVA/VO_2 Laminate Films for Smart Window Applications and Energy Efficiency in Building. *Materials* **2017**, *10*, 53. [CrossRef] [PubMed]

27. Yang, L.; Zhi-Min, L. The Research of Temperature Indicating Paints and its Application in Aero-engine Temperature Measurement. *Procedia Eng.* **2015**, *99*, 1152–1157. [CrossRef]

28. Popescu, M.; Serban, L. Thermo-indicating paint for damage warning. *J. Therm. Anal. Calorim.* **1996**, *46*, 317–321. [CrossRef]

29. Nguyen, D.K.; Bach, Q.-V.; Lee, J.-H.; Kim, I.-T. Synthesis and Irreversible Thermochromic Sensor Applications of Manganese Violet. *Materials* **2018**, *11*, 1693. [CrossRef]

30. Lee, J.D.; Browne, L.S. The nature and properties of manganese violet. *J. Chem. Soc. A* **1968**, 559–561. [CrossRef]

31. Begum, Y.; Wright, A.J. Relating highly distorted Jahn–Teller MnO_6 to colouration in manganese violet pigments. *J. Mater. Chem.* **2012**, *22*, 21110. [CrossRef]

32. Moresova, K.; Skvara, F. White cement-properties, manufacture, prospects. *Ceram. Silikaty* **2001**, *45*, 158–163.

33. Tkaleie, M.; Tasie, J.F. *Color Spaces-Perceptual, Historical and Applicational Background*; IEEE: Ljubljana, Slovenia, 2003; pp. 304–308.

34. Manganese Violet Pigment. Available online: https://www.kremer-pigmente.com (accessed on 1 January 2018).

35. *ACI Committee 212*; Report on Chemical Admixtures for Concrete; American Concrete Institute: Farmington Hills, MI, USA, 2010.

Article

The Influence of Selected Local Phenomena in CFRP Laminate on Global Characteristics of Bolted Joints

Krzysztof Puchała [1,*], Elżbieta Szymczyk [1], Jerzy Jachimowicz [1], Paweł Bogusz [1] and Michał Sałaciński [2]

[1] Institute of Mechanics and Computational Engineering, Military University of Technology, gen. Sylwestra Kaliskiego Street 2, 00-908 Warsaw, Poland; elzbieta.szymczyk@wat.edu.pl (E.S.); jerzy.jachimowicz@wat.edu.pl (J.J.); pawel.bogusz@wat.edu.pl (P.B.)

[2] Air Force Institute of Technology, Ksiecia Boleslawa Street 6, 01-494 Warsaw, Poland; michal.salacinski@itwl.pl

* Correspondence: krzysztof.puchala@wat.edu.pl; Tel.: +48-261-839-039

Received: 25 October 2019; Accepted: 9 December 2019; Published: 10 December 2019

Abstract: High specific mechanical properties of composites are the reason for their use in various fields, e.g., the aerospace industry. Mechanical joints are still used in the aerospace industry to assembly large aircraft structures. The properties of laminate around the hole can be, however, weakened, compared to their nominal values as a result of a drilling process or cyclic loading. This paper aims at the classification and analysis of imperfections affecting mechanically fastened joints in a laminate structure. A method of modeling the hole vicinity, a gradient material model, as well as the numerical and experimental estimation of laminate deterioration in this area, were proposed and analyzed. Comparative analysis of numerical and experimental results based on displacements of the testing machine grip and the extensometer length confirmed the aforementioned results as consistent in linear ranges. Therefore, joint characteristics obtained based upon measurement of the grip displacement and the ratio of stiffness in linear ranges are sufficient to determine the parameters of a gradient material model. Some imperfections resulting from, e.g., asymmetry, were included in the gradient material model; thus, the obtained weakening of laminate properties in the hole vicinity can be overestimated. Therefore, further analyses of the gradient material model for laminate structures are necessary.

Keywords: CFRP laminate; mechanically fastened joints; gradient material model

---

1. Introduction

Advantageous mechanical properties of composites together with their lower mass in comparison to conventional materials [1–3] are the reasons for their use in various branches of industry, e.g., aerospace, military, high-tech car, energy turbines and other domains, such as pipeline repairs [4,5] (Figure 1a). Beneficial features of laminates, including properties tailoring, i.e., adapting their configuration to service load, create a possibility to use them for primary structures, e.g., frames, longerons, stringers, ribs and coverings (Figure 1b). Thin-walled aircraft elements, including laminate panels, are connected and reinforced with stiffeners in order to build an airframe. Assembling a whole structure from simpler elements also creates a possibility to repair and replace a selected part if it is necessary.

Mechanically fastened joints (riveted, bolted, etc.) belong to the most reliable methods of connection, and despite some disadvantages, especially in laminate structures, they are still used beside adhesive and hybrid joints [6,7]. However, the holes, necessary for mechanical joints, constitute free edges and lead to stress concentration. Additionally, they are subjected to point loads. These phenomena lead to severe stress concentrations in the area of mechanical joints, and therefore, they demand special attention during the design, manufacturing and service stages.

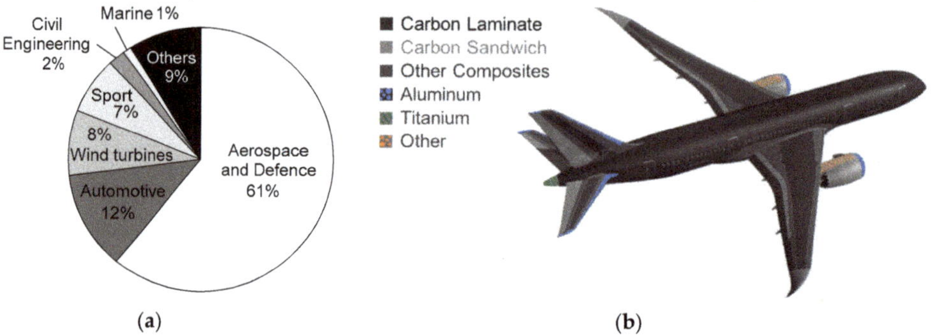

Figure 1. Use of composites: (**a**) Carbon Composite material market (redrawn from [8]); (**b**) Materials used in a Boeing 787 (redrawn from [9])

Many papers deal with mechanical joints, especially in a laminate structure [10–17], and conclude that mechanical joints still undergo both the experimental and numerical studies in order to improve their behavior. Various methods for modification of the hole vicinity are presented and analyzed [18–21]. Local FML, analyzed in refs. [3,22–24], is one of the aforementioned methods. In the paper, the comparison of both the experimental and the numerical results of CFRP laminate in mechanical joints is performed in the aspect of laminate modeling and a possible further use in the modeling of FML.

There are different methods and levels for laminate modeling, due to a complex laminate structure and different aims of analysis [25–27]. Despite a continuous increase in the computing power and performance (calculation ability) of numerical algorithms, the modeling of mechanically fastened joints (riveted, bolded, etc.) of CFRP laminate is still a challenging task. On the one hand, this difficulty is caused by a complex laminate structure (different stiffness of components, various modes of laminate failure, as well as notch and point load sensitivity of laminates [1,28,29]). On the other hand, it results from the nonlinearity of the contact problem and size of the connected parts of the structure.

The development of a suitable model of the joint requires identification of parameters affecting its stiffness and load carrying capacity (i.e., strength of the joint) as well as classification of those parameters with respect to an adequate criterion (adequately to the aim of analysis). The paper presents a classification and discussion of the selected parameters.

In order to obtain materials of high quality, characterized by repeatable properties, composite aircraft structures are made of preimpregnates (prepregs) using the autoclave technology [30,31]. The stiffness and strength of large structural components assembled with a set of mechanically fastened parts are dependent on the joint quality and influenced by the drilling technology. The properties of laminate around the hole can be weakened/deteriorated compared to their nominal values. This deterioration can be a result of the drilling technology, and on the other hand, the cyclic load during the service of the structure [32–34]. Analysis of the hole machining effects can be found in refs. [35–41]. A literature review on the drilling of composites and damage connected with it is presented in refs. [42–46]. The reverse topography obtained for the hole surface with computer tomography can be found in ref. [47]. The damage analysis, by means of acoustic emission during drilling, is presented in ref. [48]. Furthermore, since the modern aircrafts consist of different materials (e.g., composites and metal alloys), there is a need of drilling the composite–metal stack [49,50].

Delamination is the most studied damage form connected with hole drilling in laminates. However, it can be concluded based on refs. [51–53] that other forms of damage also occur and influence mechanical properties of laminate around the hole.

The fundamental methods for determining the material properties and characteristics of structural components (including joints) are experimental tests. They are especially significant in the case of laminates, due to their sensitivity to the manufacturing technology [54], as well as notch and point load sensitivity [55–58].

As it was mentioned above, development of an adequate model of a mechanically fastened composite structure requires an identification of material parameters. Usually, the properties of a single lamina (in the case of an orthotropic material model—nine stiffness and nine stress components) are experimentally determined. In the next step, the properties of laminate, for the selected stacking sequence, can be established based on a Classical Laminate Theory (CLT). Since the actual and calculated properties of laminate usually differ (actual values are usually several percent lower than theoretical ones), they should be experimentally verified and corrected, if necessary, in order to develop the numerical model [59]. Moreover, for the selected stacking sequence of laminate, it is desirable to determine its bearing properties (stiffness and strength) and/or the tensile characteristics of a simple joint. The latter characteristics can be of greater value due to the coupling effects of bearing, by-pass loads and secondary bending in the case of single-lap joints [60,61].

The aim of a series of papers (including refs. [53,62] and this paper) is analysis of the parameters determining the stiffness and strength of laminate in the area of mechanically fastened joints to develop a gradient material model.

2. Object of Analysis

A double-shear bolted joint with four steel fasteners (Figure 2) is analyzed in the paper. Development of the joint is presented in ref. [63]. The outer sheets are made of 2024T3 aluminum alloy, and the inner element is made of quasi-isotropic CFRP laminate (thermoset epoxy resin reinforced with carbon fibers). CFRP laminate consists of UD layers (HTA/913) and external fabric layers (TR30S twill woven). A stacking sequence of this laminate part is $[(0)/0/45/90/45/0/45/90/45/0/90]_s$. The specimen length L is 300 mm. The nominal diameter d of both the bolt and the hole is 6 mm. Other dimensions of the joint are as follows: The pitch length is equal to $5d$ (30 mm), the specimen width w is equal to 70 mm, and the thickness of the aluminum alloy sheet is 2 mm.

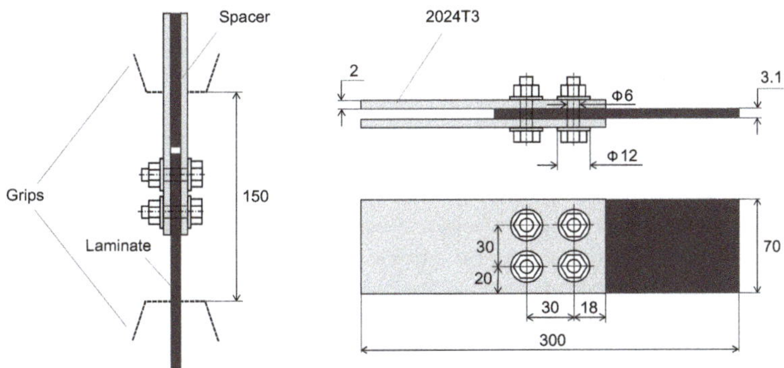

Figure 2. The analyzed double-shear bolted joint.

Specimens were manufactured by the Air Force Institute of Technology (Warsaw, Poland). The laminate panels were made of preimpregnates in the autoclave technology. The average thickness of a laminate element is 3.1 mm and the standard deviation (SD) is 0.02 mm. Parameters of HTA/913 lamina, TR30S lamina and interfaces are presented in ref. [53]. Pilot holes were made in each part. In the non-cured laminate the initial pilot holes were perforated by the 2.8 mm pins fixed to the mold. Laminate panels (after curing in the autoclave) and aluminum sheets were cut to the specified dimensions with the use of the water-jet. Subsequently, the elements were collected together (as in Figure 2) and pilot holes of 3 mm were drilled.

Final holes were made in stack of elements in two stages. In the first stage, a 5.8 mm drill was used. In the second stage the holes were reamed with a 6H7 hand reamer. The specimens were assembled with the bolt torque of 10.6 Nm.

The present paper is part of a work intended to improve the performance of mechanically fastened laminates by local reinforcement/hybridization with metal (titanium alloy) sheets. The analyzed joint is a basis to evaluate the reinforcement; therefore, both the basic and reinforced joints should be made with as similar as possible technology. Taking into account the fact that local reinforcement was supposed to be made of titanium alloy sheets, cobalt drills were chosen. A point angle, lip relief angle and a helix angle of the used drills are 140°, 12° and 30°, respectively. The rotational speed was 750 rpm and the feed was 0.1 mm/rev. Dry, compressed air was used as a cooling agent. An average hole diameter measured in the specimens is 6.00 and 6.06 mm, whereas its standard deviation is 0.02 and 0.06 mm for aluminum alloy and laminate, respectively.

The areas of the initial material failure around the hole were detected with non-destructive testing. An ultrasonic method with C-scan imaging was used with parameters based on refs. [64–66]. An average diameter of those areas is 11.25 mm and standard deviation is 1.26 mm [53].

3. Numerical Model of Mechanical Joint

Numerical models were developed and analyzed with Mentat® and Marc® code using standard (build in Marc code), finite elements, material models and failure procedures. A nominal (base) numerical model of the joint and its modification leading to satisfactory results (good agreement with the experiment) is presented in refs. [53,62]. The main features of the numerical model are described below. A solid element was used to model all of the components (aluminum sheets, laminate part and fasteners). The results of numerical analysis can be strongly influenced by the mesh density. The proper choice of an element size is especially important in the areas of stress concentration, e.g., holes and point loads. The hole in the joint satisfies both of these conditions. It was noticed that division of the hole circumference into 32 elements is satisfactory for mechanical joints, i.e., further refinement practically does not influence the results [67]. Two times denser mesh (64 elements in the circumferential direction) was used in the analyzed model. The size of elements in the radial direction was chosen so as to achieve proportional discretization in the hole vicinity and a gradual increase along with the radial distance from the hole. In ref. [59], a finite element mesh arrangement and density are also discussed for a single-lap joint, and similar conclusions, concerning the fine mesh around the hole and under the washer, are presented. Additionally, refining of the mesh in the non-overlap region is proposed, due to secondary bending of a single-lap joint. However, the latter effect does not occur in a double-lap joint analyzed in the paper.

Metal components (aluminum alloy sheets and bolts) were modeled as an isotropic elastic –plastic material. Each lamina was modeled as an orthotropic elastic–brittle material using one layer of finite elements. Connections of laminate plies were described using a cohesive material model (zero thickness cohesive elements). Various descriptions of contact were compared in ref. [27], and it was found that a segment to segment analytical contact built in Marc® code gave satisfactory results; therefore, this procedure was taken for the further analyses.

The mesh, boundary and symmetry conditions are presented in Figure 3. The left grip edge is fixed and the right grip edge is pulled (displacement u is enforced). Nonlinear, quasistatic analysis was performed using the incremental technique with a constant step size (increment of displacement) equal to 0.01 of the total displacement. It was assumed, due to the joint symmetry, that only a quarter of the joint is necessary to be modeled. Symmetry conditions are frequently used in analyses, e.g., the quarter of the model is presented in refs. [68–70]. Taking into account a large number of models and calculations to be carried out for sensitivity analysis and the determining parameters of the gradient material model [53,62], a simplified symmetrical model presented in Figure 3 is also used in the paper.

On the other hand, assumption of symmetry is fully justified for a nominal model only, since potential imperfections in the real specimen lead to its asymmetry, which was recognized in the experimental tests. Therefore, an influence of asymmetry on the joint behavior is a subject of further analyses on the development of a gradient material model.

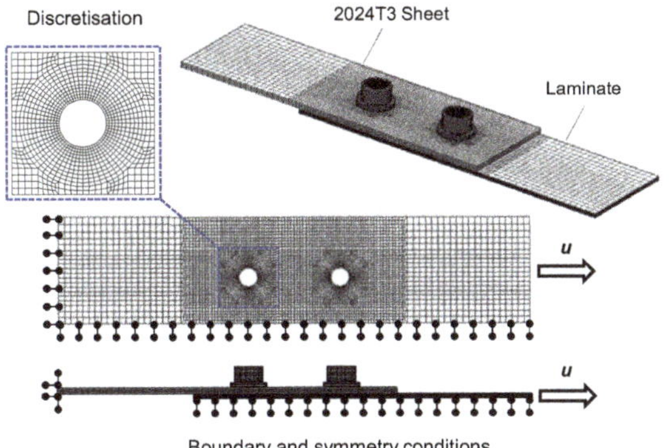

Figure 3. Finite element model—mesh, boundary and symmetry conditions.

4. Analysis of Joint Behavior in the Aspect of Numerical Model Validation

The typical characteristics of the joint (applied load vs. grip displacement curve) is presented in Figure 4.

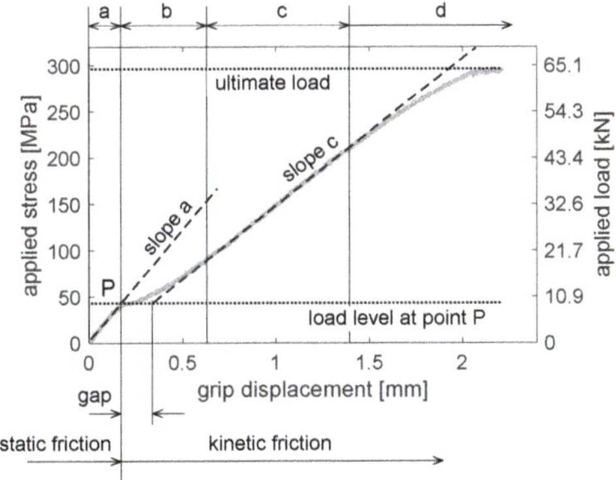

Figure 4. Joint characteristics—load vs. displacement curve: a, proportional range; b, imperfection range; c, main (work) range; d, degradation range.

The applied stress is calculated as a ratio of the applied load to the cross section of the laminate part. A similar graph is presented, e.g., in ref. [71].

Four ranges *a-b-c-d* can be identified on the graph. In the first proportional/linear range (marked with the letter *a*), the load between parts of the joint is transferred mainly by static friction. In point P, friction becomes kinetic and stiffness rapidly decreases. An imperfection range (marked with the letter *b*) is characterized by a gradual stiffness increase. In this range a joint is straightened, clearance is taken up and other imperfection are levelled (e.g., lack of symmetry, lack of holes concentricity). In the main range (marked with the letter *c*), the curve is almost linear as at the beginning. The load in the connected parts is transferred by both bearing and by-pass stresses.

In the degradation range (marked with the letter *d*), an evident failure of composite elements occurs, firstly in the bearing area, causing a gradual decrease in the joint stiffness (change of curve slope), and subsequently, in the net cross-section, leading to tearing of the composite element. Simultaneously, the yield stress state can expand in the bearing area of the aluminum alloy sheets. A sequence of the failure of specimen components depends on the joint dimensions and the materials used [63].

Parameters of the joint characteristics, i.e., slope in linear ranges *a* and *c*, size of ranges *b* and *d*, depend upon many factors. In general, due to the assumed permissible tolerances of the manufacturing process, as well as a different kind of imperfections, the results obtained for the nominal model, built using nominal sizes of a joint and nominal material data, do not lead to a satisfactory agreement with the experimental results [53,59,62].

In order to study the joint behavior, experimental and numerical analysis was performed. The main stages of analysis are presented in Figure 5.

After the initial analysis, the following parameters (imperfections), which should be taken into consideration at a validation stage, were indicated (Figure 6):

- geometrical parameters,
- stiffness parameters,
- material failure parameters,
- stress parameters (initial stress).

In the above classification, the shape and material imperfections of a real specimen, including the initial laminate failure caused by drilling, as well as numerical implementation of this failure, i.e., parameters of the gradient material model, are mainly taken into account.

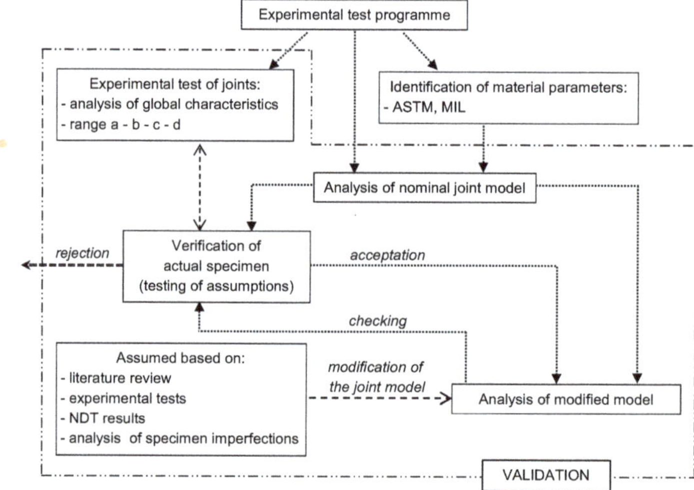

Figure 5. Algorithm of mechanical joint analysis.

In the group of geometrical parameters, manufacturing tolerances (allowable changes in dimensions of the joint and its components, as well as in the shape and in positional hole tolerances describing fits) are distinguished. Manufacturing tolerances can result in relative displacements of the connected parts, lack of symmetry, non-uniform load distribution, etc. [72].

The joint stiffness depends on the stiffness of elements (components) and their interactions. In ref. [59], different stiffness for the tension and compression of unidirectional CFRP laminate is reported and implemented in the user-defined subroutine in order to improve a numerical model of a single-lap joint. The sensitivity to stiffness components, in the case of a double-lap joint, is analyzed in refs. [53,62]. However, changes in the nominal stiffness by 10% only slightly influence the global response of the joint (global curves); therefore, there should be other reasons for the difference in stiffness between the actual specimen and the nominal model reported in both ref. [53] and ref. [59]. The area of the deteriorated/weakened material around the hole, defined with significantly lower stiffness (compared to the nominal stiffness of the laminate), seems to be a good explanation of the aforementioned difference [53,62]. Both a number and a size of the gradient material model zones are treated as geometrical parameters.

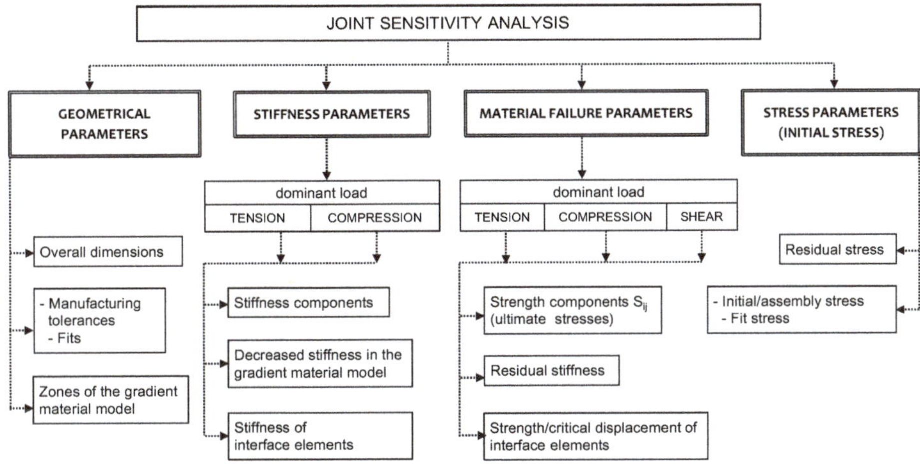

Figure 6. Parameters of joint sensitivity analysis.

Load carrying capacity of the joint depends on the strength components of CFRP lamina as well as on the yield and ultimate stress of the aluminum alloy. In the numerical model, lamina failure is defined with a failure criterion and a degradation procedure. Failure criteria compare the appropriate components of the stress or strain tensor in the system of material coordinates or their combination with the corresponding values of strength or ultimate strain. When a failure index (a stress or strain ratio of an actual to ultimate value) reaches unity, the degradation of material properties is employed. The residual stiffness defines the ratio of a degraded to a nominal stiffness value. In the paper, the selective gradual degradation procedure (built in Marc® code) is used when the failure, according to the Hashin criterion, occurred. This procedure involves a gradual reduction of the stiffness components to the residual value according to an exponential function of the current failure index [73]. The Hashin criterion and other physically-based criteria (e.g., Puck) allows for the identification of the failure mechanism important for analysis of the complex stress state occurring in a hole vicinity of mechanically fastened joints. However, this paper deals with global, not local, analysis.

In mechanically fastened joints, bearing is coupled with tension in the net section. Thus, forms of joint failure specific for bearing and for tension can be distinguished. In the tension area, stiffness and load carrying capacity after the material failure equals to zero (due to disintegration of material).

Whereas, in the bearing area, the stiffness of the failed/degraded material can be substantially greater than in the tension area, especially if the displacement of material in the transverse direction is constrained (Figure 7).

Interfaces that undergo mainly shear stresses were modeled with the cohesive elements. This created a possibility to perform analysis of the interlaminar shear stress state and simulation of the delamination process. A decrease in cohesive energy (cohesive energy release rate) for the elements placed around the holes was used to describe initial delamination in this area.

In the group of stress parameters, residual and initial stresses were considered. These stresses occur at the manufacturing and assembly stages, including sheet rolling, laminate curing, hole drilling and elements connecting. Manufacturing/assembly stresses can be divided into intentional (desired) and unintentional (undesired). The stresses caused by bolt-torque (and thus washer/nut pressure) belong to the first group. Undesired effects are the result of technological imperfections such as clearances (interferences), lack of joint symmetry, lack of holes concentricity and uneven bearing stress distribution.

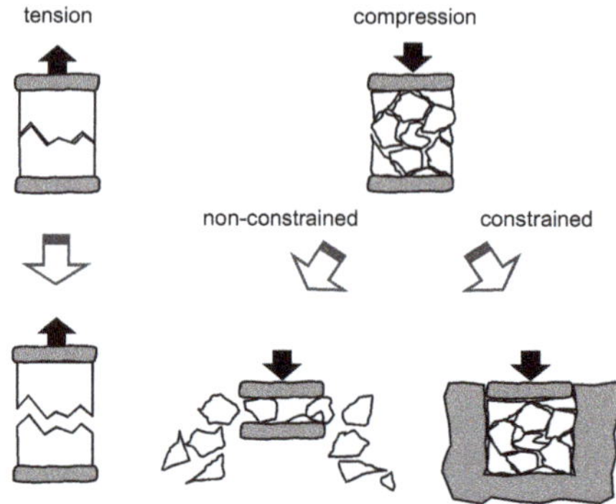

Figure 7. Material behavior in the case of tension and bearing/compression.

The residual stresses remained after the stage of curing the laminate are a result of different thermal expansion coefficients. This phenomenon concerns all laminates consisting of orthotropic plies aligned in different directions; however, it is especially visible for laminates co-curing with metal alloys (typically aluminum or titanium alloys in aircraft structures) [74–76]. This effect was not directly involved in the analyzed models; however, it can influence the stiffness and strength of a real laminate specimen. It was observed, during material identification tests, that the longitudinal stiffness of quasi-isotropic laminate is several percent lower than that calculated by means of Classical Laminated Theory (Figure 8).

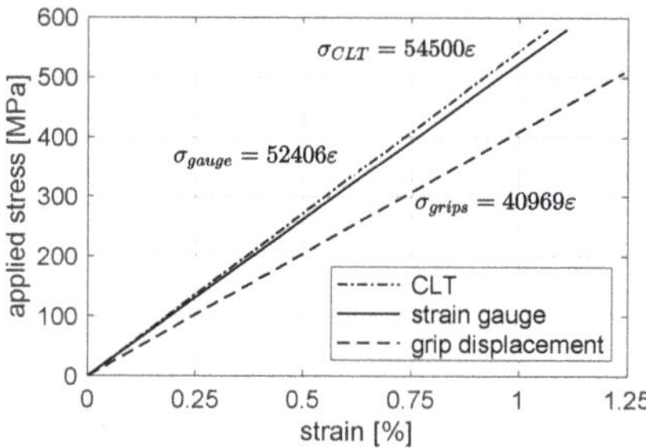

Figure 8. Comparison of stiffness of quasi-isotropic laminate.

It is worth mentioning that parameters in the aforementioned four groups are not independent, e.g., stress parameters depend upon geometrical ones (i.e., bolt preload stresses depend on fits); residual stiffness depends on both nominal stiffness and dominant load. Therefore, sensitivity analysis was a time-consuming process, due to couplings among the parameters.

A detailed validation of a numerical model of a bolted joint in the laminate structure, involving clearances, is presented in ref. [77]. Whereas, a literature review shows that parameters of mechanically fastened laminates are still widely studied. The reliability analysis concerning stacking sequence is presented in ref. [78]. An influence of the variation of several parameters on the hybrid bolted-bonded joint behavior is analyzed in ref. [79]. The analyses of the bolt preload as well as of fits including clearances and friction phenomenon can be found in refs. [13,80–83]. The influence of a hole perpendicularity error on the performance of the composite mechanical joint is presented in ref. [84].

In ref. [59] it is reported that modification of the laminate compression stiffness (from 140 MPa to 130 MPa), mesh refinement in the non-overlap region of the single lap joint, use of the assumed strain formulation with the first order finite elements, as well as modification of boundary conditions by fixing the surface of the clamped region, result in a better agreement between the numerical model and the experimental specimen. Additionally, refinement of the mesh around the hole (under the washer) is applied to improve strain and stress fields. The difference in stiffness decreases from 23.6%, for the base model, to 12.6% for the improved one. The latter difference is explained in ref. [59] mainly by difficulties with a precise measurement of displacement or a nonlinear character of the resin-rich clamped outer surface of the specimen. In the authors' opinion, this conclusion can be arguable, since various techniques for the measurement of displacement were considered to find the most appropriate solution independent of the clamping effect. Moreover, analysis of the sensitivity of the mechanically fastened CFRP laminate to a number of mechanical and numerical parameters presented in refs. [53,62] shows that deterioration of laminate around the holes can contribute to the characteristics of the joint more than the clamping effect. This conclusion is consistent with the observations presented in refs. [51,52].

Recently, a probabilistic approach to sensitivity analysis has been commonly used. A probabilistic model and sensitivity analysis of several factors of a double-lap, single bolt joint is presented in ref. [85]. Askri et al. [72] proposed a probabilistic analysis of uncertainties (hole-location errors, clearance size and bolt preload) and worst-case analysis of a mechanical four-fastener single-lap joint with the use of a simplified description of the joint and a genetic algorithm. Probabilistic analysis seems to be an appropriate approach to a further development of the weakened material model in the hole vicinity.

Experimental tests are the basis for the identification of material parameters and validation of numerical models. It is worth mentioning that results of experimental tests are dependent on nominal dimensions and material parameters, imperfections of a real specimen, as well as on environmental conditions and measurement techniques.

The object of analysis and nominal parameters are presented in the previous section. Possible imperfections of a real specimen (parameters of joint sensitivity analysis) are listed in Figure 6 and discussed above.

The simplest and most popular method of determining the global characteristics of a tensile loaded joint is to measure the applied force and displacement of the testing machine grip. However, in this case, the total displacement of the grip is a result of specimen compliance as well as compliance (and imperfections) of the fixture (grips compliance, slips and deformation of the specimen in the grip area). An exemplary comparison of the quasi-isotropic laminate stiffness estimated using displacement of the testing machine grip and the strain gauge technique is presented in Figure 8. A similar effect (stiffness 5% lower than nominal) was observed in the laminate part of the tensile loaded joint using a digital image correlation (DIC) technique. This effect is also consistent with the results presented in ref. [59].

In ref. [59], several techniques used to estimate the elongation of the free length of the single-lap joint, including determination of a testing machine grip compliance in order to correct stiffness of a numerical model, as well as the use of extensometers, are compared.

It is reported that the use of linear variable displacement transducers to obtain displacement between steel blocks glued to the side of the specimen at the ends of its free/measurement length was found to be the most appropriate measurement technique.

In ref. [53], a gradient material model is developed based on the global characteristics of the joint, i.e., the applied load versus the displacement of the testing machine grip. In this case, the nominal model does not lead to a satisfactory agreement with the experimental results, mainly due to a grip compliance. Therefore, the numerically obtained characteristics of the joint were scaled/calibrated. The calibration factor was calculated as a stiffness ratio of an experimental to a numerical curve in the proportional range (marked with the letter *a* in Figure 4). Despite the aforementioned calibration, a slope in range *c* was still different for a numerical and an experimental curve (an experimental slope was lower than the numerical one). Therefore, a gradient material model, with an average stiffness significantly lower than the nominal stiffness of the laminate, was defined around the hole. Parameters of the weakened area were estimated with a series of simulations performed to obtain a consistent slope in the range c. In other words, in order to exclude the grip compliance and efficiently compare the results of experimental and numerical analysis, two parameters were taken into account in refs. [53,62], i.e., experimentally obtained global characteristics of the joint (based on displacement of the testing machine grip) and a ratio of stiffness in linear ranges c and a (Figure 4).

The aforementioned scaling/calibration of the joint characteristics does not affect the stiffness ratio c/a, and should not affect the parameters of the gradient material model as well. However, a shape of the curve, especially in the ranges b and d, is significantly dependent on calibration, since it is highly nonlinear in these ranges.

Taking into account the aforementioned conditions (deficiency of a curve based on displacement of the testing machine grip), standard and virtual extensometers were used to measure deformation of the joint in the overlap region.

The non-contact optical strain and displacement measurement system Aramis® was involved to obtain a displacement field of selected tensile loaded joints. The measuring area of a specimen was marked with a stochastic black and white pattern. Then, relative displacements of two selected points within this area, used as a virtual extensometer with a 50 mm base, were measured, as shown here in Figure 15c.

The Aramis® system is designed for measurement of deformation on a surface of material during loading. The equipment uses two high resolution CCD cameras of 2358×1728 pixels and the digital image correlation (DIC) method to obtain a three-dimensional image sequence.

The aim of the paper is to confirm the correctness of parameters obtained in refs. [53,62] for a gradient material model and in-depth verification of a numerical model of a mechanically fastened joint. Comparative analysis of the numerical and experimental results based on both the displacement of the testing machine grip and length of the extensometer was performed in order to reach this aim.

5. Gradient Material Model

It was identified using the NDT ultrasonic method that the drilling process caused deterioration/changes in the material around the hole. The shape of the area of the deteriorated material (ADM) is irregular; however, it can be approximated using a ring shape with the inner diameter coincident with the hole diameter (Figure 9). Such approximation is more suitable for numerical implementation. The average outer diameter of those areas in the analyzed laminate coupons is equal to about 11 mm. The exemplary results of the NDT tests and corresponding ADM in the finite element model are presented in ref. [53].

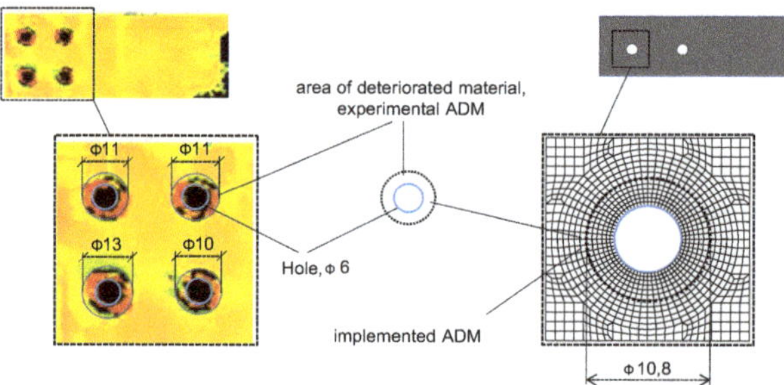

Figure 9. NDT results (C-Scan images) and implemented area of deteriorated material in numerical model [53].

In refs. [53,62], it was found that the initial delamination only (without any changes of intralaminar properties) did not sufficiently affect the numerical results. This leads to the conclusion that the drilling process caused significant changes of laminae in the hole vicinity, i.e., changes in material properties. In order to describe the laminate deterioration, a gradient material model is proposed in ref. [53]. The function of gradient material properties is unknown and cannot be determined experimentally, therefore, sample functions were proposed to estimate it. The concept of a material gradient model is presented in Figure 10.

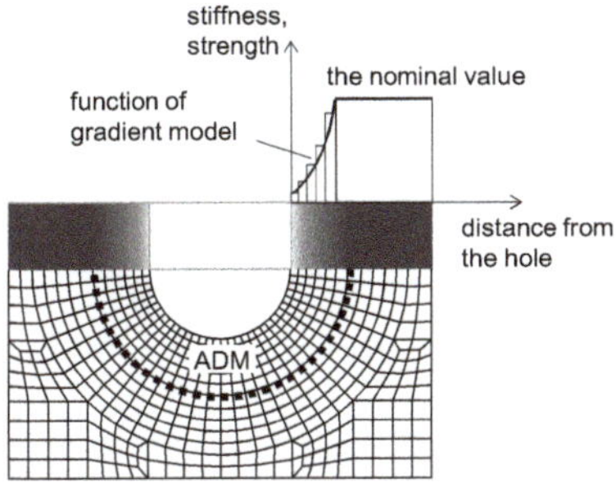

Figure 10. The concept of gradient material model in the hole vicinity.

Five sample functions were analyzed: cube root, square root, linear, quadratic and cubic function (Figure 11). The sample function describes a stiffness/strength ratio of a current to a nominal value versus the normalized radial distance from the edge of the hole. The normalizing factor is equal to the width of the ADM ring. The minimum value of stiffness/strength at the edge of the hole was established as 7% of its nominal value. This value is both close to zero and high enough to ensure convergence of the numerical calculations. The material parameter at the outer diameter of ADM and beyond this area is equal to its nominal value.

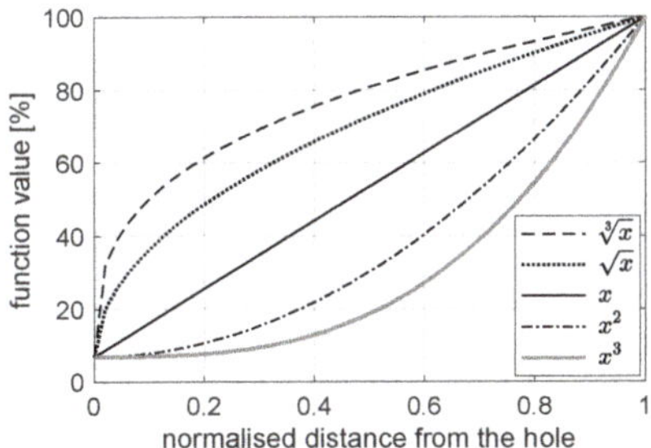

Figure 11. Analyzed sample functions.

A sample function defines a decrease in the material stiffness/strength compared to its nominal value. An average decrease corresponding to each sample function is shown in Table 1. An average decrease in stiffness/strength (deficiency parameter) can be a measure of the laminate weakness in the hole vicinity.

A series of simulations were performed [53,62] to obtain a satisfactory agreement between the numerical and the experimental results (to estimate stiffness in the hole vicinity).

ADM was implemented in a discrete manner; i.e., the total area was divided into zones with different values of material properties. A value of a sample function corresponding to the middle of a selected zone (of gradient model) was assigned to the whole zone. Several cases of ADM division (in the hole vicinity) were tested [53], and it was found that three zones with a radially increased number of elements (shown in Figure 12) led to the most appropriate results [62], and therefore this case was adopted to further analyses.

Table 1. Comparison of sample functions' deficiency parameters

No.	Sample Function	Deficiency Parameter (%)
1	cubic root	75.0
2	quadratic root	66.7
3	linear	50.0
4	square	33.3
5	cubic	25.0

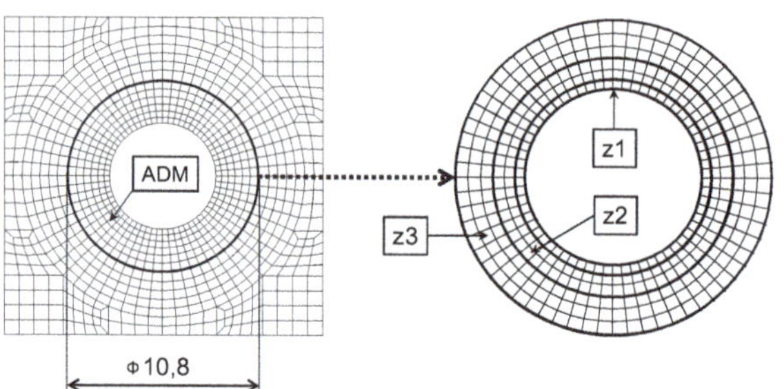

Figure 12. Gradient material model—ADM zones (uneven division).

In order to adequately describe the joint failure, two regions, subjected to tension (including tearing) and bearing/compression, were distinguished around the hole (Figure 13).

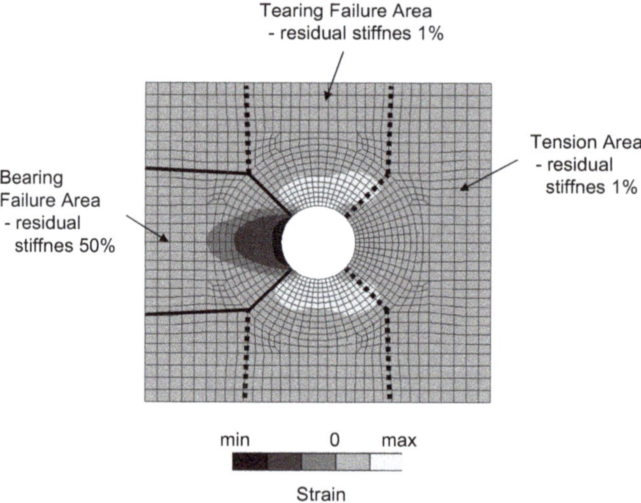

Figure 13. Strain state around the hole. Tension and bearing areas distinguished in the model.

The residual stiffness for the tension area was assumed as 1% of the nominal stiffness. This value is small enough to be treated as zero. The value of the residual stiffness in the bearing area should be substantially greater, as it is illustrated in Figure 7. This value was found after several trials to gain good agreement with the experimental results in the degradation range *d*.

6. Results of Experimental Tests

Experimental data were collected via a force transducer installed on the transverse beam of the machine, standard extensometers or the digital image correlation (DIC) system Aramis® (serving as virtual extensometer). The results of experimental tests, i.e., characteristics of tensile loaded, mechanically fastened joints, are presented in Figure 14a,b based on displacement of the testing machine grip and length of the extensometer gauge.

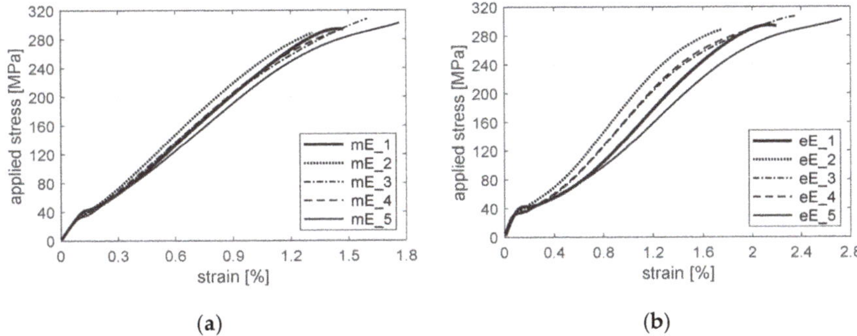

Figure 14. Characteristics of mechanically fastened joints based on: (**a**) displacement of testing machine grip; (**b**) length of extensometer gauge; where prefix m/e—grip displacement/extensometer length, E—experimental results, 1–5—specimen numbers.

Characteristics of the joint are presented as applied stress vs. strain in order to easily compare the different bases for measuring the displacement. The free length of the specimen fixed in the grips of the testing machine is 150 mm, and the length of the extensometer base is 50 mm (Figure 15).

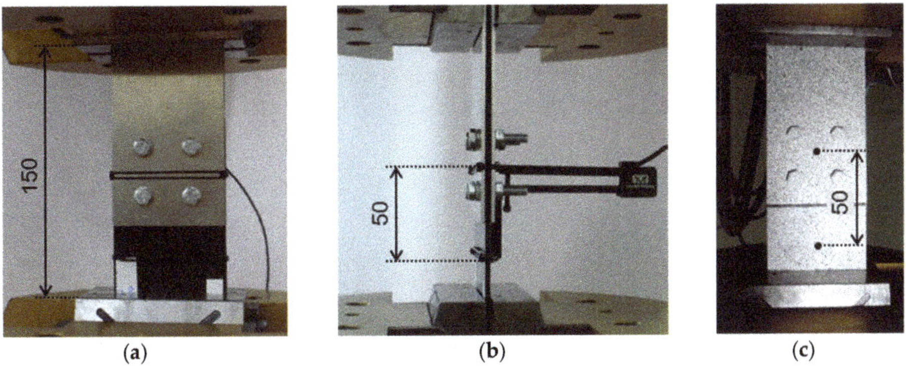

Figure 15. Specimen fixed in the testing machine grips with (**a**), (**b**) extensometer attached across the overlap area and the laminate part; (**c**) virtual extensometer based on Aramis® field

Comparison of the experimental results, i.e., the load level and the gap (clearance) at point P (Figure 4), as well as stiffness in the ranges *a* and *c*, is presented in Table 2. Specimens numbered as 2, 3 and 4 were initially loaded (preloaded) to 100 MPa, which resulted in smaller imperfections (smaller size of range b) compared to specimens numbered as 1 and 5 (without preload). Those imperfections caused larger deformation of the joint in the overlap region. Strain of the joint obtained, which strain was based on the length of the extensometer, is almost twice as large as the one based on the displacement of the grip (Figure 14a,b). This effect results from both different measurement bases (150 mm and 50 mm) and approximately the same region that is mainly deformed. Stiffness always depends on the measurement base of displacement. It results from a strongly non-uniform strain state in the overlap region.

Analysis presented in the paper is focused mainly on range *c*—the main range of the joint work. In this range, values of average slopes are 23,500 and 19,200 MPa, whereas standard deviations are 1130 and 1948 MPa based on displacement of the grip and length of the extensometer, respectively (Table 2).

Table 2. Comparison of the joint stiffness in ranges *a* and *c*.

Specimen No.	Stress (MPa)	Grip Displacement *			Extensometer Length		
		Stiffness (MPa)		Gap (mm)	Stiffness (MPa)		Gap (mm)
	Point P	Range *a*	Range *c*	Point P	Range *a*	Range *c*	Point P
No. 1	42	36100	23900	0.163	39500	18500	0.181
No. 2	31	35700	25100	0.083	42300	21600	0.103
No. 3	30	36500	23200	0.096	39800	19600	0.124
No. 4	31	36900	23300	0.094	35700	19800	0.122
No. 5	31	37400	22000	0.133	38300 **	16300 **	0.170 **
mean		36500	23500	0.114	39100	19200	0.140
std. dev.		665	1130	0.033	2402	1948	0.034
difference		1700	3100	0.080	6600	5300	0.078

* results obtained directly with measurement of the grip displacement (without calibration). ** results obtained using a virtual extensometer based on Aramis displacement field.

7. Discussion

As it was mentioned above, displacement of the testing machine grip depends on both joint and fixture compliance, and therefore experimental stiffness was scaled/calibrated (with coefficient 5/3, obtained as a ratio of the numerical to the experimental stiffness in range *a*) in order to compare it to the numerical result. The maximum difference (scatter) of slopes obtained experimentally in range *c* is 5170 (3100 before scaling/calibration) and 5300 MPa, based on displacement of the grip and length of the extensometer, respectively (row difference in Table 2).

Comparison of joint stiffness obtained both numerically and experimentally in range *c*, based on displacement of the grip and length of the extensometer, is shown in Figure 16. Numerical slope for the nominal model of the joint (N_0) in range *c* is 28,700 MPa. Whereas, a difference of slopes in range *c* between the nominal model (N_0) and the average experimental result (E_av) is about 80% larger than the scatter of the experimental results.

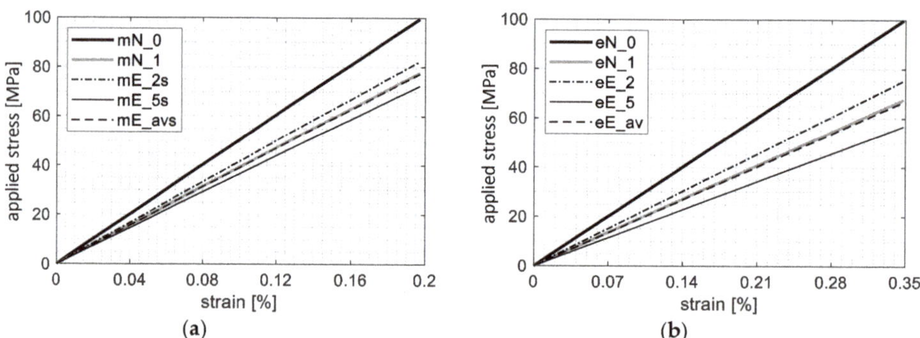

Figure 16. Comparison of the joint stiffness obtained both numerically and experimentally in range *c* based on (**a**) grip displacement, (**b**) extensometer length, where: prefix m/e—grip displacement/extensometer length; N/E—numerical/experimental results; 0—nominal model; 1, 2, 5 —specimen number; av—average stiffness; suffix s—stiffness calibrated with coefficient 5/3

The aforementioned difference between the nominal model of the joint (mN_0, eN_0) and the average (mE_avs, eE_av) or maximum experimental result (mE_2) can be explained by deterioration of the laminate around the hole, which resulted from the drilling process. In ref. [62], based on characteristics of sample No. 1, obtained using displacement of the testing machine grip, the area of the deteriorated material (ADM) around the hole was defined as a gradient material model with a linear function of stiffness and a quadratic function of strength (mN_1). In this case, the result of numerical simulation is in good agreement with both the experimental stiffness of sample No. 1 (as is shown in ref. [62]) and the average experimental value (mE_avs). The results obtained based on displacement of the grip (Figure 16a) are consistent with those based on the extensometer measurement (Figure 16b). Thus, the laminate stiffness around the hole was estimated at 50% of the nominal stiffness, i.e., the average decrease for a linear sample function (compare Table 1). This level is the upper estimation bound since other imperfections can also influence the joint stiffness.

Although the joint stiffness is different for displacement of the grip and length of the extensometer (as is shown in Figure 14 and Table 2), both of them can be used to obtain parameters of a gradient material model and to estimate laminate weakness in the hole vicinity. However, characteristics of the joint based on displacement of the testing machine grip should be scaled/calibrated in order to compare the numerical and the experimental results (Figure 17a). Differences in the joint stiffness depend on various imperfections of a real specimen and the conditions of the experimental test. In refs. [53,62], a relative shift of the hole was used to initially describe all imperfections in range b and to obtain good agreement with the experimental results (Figure 17a). Comparison of the numerical and the experimental results, based on the extensometer measurement in the overlap region, revealed that real imperfections were substantially larger than a possible clearance effect (especially for specimens without preload as in the case of specimen No. 1). It was found that in numerical model N_1, virtual clearance (gap) equal to not less than 0.22% of the extensometer base covered all imperfections in the overlap region (Figure 17b).

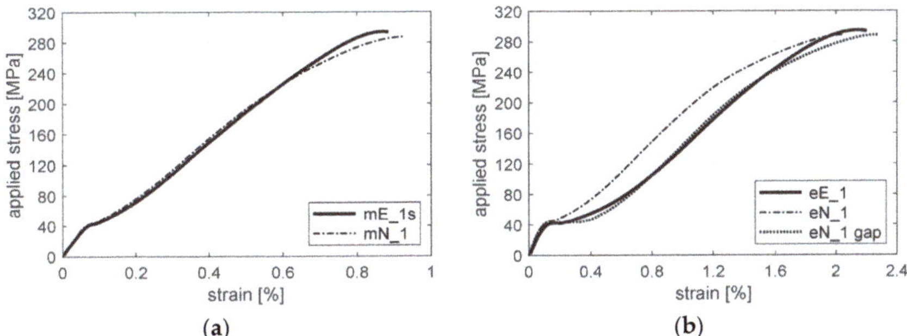

Figure 17. Comparison of the joint characteristics obtained both numerically and experimentally based on: (**a**) displacement of testing machine grip; (**b**) length of extensometer gauge; where prefix m/e—grip displacement/extensometer length; N/E—numerical/experimental results; 1—specimen number; suffix s—grip displacement calibrated with coefficient 3/5; gap—model with increased gap (virtual clearance equal to 0.22% of the extensometer base)

Summarizing, the aim of the paper is analysis of parameters determining the laminate stiffness and strength in the joint area. The paper presents classification and discussion of the selected parameters affecting the stiffness and load carrying capacity of the joint. The shape and material imperfections of a real specimen, including the initial laminate failure resulting from the drilling, and the initial stresses induced during both the manufacturing and the assemble stage, are classified in Section 4.

Delamination is the most studied damage form connected with drilling a hole in composites/laminates. However, other forms of damage also occur and influence the mechanical properties of laminate around the hole. Therefore, a method for modeling the hole vicinity, namely, a gradient material model, as well as the numerical and experimental estimation of laminate deterioration in this area, was proposed and analyzed.

The laminate stiffness beside the joint area, determined with the DIC technique in the Aramis® system, is 5% lower than the nominal (theoretical) stiffness calculated according to Classical Laminate Theory. A similar result was obtained for the quasi-isotropic laminate during material identification tests, and this effect is consistent with the results presented in literature. Therefore, laminate properties, several percent lower than nominal values, can be taken into account in the numerical model. However, this modification did not cover differences between the nominal model and the real specimen equal to about 50% (Figure 16b).

Weakness of laminate (deterioration of material properties) in the hole vicinity, caused by the drilling process or cyclic loading during the service life of the structure, can be described in the numerical model using a gradient material model. The function of gradient material properties is unknown and cannot be determined experimentally, therefore, the sample functions were used to describe changes in material parameters in the hole vicinity (Figure 11, Table 1). The aforementioned approach is a novelty in the field of the analysis of mechanically connected composite elements.

The results of the experimental tests are dependent on imperfections of the real specimen, test conditions, as well as measurement techniques. Despite nonlinearity of the characteristics of the tensile loaded joint, comparative analysis of the numerical and the experimental results, based on both displacement of the testing machine grip and length of the extensometer, confirmed that the characteristics of the joint, obtained from the measurement of the grip displacement and the ratio of stiffness in the linear ranges c and a, are sufficient to determine the parameters of the gradient material model and to estimate the weakness of laminate (to find the upper estimation of the laminate weakness) in the hole vicinity.

A simplified symmetrical model used in the paper was sufficient to analyze the global characteristics of the joint; however, it is not able to capture all of the imperfections that resulted from asymmetry of the real joint, particularly in the overlap region (e.g., the clearance shown in range b in Figure 17b). Those imperfections can affect to some extent also the joint stiffness in ranges a and c. If their influence is not negligible, they are included in the gradient material model, and deterioration of laminate properties in the hole vicinity is overestimated.

Analysis of the full model creates a possibility to study the influence of asymmetry on the joint behavior, and is a subject of further works on development of the gradient material model.

8. Conclusions

Four ranges *a-b-c-d* were identified on the experimental graphs, i.e., proportional range (marked with the letter a), imperfection range (b), main work range (c) and degradation range (d). Numerical slope for the nominal model in range c (28,700 MPa) was found about 50% larger than the average experimental result (19,200 MPa). Moreover, the difference of slopes in range c (9500 MPa) between the nominal model of the joint and the average experimental result was found about 80% larger than the scatter of the experimental results (5300 MPa). The aforementioned difference can be explained by the laminate weakness/deterioration around the hole resulting from the drilling process.

Weakness of laminate in the hole vicinity was described in the numerical model using a gradient material model with the linear function; i.e., the average laminate stiffness around the hole was estimated at 50% of the nominal stiffness.

Grip displacements are the most easily obtained measurement data, and they are less sensitive to local imperfections than the extensometer data. In the case of the grip displacements, standard deviation of the joint stiffness was about 60% lower than in the case of the extensometer data. Parameters of the gradient material model obtained based on displacement of the testing machine grip and length of the extensometer were almost the same. Therefore, they were successfully verified.

Author Contributions: Conceptualization, K.P. and J.J.; methodology, E.S., K.P.; validation, E.S., K.P. and P.B.; formal analysis, E.S.; investigation, K.P., E.S. and M.S.; resources, K.P. and M.S.; data curation, K.P., E.S.; writing—original draft preparation, K.P., E.S.; writing—review and editing, K.P., E.S. and P.B.; visualization, K.P., E.S.; supervision, J.J.; project administration, E.S.; funding acquisition, E.S.

Funding: This research has been funded from the Polish-Norwegian Research Programme under the Norwegian Financial Mechanism 2009–2014 within Project Contract Pol-Nor/210974/44/2013 and the APC was funded from statutory activities 2019.

Conflicts of Interest: The authors declare no conflict of interest.

References

1. Niu Michael, C.Y. *Composite Airframe Structures*; Conmilit Press Ltd.: Hong Kong, China, 1992.
2. Ashby, M.F. *Materials Selection in Mechanical Design*; Butterworth-Heinemann: Oxford, UK, 2010.
3. Jachimowicz, J.; Szymczyk, E.; Puchała, K. Study of material mass efficiency and numerical analysis of modified CFRP laminate in bearing conditions. *Compos. Struct.* **2015**, *134*, 114–123. [CrossRef]
4. Mazurkiewicz, L.; Tomaszewski, M.; Malachowski, J.; Sybilski, K.; Chebakov, M.; Witek, M.; Yukhymets, P.; Dmitrienko, R. Experimental and numerical study of steel pipe with part-wall defect reinforced with fibre glass sleeve. *Int. J. Press. Vessel. Pip.* **2017**, *149*, 108–119. [CrossRef]
5. Mazurkiewicz, Ł.; Malachowski, J.; Tomaszewski, M.; Baranowski, P.; Yukhymets, P. Performance of steel pipe reinforced with composite sleave. *Compos. Struct.* **2018**, *183*. [CrossRef]
6. U.S. Department of Transportation. *Federal Aviation Administration (FAA) Advisory Circular—Composite Aircraft Structure*; U S. Department of Transportation: Washington, DC, USA, 2009.
7. Schmid Fuertes, T.A.; Kruse, T.; Körwien, T.; Geistbeck, M. Bonding of CFRP primary aerospace structures—Discussion of the certification boundary conditions and related technology fields addressing the needs for development. *Compos. Interfaces* **2015**, *22*, 795–808. [CrossRef]
8. Witten, E.; Kraus, T.; Kühnel, M. *Composites Market Report 2016. Market Developments, Trends, Outlook and Challenges*; AVK Federation of Reinforced Plastics: Frankfurt, Germany, 2016.
9. Boeing: 787 by Design. Available online: https://www.boeing.com/commercial/787/by-design/#/advanced-composite-use (accessed on 8 October 2019).
10. Nerilli, F.; Vairo, G. Progressive damage in composite bolted joints via a computational micromechanical approach. *Compos. Part B Eng.* **2017**, *111*, 357–371. [CrossRef]
11. Xiang, J.; Zhao, S.; Li, D.; Wu, Y. An improved spring method for calculating the load distribution in multi-bolt composite joints. *Compos. Part B Eng.* **2017**, *117*, 1–8. [CrossRef]
12. Yazdani Nezhad, H.; Egan, B.; Merwick, F.; McCarthy, C.T. Bearing damage characteristics of fibre-reinforced countersunk composite bolted joints subjected to quasi-static shear loading. *Compos. Struct.* **2017**, *166*, 184–192. [CrossRef]
13. Chen, C.; Hu, D.; Liu, Q.; Han, X. Evaluation on the interval values of tolerance fit for the composite bolted joint. *Compos. Struct.* **2018**, *206*, 628–636. [CrossRef]
14. Hu, X.F.; Haris, A.; Ridha, M.; Tan, V.B.C.; Tay, T.E. Progressive failure of bolted single-lap joints of woven fibre-reinforced composites. *Compos. Struct.* **2018**, *189*, 443–454. [CrossRef]
15. Zhuang, F.; Arteiro, A.; Furtado, C.; Chen, P.; Camanho, P.P. Mesoscale modelling of damage in single- and double-shear composite bolted joints. *Compos. Struct.* **2019**, *226*. [CrossRef]
16. Hu, J.; Zhang, K.; Xu, Y.; Cheng, H.; Xu, G.; Li, H. Modeling on bearing behavior and damage evolution of single-lap bolted composite interference-fit joints. *Compos. Struct.* **2019**, *212*, 452–464. [CrossRef]
17. Grzejda, R. Determination of Bolt Forces and Normal Contact Pressure between Elements in the System with Many Bolts for its Assembly Conditions. *Adv. Sci. Technol. Res. J.* **2019**, *13*, 116–121. [CrossRef]
18. Jancelewicz, B.; Mądry, W. *Złącze Do Wprowadzania Siły Skupionej W Powłokę Kompozytową*; Urząd Patenowy PRL: Warszawa, Polska, 1988.
19. Camanho, P.P.; Tavares, C.M.L.; de Oliveira, R.; Marques, A.T.; Ferreira, A.J.M. Increasing the efficiency of composite single-shear lap joints using bonded inserts. *Compos. Part B Eng.* **2005**, *36*, 372–383. [CrossRef]
20. Crosky, A.; Kelly, D.; Li, R.; Legrand, X.; Huong, N.; Ujjin, R. Improvement of bearing strength of laminated composites. *Compos. Struct.* **2006**, *76*, 260–271. [CrossRef]
21. Asi, O. An experimental study on the bearing strength behavior of Al_2O_3 particle filled glass fiber reinforced epoxy composites pinned joints. *Compos. Struct.* **2010**, *92*, 354–363. [CrossRef]
22. Kolesnikov, B.; Herbeck, L.; Fink, A. CFRP/titanium hybrid material for improving composite bolted joints. *Compos. Struct.* **2008**, *83*, 368–380. [CrossRef]

23. Fink, A.; Camanho, P.P. 1—Reinforcement of composite bolted joints by means of local metal hybridization. Woodhead Publishing Series in Composites Science and Engineering. In *Composite Joints and Connections*; Camanho, P., Tong, L., Eds.; Woodhead Publishing: Cambridge, UK, 2011; pp. 3–34.

24. Gerendt, C.; Dean, A.; Mahrholz, T.; Rolfes, R. On the progressive failure simulation and experimental validation of fiber metal laminate bolted joints. *Compos. Struct.* **2019**, *229*. [CrossRef]

25. Vasiliev, V.V.; Morozov, E.V. *Mechanics and Analysis of Composite Materials*; Elsevier Science: Oxford, UK, 2001.

26. Kwon, Y.; Allen, D.H.; Talreja, R.R. *Multiscale Modeling and Simulation of Composite Materials and Structures*; Springer: New York, NY, USA, 2008.

27. Szymczyk, E.; Puchała, K.; Jachimowicz, J. About numerical analysis of pin loaded joints in laminate structure. *AIP Conf. Proc.* **2019**, *2078*. [CrossRef]

28. Jones, R.M. *Mechanics of Composite Materials*, 2nd ed.; Taylor & Francis, Inc.: Philadelphia, PA, USA, 1998.

29. Baker, A.A.; Scott, M.L. *Composite Materials for Aircraft Structures*, 3rd ed.; American Institute of Aeronautics and Astronautics, Inc.: Washington, DC, USA, 2016.

30. Mikulik, Z.; Haase, P. *CODAMEIN—Composite Damage Metrics and Inspection. EASA.2010.C13 Final Report*; Bishop GmbH—Aeronautical Engineers: Hamburg, Germany, 2012.

31. Breuer, U.P. *Commercial Aircraft Composite Technology*; Springer: Berlin, Germany, 2016.

32. Seike, S.; Takao, Y.; Wang, W.X.; Matsubara, T. Bearing damage evolution of a pinned joint in CFRP laminates under repeated tensile loading. *Int. J. Fatigue* **2010**, *32*, 72–81. [CrossRef]

33. Persson, E.; Eriksson, I.; Hammersberg, P. Propagation of Hole Machining Defects in Pin-Loaded Composite Laminates. *J. Compos. Mater.* **1997**, *31*, 383–408. [CrossRef]

34. Starikov, R.; Schön, J. Local fatigue behaviour of CFRP bolted joints. *Compos. Sci. Technol.* **2002**, *62*, 243–253. [CrossRef]

35. Persson, E.; Eriksson, I.; Zackrisson, L. Effects of hole machining defects on strength and fatigue life of composite laminates. *Compos. Part A Appl. Sci. Manuf.* **1997**, *28*, 141–151. [CrossRef]

36. Durão, L.M.P.; Tavares, J.M.R.S.; de Albuquerque, V.H.C.; Marques, J.F.S.; Andrade, O.N.G. Drilling Damage in Composite Material. *Materials* **2014**, *7*, 3802–3819. [CrossRef] [PubMed]

37. Bhatnagar, D.N.; Singh, I.; Nayak, D. Damage Investigation in Drilling of Glass Fiber Reinforced Plastic Composite Laminates. *Mater. Manuf. Process.* **2004**, *19*, 995–1007. [CrossRef]

38. Tsao, C.C.; Hochengb, H. Computerized tomography and C-Scan for measuring delamination in the drilling of composite materials using various drills. *Int. J. Mach. Tools Manuf.* **2005**, *45*, 1282–1287. [CrossRef]

39. Davim, J.P.; Rubio, J.C.; Abrao, A.M. A novel approach based on digital image analysis to evaluate the delamination factor after drilling composite laminates. *Compos. Sci. Technol.* **2007**, *67*, 1939–1945. [CrossRef]

40. Grilo, T.J.; Paulo, R.M.F.; Silva, C.R.M.; Davim, J.P. Experimental delamination analyses of CFRPs using different drill geometries. *Compos. Part B Eng.* **2013**, *45*, 1344–1350. [CrossRef]

41. Gemi, L.; Morkavuk, S.; Köklü, U.; Gemi, D.S. An experimental study on the effects of various drill types on drilling performance of GFRP composite pipes and damage formation. *Compos. Part B Eng.* **2019**, *172*, 186–194. [CrossRef]

42. Liu, D.; Tang, Y.; Cong, W.L. A review of mechanical drilling for composite laminates. *Compos. Struct.* **2012**, *94*, 1265–1279. [CrossRef]

43. Kavad, B.V.; Pandey, A.B.; Tadavi, M.V.; Jakharia, H.C. A Review Paper on Effects of Drilling on Glass Fiber Reinforced Plastic. *Procedia Technol.* **2014**, *14*, 457–464. [CrossRef]

44. Panchagnula, K.K.; Palaniyandi, K. Drilling on fiber reinforced polymer/nanopolymer composite laminates: A review. *J. Mater. Res. Technol.* **2018**, *7*, 180–189. [CrossRef]

45. Karataş, A.M.; Gökkaya, H. A review on machinability of carbon fiber reinforced polymer (CFRP) and glass fiber reinforced polymer (GFRP) composite materials. *Def. Technol.* **2018**, *14*, 318–326. [CrossRef]

46. Geng, D.; Liu, Y.; Shao, Z.; Lu, Z.; Cai, J.; Li, X.; Jiang, X.; Zhang, D. Delamination formation, evaluation and suppression during drilling of composite laminates: A review. *Compos. Struct.* **2019**, *216*, 168–186. [CrossRef]

47. Kourra, N.; Warnett, J.M.; Attridge, A.; Kiraci, E.; Gupta, A.; Barnes, S.; WIlliams, M.A. Metrological study of CFRP drilled holes with x-ray computed tomography. *Int. J. Adv. Manuf. Technol.* **2015**, *78*, 2025–2035. [CrossRef]

48. Karimi, Z.N.; Minak, G.; Kianfar, P. Analysis of damage mechanisms in drilling of composite materials by acoustic emission. *Compos. Struct.* **2015**, *131*, 107–114. [CrossRef]

49. Zitoune, R.; Krishnaraj, V.; Collombet, F. Study of drilling of composite material and aluminium stack. *Compos. Struct.* **2010**, *92*, 1246–1255. [CrossRef]
50. Wang, G.D.; Melly, S.K.; Li, N. Experimental studies on a two-step technique to reduce delamination damage during milling of large diameter holes in CFRP/Al stack. *Compos. Struct.* **2018**, *188*, 330–339. [CrossRef]
51. Tagliaferri, V.; Caprino, G.; Diterlizzi, A. Effect of drilling parameters on the finish and mechanical properties of GFRP composites. *Int. J. Mach. Tools Manuf.* **1990**, *30*, 77–84. [CrossRef]
52. Haeger, A.; Schoen, G.; Lissek, F.; Meinhard, D.; Kaufeld, M.; Schneider, G.; Schuhmacher, S.; Knoblauch, V. Non-destructive Detection of Drilling-induced Delamination in CFRP and its Effect on Mechanical Properties. *Procedia Eng.* **2016**, *149*, 130–142. [CrossRef]
53. Puchała, K.; Elżbieta, S.; Jachimowicz, J.; Bogusz, P. Gradient material model in analysis of mechanical joints of CFRP laminate. *AIP Conf. Proc.* **2018**, *1922*. [CrossRef]
54. Rueda, S.H. *Curing, Defects and Mechanical Performance of Fiber-Reinforced Composites*; Universidad Politécnica de Madrid: Madrid, Spain, 2013.
55. Awerbuch, J.; Madhukar, M.S. Notched Strength of Composite Laminates: Predictions and Experiments—A Review. *J. Reinf. Plast. Compos.* **1985**, *4*, 3–159. [CrossRef]
56. Xiao, Y.; Ishikawa, T. Bearing strength and failure behavior of bolted composite joints (part I: Experimental investigation). *Compos. Sci. Technol.* **2005**, *65*, 1022–1031. [CrossRef]
57. Shah, P.D.; Melo, J.D.D.; Cimini, C.A.; Ridha, M. Evaluation of Notched Strength of Composite Laminates for Structural Design. *J. Compos. Mater.* **2010**, *44*, 2381–2392. [CrossRef]
58. Camanho, P.P.; Hallett, S. *Composite Joints and Connections: Principles, Modelling and Testing*; Woodhead Publishing: Oxford, UK, 2011.
59. McCarthy, M.A.; McCarthy, C.T.; Lawlor, V.P.; Stanley, W.F. Three-dimensional finite element analysis of single-bolt, single-lap composite bolted joints: Part I—model development and validation. *Compos. Struct.* **2005**, *71*, 140–158. [CrossRef]
60. Crews, J.H., Jr.; Naik, R.A. *Bearing-Bypass Loading on Bolted Composite Joints*; NASA: Washington, DC, USA, 1987.
61. Hung, C.L.; Chang, F.K. Strength Envelope of Bolted Composite Joints under Bypass Loads. *J. Compos. Mater.* **1996**, *30*, 1402–1435. [CrossRef]
62. Puchała, K.; Szymczyk, E.; Jachimowicz, J. Sensitivity analysis of numerical model of CFRP mechanical joint. *AIP Conf. Proc.* **2019**, *2078*, 020096. [CrossRef]
63. Puchała, K.; Szymczyk, E.; Jachimowicz, J. FEM design of composite—metal joint for bearing failure analysis. *Przegląd Mech.* **2015**, *2*, 33–41.
64. Dragan, K.; Bieniaś, J.; Sałaciński, M.; Synaszko, P. Inspection methods for quality control of fibre metal laminates in aerospace components. *Composites* **2011**, *11*, 130–135.
65. Wronkowicz-Katunin, A.; Katunin, A.; Dragan, K. Ultrasonic C-Scan Image Processing Using Multilevel Thresholding for Damage Evaluation in Aircraft Vertical Stabilizer. *Int. J. Image Graph. Signal Process.* **2015**, *11*, 1–8. [CrossRef]
66. Sałaciński, M.; Synaszko, P.; Olesiński, D.; Samoraj, P. Approach to evaluation of delamination on the MiG-29's vertical stabilizers composite skin. In *The ICAF 2019—Structural Integrity in the Age of Additive Manufacturing*; Niepokolczycki, A., Komorowski, J., Eds.; Springer: Berlin, Germany, 2019; pp. 865–873.
67. Szymczyk, E. *Numeryczna Analiza Zjawisk Lokalnych w Połączeniach Nitowych Konstrukcji Lotniczych*; Wojskowa Akademia Techniczna: Warsaw, Poland, 2013.
68. Kelly, G. Quasi-static strength and fatigue life of hybrid (bonded/bolted) composite single-lap joints. *Compos. Struct.* **2006**, *72*, 119–129. [CrossRef]
69. Dano, M.-L.; Kamal, E.; Gendron, G. Analysis of bolted joints in composite laminates: Strains and bearing stiffness predictions. *Compos. Struct.* **2007**, *79*, 562–570. [CrossRef]
70. Du, A.; Liu, Y.; Xin, H.; Zuo, Y. Progressive damage analysis of PFRP double-lap bolted joints using explicit finite element method. *Compos. Struct.* **2016**, *152*, 860–869. [CrossRef]
71. Stocchi, C.; Robinson, P.; Pinho, S.T. A detailed finite element investigation of composite bolted joints with countersunk fasteners. *Compos. Part A Appl. Sci. Manuf.* **2013**, *52*, 143–150. [CrossRef]
72. Askri, R.; Bois, C.; Wargnier, H.; Gayton, N. Tolerance synthesis of fastened metal-composite joints based on probabilistic and worst-case approaches. *Comput.-Aided Des.* **2018**, *100*, 39–51. [CrossRef]

73. MSC. Marc 2013 Documentation vol. A. In *Theory and User Information*; MSC Corporation: Newport Beach, CA, USA, 2013.

74. Kim, K.S.; Hahn, H.T. Residual stress development during processing of graphite/epoxy composites. *Compos. Sci. Technol.* **1989**, *36*, 121–132. [CrossRef]

75. Cho, J.; Sun, C.T. Lowering Thermal Residual Stresses in Composite Patch Repairs in Metallic Aircraft Structure. *AIAA J.* **2001**, *39*, 2013–2018. [CrossRef]

76. Mulle, M.; Collombet, F.; Olivier, P.; Grunevald, Y.-H. Assessment of cure residual strains through the thickness of carbon–epoxy laminates using FBGs, Part I: Elementary specimen. *Compos. Part A Appl. Sci. Manuf.* **2009**, *40*, 94–104. [CrossRef]

77. McCarthy, C.T.; McCarthy, M.A. Three-dimensional finite element analysis of single-bolt, single-lap composite bolted joints: Part II—Effects of bolt-hole clearance. *Compos. Struct.* **2005**, *71*, 159–175. [CrossRef]

78. Khashaba, U.A.; Sebaey, T.A.; Alnefaie, K.A. Failure and reliability analysis of pinned-joints composite laminates: Effects of stacking sequences. *Compos. Part B Eng.* **2013**, *45*, 1694–1703. [CrossRef]

79. Lopez-Cruz, P.; Laliberté, J.; Lessard, L. Investigation of bolted/bonded composite joint behaviour using design of experiments. *Compos. Struct.* **2017**, *170*, 192–201. [CrossRef]

80. Askri, R.; Bois, C.; Wargnier, H. Effect of Hole-location Error on the Strength of Fastened Multi-Material Joints. *Procedia CIRP* **2016**, *43*, 292–296. [CrossRef]

81. Lü, X.; Zhao, J.; Hu, L.; Wang, H. Effect of interference fits on the fatigue lives of bolted composite joints. *J. Shanghai Jiaotong Univ. Sci.* **2016**, *21*, 648–654. [CrossRef]

82. Giannopoulos, I.K.; Doroni-Dawes, D.; Kourousis, K.I.; Yasaee, M. Effects of bolt torque tightening on the strength and fatigue life of airframe FRP laminate bolted joints. *Compos. Part B Eng.* **2017**, *125*, 19–26. [CrossRef]

83. Mandal, B.; Chakrabarti, A. Numerical failure assessment of multi-bolt FRP composite joints with varying sizes and preloads of bolts. *Compos. Struct.* **2018**, *187*, 169–178. [CrossRef]

84. Liu, X.; Yang, Y.; Wang, Y.; Bao, Y.; Gao, H. Effects of Hole Perpendicularity Error on Mechanical Performance of Single-Lap Double-Bolt Composite Joints. *Int. J. Polym. Sci.* **2017**, *2017*. [CrossRef]

85. Zhao, L.; Shan, M.; Liu, F.; Zhang, J. A probabilistic model for strength analysis of composite double-lap single-bolt joints. *Compos. Struct.* **2017**, *161*, 419–427. [CrossRef]

 *materials* 

Article

Experimental Investigation of a Slip in High-Performance Steel-Concrete Small Box Girder with Different Combinations of Group Studs

Bishnu Gupt Gautam [1], **Yiqiang Xiang** [1,*], **Xiaohui Liao** [2], **Zheng Qiu** [1] **and Shuhai Guo** [1]

1   College of Civil Engineering and Architecture, Zhejiang University, Hangzhou 310058, China
2   College of Civil Engineering and Architecture, Quzhou University, Quzhou 324000, China
*   Correspondence: xiangyiq@zju.edu.cn; Tel.: +86-571-8820-8700

Received: 11 August 2019; Accepted: 26 August 2019; Published: 29 August 2019

Abstract: Due to the significant advantages of steel-concrete composite beams, they are widely used for accelerated bridge construction (ABC). However, there is still a lack of experimental research on the proper design of ABC, especially in the slip with a different group of shear connectors. As a component of steel-concrete composite structure, shear studs play a vital role in the performance of composite structures. This paper investigates the influence of group studs in simply supported and continuous box girders. To this end, three sets of simply supported steel-concrete composite small box girders and two continuous steel-concrete composite small box girders were made with different groups of shear studs, and the slip generated along the beams was recorded under different caseloads. The results were then compared with the proposed simplified equations. The results show that the slip value of the test beam is inversely proportional to the degree of shear connection. The slip of Simply Supported Prefabricated Beam-3 (SPB3) is 1.247 times more than Simply Supported Prefabricated Beam-1 (SPB1), and 2.023 times more than Simply Supported Prifabricated Beam-2 (SPB2). Also, the slip value of Experimental Continuous Beam-1 (ECB1) is 1.952 times more than Experimental Continuous Beam-2 (ECB2). The higher the degree of shear connection, the smaller the maximum slip value.

Keywords: slip; group studs; composite beam; accelerated bridge construction; steel fiber

---

1. Introduction

With the development of urbanization and the rapid progress of civil engineering, the problem of urban congestion has become increasingly prominent. Therefore, it is necessary to improve the construction of urban transportation infrastructure methods and expand urban expressway networks. In order to build, replace, and repair a bridge, or a series of bridges, in areas with heavy traffic, traditional construction methods are usually used in current engineering projects. These methods generally include field activities, such as the installation of supports, formwork, binding of reinforcement bars, pouring and curing of concrete, etc. These onsite construction activities not only consume a great deal of time but also lead to traffic congestion, which weakens the safety and efficiency of the traffic network. On the other hand, due to the constraints of the construction site and climatic conditions, the components of onsite construction are prone to quality problems, which may potentially affect the durability of the structure. Accelerated bridge construction (ABC) is a promising approach to reduce the impact of construction on traffic, improve the quality of materials and durability of products, and minimize the construction time [1–3].

A combination of steel beams and a concrete deck with shear connectors, known as steel-concrete composite beams, are widely used in the ABC approach. Due to the use of shear stud tied together in the upper part of steel beam, this type of structure can give full strength to the respective properties of the two materials, especially for short and medium-span urban areas and highways [4–7]. The headed stud shear connectors are usually used to resist longitudinal slip and vertical separation between the steel beam and concrete slab. In general practice, construction of steel-concrete composite girder bridge, I-shaped steel beams, or a small steel box girder flange of the steel beam structure are used [8,9]. Compared with I-shaped steel girder, the small box section has better stability and torsional resistance. Moreover, compared with large box girders, small box girders have a flexible cross-section, lightweight, easy fabrication, and assembly, and thus, are more suitable for accelerated construction of the bridges [10].

The characteristics of concrete play a vital role in all kinds of structure for overall strength. Concrete is brittle and has partial ductile behavior. Therefore, a different form of reinforcement is needed to boost structural stability. Steel bars are used as reinforcement in concrete structures; however, there's still a chance of crack, deflection and slip formation, which causes a big problem within the overall stability of the structure [11–17].

Concrete is a comparatively widely used material. However, it will run into huge issues and defects if it's not prepared correctly. The defects are started from the concrete parts of reinforcement concrete members and extended until encountering a reinforcing bar in which yield to significant problems. We must control these defects to prolong the life period of concrete structures. Hence, according to the application of the concrete in the structure, a multi-directional and closely spaced reinforcement may need to be utilized. The use of fibers is another solution to enhance the ductility of concrete material. Fiber-reinforced concrete (FRC) consists of short distinct fibers that are uniformly distributed and at random bound among the concrete matrix. The fibers are mostly classified into four categories, namely—steel fiber, glass fiber, natural fiber, and synthetic fiber, all of which have their own respective properties [18–22].

The use of fibers in concrete components causes the increase of structural integrity, provides high tensile strength to plain concrete, decreases the permeableness of concrete, and increases the resistance to impact load. Fibers reduce the number of rebars without affecting the strength. It will additionally eliminate the defects by bridging action. Flexural behavior, bond strength, and particularly toughness of SFRC (steel fiber reinforced concrete) increases by increasing fiber content. Carbon or steel fibers are added to a cement matrix at a high volume fraction (0.5%–3%) to extend the conduction of the composite. The properties of fiber concrete depend on the quantity of fibers used [23–28].

Various problems are occurred in composite materials in terms of slip deflection and crack. The crack in a composite structure could decrease the stiffness and strength, which may accelerate the failure of the structure. Due to the inherent material properties of fiber concrete, the presence of fiber improves the resistance of conventionally reinforced structural members to serviceability deflection condition. Although the slip is not the main reason for a structural failure, it is part of the failure process due to losses in the structural integrity. Therefore, controlling slip plays an essential role in the failure mechanism of steel-concrete composite structures [29–35].

Generally, the concrete grade is below C50 in traditional construction, while high-performance concrete with a grade higher than C60 is used in the ABC process. Therefore, due to the high mechanical strength and durability, high-performance concrete in the composite beam receives much attention [36]. In the steel-concrete composite beams, there is a sliding problem in the interface layer between two materials, which is usually appeared by increasing load intensity. It can increase damping, which is beneficial for dynamic loading conditions [37,38]. The existence of such an interface slip can reduce the stiffness, increase the curvature and the deformation of composite beams [39–45]. Therefore, the actual working condition of composite beams cannot be precisely determined without considering the interface slip effect. In the past few years, many studies attempted to address the slip problem; however, few of them, if any, considered the group studs in the beam [46–52]. So, this research is only focused on the slip tests of a steel-concrete composite small box girder (SCCSBG) in the simply supported and continuous beams.

In this research, two equations were proposed, one for the simply supported beam and another for continuous beams to consider the influence of group studs and compare their performance. For this purpose, three simply supported SCCSBG (steel-concrete composite small box girder), and two continuous SCCSBG were experimentally tested under the action of the vertically applied load. The slip values were observed throughout the full span of the beam. The details of the experiments and their behavior are discussed in the following sections.

2. Design and Fabrication of Experimental Beams

The experimental SCCSBG beams, which were made using accelerated construction techniques, consist of an open steel box girder, diaphragm, welded stud, and concrete slab. Due to difficulties in the process of making a real-size scale model, especially in the case of the concrete bridge deck and the thickness of the steel plate, it followed the design specification. Based on the 25 m span prototype bridge, a stereotype model of steel concrete steel fibrous small box girder was prepared to the actual length ratio of 1:4 and 1:6 for continuous and simply supported beams, respectively. The ECB-1 and ECB-2 were used for the continuous beam, whereas SPB1 to SPB3 were used for the simply supported beam. The material details about both types of the beams are listed in Table 1.

Table 1. Material details of experimental beams.

Material	Parameter	Value
Concrete	Density, (kg/m^3)	2400
	Elastic modulus, (MPa)	36,400
	Poisson's ratio	0.167
	Design strength of bridge deck concrete, (MPa)	C60
	Reserved hole concrete design strength, (MPa)	C80
	Yield strength, (MPa)	76.24
Steel (Q345qc)	Density, (kg/m^3)	7850
	Elastic modulus, (GPa)	210
	Poisson's ratio	0.3
	Yield strength, (MPa)	421
steel bar (HPB 300 and HRB 400)	Density, (kg/m^3)	7800
	Elastic modulus, (GPa)	206
	Poisson's ratio	0.3
	Yield strength, HRB400, (MPa)	445
	Yield strength, HPB300,(MPa)	363
Stud	Density, (kg/m^3)	7800
	Elastic modulus, (GPa)	210
	Poisson's ratio	0.3
	Yield strength, (MPa)	360

In the accelerated construction of SCCSBG, the spacing of studs, group arrangement of studs, and concrete materials are the three main factors. Using the high-performance concrete reduces the structural weight and cost and increases the durability. Therefore, recently, it is more widely used in bridge engineering. The spacing and arrangement of the studs should be determined using the degree of shear connection. The degree of shear connection is the ratio of the number of actual welded studs in the shear span to the number of welded studs required for the complete shear connection. In the designing procedures, while considering an arrangement for studs, a minimum space according to the degree of shear connection is defined to ensure the mechanical performance of the structure. However, in term of construction, having more space between the group studs is preferred. To design the test beams, the degree of the shear connection was determined based on the designing of steel and concrete composite bridges code GB50917-2013 [53]. However, there is no code available to specify the arrangement and spacing of group studs. Therefore, the formula of bearing capacity for a single stud defined in this code is used here. EC4 [54] requires that the maximum spacing of stud uniformly distributed along the steel-concrete composite beams should not exceed four times of the concrete slab thickness or the value of 800 mm.

Based on the steel structure design code GB50017-2017 [55], if the strength and deformation are satisfied, the longitudinal shear capacity of the shear connectors at the interface of composite beams guarantees the full flexural capacity of the beam. Hence, the beam can be designed according to the partial shear connection. However, the partial shear connection is limited to the composite beams with equal cross-section and the spans less than 20 m. The American Association of State Highway and Transportation Officials (AASHTO) [56] bridge design code stipulates that the spacing of reserved holes is not more than 610 mm.

In the fabrication, two different lengths of girders are used to manufacture the beams. The smaller one is used in the SPBs, and the larger one is used in the ECBs. The SPBs are 4.5 m in length, 4.2 m in support spacing, 258 mm in depth of the composite beam and 70 mm in the thickness of the precast concrete slab. For the steel U-type girder, the depth is 0.187 m, the top width is 180 mm, and the bottom width is 150 mm. The upper flange is 6 mm thick and 100 mm wide. The web is 6 mm thick, and the bottom is 8 mm thick. A solid web diaphragm is set every 700 mm from support. The diaphragm is 6 mm thick and 160 mm high. The steel U-type girder is made of Q345qc steel. The group studs are welded to the upper flange of the steel beam. The dimension and materials of welding for studs are 13 mm × 45 mm and ML15AL, respectively. The SPBs were made of C60 high-performance concrete with a width of 600 mm and a thickness of 70 mm. The center distance of the group studs varies from 350 mm to 420 mm from support to midspan. Two layers of steel bars were arranged longitudinally. The upper layer of steel bars is HRB400 with a diameter of 10 mm, and the lower layer of steel bars and stirrups are HPB300 with a diameter of 8 mm. The spacing of stirrups is 60 mm, and no stirrups are arranged in the reserved hole position. The details about stud arrangement and cross-section of the SPBs is shown in Figure 1a. The design parameters of SPBs and ECBs are presented in Table 2.

ECBs were made of two spans, each span is 3 m in length, with a total length of 6.3 m long. The total depth of the composite box girder was 327 mm with 70 mm slab thickness and 257 mm depth of steel U-type girders. The upper flange was 6 mm thick and 135 mm wide. The essential of the web was 6 mm, whereas the bottom plate was 8 mm. The diaphragm was set to every 600 mm from support to throughout the box girder. The thickness and height of the diaphragm were 6 mm and 220 mm, respectively. The construction of prefabricated concrete slabs, welding steel beams, and welding studs was in accordance with the requirements of design drawings and technical construction specifications. The C60 high-performance concrete and C80 steel fiber high-performance concrete are used for prefabricated slab and reserved hole filling, respectively. The detail configuration of both types of box girder is shown in Figure 1, and the main fabrication process is shown in Figure 2.

For the high-performance concrete element, copper-plated steel fiber with a diameter of 0.2 mm and length of 13 mm, tensile strength of 2000 MPa, and a volume fraction of 1.5% was used. The details about the physical properties of the steel fiber are listed in Table 3. The obtained physical properties by the use of cement, fine aggregate, coarse aggregate, water, and chemicals are based on different codes [57–59]. The information about different of properties and materials uses are described in Appendix A. The arrangement of bars in the ECB's are the same as SPB's. However, for the stud arrangement, the hole with a plane size of 160 mm × 100 mm is reserved and filled with C80 steel fiber concrete. The size of the ECB's welding stud is 13 mm × 50 mm.

Table 2. The design parameters of SPBs and ECBs.

Parameter	SPB1	SPB2	SPB3	ECB1	ECB2
Size of stud (mm)	$\Phi13 \times 45$	$\Phi13 \times 45$	$\Phi13 \times 55$	$\Phi13 \times 50$	$\Phi13 \times 50$
Distribution of studs (Tr. × L.)	2×2	2×2	2×3	2×3	3×3
Longitudinal spacing between adjacent group of studs (mm)	350	350–420	420	600	600
Total number of studs	104	144	88	132	198
Degree of shear connection	1.13	1.57	0.96	1.02	1.53
Transverse c/c spacing of studs	50	50	50	50	50
Longitudinal c/c spacing of studs	65	65	65	65	65
Beam total length (m)	4.5	4.5	4.5	6.3	6.3
Combined beam width (mm)	600	600	600	700	700
Beam height (mm)	257	257	257	327	327
Thickness of concrete deck (mm)	70	70	70	70	70

Table 3. Physical properties of steel fibers.

Fiber Type	Length (mm)	Diameter (mm)	Aspect Ratio	Tensile Strength (Mpa)	Modulus of Elasticity (GPa)	Density (Kg/m^3)
Steel fiber	13	0.2	65	2000	210	7800

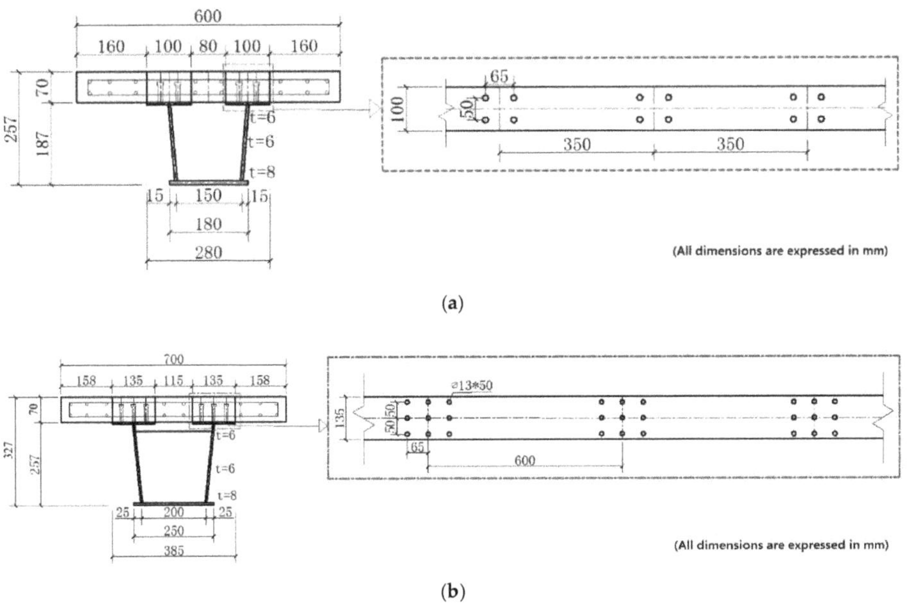

Figure 1. Details of stud arrangement cross-section of experimental accelerated construction steel-concrete steel fibrous high-performance composite box girder bridge. (**a**)Typical arrangement of a simply supported beam (SPB1); (**b**) Typical arrangement of a continuous beam (ECB-2).

Figure 2. Fabrication process of the experimental beams: (**a**) formwork erection and reinforcing case binding; (**b**) casting and curing of concrete slab; (**c**) stud welding of test specimens; (**d**) completion of the concrete slab and steel U-type girder; (**e**) placing of concrete slab on steel box girder to make it composite; (**f**) reserve hole filling.

3. Experimental Test

3.1. Test Method and Instrumentation

The relative slip between the concrete, slab, and steel U-type girder was measured by means of the dial gauge (Wanmu, Chengdu, China). The dial gauges were fixed at the lower edge of the concrete slab, and free end contacts with the aluminum block extending from the bottom of the upper flange of the steel U-type girder. The dial gauges were installed along the experimental beams under the group studs, as shown in Figure 3c. Each beam is placed with a dial gauge at the position of group studs from the support to the throughout the span. Due to have a symmetric shape, the dial gauges are mainly arranged in only one side.

Figure 3. *Cont.*

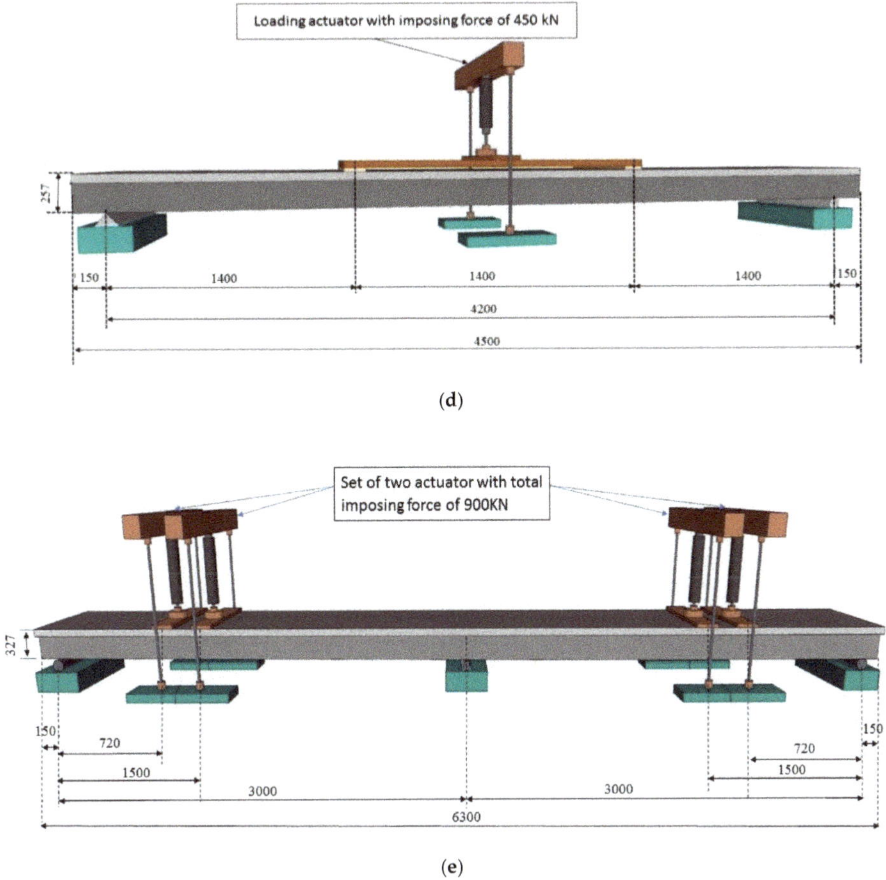

Figure 3. Experimental setup for slip measurements (m); (**a**) Loading system of SPBs (**b**) Loading system of ECBs; (**c**) Arrangement of dial gauges; (**d**) Schematic view of SPBs (mm); (**e**) Schematic view of ECBs (mm).

3.2. Loading Procedure

The 500 kN servo loading system (Popwil, Hangzhou, China was used for the test loading. The boundary conditions of the beams were roller and hinge bearing at both ends of the beam. The experiment was performed in the structural lab of Quzhou University in China. The loading was directly exerted to an area of 60 cm × 20 cm of the experimental beam. For SPBs, the loading was symmetrically applied at two areas within the full span. The distance between the loading areas is 1.4 m. A layer of fine sand is cushioned at the loading area to ensure that the load of the test beam is uniformly applied to the beam. The loading, as mentioned above procedure, was performed for SPBs. Whereas, for the ECBs, a set of two actuators with the total loading capacity of 1000 kN was employed. The loading was applied at each mid-span of a two-span continuous SCCSBG with a loading area of 70 cm × 20 cm. The ECB's were supported by the roller bearing at both ends and a hinge bearing at mid support. The loading increment applied in the laboratory model was about 5% of the yield load (Py) in each load step. The slip values of test beams were carefully measured throughout the beams, which is shown in Figure 3c. The loading arrangement of the test beams are shown in Figure 3d,e.

The full experimental detail is shown in Figure 3. Before the formal loading, two preloads were carried out to verify the normal working of test instruments and meet the quality test requirements.

3.3. Slip Measurement

The slip mainly occurs in the shear span between support and the loading point. The slip values in the pure bending segment of the beams and the midspan are minimal. There is no shear force at the section of the steel beam and concrete slab in the pure bending segment in the middle of the experimental beam. In the general concept, there should be no slip, but as the deflection of the beam increases, the concrete slab near the loading point has an oblique extrusion effect on the welding studs, which makes the relative displacement between the steel beam and a concrete slab in a longitudinal direction. Because of the boundary condition, the maximum slip value does not appear in the support region, but it occurs at the shear span segment. Moreover, because of the different restrained strength of the support at both ends, the slip of the supports is asymmetric. The arrangement of dial gauges to measure slip values are shown in Figure 4. The experimental slip results in the longitudinal direction of the SPBs and ECBs under the action of vertical load are shown in Figure 5.

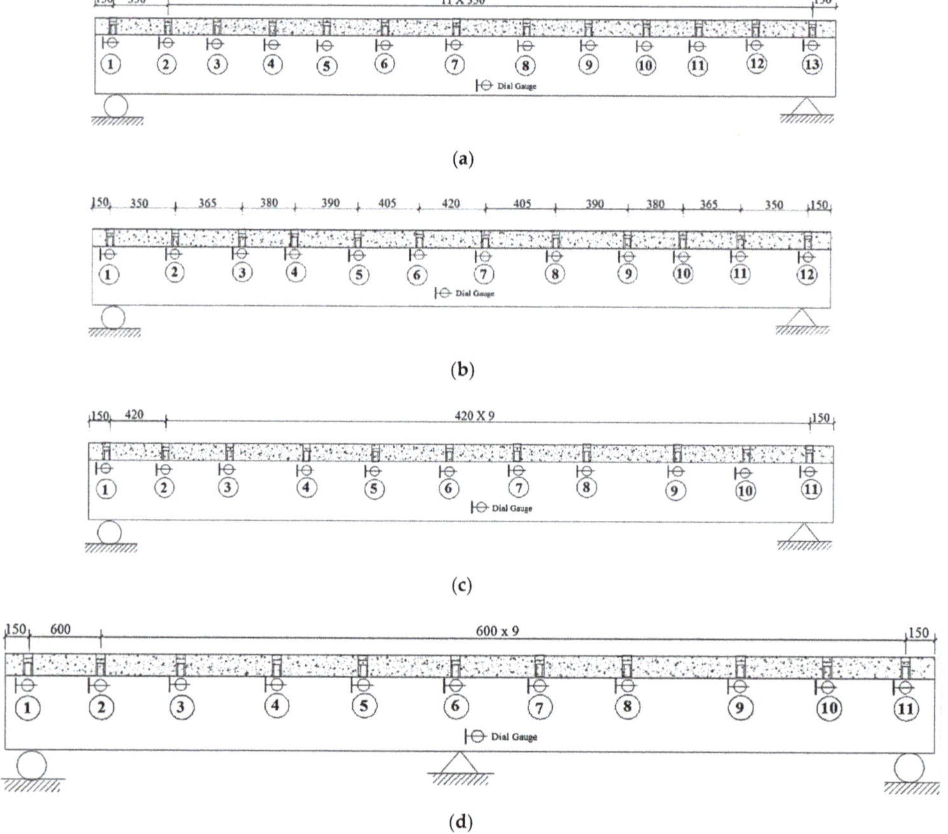

Figure 4. Location of dial gauges (mm); (**a**) SPB1; (**b**) SPB2; (**c**) SPB3; (**d**) ECB-1 and ECB-2.

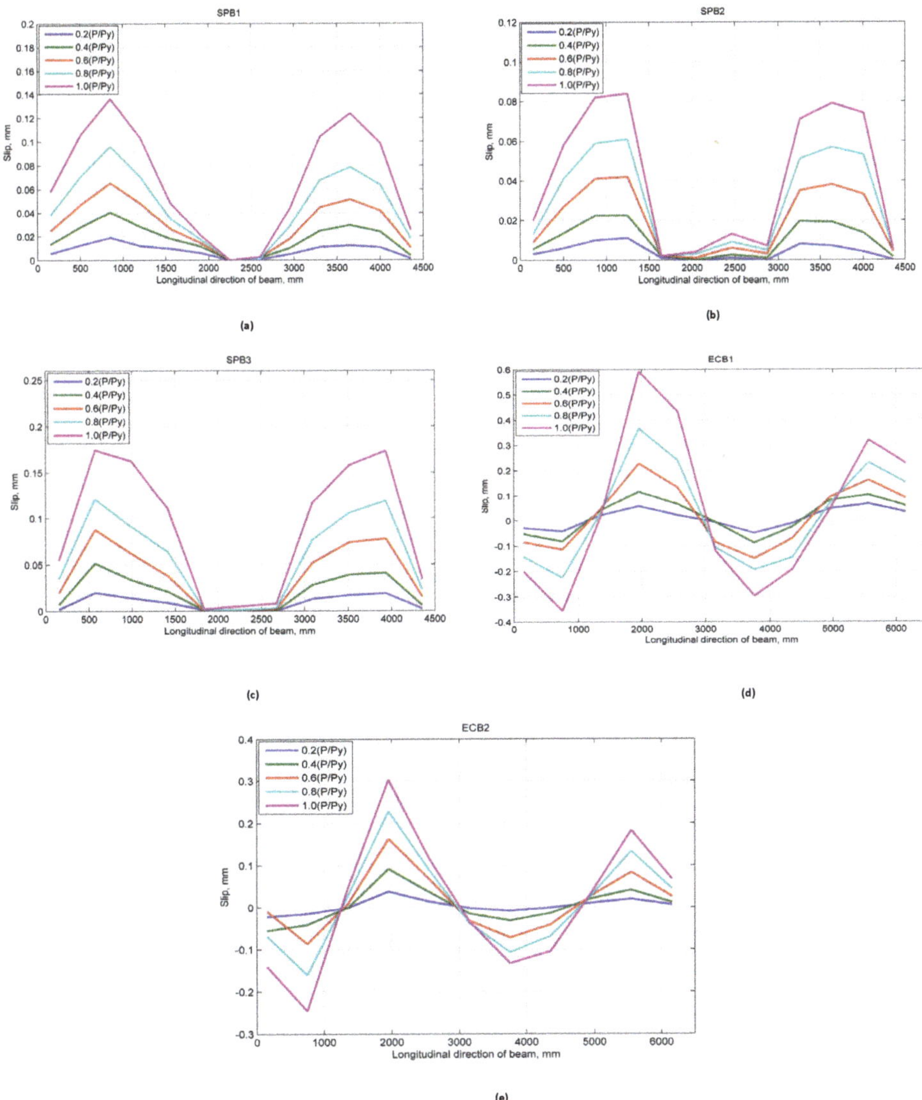

Figure 5. Variation of measured interface slip with a longitudinal direction of the beam under the action of vertical loads: (**a**) SPB1; (**b**) SPB2; (**c**) SPB3; (**d**) ECB1; (**e**) ECB2.

4. Simplified Model

The slip was observed under 24 kN of load where the values of SPB1, SPB2, and SPB3 were 0.009, 0.004 and 0.008 mm, respectively. The slip values of the SPBs increases with the increase of the load. As shown in Figure 5, the relationship between the growth rate of the SPBs slip values are SPB3> SPB1 > SPB2, and it does not increase linearly with the increase of the load. The inconsistency growth rates of slip values are mainly affected by the degree of shear connection. Therefore, as the load increases, the welding studs gradually entered into yielding. So, the growth rate of slip values increases continuously, and the slip value of each stud group in the shear span tends to be uniform.

The maximum slip values of SPB1, SPB2, and SPB3 under 270 kN load are 0.136, 0.084, and 0.170 mm, respectively. The maximum slip values have an inverse relationship with the degree of shear connection. The larger the degree of shear connection, the smaller the slip value. The main reason for this is the local effect of group studs shear connectors. The location of reserved holes is being centralized, and the shear stiffness of this location is becoming more significant than the traditional uniformly distributed studs.

There are so many conditions which need an ideal mathematical solution to represent a set of data. Using the N-order polynomial regression model, a line can be fitted to the experimental scatter data, and formula can be extracted to represent/estimate the data [60]. Therefore, two quadratic equations extracted using N-order polynomial regression are proposed in the form of Equations (1) and (2). The residual value for each data point is the distance from the target point to the regression line, which is error in prediction. The norm of residual is a measure of goodness of fit, so that a value centered around zero indicates a better fit [61]. Equation (1) is for SPBs, while Equation (2) addresses ECBs. The norm of residuals for Equations (1) and (2) are 0.027 and 0.112, respectively.

$$P/Py = -14 \times s^2 + 8.1 \times s + 0.014 \tag{1}$$

$$P/Py = -2.5 \times s^2 + 3.1 \times s + 0.031 \tag{2}$$

where P/Py is the ratio of an applied load to yield load, and s is the slip of the beam at corresponding load. The proposed equation can be used for the prediction of the slip in SCCSBG of simply supported and continuous beam with fully composite action. When the degree of shear connection is equal, these equations can predict the slip value of that beam. Whereas, with a different degree of shear connection, it can relate the change in slip values between them. However, the limitation of the proposed equation regarding the dimension and construction size should also be considered while applying it. Figures 6 and 7 show the proposed equations with the experimental data and corresponding residual values. The residual (%) in the fitting process of SPB3 and ECB1 load-slip curves are presented in Tables 4 and 5, respectively.

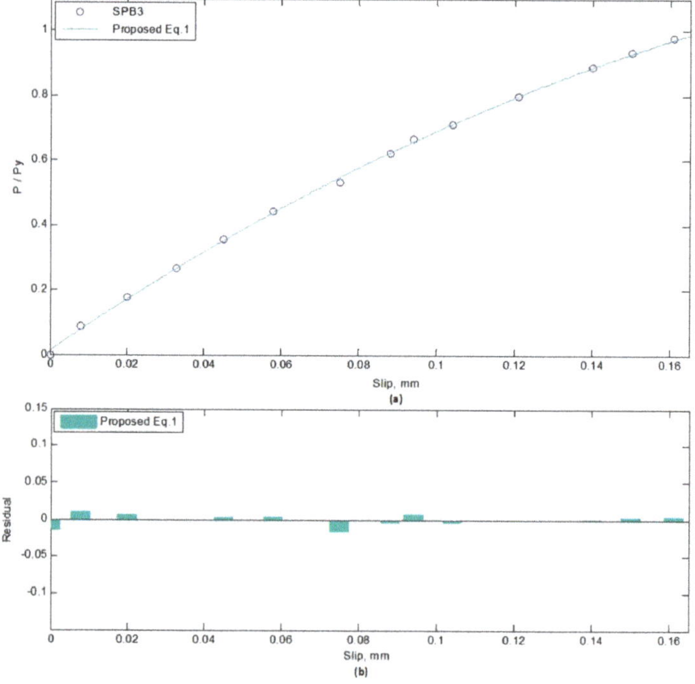

Figure 6. Fitting of curve, SPB3: (**a**) Load-slip; (**b**) Residual-slip.

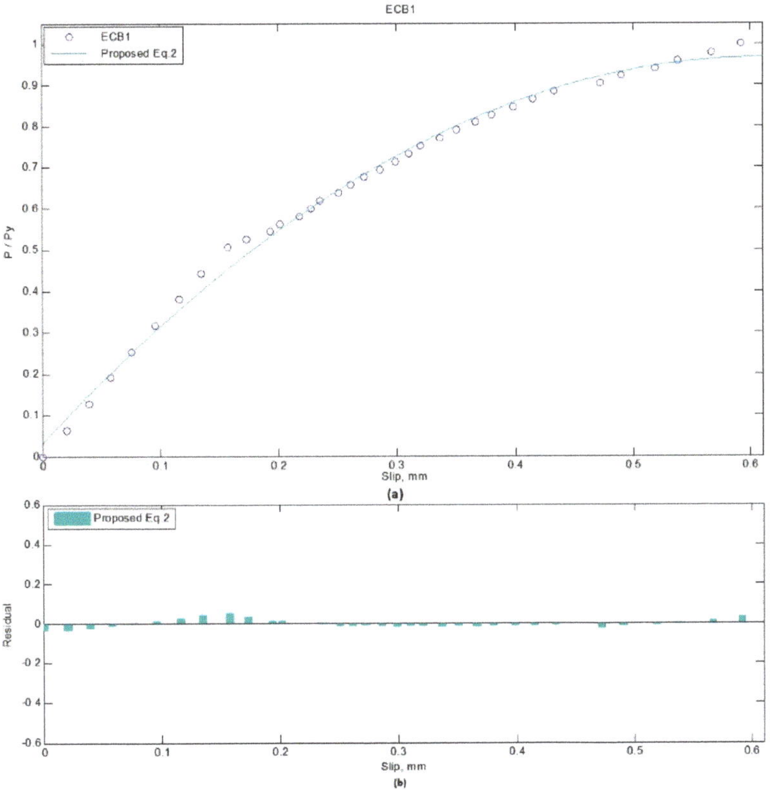

Figure 7. Fitting of curve, ECB1: (**a**) Load-slip; (**b**) Residual-slip.

5. Discussion

Based on the experimental results, the relationship between applied load and interface slip was studied, and it is noteworthy that all the provided relationships are limited to the upper yield (P_y) point of the elastic stage. The upper yield load is around 270 kN for the SPBs and 900 kN for the ECBs. To examine the characteristics performance of the different types of group studs by using the proposed equations, the comparison between all the SPBs and ECBs were performed. The comparison of the proposed equations and experimental results of the SPBs are presented in Table 4. The same comparison for ECB's are also reported in Table 5. With regard to the different group studs arrangement, by increasing the intensity of the load ratio (P/P_y), the smaller group size with a more number of welding studs has a lower value for a slip and vice versa. The diagram of the measured yield load (P_y) to the corresponding slip for SPBs and ECBs are illustrated in Figures 8 and 9, respectively. The higher degree of shear connection for simply supported beam in the SPB2 and for continuous beam in ECB2 causes less slip value. The proposed Equation (1) and SPB3 have an almost similar value of slip, so it is considered as Equation (1) with a degree of connection similar to SPB3. Equation (2) also has a degree of shear connection similar to ECB1. The slip value of SPB3 is 1.250 times more than SPB1, and 2.023 times more than SPB2. Also, the slip value of ECB1 is 1.952 times more than ECB2.

Table 4. Comparison of slip between the test results of SPBs and the proposed equation.

Measured P/Py	Slip, mm				Residual (%) SPB3
	SPB1	**SPB2**	**SPB3**	**Proposed Equation (1)**	
0.2	0.019	0.011	0.020	0.020	0
0.4	0.040	0.022	0.051	0.051	1.360
0.6	0.065	0.042	0.088	0.088	0.113
0.8	0.096	0.061	0.121	0.122	1.150
1.0	0.136	0.084	0.170	0.166	2.264

Table 5. Comparison of slip between the test results of ECBs and the proposed equation.

Measured P/Py	Slip, mm			Residual (%) ECB-1
	ECB-1	**ECB-2**	**Proposed Equation (2)**	
0.2	0.057	0.038	0.045	22.393
0.4	0.116	0.092	0.103	10.966
0.6	0.228	0.163	0.221	3.027
0.8	0.367	0.229	0.378	3.217
1.0	0.591	0.303	0.566	4.226

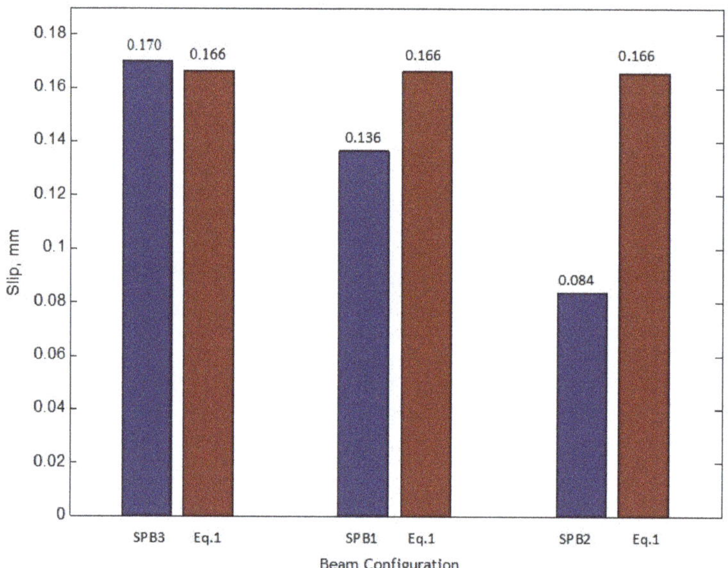

Figure 8. Comparison of proposed Equation (1) with SPB1, SPB2, and SPB3.

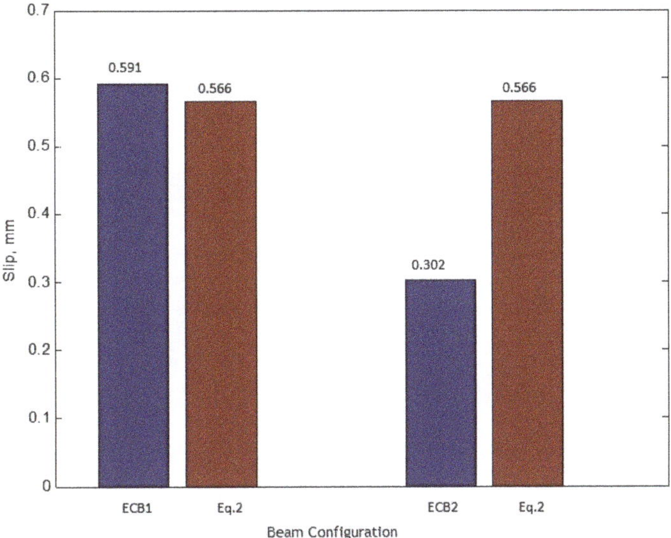

Figure 9. Comparison of proposed Equation (2) with ECB1 and ECB2.

6. Conclusions

This study focused on the static behavior of the accelerated construction steel fibrous high-performance steel-concrete composite small box girder bridge. To achieve this, two different types of simply supported and continuous beams have been tested with trapezoidal steel beams and different shear stud arrangements. Each sample was tested under a progressive applied vertical loading, and the maximum slip was recorded at key points along the span for both simply supported and continuous beams. Using the dataset obtained from laboratory tests, simplified empirical relationships were developed to approximate the slip of both simply supported and continuous beams with a different degree of shear connections.

The following main conclusions are drawn according to investigation in this paper.

- It was observed that the measured slip values for SPB1, SPB2, and SPB3 samples of simply supported beams at 0.2 (P/Py) were 0.019 mm, 0.011 mm, and 0.020 mm, respectively. While the slip values for ECB-1 and ECB-2 samples of continuous beams were 0.057 mm and 0.038 mm, respectively.
- The integrity and load-carrying capacity of accelerated construction high-performance SCCSBG are found to be more than traditional construction because of the use of steel fibers in concrete. Therefore, for SPB1, SPB2 and SPB3, the slip values of simply supported beams at 1.0 (P/Py) are bounded to 0.136 mm, 0.084 mm and 0.174 mm, respectively. In addition, for continuous beams, the sliding values measured by ECB-1 and ECB-2 at 1.0 (P/Py) are limited to 0.591 mm and 0.303 mm, respectively.
- Because of the higher degree of shear connection, the slip of SPB3 is 1.247 times more than SPB1, and 2.023 times more than SPB2. Also, the slip value of ECB1 is 1.952 times greater than ECB2. That is, the higher the degree of shear connection is, the smaller the maximum slip value becomes.

Since the proposed equations are presented in an explicit form, they can be practically used to predict the slip of group studs in either simply supported or continuous beam. As the experimental model was designed with ratios of 1:6 and 1:4 for simply supported and continuous beams, respectively, the application of the proposed model is limited to cases with a similar range of parameters. Other sizes and forms of accelerated construction beams can be investigated by the authors in future research.

Author Contributions: Conceptualization, B.G.G. and Y.X.; formal analysis, B.G.G.; data curation, B.G.G., Y.X., X.L., Z.Q. and S.G.; investigation, B.G.G., Y.X., X.L. and S.G.; methodology, B.G.G.; project administration, Y.X. and B.G.G.; resources, Y.X., B.G.G. and X.L.; validation, B.G.G., Y.X., S.G. and Z.Q.; visualization, B.G.G.; writing—original draft, B.G.G.; writing—review and editing, B.G.G.; funding acquisition, Y.X.; supervision, Y.X.

Funding: This research was funded by Zhejiang University, the Cyrus Tang Foundation in China, and National Natural Science Foundation of China (NSFC, No. 51541810).

Acknowledgments: The authors appreciate the cooperation of staff members of the structural laboratory of Quzhou University and the help from Ying Yang in the experimental research are highly appreciated.

Conflicts of Interest: The authors declare no conflict of interest. The funding sponsors had no role in the design of the study; in the collection, analyses, or interpretation of data; in the writing of the manuscript; or in the decision to publish the results.

Appendix A

A local brand cement of 52.5 grade OPC of and medium sand from Ganjiang river was used for fine aggregate with a fineness modulus of 2.8. For coarse aggregate, well-graded basalt was used with a maximum particle size less than 16 mm. The main chemical composition for bonding properties of the constituent materials are shown in Table A1. The average fineness of the slag powder, density, water reduction rate, activity index on 28d, and activity index on 56d were 3 microns, 2.40 g/cm^3, 10–15%, 105–110% on 28d, and 110–125%, respectively.

Table A1. Chemical constituents of Diatomite, SiO$_2$ and mineral powder (%).

Component	SiO$_2$	Al$_2$O$_3$	MgO	CaO	f-CaO	SO$_3$	MnO	Density	Loss
Diatomite	76.11	11.21	3.5	3.8	-	-	-	2.6	<1
Sub-nano SiO$_2$	80.11	1.49	2.69	4.64	1.98	0.65	-	-	<1.5
Micron grade Mineral powder	33.84	11.68	10.61	38.13	-	-	0.34	-	<3

Silicon Powder was made by local Building Materials Co., Ltd. according to GB/T176-1996 [56]. The test results of Silicon powder measured by GB/T18736/2002 [57] is reported in Table A2.

Table A2. Silicon powder performance index.

Item	SiO$_2$ (%)	Moisture Content (%)	Ignition Loss (%)	Water Demand Ratio (%)	Fineness (45 um) (%)	28d Activity Index
Control index	≥85	≤3.0	≤6	≤125	-	≥85
Test Result	95.3	0.90	2.1	119.5	0.9	90

Steel fibers: Straight copper-plated steel fibers with a length of 13 mm, the diameter of 0.2 mm, the volume fraction of 1.5%, the tensile strength of 2000 MPa, and were used as steel fibers. Figure A1 shows the steel fibers mixed with C80 high-performance concrete.

Figure A1. Steel fiber in concrete.

Water reducer: Superplasticizer of a local chemical company was used. The index parameters are listed in Table A3.

Table A3. Performance index of water reducer.

Item	PH Value	Density (g/mL)	Solid Content (%)	Chloride Ion Content (%)	Alkali Content (%)	1 h Loss	Water Reduction Rate.
Control Index	6.5 ± 0.02	1.04 ± 0.02	21 ± 1	≤0.2	≤3.0	40	≥25
Test result	6.5	1.046	21.5	0.05	0.12	23	30.4

Water: Ordinary tap water was used in the experiment.

Each concrete was designed with three mixing ratios and after that, C60 concrete, and C80 concrete was tested. Three sets of test pieces were made for each combination, and each group had three cube test pieces. After conducting the strength test, the design mix ratio was optimized. The concrete mix of C60 and C80 is shown in Table A4.

Table A4. Benchmark mixture proportion of C60 and C80 high-performance concrete.

Concrete	Water Binder Ratio	Sand Rate (%)	The Amount of Raw Material Used per Concrete							
			Cement	Slag Powder	Fine Aggregate	Coarse Aggregate	Water	Water Reducer	Silica Fume	Steel Fiber
C60 *	0.3	40	388	129	699	1049	155	6.2	-	-
C80 **	0.25	38	435	87	652	1063	157	10.44	58	87

* Cement = 0.75, slag powder = 0.25, fine aggregate = 1.35, coarse aggregate = 2.03, water = 0.3, Admixture = 0.012.
** Cement = 0.75, slag powder = 0.15, Silica powder = 0.1, fine aggregate = 1.124, coarse aggregate = 1.83, steel fiber = 0.15, water = 0.27, and water reducer = 0.018.

C60 and C80 high-performance concrete were used in the test of the beam. For C60, eight groups of 150 mm cubic test specimen and two groups of 150 mm × 150 mm × 300 mm prism specimen were made. Four groups of cube specimen were tested with the same curing condition as of beam. Where, with two groups of the cube, compressive strength was measured with a temperature of 600 °C. One group of the specimen was tested on the day of testing of the beam, and other was preserved. The four groups which remained was used for standardization; among them, two groups and one group were used separately for cube test of 7 days and 28 days compressive strength respectively, and last one group was used for testing splitting strength. C80 steel fiber high-performance concrete was used to pour in the reserved hole, and at the same time, three cubic specimen and two groups of prism were made. Among the cubic specimen, two groups were used to obtain 28-days cubic strength and splitting strength, and the remaining one group was preserved. Simultaneously, the remaining two groups of prism specimens were used to test 28-day compressive strength and elastic modulus of concrete. This both types, cubic and prismatic specimens were tested in a universal testing machine. The test results and elastic modulus of C60 high-performance concrete cube under standard curing and same curing condition with tested beam are shown in Tables A5–A7. The test results of C80 steel fiber high-performance concrete cube and elastic modulus are shown in Tables A8 and A9.

Table A5. Concrete strength obtained from the experimental test on standard C60 HPC cubic specimen.

Test Block Number	Standardized 7-Day Cube Strength (MPa)		Standardize 28-Day Cube Strength (MPa)	Splitting Strength (MPa)
Test block 1	63.4	55.9	77.8	5.63
Test block 2	55.2	54.1	69.9	5.08
Test block 3	48.9	57.9	67.2	4.56
Average value		55.9	71.6	5.09

Table A6. Concrete strength obtained from experimental test on C60 HPC cubic specimen at the same condition with bridge deck.

Test Block Number	Compressive Strength (MPa) *		Strength Measured at the Test Day (MPa)
Test block 1	71.3	64.9	77.87
Test block 2	71.4	61.3	75.37
Test block 3	64.7	69.3	75.47
Average value	67.15		76.24

* Strength measured at the same curing condition with bridge deck.

Table A7. Elastic modulus obtained from the experimental test on C60 HPC.

Item	First Group	Second Group	Third Group	Average Value
Modulus of Elasticity (GPa)	36.0	36.7	36.5	36.4

Table A8. Concrete strength obtained from the experimental test on standard C80 HPC cubic specimen.

Test Block Number	Standard Curing 28-Day Strength (MPa)	Splitting Strength (MPa)
Test block 1	90.4	8.27
Test block 2	87.5	9.91
Test block 3	84.4	8.83
Average value	87.4	9

Table A9. Elastic modulus obtained from the experimental test on C80 HPC.

Item	First Group	Second Group	Third Group	Average Value
Modulus of elasticity (GPa)	37.1	36.1	36.3	36.5

References

1. Khan, M.A. *Accelerated Bridge Construction: Best Practices and Techniques*; Elsevier/BH: Amsterdam, The Netherlands; Boston, MA, USA, 2015; pp. 3–4, ISBN 978-0-12-407224-4.
2. Xiang, Y.Q.; Zhu, S.; Zhao, Y. Research and development on Accelerated Bridge Construction Technology. *China J. Highw. Transp.* **2018**, *31*, 1–27.
3. Gautam, B.G.; Xiang, Y.Q.; Qiu, Z.; Guo, S.-H. A semi-empirical deflection-based method for crack width prediction in accelerated construction of steel fibrous high-performance composite small box girder. *Materials* **2019**, *12*, 964. [CrossRef] [PubMed]
4. Brozzetti, J. Design development of steel-concrete composite bridges in France. *J. Constr. Steel Res.* **2000**, *55*, 229–243. [CrossRef]
5. Nakamura, S.-I.; Momiyama, Y.; Hosaka, T.; Homma, K. New technologies of steel/concrete composite bridges. *J. Constr. Steel Res.* **2002**, *58*, 99–130. [CrossRef]
6. Hanswille, G. Composite bridges in Germany designed according to Eurocode 4-2. In *Composite Construction in Steel and Concrete VI*; American Society of Civil Engineers: Reston, VA, USA, 2011; pp. 391–405.
7. Akbari, R. Accelerated construction of short span railroad bridges in Iran-case study. *Pract. Period. Struct. Des. Constr.* **2019**, *24*, 05018004. [CrossRef]
8. Wang, Y.-H.; Yu, J.; Liu, J.-P.; Frank Chen, Y. Shear behavior of shear stud groups in precast concrete decks. *Eng. Struct.* **2019**, *187*, 73–84. [CrossRef]
9. Xiang, Y.Q.; He, X.Y. Short- and long-term analyses of shear lag in RC box girders considering axial equilibrium. *Struct. Eng. Mech.* **2017**, *62*, 725–737.
10. Xiang, Y.Q.; Guo, S.H.; Qiu, Z. Influence of group studs layout style on static behavior of the steel-concrete composite small box girder models. *J. Build. Struct.* **2017**, *38*, 376–383.
11. Gribniak, V.; Arnautov, A.K.; Norkus, A.; Tamulenas, V.; Gudonis, E.; Sokolov, A. Experimental investigation of the capacity of steel fibers to ensure the structural integrity of reinforced concrete specimens coated with CFRP sheets. *Mech. Compos. Mater.* **2016**, *52*, 401–410. [CrossRef]

12. Gribniak, V.; Kaklauskas, G.; Kliukas, R.; Meskenas, A. Efficient technique for constitutive analysis of reinforced concrete flexural members. *Inverse Probl. Sci. Eng.* **2016**, *25*, 27–40. [CrossRef]

13. Chalioris, C.E.; Kosmidou, P.-M.K.; Karayannis, C.G. Cyclic Response of Steel Fiber Reinforced Concrete Slender Beams: An Experimental Study. *Materials* **2019**, *12*, 1398. [CrossRef] [PubMed]

14. Gribniak, V.; Tamulenas, V.; Ng, P.-L.; Arnautov, A.K.; Gudonis, E.; Misiunaite, I. Mechanical Behavior of Steel Fiber-Reinforced Concrete Beams Bonded with External Carbon Fiber Sheets. *Materials* **2017**, *10*, 666. [CrossRef] [PubMed]

15. Gribniak, V.; Arnautov, A.K.; Norkus, A.; Kliukas, R.; Tamulenas, V.; Gudonis, E.; Sokolov, A. Steel Fibres: Effective Way to Prevent Failure of the Concrete Bonded with FRP Sheets. *Adv. Mater. Sci. Eng.* **2016**, *2016*, 4913536. [CrossRef]

16. Smarzewski, P. Analysis of Failure Mechanics in Hybrid Fibre-Reinforced High-Performance Concrete Deep Beams with and without Openings. *Materials* **2019**, *12*, 101. [CrossRef] [PubMed]

17. Ma, K.; Qi, T.; Liu, H.; Wang, H. Shear Behavior of Hybrid Fiber Reinforced Concrete Deep Beams. *Materials* **2018**, *11*, 2023. [CrossRef] [PubMed]

18. Chalioris, C.E. Analytical approach for the evaluation of minimum fibre factor required for steel fibrous concrete beams under combined shear and flexure. *Constr. Build. Mater.* **2013**, *43*, 317–336. [CrossRef]

19. Chalioris, C.E.; Karayannis, C.G. Effectiveness of the use of steel fibres on the torsional behavior of flanged concrete beams. *Cem. Concr. Compos.* **2009**, *31*, 331–341. [CrossRef]

20. Campione, G. Analytical prediction of load deflection curves of external steel fibers R/C beam–column joints under monotonic loading. *Eng. Struct.* **2015**, *83*, 86–98. [CrossRef]

21. Guerini, V.; Conforti, A.; Plizzari, G.; Kawashima, S. Influence of Steel and Macro-Synthetic Fibers on Concrete Properties. *Fibers* **2018**, *6*, 47. [CrossRef]

22. Chalioris, C.E.; Sfiri, E.F. Shear Performance of Steel Fibrous Concrete Beams. *Procedia Eng.* **2011**, *14*, 2064–2068. [CrossRef]

23. Chalioris, C.E. Steel fibrous RC beams subjected to cyclic deformations under predominant shear. *Eng. Struct.* **2013**, *49*, 104–118. [CrossRef]

24. Chalioris, C.E.; Kosmidou, P.M.K.; Papadopoulos, N.A. Investigation of a New Strengthening Technique for RC Deep Beams Using Carbon. *Fibers* **2018**, *6*, 52. [CrossRef]

25. Juárez, C.; Valdez, P.; Durán, A.; Sobolev, K. The diagonal tension behavior of fiber reinforced concrete beams. *Cem. Concr. Compos.* **2007**, *29*, 402–408. [CrossRef]

26. Lin, W.T.; Wu, Y.C.; Cheng, A.; Chao, S.J.; Hsu, H.M. Engineering Properties and Correlation Analysis of Fiber Cementitious Materials. *Materials* **2014**, *7*, 7423–7435. [CrossRef] [PubMed]

27. ACI Committee 544. *Report on the Physical Properties and Durability of Fiber-Reinforced Concrete*; ACI Manual of Concrete Practice (MCP), American Concrete Institute: Farmington Hills, MI, USA, 2010.

28. Mirza, F.A.; Soroushian, P. Effects of alkali-resistant glass fiber reinforcement on crack and temperature resistance of lightweight concrete. *Cem. Concr. Compos.* **2002**, *24*, 223–227. [CrossRef]

29. Ryua, H.K.; Changa, S.P.; Kimb, Y.J.; Kim, B.S. Crack control of a steel and concrete composite plate girder with prefabricated slabs under hogging moments. *Eng. Struct.* **2005**, *27*, 1613–1624. [CrossRef]

30. Muttoni, A.; Ruiz, M.F. Concrete Cracking in Tension Members and Application to Deck Slabs of Bridges. *J. Bridge Eng.* **2007**, *12*, 646–653. [CrossRef]

31. Xing, Y.; Han, Q.H.; Xu, J.; Guo, Q.; Wang, Y.H. Experimental and numerical study on static behavior of elastic concrete-steel composite beams. *J. Constr. Steel Res.* **2016**, *123*, 79–92. [CrossRef]

32. Lin, W.W.; Yoda, T.; Taniguchi, N. Application of SFRC in steel–concrete composite beams subjected to hogging moment. *J. Constr. Steel Res.* **2014**, *101*, 175–183. [CrossRef]

33. Ge, W.J.; Ashour, A.F.; Yu, J.; Gao, P.; Cao, D.F.; Cai, C.; Ji, X. Flexural Behavior of ECC–Concrete Hybrid Composite Beams Reinforced with FRP and Steel Bars. *J. Compos. Constr.* **2019**, *23*, 04018069. [CrossRef]

34. Shamass, R.; Cashell, K.A. Analysis of stainless steel-concrete composite beams. *J. Constr. Steel Res.* **2019**, *152*, 132–142. [CrossRef]

35. Xu, L.Y.; Nie, X.; Tao, M.X. Rational modeling for cracking behavior of RC slabs in composite beams subjected to a hogging moment. *Constr. Build. Mater.* **2018**, *192*, 357–365. [CrossRef]

36. Tayeh, B.A.; Abu Bakar, B.H.; Megat Johari, M.A.; Voo, Y.L. Utilization of ultra-high performance fibre concrete(UHPFC) for rehabilitation—A review. *Procedia Eng.* **2013**, *54*, 525–538. [CrossRef]

37. Elnashai, A.S.; Di Sarno, L. *Fundamentals of Earthquake Engineering*; Wiley and Sons Ltd.: Chichester, UK, 2008.

38. Di Sarno, L.; Elnashai, A.S. Innovative strategies for seismic retrofitting of steel and composite structures. *J. Prog. Struct. Eng. Mater.* **2003**, *7*, 115–135. [CrossRef]
39. Cuil, Y.; Nakashima, M. Application of headed studs in steel fiber reinforced cementitious composite slab of steel beam-column connection. *Earthq. Eng. Eng. Vib.* **2012**, *11*, 11–21. [CrossRef]
40. Lin, Z.; Liu, Y.; Roeder, C.W. Behavior of stud connections between concrete slabs and steel girders under transverse bending moment. *Eng. Struct.* **2016**, *117*, 130–144. [CrossRef]
41. Wang, Y.-H.; Yu, J.; Liu, J.; Chen, Y.F. Experimental and Numerical Analysis of steel-Block shear connectors in Assembled monolithic steel-concrete composite beams. *J. Bridge Eng.* **2019**, *24*, 04019024. [CrossRef]
42. Ovuoba, B.; Prinz, G.S. Headed shear stud fatigue demands in composite bridge girders having varied stud pitch, girder, depth and span length. *J. Bridge Eng.* **2018**, *23*, 04018085. [CrossRef]
43. Yang, F.; Liu, Y.; Jiang, Z.; Xin, H. Shear performance of a novel demountable steel-concrete bolted connector under static push-out tests. *Eng. Struct.* **2018**, *160*, 133–146. [CrossRef]
44. Hicks, S.J. Stud shear connectors in composite beam that support slabs with profiled steel sheeting. *Struct. Eng. Int.* **2014**, *24*, 246–253. [CrossRef]
45. Zeng, X.; Jiang, S.-F.; Zhou, D. Effect of shear connector layout on the behavior of steel-concrete composite beams with interface slip. *Appl. Sci.* **2019**, *9*, 207. [CrossRef]
46. Abada, G.; Bernrd, F.; Lim, S.; Tehami, M. Simulstion of push-out tests: Influence of several parameters and structural arrangements. *Proc. Inst. Civ. Eng. Struct. Build.* **2019**, *172*, 340–357. [CrossRef]
47. Xu, C.; Sugiura, K.; Wu, C.; Su, Q. Parametrical static analysis on group studs with typical push-out tests. *J. Constr. Steel Res.* **2012**, *72*, 84–96. [CrossRef]
48. Spremic, M.; Markovic, Z.; Dobric, J.; Vejikovic, M.; Budevac, D. Shear connection with groups of headed studs. *Gradevinar* **2017**, *69*, 379–386. [CrossRef]
49. Xu, C.; Su, Q.; Sugiura, K. Mechanism study on the low cycle fatigue behavior of group studs shear connectors in steel-concrete composite bridges. *J. Constr. Steel Res.* **2017**, *138*, 196–207. [CrossRef]
50. Wang, J.-Y.; Guo, J.-Y.; Jia, L.-J.; Chen, S.-M.; Dong, Y. Push-out tests of demountable headed stud shear connectors in steel-UHPC composite structures. *Compos. Struct.* **2017**, *170*, 69–79. [CrossRef]
51. Qureshi, J.; Lam, D.; Ye, J. Effect of shear connector spacing and layout on the shear connector capacity in composite beams. *J. Constr. Steel Res.* **2011**, *67*, 706–719. [CrossRef]
52. Wang, J.; Xu, Q.; Yao, Y.; Qi, J.; Xiu, H. Static behavior of grouped large headed stud-UHPC shear connectors in composite structures. *Compos. Struct.* **2018**, *206*, 202–214. [CrossRef]
53. Ministry of Housing and Urban-Rural Development of the People's Republic of China. *Code for Design of Steel and Concrete Composite Bridges*; GB50917-2013; China Planning Press: Beijing, China, 2014.
54. The European Union. *Design of Composite Steel and Concrete Structures—Part 1-1: General Rules and Rules for Building*; Eurocode 4, EN, 1994-1-1; European Publication: London, UK, 2004.
55. Ministry of Housing and Urban-Rural Development of the People's Republic of China. *Code for Design of Steel Structures*; GB 50017-2017; China Building Industry Press: Beijing, China, 2018.
56. American Association of State Highway and Transportation Officials. *AASHTO LRFD Bridge Design Specifications*, 4th ed.; American Association of State Highway and Transportation Officials: Washington, DC, USA, 2007.
57. Ministry of Housing and Urban-Rural Development of the People's Republic of China. *Code for Design of Concrete Structures*; GB50010-2010; Chemical Industry Press: Beijing, China, 2015.
58. General Administration of Quality Supervision, Inspection and Quarantine of the People's Republic of China. *Method for Chemical Analysis of Cement*; GB/T 176-1996; Biaoxin Technology (Beijing) Co. Ltd.: Beijing, China, 1996.
59. General Administration of Quality Supervision, Inspection and Quarantine of the People's Republic of China. *Mineral Admixtures for High Strength and High Performance Concrete*; GB/T 18736/2002; Biaoxin Technology (Beijing) Co. Ltd.: Beijing, China, 2002.
60. Rudziewicz, M.; Bosse, M.J.; Marland, E.S.; Rhoads, G.S. Visualisation of lines of best fit. *Int. J. Math. Teach. Learn.* **2017**, *18*, 359–382.
61. Shampine, L.F. solving ODEs and DDEs with residual contro. *Appl. Numer. Math.* **2005**, *52*, 113–127. [CrossRef]

Article

Sound Radiation of Orthogonal Antisymmetric Composite Laminates Embedded with Pre-Strained SMA Wires in Thermal Environment

Yizhe Huang, Zhifu Zhang, Chaopeng Li, Jiaxuan Wang, Zhuang Li and Kuanmin Mao *

State Key Laboratory of Digital Manufacturing Equipment and Technology, Huazhong University of Science and Technology, Wuhan 430074, China; yizhehuang@hust.edu.cn (Y.H.); jeff.zfzhang@foxmail.com (Z.Z.); 15927673439@163.com (C.L.); wjx@hust.edu.cn (J.W.); lz_mse@hust.edu.cn (Z.L.)
* Correspondence: maokm@hust.edu.cn

Received: 9 July 2020; Accepted: 17 August 2020; Published: 19 August 2020

Abstract: The interest of this article lies in the sound radiation of shape memory alloy (SMA) composite laminates. Different from the traditional method of avoiding resonance sound radiation of composite laminates by means of structural parameter design, this paper explores the potential of adjusting the modal peak of the resonant acoustic radiation by using material characteristics of shape memory alloys (SMA), and provides a new idea for avoiding resonance sound radiation of composite laminates. For composite laminates embedded with pre-strained SMA, an innovation of vibration-acoustic modeling of SMA composite laminates considering pre-stain of SMA and thermal expansion force of graphite-epoxy resin is proposed. Based on the classical thin plate theory and Hamilton principle, the structural dynamic governing equation and the frequency equation of the laminates subjected to thermal environment are derived. The vibration sound radiation of composite laminates is calculated with Rayleigh integral. Effects of ambient temperature, pre-strain, SMA volume fraction, substrate ratio, and geometrical parameters on the sound radiation were analyzed. New laws of SMA material and pre-strain characteristics on sound radiation of composite laminates under temperature environment are revealed, which have theoretical and engineering functional significance for vibration and sound radiation control of SMA composite laminates.

Keywords: shape memory alloys; thermal environment; composite laminates; sound radiation

---

1. Introduction

Composite laminates as structural components have been extensively implemented in various fields of mechanical engineering such as aeronautics, astronautics, naval architecture, and automotive engineering. Composite plate offers advantages such as lightweight, high strength; but resulting in considerable vibration noise. Especially when the external excitation frequency is equivalent to or close to the natural frequency of the composite laminates, the strong structural resonance leads to more radiated noise. Shape memory alloy (SMA) reinforced composites are an extremely versatile class of materials which have the characteristics of larger internal forces, unique ability of changing its material properties, wide range of operational temperature, and high durability [1]. Though embedded with SMA wire to the composite material, the modal performance of composite laminates can be adjusted by the volume fraction and pre-strain of SMA without modifying the original shape dimensions and boundary conditions. This modification can effectively avoid the structure resonance and minimize the noise radiated into the surroundings.

Acoustic radiation of composite laminates is intimately related with the structural mode and vibration response. Over recent decades, the vibration of SMA composite plates has attracted the attention of some scholars. Rogers et al. [2] proposed utilizing SMA fibers modified structural performance of laminated plates via theoretically analyzed natural frequencies and mode shapes. Ostachowicz et al. [3] established a finite element model for predicting the first three natural frequencies subjected to several kinds of pre-strain. Malekzadeh et al. [4–7] developed the linear free vibration of hybrid laminated composite plates embedded with SMA fibers, employing the Ritz method, which lead to extracting the fundamental natural frequency, and investigated the dynamic behavior on detailed parametric analyses such as volume fraction, pre-strain, and aspect ratio. Park et al. [8] studied the vibration behavior of thermally buckled SMA composite plates using the von Karman nonlinear incremental strain–displacement relation. Furthermore, there are some investigations of the SMA composite plates with regard to vibration response, such as nonlinear vibration analysis [9–11], prediction of thermo-mechanical response under the combined action of thermal and mechanical loads [12,13], active vibration control of structures [14–17], and the experiment of structural acoustic characteristics [18].

Few studies about the acoustic radiation of SMA composite laminates are presented in the reported researching. However, research on vibration and acoustic radiation of conventional composite laminates has been very active. The design of material and structure parameters is an important means to improve the acoustic radiation. Comparing and analyzing sound radiation from material parameters of laminates, the conspicuousness of material composition on response, and the coupling effect of material proportion on sound radiation is often of great significance [19–21]. In terms of structural parameters, the research on sound radiation has developed from the design of normal geometrical parameters (aspect ratio, thickness) of laminates to more complex structural forms such as periodically reinforced structures and non-uniform laying [22–25]. In addition, as a current hotspot, the investigation of acoustic radiation of functionally graded laminates comprehensively considers the effects of materials, structures, external excitations, and the environment [26–28]. For this study, a new method to estimate sound power of orthogonal antisymmetric composite laminates embedded with SMA considering the pre-strain of SMA and thermal environment was presented and so the thermal expansion force and recovery force are added to the constitutive model of composite laminates. Based on the Hamilton principle and Rayleigh integration, the vibration differential equation and sound radiation calculation model of SMA composite laminates are established. The influence of SMA wires on the sound radiation power of laminates was examined by varying the pre-strain, SMA volume fraction, substrate ratio, and structural size parameters with temperature. The sound radiation results in the study provide a theoretical basis for vibration and sound radiation control of embedded SMA composite laminates.

2. Material Properties of SMA in Composite Laminates

Consider a rectangular composite laminates of length L_a, width L_b, and thickness h in a Cartesian coordinate system as shown in Figure 1.

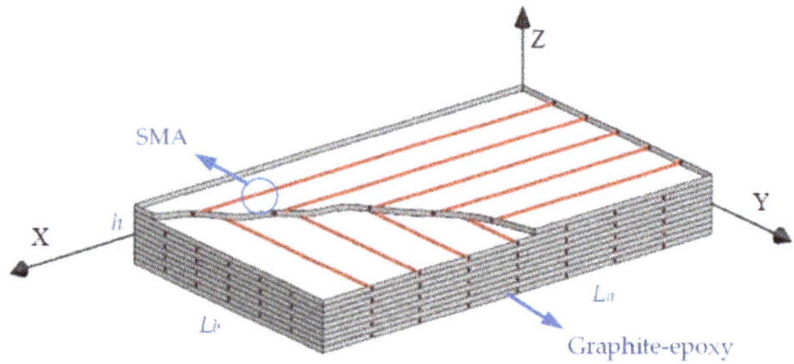

Figure 1. Schematic of orthogonal antisymmetric composite laminates embedded with shape memory alloy (SMA) wires.

Each layer of material is composed of SMA (nickel-titanium alloy), graphite, and epoxy in which SMA fibers are distributed in laminates in the form of orthogonal anti-symmetry. The elastic modulus of the SMA appears to have a strong temperature dependence [29]. From Figure 2, the modulus is 25 to 30 GPa when the temperature is below 40 °C as SMA in the martensitic phase.

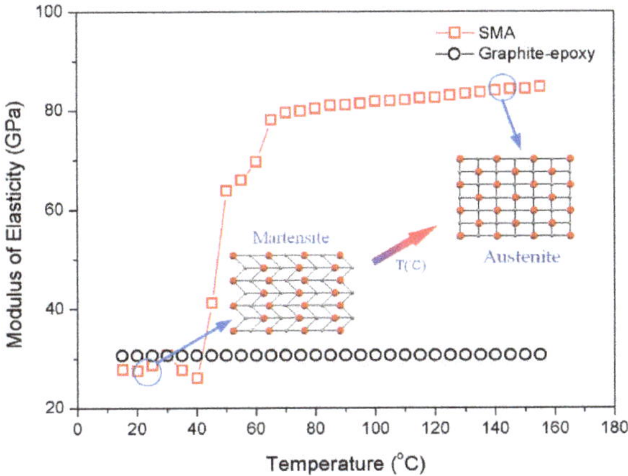

Figure 2. SMA modulus of elasticity vs. temperature.

As the temperature increases, the modulus of SMA increases rapidly when it transforms from the martensitic phase to austenite phase. Most of SMA are in the austenite phase when the temperature reaches 60 °C, and its modulus reaches about 80 GPa. After that, the modulus increases slowly with the rising temperature.

Recovery stress produced by SMA is not only temperature dependent, but also tensile pre-strain dependent. It can be seen from Figure 3 that the recovery stress increases with the increase of pre-strain after the temperature is higher than 60 °C [29].

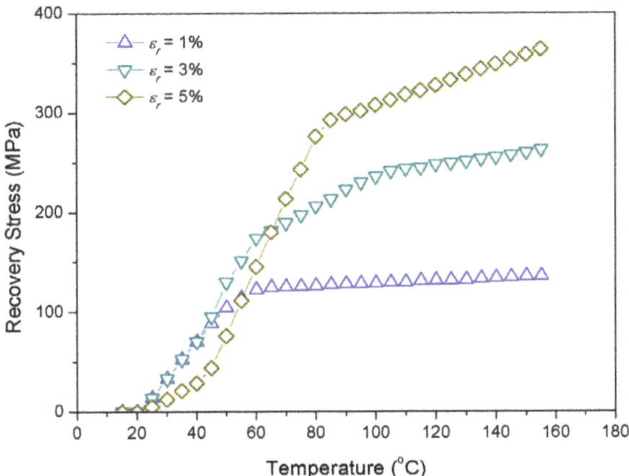

Figure 3. SMA recovery stress vs. temperature.

Furthermore, the effect of pre-strain is not significant for normal or low temperature environment. To composite laminates embedded with SMA, owing to its temperature dependent material properties and pre-strain characteristics, the thermal environment should be fully considered in the process of its application. Especially through the control of material properties by adjusting the temperature, which has great potential for improving the vibration and sound radiation of the composite laminates.

3. Structural Dynamic Response

Assuming that the SMA orthogonal anti-symmetric composite laminate shown in Figure 1 contains N layers, the stress–strain relation of the kth layer can be given as [30], Equation (1):

$$
\left\{\begin{array}{c} \sigma_x \\ \sigma_y \\ \tau_{xy} \end{array}\right\}^{(k)} = \left[\begin{array}{ccc} \widetilde{Q}_{11}^{(k)} & \widetilde{Q}_{12}^{(k)} & 0 \\ \widetilde{Q}_{12}^{(k)} & \widetilde{Q}_{22}^{(k)} & 0 \\ 0 & 0 & \widetilde{Q}_{66}^{(k)} \end{array}\right] \left\{\begin{array}{c} \varepsilon_x \\ \varepsilon_y \\ \gamma_{xy} \end{array}\right\}^{(k)} - \left[\begin{array}{ccc} \overline{Q}_{11}^{(k)} & \overline{Q}_{12}^{(k)} & 0 \\ \overline{Q}_{12}^{(k)} & \overline{Q}_{22}^{(k)} & 0 \\ 0 & 0 & \overline{Q}_{66}^{(k)} \end{array}\right] \left\{\begin{array}{c} \alpha_{11} \\ \alpha_{22} \\ 0 \end{array}\right\}^{(k)} \Delta T + \left\{\begin{array}{c} \sigma_x^r \\ \sigma_y^r \\ 0 \end{array}\right\}^{(k)}, \quad (1)
$$

where $\widetilde{Q}^{(k)}$ is the transformed reduced stiffness of the kth layer of composite laminate embedded with SMA and $\overline{Q}^{(k)}$ is the transformed reduced stiffness of the kth layer of the substrate which is graphite-epoxy. The term σ^r denotes the recovery stress in the SMA fiber direction. Other equations and terms in Equation (1) are completed in Appendix A. Utilizing the classic thin plate theory, the strains associated with the displacements are, Equation (2):

$$
\left\{\begin{array}{c} \varepsilon_x \\ \varepsilon_y \\ \gamma_{xy} \end{array}\right\} = \left\{\begin{array}{c} \frac{\partial u_0}{\partial x} \\ \frac{\partial v_0}{\partial y} \\ \frac{\partial v_0}{\partial x} + \frac{\partial u_0}{\partial y} \end{array}\right\} - z \left\{\begin{array}{c} \frac{\partial^2 w}{\partial x^2} \\ \frac{\partial^2 w}{\partial y^2} \\ 2\frac{\partial^2 w}{\partial x \partial y} \end{array}\right\}, \quad (2)
$$

where 'u_0', 'v_0' are the displacement components of the middle plane along the x, y coordinates, respectively and 'w' is the displacement components of along z coordinate.

Then, the strain energy U_1 is obtained as, Equation (3):

$$
U_1 = \frac{1}{2} \sum_{k=1}^{N} \int_{h_{k-1}}^{h_k} \iint_S \left\{ \sigma_x \varepsilon_x + \sigma_y \varepsilon_y + \tau_{xy} \gamma_{xy} \right\} dx dy dz. \quad (3)
$$

The additional strain energy U_2 which consists of the strain energy caused by recovery stress and the thermal deformation energy can be expressed as, Equation (4):

$$
U_2 = \frac{1}{2} \iint_S \left\{ \left(N_x^r - N_x^T\right)\left(\frac{\partial w}{\partial x}\right)^2 + 2\left(N_{xy}^r - N_{xy}^T\right)\left(\frac{\partial w}{\partial x}\right)\left(\frac{\partial w}{\partial y}\right) + \left(N_y^r - N_y^T\right)\left(\frac{\partial w}{\partial y}\right)^2 \right\} dx dy, \quad (4)
$$

where Equations (5) and (6):

$$
\left\{\begin{array}{c} N_x^T \\ N_y^T \\ N_{xy}^T \end{array}\right\} = \sum_{k=1}^{N} \left[\begin{array}{ccc} \overline{Q}_{11}^{(k)} & \overline{Q}_{12}^{(k)} & 0 \\ \overline{Q}_{12}^{(k)} & \overline{Q}_{22}^{(k)} & 0 \\ 0 & 0 & \overline{Q}_{66}^{(k)} \end{array}\right] \left\{\begin{array}{c} \alpha_{11} \\ \alpha_{22} \\ 0 \end{array}\right\} \Delta T dz, \quad (5)
$$

$$
\left\{\begin{array}{c} N_x^r \\ N_y^r \\ N_{xy}^r \end{array}\right\} = \sum_{k=1}^{N} \left\{\begin{array}{c} \sigma_x^r \\ \sigma_y^r \\ 0 \end{array}\right\}^{(k)} dz. \quad (6)
$$

The damping deformation energy U_3 can be given as, Equation (7):

$$
U_3 = \iint_S F_d w(x, y, t) dx dy, \quad (7)
$$

in which the damping is $F_d = \lambda \dot{w}$ and λ is the viscous damping coefficient. The total potential energy of composite laminates can be written as, Equation (8):

$$U = U_1 + U_2 + U_3. \tag{8}$$

The kinetic energy of composite laminates can be calculated as, Equation (9):

$$T = \frac{1}{2}\sum_{k=1}^{N}\int_{h_{k-1}}^{h_k}\iint_{S}\overline{\rho}^{(k)}\left(\frac{\partial w}{\partial t}\right)^2 dxdydz. \tag{9}$$

It is noteworthy in Equation (9) that the sound radiation of the composite laminates is mainly generated by bending vibration, so that the kinetic energy in the x and y directions can be neglected. For a transverse load $q(x, y, t)$ on the surface of the laminates, the work it has done on the system can be obtained as, Equation (10):

$$W = \iint_{S} q(x, y, t)w(x, y, t)dxdy. \tag{10}$$

According to the Hamilton's principle, the dynamic equations of the laminates can be derived by Equation (11):

$$\delta\Pi = \int_{t_0}^{t_1}(\delta T - \delta U + \delta W)dt = 0. \tag{11}$$

Substituting Equations (8)–(10) into Equation (11), meanwhile considering the condition of orthogonal antisymmetric composite laminates, only B_{11} and B_{22} are not zero in B_{ij}, and $B_{22} = -B_{11}$. Then, the vibration equation of an orthogonal antisymmetric composite laminate embedded in SMA can be derived as, Equations (12)–(14):

$$A_{11}\frac{\partial^2 u_0}{\partial x^2} + A_{66}\frac{\partial^2 u_0}{\partial y^2} + (A_{12} + A_{66})\frac{\partial^2 v_0}{\partial x \partial y} - B_{11}\frac{\partial^3 w}{\partial x^3} = 0, \tag{12}$$

$$(A_{12} + A_{66})\frac{\partial^2 u_0}{\partial x \partial y} + A_{66}\frac{\partial^2 v_0}{\partial x^2} + A_{22}\frac{\partial^2 v_0}{\partial y^2} + B_{11}\frac{\partial^3 w}{\partial y^3} = 0, \tag{13}$$

$$B_{11}\left(-\frac{\partial^3 u_0}{\partial x^3} + \frac{\partial^3 v_0}{\partial y^3}\right) + D_{11}\frac{\partial^4 w}{\partial x^4} + 2(D_{12} + 2D_{66})\frac{\partial^4 w}{\partial x^2 \partial y^2} + D_{22}\frac{\partial^4 w}{\partial y^4} - \left(N_x^r - N_x^T\right)\frac{\partial^2 w}{\partial x^2}$$
$$- \left(N_y^r - N_y^T\right)\frac{\partial^2 w}{\partial y^2} + \lambda\frac{\partial w}{\partial t} + I\frac{\partial^2 w}{\partial t^2} = q, \tag{14}$$

where Equations (15)–(18):

$$A_{ij} = \sum_{k=1}^{N}\widetilde{Q}_{ij}^{(k)}(h_k - h_{k-1}), \tag{15}$$

$$B_{ij} = \frac{1}{2}\sum_{k=1}^{N}\widetilde{Q}_{ij}^{(k)}(h_k^2 - h_{k-1}^2), \tag{16}$$

$$D_{ij} = \frac{1}{3}\sum_{k=1}^{N}\widetilde{Q}_{ij}^{(k)}(h_k^3 - h_{k-1}^3), \tag{17}$$

$$I = \sum_{k=1}^{N}\overline{\rho}^{(k)}(h_k - h_{k-1}). \tag{18}$$

In this paper, mechanical boundary conditions of composite laminates are intended to be simply supported, Equation (19):

$$\begin{aligned}
w = v_0 = 0 \ at \ x = 0, L_a \\
w = u_0 = 0 \ at \ y = 0, L_b.
\end{aligned} \tag{19}$$

The state variables satisfying the simply supported boundary conditions of Equation (19) are assumed as Equation (20):

$$
\begin{cases}
u(x,y,t) = \sum\limits_{m=1}^{\infty}\sum\limits_{n=1}^{\infty} U_{mn}\cos\frac{m\pi x}{L_a}\sin\frac{n\pi y}{L_b}e^{j\omega t} \\
v(x,y,t) = \sum\limits_{m=1}^{\infty}\sum\limits_{n=1}^{\infty} V_{mn}\sin\frac{m\pi x}{L_a}\cos\frac{n\pi y}{L_b}e^{j\omega t} \quad , \\
w(x,y,t) = \sum\limits_{m=1}^{\infty}\sum\limits_{n=1}^{\infty} W_{mn}\sin\frac{m\pi x}{L_a}\sin\frac{n\pi y}{L_b}e^{j\omega t}
\end{cases}
\tag{20}
$$

where U_{mn}, V_{mn}, and W_{mn} are the Fourier expansion coefficients of the solution, m and n are the half wave numbers in x and y directions, and ω is the circular frequency. Similarly, the load $q(x,y,t)$ is assumed to be expanded into a double Fourier series form, Equation (21):

$$
q(x,y,t) = \hat{q}(x,y)e^{j\omega t} = \sum\limits_{m=1}^{\infty}\sum\limits_{n=1}^{\infty} q_{mn}\sin\frac{m\pi x}{L_a}\sin\frac{n\pi y}{L_b}e^{j\omega t},
\tag{21}
$$

where q_{mn} as the load coefficient can be given as, Equation (22):

$$
q_{mn} = \frac{4}{L_a L_b}\left(\sum\limits_{m=1}^{\infty}\sum\limits_{n=1}^{\infty}\int\limits_{0}^{L_a}\int\limits_{0}^{L_b}\hat{q}(x,y)\sin\frac{m\pi x}{L_a}\sin\frac{n\pi y}{L_b}dxdy\right).
\tag{22}
$$

Substituting Equations (20) and (21) into Equations (12)–(14) and the vibration equation can be rewritten into a form as, Equation (23):

$$
\left([K] + j\omega[C] - \omega^2[M]\right)\{X_{mn}\} = \{F_{mn}\},
\tag{23}
$$

where $\{X_{mn}\} = \{U_{mn}, V_{mn}, W_{mn}\}^T$ and $\{F_{mn}\} = \{0, 0, q_{mn}\}^T$ are displacement amplitude vector and load amplitude vector, respectively. Moreover, the mass matrix, damping matrix, and stiffness matrix is respectively expressed as, Equation (24):

$$
[M] = \begin{bmatrix} 0 & & \\ & 0 & \\ & & I \end{bmatrix}, \quad [C] = \begin{bmatrix} 0 & & \\ & 0 & \\ & & \eta\omega_{mn} \end{bmatrix}, \quad [K] = \begin{bmatrix} K_{11} & K_{12} & K_{13} \\ K_{12} & K_{22} & K_{23} \\ K_{13} & K_{23} & K_{33} \end{bmatrix},
\tag{24}
$$

in which the damping loss factor is, Equation (25):

$$
\eta = \lambda / \left(\omega_{mn}\sum\limits_{k=1}^{N}\overline{\rho}^{(k)}\right),
\tag{25}
$$

and K_{ij} are given as, Equations (26)–(31):

$$
K_{11} = -A_{11}\left(\frac{m\pi}{L_a}\right)^2 - A_{66}\left(\frac{n\pi}{L_b}\right)^2,
\tag{26}
$$

$$
K_{12} = -(A_{12} + A_{66})\left(\frac{m\pi}{L_a}\right)\left(\frac{n\pi}{L_b}\right),
\tag{27}
$$

$$
K_{13} = B_{11}\left(\frac{m\pi}{L_a}\right)^3,
\tag{28}
$$

$$
K_{22} = -\left[A_{66}\left(\frac{m\pi}{L_a}\right)^2 + A_{11}\left(\frac{n\pi}{L_b}\right)^2\right],
\tag{29}
$$

$$K_{23} = -B_{11}\left(\frac{n\pi}{L_b}\right)^3,$$ (30)

$$K_{33} = D_{11}\left[\left(\frac{m\pi}{L_a}\right)^4 + \left(\frac{n\pi}{L_b}\right)^4\right] + 2(D_{12} + 2D_{66})\left(\frac{m\pi}{L_a}\right)^2\left(\frac{n\pi}{L_b}\right)^2 + \left(N_x^r - N_x^T\right)\left(\frac{m\pi}{L_a}\right)^2 + \left(N_y^r - N_y^T\right)\left(\frac{n\pi}{L_b}\right)^2.$$ (31)

Multiply both sides of the Equation (23) by the inverse matrix of the coefficient matrix, Equation (23):

$$\{X_{mn}\} = \left([K] + j\omega[C] - \omega^2[M]\right)^{-1}\{F_{mn}\}.$$ (32)

Solving Equation (32) obtains displacement amplitude in z direction of composite laminates, Equation (33):

$$W_{mn} = q_{mn}/\left\{\overline{\rho}\left[\left(\omega_{mn}^2 - \omega^2\right) + j\eta\omega\omega_{mn}\right]\right\},$$ (33)

where ω_{mn} is the natural frequency which can be solved by the following frequency Equations (34):

$$det\left([K] - \omega^2[M]\right) = 0.$$ (34)

Substituting Equation (33) into the third equation of Equation (20), it can be obtained that the displacement components along z coordinate, Equation (35):

$$w(x,y,t) = \frac{\sum_{m=1}^{\infty}\sum_{n=1}^{\infty}q_{mn}}{\left\{\overline{\rho}\left[\left(\omega_{mn}^2 - \omega^2\right) + j\eta\omega\omega_{mn}\right]\right\}}sin\frac{m\pi x}{L_a}sin\frac{n\pi y}{L_b}e^{j\omega t}.$$ (35)

The derivative of the displacement in Equation (35) with respect to time is utilized to obtain the amplitude of the (m, n) order modal vibration velocity of composite laminates, Equations (36):

$$\hat{U}_{mn}(x,y) = \frac{j\omega q_{mn}}{\left\{\overline{\rho}\left[\left(\omega_{mn}^2 - \omega^2\right) + j\eta\omega\omega_{mn}\right]\right\}}sin\frac{m\pi x}{L_a}sin\frac{n\pi y}{L_b}.$$ (36)

4. Sound Radiation Power

For rectangular composite laminates embedded with SMA, each unit in the laminates can be regarded as a source of external vibrational acoustic radiation. The radiation sound pressure of the laminates is obtained by radiating the sound pressure of each unit, and then performing Rayleigh integration. Assuming that the amplitude of the vibration velocity of the ds' in the laminates is $\hat{U}_{mn}(x',y')$, the amplitude of the sound pressure at ds can be formulated as, Equation (37):

$$dP_{mn}(x,y) = j\frac{k\rho_0 c_0}{2\pi r}\hat{U}_{mn}(x',y')e^{-jkr}ds',$$ (37)

where k, ρ_0, and c_0 are the sound wave number, the density of the air, and the sound velocity, respectively. r is the distance from unit ds' to unit ds. The sound pressure and sound intensity of all panel elements are radiated as being defined by [30], Equations (38) and (39):

$$P_{mn}(x,y) = j\frac{k\rho_0 c_0}{2\pi}\iint_S \frac{e^{-jkr}}{r}\hat{U}_{mn}(x',y')ds',$$ (38)

$$\hat{I}_{mn}(x,y) = \frac{1}{2}Re\left[P_{mn}(x,y)\hat{U}_{mn}^*(x,y)\right],$$ (39)

where $\hat{U}_{mn}^{*}(x, y)$ is the conjugate of $\hat{U}_{mn}(x, y)$. Using Equation (39), the radiated sound power of the composite laminate can be formulated as [26], Equation (40):

$$\hat{W} = \frac{k\rho_0 c_0}{4\pi} \sum_{m,n}^{\infty} \iint_{S'} \int_{S} R_e\left[\hat{U}_{mn}(x', y')\frac{e^{-jkr}}{r} \hat{U}_{mn}^{*}(x, y)\right] ds' ds. \tag{40}$$

The radiated sound power is usually written in the form of sound power level in decibel, which is defined by, Equation (41):

$$L_{\hat{W}} = 10\lg\left(\hat{W}/\hat{W}_0\right), \tag{41}$$

where $\hat{W}_0$ is the reference power and $\hat{W}_0 = 1 \times 10^{-12}$W.

5. Results and Discussion

5.1. Materials and Verification

The analytical method is thus deployed to carry out several parametric studies to explore the acoustic response of composite laminates in thermal environments. The rectangular composite laminates are composed of ten layers, simply supported on all edges with dimensions of 0.4 m × 0.3 m × 0.008 m are considered as shown in Figure 1. Young's modulus of elasticity and recovery stress for SMA were taken from Figures 2 and 3, respectively. Poisson's ratio, density, and thermal expansion coefficient of SMA are 0.3, 645 kg·m^3, and 10.26 × 10^{-6} 1/°C, respectively. SMA is embedded in each layer with a volume fraction of 20%; graphite and epoxy are invoked as the substrate for a ratio of 1/9. The material properties of the substrate are given in Table 1.

Table 1. Material properties of the substrate.

Substrate	E_{1mf} (GPa)	E_{2mf} (GPa)	G_{mf} (GPa)	v_{12mf}	α_{11} (1/°C)	α_{22} (1/°C)	ρ_{mf} (kg·m^3)
graphite-epoxy	30.65	3.81	1.41	0.34	80.26 × 10^{-6}	60.76 × 10^{-6}	1315

In addition, the stacking sequence is $[0°/90°/0°/90°/0°]_a$ and the damping loss factor of the laminates is set at 0.01. To verify the code of sound radiation characteristics of composite laminates embedded with SMA, the natural frequency and the sound radiation were calculated by the present method. The first example is the natural frequency from a rectangular laminated composite plate with SMA, which is taken from Park et al. [8]. The size of the plate is 0.38 m × 0.305 m × 0.002 m with simply supported boundary conditions for all edges. Table 2 shows the comparison of the first 3 natural frequencies of the laminated composite plate subjected to temperature at 25, 45, 65, 100, and 120 °C.

Table 2. Comparison of natural frequencies (Hz) with Park et al.

Modal Indices	Park et al.					Present				
	25 °C	45 °C	65 °C	100 °C	120 °C	25 °C	45 °C	65 °C	100 °C	120 °C
(1,1)	109.83	119.32	107.12	81.36	54.24	110.25	121.45	108.92	82.07	54.89
(2,1)	221.02	234.58	215.59	181.70	150.51	222.76	235.87	216.15	183.66	152.01
(1,2)	241.36	252.20	238.64	214.24	193.90	239.89	249.22	238.03	212.27	193.11

As can be seen from Table 2, the comparison certifies the correctness of natural frequencies that was calculated by the present method. The second example is sound radiation from antisymmetric cross-ply $[0°/90°]_2$ laminated composite flat panel (0.5 m × 0.5 m × 0.02 m) subjected to unit harmonic point load at x = 0.125 m and y = 0.125 m which is taken from Sharma et al. [20]. Table 3 represents the material properties of the graphite-epoxy and glass-epoxy.

Table 3. Composite material properties of graphite-epoxy and glass-epoxy.

Properties	Graphite-Epoxy	Glass-Epoxy
E_1 (GPa)	137	38.6
$E_2 = E_3$ (GPa)	8.9	8.2
$v_{12} = v_{23} = v_{13}$	0.28	0.26
$G_{12} = G_{13}$ (GPa)	7.1	4.2
G_{23} (GPa)	$0.5G_{12}$	$0.5G_{12}$
ρ (kg·m^3)	1600	1900

As shown in Figure 4, the calculated result has an excellent agreement with Sharma et al. [20]. Thus, the present analytical method was a sound choice for the study.

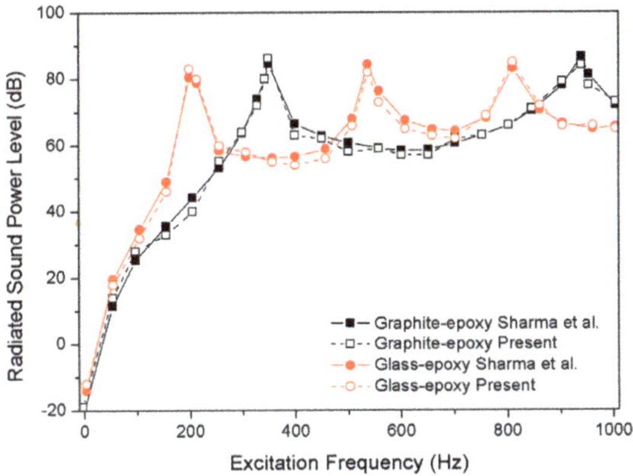

Figure 4. Comparison of sound power level with Sharma et al.

5.2. Investigating the Influence of Temperature Dependent Material Properties

Composite laminates embedded with SMA are temperature dependent. In order to depict the acoustic radiation power of the composite laminates in the thermal field, it assumes to be a uniform heating process. Figure 5 shows a three-dimensional diagram of the acoustic radiation power of composite laminates in the frequency range of 10 to 1000 Hz during the temperature rise from 15 to 155 °C.

As shown in Figure 5, it can be observed that there are four to five peaks of the sound power level corresponding to the natural frequencies. First and second modes shift to lower frequency and approach to each other with the heating case. Simultaneously, the frequency gap between the second mode and the third mode is widened. Moreover, the value of the sound radiation power level also varies and fewer peaks occur in the frequency range showed owing to the effect of heating. These significant effects stem from the temperature dependent material properties of the modulus and pre-strain of the SMA. To investigate further sound radiation characteristics of composite laminates embedded with SMA, representative temperatures such as 25 °C (SMA in martensitic phase) and 100 °C (SMA in austenite phase) were selected as conditions for subsequent study.

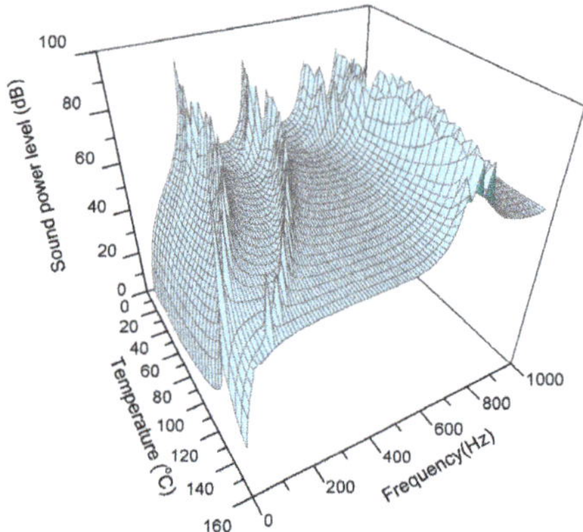

Figure 5. Sound radiation power of composite laminates embedded with SMA in the thermal environment.

5.3. Investigating the Influence of SMA Pre-Strain

The influence of SMA pre-strain on the acoustic radiation of composite laminates is studied. Ensure that the substrate material ratio and SMA volume fraction are constant. The tensile pre-strains of SMA are treated as 1%, 3%, and 5%, respectively, to calculate the acoustic radiation power of composite laminates.

Figures 6 and 7 show the effects of pre-strain on the sound radiation of the composite laminates with SMA in 25 and 100 °C, respectively.

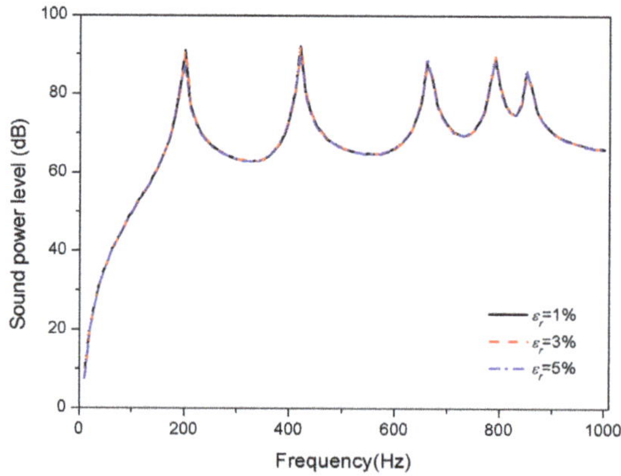

Figure 6. Sound radiation power of composite laminates embedded with SMA subjected to different pre-strain at 25 °C.

No considerable changes are seen in the sound power level under different pre-strain at 25 °C; however, the peaks of sound power level shift towards high frequencies with the increase of pre-strain at 100 °C. From Figures 2 and 3, SMA in the austenite phase has higher modulus and recovery stress than that in the martensitic phase, which can explain the fact that extremely small recovery stress is generated by pre-strain at normal temperature since SMA is in the martensitic phase; but recovery stress increases gradually with rising temperature, which is due to the transformation of the martensitic phase to austenite phase. Moreover, it can be known from Equation (31) that the recovery stress has a positive correlation with the coefficient K_{33} in the stiffness matrix. Therefore, the influence of recovery stress produced by tensile pre-strain of SMA at 25 °C can be ignored; nevertheless, at 100 °C, the increase of the stiffness of the composite laminates with the increase of recovery stress, results in the increase of the corresponding natural frequencies.

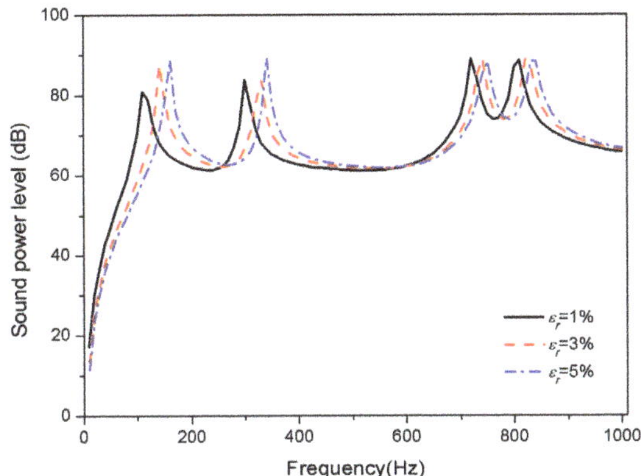

Figure 7. Sound radiation power of composite laminates embedded with SMA subjected to different pre-strain at 100 °C.

5.4. Investigating the Influence of SMA Volume Fraction

Materials volume fraction is an important component parameter of the plate materials for the acoustic properties. This section considers the effect of SMA volume fractions of 10%, 20%, 30%, and 40% on the sound radiation of composite laminates.

The variation of sound radiation power with different SMA volume fraction of the composite laminates is depicted in Figures 8 and 9.

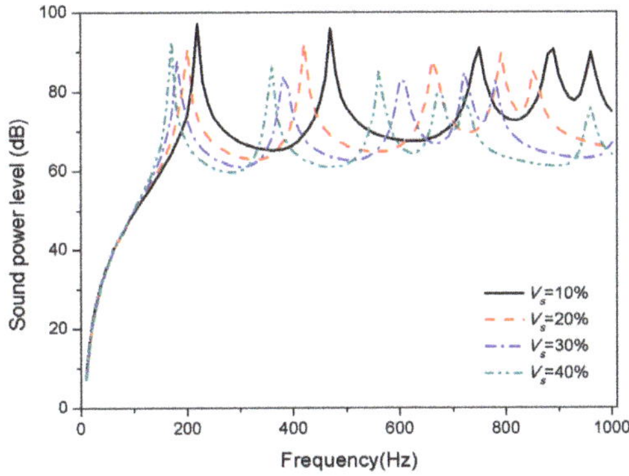

Figure 8. Sound radiation power of composite laminates embedded with SMA subjected to different SMA volume fraction at 25 °C.

The plot Figure 8 reveals that at 25 °C, the modes shift towards low frequencies with an increase of SMA volume fraction. Modulus of SMA in the martensitic phase at ordinary temperature is less than that of substrate modulus as shown in Figure 2. The increase in SMA volume fraction leads to a decrease in the overall stiffness of the laminates, which bring about the decrease of natural frequencies. When the temperature is 100 °C, the modulus of SMA in austenite phase is greater than the substrate. As the SMA volume fraction increases, the stiffness of the laminates increases, so the first two modes shift towards high frequencies. Besides, third and fourth modes shift slightly towards low frequencies.

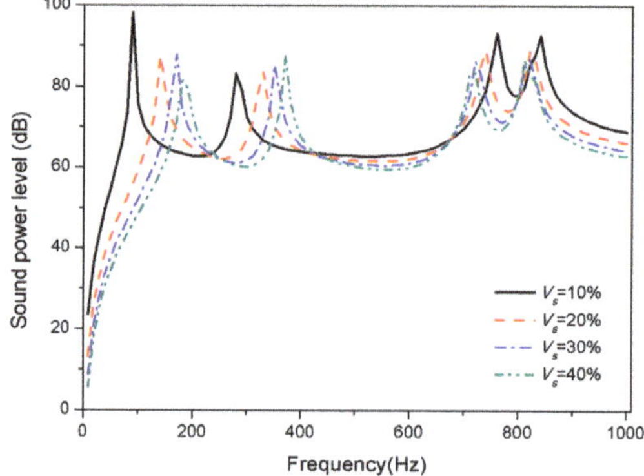

Figure 9. Sound radiation power of composite laminates embedded with SMA subjected to different SMA volume fraction at 100 °C.

5.5. Investigating the Influence of Substrate Ratio

Considering the larger modulus of graphite in the substrate, the volume fraction ratio of graphite to epoxy studied in this section determines the modulus of the substrate. The volume fraction of SMA is 20%, so the volume fraction of substrate is 80%. A comparative study was carried out with the volume fraction ratio of graphite to epoxy (V_f/V_m) as 1/9, 2/8, 3/7, and 4/6.

The result in Figures 10 and 11 illustrate influence of the substrate ratio on the sound radiation of composite laminates.

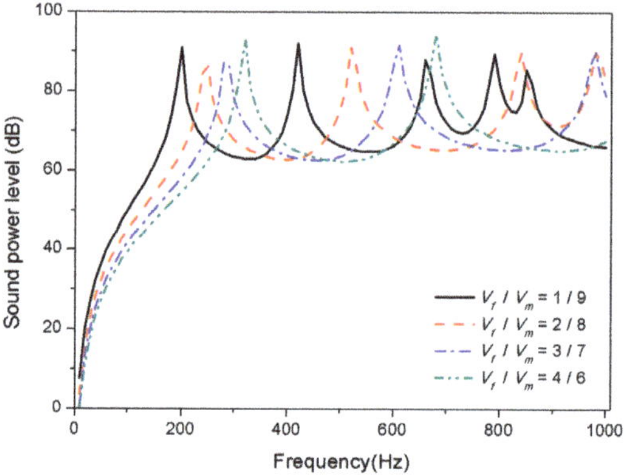

Figure 10. Sound radiation power of composite laminates embedded with SMA subjected to different substrate ratio at 25 °C.

It is observed that the corresponding peaks of the sound power level shift towards the higher frequency domain when the volume fraction of graphite increases in Figure 10. The reason is that graphite has a larger modulus than that SMA in the martensitic phase which significantly affects the stiffness of the laminates at 25 °C. As can be seen from Figure 11, at 100 °C, increasing the volume fraction of graphite in the substrate has little effect on the first mode of the composite laminates, which is due to the influence of the modulus of SMA in the austenite phase on the stiffness increase. However, after the second mode, as the proportion of graphite increases, the peaks of sound power gradually shift towards the high frequency, i.e., the higher the volume fraction of graphite in the substrate, the higher the high-order modal frequency can be increased.

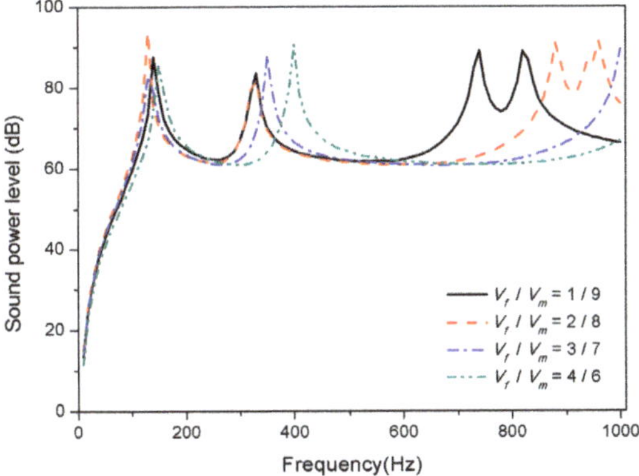

Figure 11. Sound radiation power of composite laminates embedded with SMA subjected to different substrate ratio at 100 °C.

5.6. Investigating the Influence of Geometrical Properties

The geometric properties of composite laminates are studied from the thickness and aspect ratio. Figures 12 and 13 display the variation of sound radiation power with constant length and width (L_a = 0.4 m, L_b = 0.3 m) but different thickness, that is, h = 0.008 m, h = 0.012 m, h = 0.016 m, and h = 0.02 m are considered.

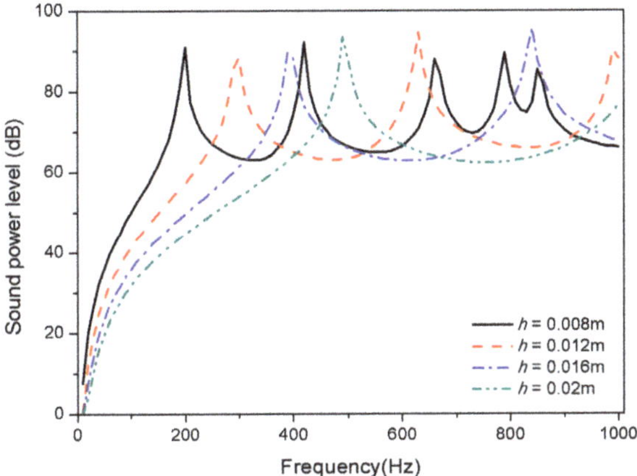

Figure 12. Sound radiation power of composite laminates embedded with SMA subjected to different thickness at 25 °C.

From Figures 12 and 13, the trend of mode shift caused by thickness variation is consistent for 25 and 100 °C. With the increase of thickness, the first and second mode shifts towards high frequencies; fewer peaks occur in the frequency range showed for thickness h > 0.012 m in both normal and high temperature, which is due to the fact that some higher-order modes have shifted above 1000 Hz. In addition, the increase of the thickness of composite laminates leads to a reduction of sound radiation, which is significantly below 400 Hz; however, the thickness has no significant effect on sound radiation in the high frequency region.

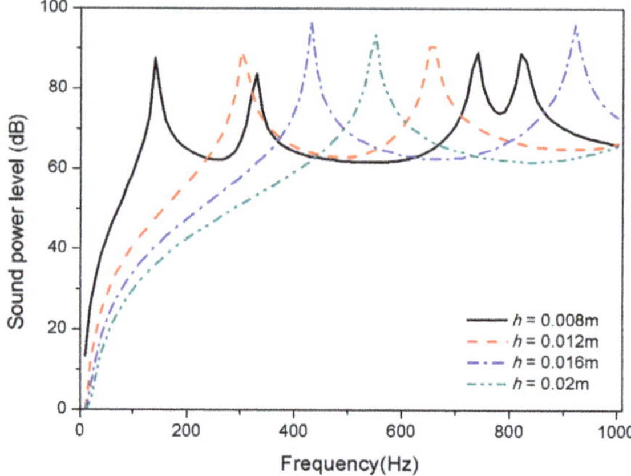

Figure 13. Sound radiation power of composite laminates embedded with SMA subjected to different thickness at 100 °C.

Similar to the plate thickness, aspect ratio is another structural geometric parameter of composite laminates. In order to study the effect of increasing aspect ratio on sound radiation, the length of the laminates is increased and the width is shortened while maintaining a constant area, i.e., 0.4 m × 0.3 m, 0.425 m × 0.283 m, 0.490 m × 0.254 m, and 0.6 m × 0.2 m. The results presented in Figures 14 and 15 indicate that the aspect ratio has a significant effect on the acoustic radiation, in which the first mode shifts towards high frequencies with the increase of the aspect ratio.

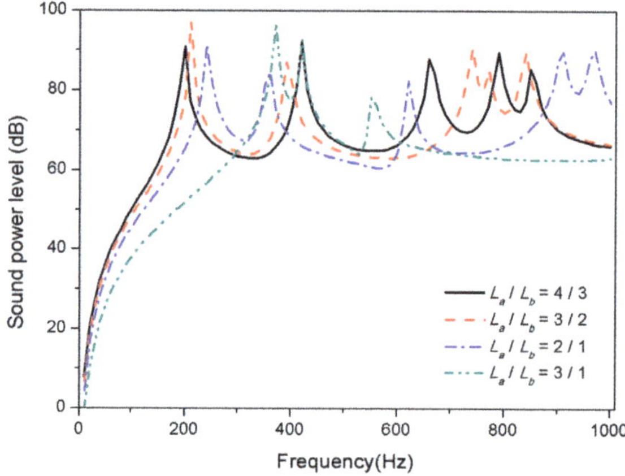

Figure 14. Sound radiation power of composite laminates embedded with SMA subjected to different aspect ratio at 25 °C.

The foremost reason is that the bending stiffness of composite laminates increases with the ratio of length to width under the condition of a certain area, which increases the modal frequency.

As mentioned above, the observations that the variation of thickness and aspect ratio affects sound radiation is determined by the variation of structural stiffness. However, temperature changes stiffness of laminates by affecting material properties, which is not a factor influencing the stiffness from structural geometric dimension. Therefore, the variation of structural geometric parameters has no remarkable law on acoustic radiation between normal and high temperature, but the trends are comparable.

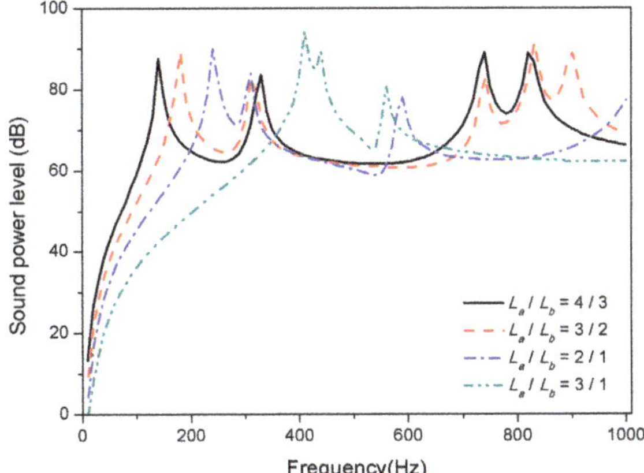

Figure 15. Sound radiation power of composite laminates embedded with SMA subjected to different aspect ratio at 100 °C.

6. Conclusions

Vibro-acoustic radiation of composite laminates which embeds SMA orthogonal antisymmetric has been developed. The constitutive equation, energy method, and Hamilton principle are utilized to derive the governing equations, and sound radiation is calculated by employing the Rayleigh integral approach. From the current investigation, the following conclusions are made:

(1) The pre-strain and volume fraction of SMA in composite laminates have a significant impact on vibro-acoustic radiation. The higher the pre-strain of the SMA at high temperatures, the modes shift towards high frequencies. However, the volume fraction of SMA has opposite effects on the modalities of acoustic radiation between low and high temperatures. That is, as the volume fraction of SMA increases, modes shift towards high frequencies at elevated temperatures, and the opposite occurs at low temperatures.

(2) Graphite in the base material has a larger modulus which affects the stiffness of the laminates. The effect of increasing volume fraction of graphite at low temperature is opposite to that of SMA on sound radiation. At this point, the modulus of graphite is larger than that of SMA, at which point the graphite content dominates the overall stiffness of the laminates.

(3) Several geometric parameter studies on the effects of geometrical properties on sound radiation of composite laminates are carried out. It is noticed that composite laminates embedded SMA have similar characteristics of sound radiation with homogeneous laminates in the temperature range. Influence of geometric parameters on sound radiation is mainly dependent on structural stiffness rather than material properties, resulting in insensitivity to temperature.

The above conclusion fully demonstrates that the composite laminates embedded with SMA wires have excellent ability to adjust the mode frequency and change sound radiation characteristics. This study provides a basis for avoiding modal resonance and reducing vibration and sound radiation of composite laminates from the point of view of material design.

Author Contributions: Conceptualization, Y.H., Z.Z. and K.M.; methodology, Y.H. and C.L.; characterization Y.H. and J.W.; writing—original draft preparation, Y.H. and Z.L.; writing review and editing, Y.H., Z.Z. and K.M. All authors have read and agreed to the published version of the manuscript.

Funding: This research was funded by National Natural Science Foundation of China, grant number 51575201 and the APC was funded by China Postdoctoral Science Foundation, grant number 2020M672360.

Conflicts of Interest: The authors declare no conflict of interest.

Appendix A

The engineering parameters of the substrate are related to the volume fraction ratio of graphite and epoxy. According to the material mixing law [6], the engineering parameters of the base material can be written as:

$$E_{1mf} = E_m V_m + E_f V_f, \tag{A1}$$

$$E_{2mf} = \frac{E_m E_f}{E_m V_f + E_f V_m}, \tag{A2}$$

$$v_{12mf} = v_m V_m + v_f V_f, \tag{A3}$$

$$v_{21mf} = v_{12mf} E_{2mf} / E_{1mf}, \tag{A4}$$

$$G_{12mf} = \frac{G_m G_f}{G_m V_f + G_f V_m}, \tag{A5}$$

$$\alpha_{11} = \frac{E_{1mf} \alpha_m V_m + E_f \alpha_f V_f}{E_{1mf}}, \tag{A6}$$

$$\alpha_{22} = \alpha_m V_m + \alpha_f V_f, \tag{A7}$$

$$\rho_{mf} = \rho_m V_m + \rho_f V_f. \tag{A8}$$

The engineering parameters of SMA fiber embedded in the substrate can be expressed as:

$$E_{11} = E_{1mf}(1 - V_s) + E_s V_s, \tag{A9}$$

$$E_{22} = \frac{E_s E_{1mf}}{E_s(1 - V_s) + E_{1mf} V_s}, \tag{A10}$$

$$v_{12} = v_{12mf}(1 - V_s) + v_s V_s, \tag{A11}$$

$$v_{21} = v_{12} E_{22} / E_{11}, \tag{A12}$$

$$G_{12} = \frac{G_{12mf} E_s}{2 G_{12mf} V_s (1 + v_s) + E_s (1 - V_s)}, \tag{A13}$$

$$\bar{\rho} = \rho_{mf}(1 - V_s) + \rho_s V_s. \tag{A14}$$

The coefficient expression of the transformed reduced stiffness matrix are as follows:

$$\widetilde{Q}_{11}^{(k)} = E_{11} / (1 - v_{12} v_{21}), \tag{A15}$$

$$\widetilde{Q}_{12}^{(k)} = E_{22} v_{12} / (1 - v_{12} v_{21}), \tag{A16}$$

$$\widetilde{Q}_{22}^{(k)} = E_{22} / (1 - v_{12} v_{21}), \tag{A17}$$

$$\widetilde{Q}_{66}^{(k)} = G_{12}, \tag{A18}$$

$$\overline{Q}_{11}^{(k)} = E_{1mf} / \left(1 - v_{12mf} v_{21mf}\right), \tag{A19}$$

$$\overline{Q}_{12}^{(k)} = E_{2mf} v_{12mf} / \left(1 - v_{12mf} v_{21mf}\right), \tag{A20}$$

$$\overline{Q}_{22}^{(k)} = E_{2mf} / \left(1 - v_{12mf} v_{21mf}\right), \tag{A21}$$

$$\overline{Q}_{66}^{(k)} = G_{12mf}, \tag{A22}$$

where subscript '*m*', '*f*' and '*s*' denote the epoxy, graphite, and SMA, respectively.

References

1. Ro, J.; Baz, A. Nitinol-reinforced plates, Part I—III. *Compos. Eng.* **1995**, *5*, 61–106. [CrossRef]
2. Rogers, C.A.; Liang, C.; Jia, J. Structural modification of simply supported laminated plates using embedded shape memory alloy fibers. *Comput. Struct.* **1991**, *38*, 569–580. [CrossRef]
3. Ostachowicz, W.; Krawczuk, M.; Żak, A. Natural frequencies of a multilayer composite plate with shape memory alloy wires. *Finite Elem. Anal. Des.* **1999**, *32*, 71–83. [CrossRef]
4. Malekzadeh, K.; Mozafari, A.; Ghasemi, F.A. Free vibration response of a multilayer smart hybrid composite plate with embedded SMA wires. *Lat. Am. J. Solids Struct.* **2014**, *11*, 279–298. [CrossRef]
5. Mahabadi, R.K.; Shakeri, M.; Pazhooh, M.D. Free vibration of laminated composite plate with shape memory alloy fibers. *Lat. Am. J. Solids Struct.* **2016**, *13*, 314–330. [CrossRef]
6. Huang, Y.; Zhang, Z.; Li, C.; Mao, K.; Huang, Q. Modal performance of two-fiber orthogonal gradient composite laminates embedded with SMA. *Materials* **2020**, *13*, 1102. [CrossRef]
7. Huang, Y.; Li, L.; Xu, Z.; Li, C.; Mao, K. Free vibration analysis of functionally gradient sandwich composite plate embedded SMA wires in surface layer. *Appl. Sci.* **2020**, *10*, 3921. [CrossRef]
8. Park, J.S.; Kim, J.H.; Moon, S.H. Vibration of thermally post buckled composite plates embedded with shape memory alloy fibers. *Compos. Struct.* **2004**, *63*, 179–188. [CrossRef]
9. Chowdhury, M.A.; Rahmzadeh, A.; Alam, M.S. Improving the seismic performance of post-tensioned self-centering connections using SMA angles or end plates with SMA bolts. *Smart Mater. Struct.* **2019**, *28*, 075044. [CrossRef]
10. Panda, S.K.; Singh, B.N. Nonlinear finite element analysis of thermal post-buckling vibration of laminated composite shell panel embedded with SMA fibre. *Aerosp. Sci. Technol.* **2013**, *29*, 47–57. [CrossRef]
11. Shariyat, M.; Moradi, M.; Samaee, S. Enhanced model for nonlinear dynamic analysis of rectangular composite plates with embedded SMA wires, considering the instantaneous local phase changes. *Compos. Struct.* **2014**, *109*, 106–118. [CrossRef]
12. Tehrani, B.T.; Kabir, M.Z. Non-linear load-deflection response of SMA composite plate resting on winkler-pasternak type elastic foundation under various mechanical and thermal loading. *Thin Wall. Struct.* **2018**, *129*, 391–403. [CrossRef]
13. Turner, T.L. A new thermoelastic model for analysis of shape memory alloy hybrid composites. *J. Intel. Mater. Syst. Struct.* **2000**, *11*, 382–394. [CrossRef]
14. Abdullah, E.J.; Gaikwad, P.S.; Azid, N.; Majid, D.A.; Rafie, A.M. Temperature and strain feedback control for shape memory alloy actuated composite plate. *Sens. Actuators A Phys.* **2018**, *283*, 134–140. [CrossRef]
15. Jiang, E.Y.; Zhu, X.J.; Shao, Y.; Wang, Z.L. Active vibration control of smart structure based on alternate driving SMA actuators. *Appl. Mech. Mater.* **2011**, *39*, 61–66. [CrossRef]
16. Amarante dos Santos, F.P.; Cismaşiu, C.; Pamies Teixeira, J. Semi-active vibration control device based on super-elastic NiTi wires. *Struct. Control Health Monit.* **2013**, *20*, 890–902. [CrossRef]
17. Bodaghi, M.; Shakeri, M.; Aghdam, M.M. Passive vibration control of plate structures using shape memory alloy ribbons. *J. Vib. Control.* **2017**, *23*, 69–88. [CrossRef]
18. Liang, C.; Rogers, C.A.; Fuller, C.R. Acoustic transmission and radiation analysis of adaptive shape-memory alloy reinforced laminated plates. *J. Sound Vib.* **1991**, *145*, 23–41. [CrossRef]
19. Sharma, N.; Mahapatra, T.R.; Panda, S.K. Vibro-acoustic analysis of laminated composite plate structure using structure-dependent radiation modes: An experimental validation. *Sci. Iran. B* **2018**, *25*, 2706–2721. [CrossRef]
20. Sharma, N.; Mahapatra, T.R.; Panda, S.K. Numerical study of vibro-acoustic responses of un-baffled multi-layered composite structure under various end conditions and experimental validation. *Lat. Am. J. Solids Struct.* **2017**, *14*, 1547–1568. [CrossRef]
21. Jeyaraj, P.; Ganesan, N.; Padmanabhan, C. Vibration and acoustic response of a composite plate with inherent material damping in a thermal environment. *J. Sound Vib.* **2009**, *320*, 322–338. [CrossRef]
22. Yin, X.W.; Gu, X.J.; Cui, H.F.; Shen, R.Y. Acoustic radiation from a laminated composite plate reinforced by doubly periodic parallel stiffeners. *J. Sound Vib.* **2007**, *306*, 877–889. [CrossRef]
23. Daneshjou, K.; Talebitooti, R.; Kornokar, M. Vibroacoustic study on a multilayered functionally graded cylindrical shell with poro-elastic core and bonded-unbonded configuration. *J. Sound Vib.* **2017**, *393*, 157–175. [CrossRef]

24. Santoni, A.; Schoenwald, S.; Fausti, P.; Tröbs, H.M. Modelling the radiation efficiency of orthotropic cross-laminated timber plates with simply-supported boundaries. *Appl. Acoust.* **2019**, *143*, 112–124. [CrossRef]

25. Zhang, D.; Wu, Y.; Chen, J.; Wang, S.; Zheng, L. Sound radiation analysis of constrained layer damping structures based on two-level optimization. *Materials* **2019**, *12*, 3053. [CrossRef]

26. Yang, T.; Zheng, W.; Huang, Q.; Li, S. Sound radiation of functionally graded materials plates in thermal environment. *Compos. Struct.* **2016**, *144*, 165–176. [CrossRef]

27. Yang, T.; Huang, Q.; Li, S. Three-dimensional elasticity solutions for sound radiation of functionally graded materials plates considering state space method. *Shock Vib.* **2016**, *2016*, 1403856. [CrossRef]

28. Xu, Z.; Huang, Q. Vibro-acoustic analysis of functionally graded graphene-reinforced nanocomposite laminated plates under thermal-mechanical loads. *Eng. Struct.* **2019**, *186*, 345–355. [CrossRef]

29. Kuo, S.Y.; Shiau, L.C.; Chen, K.H. Buckling analysis of shape memory alloy reinforced composite laminates. *Compos. Struct.* **2009**, *90*, 188–195. [CrossRef]

30. Xue, K.; Huang, W.; Li, Q. Three-dimensional vibration analysis of laminated composite rectangular plate with cutouts. *Materials* **2020**, *13*, 3113. [CrossRef]

Article

Selected Tribological Properties and Vibrations in the Base Resonance Zone of the Polymer Composite Used in the Aviation Industry

Aneta Krzyzak [1],*, Ewelina Kosicka [2], Marek Borowiec [3] and Robert Szczepaniak [1]

[1]   Department of Airframe and Engine, Military University of Aviation, 80-521 Dęblin, Poland; r.szczepaniak@law.mil.pl
[2]   Department of Production Engineering, Lublin University of Technology, 20-618 Lublin, Poland; e.kosicka@pollub.pl
[3]   Department of Applied Mechanics, Lublin University of Technology, 20-618 Lublin, Poland; m.borowiec@pollub.pl
*   Correspondence: a.krzyzak@law.mil.pl; Tel.: +48-261-518-185

Received: 15 February 2020; Accepted: 17 March 2020; Published: 18 March 2020

Abstract: The revolution in the global market of composite materials is evidenced by their increasing use in such segments as the transport, aviation, and wind industries. The innovative aspect of this research is the methodology approach, based on the simultaneous analysis of mechanical and tribological loads of composite materials, which are intended for practical use in the construction of aviation parts. Simultaneously, the methodology allows the composition of the composites used in aviation to be optimized. Therefore, the presented tests show the undefined properties of the new material, which are necessary for verification at the application stage. They are also a starting point for further research planned by the authors related to the improvement of the tribological properties of this material. In this article, the selected mechanical and tribological properties of an aviation polymer composite are investigated with the matrix of L285-cured hardener H286 and six reinforcement layers of carbon fabric GG 280P/T. The structure of a polymer composite has a significant influence on its mechanical properties; thus, a tribological analysis in the context of abrasive wear in reciprocating the movement for the specified polymer composite was performed. Moreover, the research was expanded to dynamic analysis for the discussed composite. This is crucial knowledge of material dynamics in the context of aviation design for the conditions of resonance vibrations. For this reason, experimental dynamical investigations were performed to determine the basic resonance of the material and its dynamics behavior response. The research confirmed the assumed hypotheses related to the abrasive wear process for the newly developed material, as well as reporting an empirical evaluation of the dependencies of the resonance zone from the fabric orientation sets.

Keywords: composite material; tribology; vibrations; resonance zone

---

1. Introduction

The multifaceted approach to conducting improvements in various industry sectors is manifested both in organizational aspects, by forcing the improvement of the quality of implemented tasks (an example may be striving to eliminate the occurrence of machinery or equipment failures [1]), as well as in design aspects [2], which includes the optimization of previously developed structures [3] and technology [4] and carrying out modifications of the materials used [5,6]. Activities related to material improvements determine the dynamics of material engineering development, while assumptions related to the operating conditions of the components made of a given material determine the direction of the changes and expectations of its final properties. These are determined after a wide range of tests (often standardized) are performed, which indicate the dependency between the preparation method of the material and its characteristics, often taking into account external factors [7] such as UV radiation [8] and humidity [9].

The role of material testing increases particularly in the case of high reliability requirements for components made of this material. Among the various industrial sectors, the aviation industry is one that requires absolute reliability. For this reason, all potential solutions used in this industry, including materials, must be thoroughly tested, which indicates the topicality of the topic taken in this article. Additionally, dynamics analysis is crucial from a system stability point of view. It is impossible to exclude complex excitations of the structure from external sources; hence, the material can manifest the existence of various nonlinearities. Moreover, the structure may behave noisily or chaotically [9] and can respond by non-trivial dependence of amplitude and frequency, or even by the occurrence of multiple solutions [10,11].

According to the hypothesis made at the planning stage of the research part for this paper on the new composite material, the authors expected the following increase in abrasive wear and the maximum roughness depth R_{max} as the number of cycles increases, as well as the observation of resonance zone movements while the structure vibrates, due to fabric orientation sets. In this article, the preliminary results are described. The research aim was to develop a new composite material characterized by high tribological resistance and required mechanical properties associated with resonance vibrations. The resin used can be used to manufacture components for gliders, powered sailplanes, and motor-powered aeroplanes. The authors did not find other publications on tribological research in connection with resonance studies of parts made of this resin, hardener, or fabric in specific configurations. The obtained results are the first stage of planned research consisting of carrying out modifications to improve the properties of the aviation composite.

1.1. Polymer Composites Used in Aviation

By using various materials for warp or reinforcement, there are unlimited possibilities to modify the properties of composites. This influences the dynamics of the increase in interest in these materials observed in global industry trends. Research carried out in the field of developing high-strength and highly modular constructions, while reducing the specific gravity of composites, opens up the perspective of increasingly bolder use of them in the construction, automotive, and aerospace industries [11].

The main motivation toward the development of new materials for the aviation industry is the reduction of aircraft weight, extending the time of reliable operation of the parts, improving fuel efficiency, and thus reducing aircraft operating costs [12–14]. These materials can be found in both internal and external aircraft constructions; in internal structures (secondary structures), they include cabins, floorboards, or armchairs, but it is the external (primary) structures that constitute the main market for composite aviation materials.

Primary examples of the implementation of composite materials in the structure of aircraft are the Airbus A380 and the Boeing 787 Dreamliner. In the former, the world's largest passenger aircraft, the composites constitute 25% of the weight of the entire aircraft, where as much as 22% are fibrous composites with an epoxy resin matrix and the reinforcement is Carbon Fiber-Reinforced Polymer (CFRP) carbon fibers. The carbon fibers used ensure high rigidity (935 GPa), and their density is 60% of the density of aluminum (achieving a 40% lighter weight than is the case with constructions made of aluminum). An additional argument in favor of the use of CFRP composites is the relative ease of obtaining the complex shapes of the components. In addition, they are characterized by good thermal and electrical conductivity and X-ray absorption, as well as the ability to dampen vibrations [15].

The second passenger aircraft, which can boast of functioning implementations of composite components, is the Boeing 787 Dreamliner (Figure 1), which has been designated as a ground-breaking application of composite materials, inspiring the use of these materials deep in the currents of commercial products [16]. This aircraft accounts for up to 50% by mass (and 80% by volume) of the composite materials applied to the main airframe structures [17]. As emphasized by Merkisz and Bajerlein in [18], this results, among other things, in a significant reduction in aircraft weight, high corrosion resistance, easier production of complex shapes, and a reduction in assembly time compared to earlier conventional constructions. However, the estimated fuel consumption during operation is about 20% lower compared to the same class of aircraft.

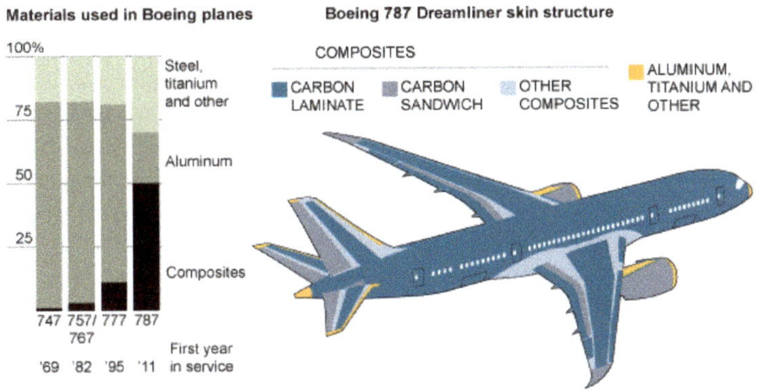

Figure 1. Materials used in the Boeing 787 Dreamliner skin structure [19].

The conducted literature analysis has allowed us to observe that, in aviation polymer composites, carbon fibers are often used as reinforcement. They are obtained mainly as a result of polyacrylonitrile pyrolysis, and their properties are shaped by the production parameters used [20]. They are characterized by good heat and chemical resistance, as evidenced by the lack of changes in the non-oxidizing atmosphere up to 2000 °C, unlike glass or aramid fibers [21]. In addition, they are characterized by low density, good thermal and electrical conductivity, and when used as friction materials, they show a low coefficient of friction. In addition, these fibers have the ability to damp vibrations and lower X-ray absorption. For this reason, their widespread use in the production of composite materials in global terms can be observed, as shown in Figure 2.

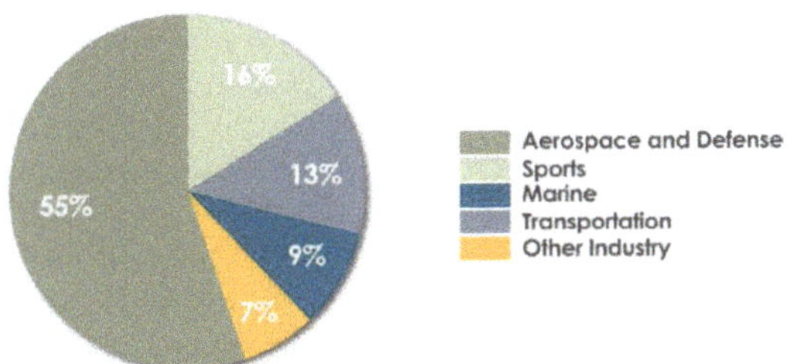

Figure 2. Areas of application for carbon fibers (based on: [22]).

Carbon fabrics, being reinforcements made on the basis of carbon fibers, have been used in aviation solutions, and it is appropriate to mention where their weaves have been used. The discussed weaves are as follows:
—Canvas (PLAIN);
—Oblique (TWILL);
—Satin (SATIN).
These weaves also have different weights and thicknesses.
It is also worth mentioning that the specifications in the context of the fabric used for the composite material produced should include the number of layers used or the way the subsequent reinforcement

layers are laid. It is obvious the composite preparation process varies and depends on the materials used, as well as the recommendation results from the intended application of a composite.

1.2. Tribological Properties

Friction is considered one of the most common phenomena occurring in nature, and there has been an intensification of research in this area aimed at eliminating, or at least reducing, its formation. The problems of the occurrence of tribological wear of machine and device components ultimately translate into performance, as well as operational, reliability. The occurrence of many machine motion units necessitates limiting the formation of this irreversible phenomenon, which is achieved, for example, by using construction materials with improved anti-wear properties [23]. Tribological research in industrial conditions is difficult and expensive; therefore, the analysis of friction and wear processes is carried out on designed friction devices (also called tribometers or tribotesters). Their task is to map the association of friction through model experiments, which is why they have different designs and modes of action.

The up-to-date analysis of issues covering tribological aspects (e.g., [24–26]) is evidenced by publications of various scientific centers, in which the results of materials engineering research are analyzed, including considerations on how to improve the tribological properties of new materials [27,28] while determining the impact on their mechanical properties. The observed intensification in the development of material engineering in aviation, apart from defining the obtained mechanical properties for a given material, also requires the determination of tribological properties, due to the further possibilities of using these materials for structural solutions in which friction occurs.

The resin used is approved by the German Federal Aviation Office and it meets the standards that allow it to be used to make parts for gliders, powered sailplanes, and motor-powered airplanes. However, no publications can be found regarding the abrasive wear of composites made of this resin.

2. Measurements Methodology

2.1. Material Tested

Polymer composites were used to conduct this research. For the preparation of composites by hand lamination, an epoxy resin with the trade name L285, together with the hardener H285, were used, as well as carbon fabric GG 280 P/T. Due to the target use of composites for the production of aircraft components, the carbon fabric GG 280 P/T (twill 2/2, fiber 3K 200 [tex], 220 [g/m^2]) used to make the composites had the appropriate certificates. The air bubbles formed in the matrix were removed prior to curing using ultrasonic waves. The fabric was arranged in the configuration: 0/22.5/45/0/22.5/45. Such fiber configurations, as well as the type of resin fabric, were determined by laminates applied in the aviation industry, especially in helicopter PZL SW-4. The composite material was manufactured by combining the laminating method, preparing the consecutive fabric layers impregnated by resin, and finally, pressing the structure by means of the hydraulic press PDM–50S Mecamaq at a pressure level of 2.5 MPa. The polymer composite was not affected by environmental conditions; it was not exposed to external factors such as changes in temperature, humidity or UV radiation. Due to the lack of standards specifying the dimensions of the samples used for the planned tests, the samples' dimensions were assumed for both tests, the tribological 125 × 25 × 2.5 mm and the dynamical (see Table 1), where the frequency of vibrations in the first resonance zone was investigated. The assumed dimensions were obtained in the process of cutting with an abrasive water jet. The research was carried out in the Department of Airframe and Engine (Military University of Aviation, Dęblin), the Department of Applied Mechanics (Lublin University of Technology, Lublin), and the Department of Production Engineering (Lublin University of Technology, Lublin).

Table 1. The parameters of the composite material beam.

Symbol and Value	Description
$\rho = 2489 \text{ kg/m}^3$	mass density of the beam
$E = 42.40 \text{ GPa}$	Young modulus of the material
$L = 200 \text{ mm}$	length of the beam
$b = 10 \text{ mm}$	width of the beam
$h = 2.4 \text{ mm}$	thickness of the beam
$A = 24 \text{ mm}^2$	cross section of the beam
$I = 22.5 \text{ mm}^4$	area moment of inertia
$N_1 = 0.05 \text{ m}$	constant no1 depends on $\psi(x)$
$N_4 = 5.80 \text{ m}^{-1}$	constant no4 depends on $\psi(x)$
$N_6 = 386.25 \text{ m}^{-3}$	constant no6 depends on $\psi(x)$

2.2. The Test Stand for Tribological Tests

The produced samples were tested in conditions of abrasive wear in reciprocating motion using a tribotester Taber Linear Abraser model 5750 (Figure 3). The abrasive stone used in the test bench had a diameter of 6.6 mm and a gradation of 200. The friction path was 101.6 mm/cycle, and the total load was 1850 g. After 100, 300, 600, 1000, and 1500 cycles, the mass of the samples was determined using a precision weight laboratory. After each measurement, the friction products were removed and, according to the manufacturer's instructions, the abrasive stone and the sample were cleaned, and then the selected parameters of the wear path roughness, which were taken by the counter sample, were measured using the MicroProf 100 FRT optical profilometer. The abrasive wear of the composite was determined as a weight loss. Basic roughness parameters were determined, such as the arithmetical mean height (R_a), the maximum height of the profile (R_z), and maximum roughness depth (R_{max}). The R_{max} parameter is understood as the largest recess in the material after the friction process. Surface parameters the arithmetical mean height (R_a) and the maximum height of the profile (R_z) were made in accordance with the norm PN-EN ISO 4287, using the following relationship [29]:

$$R_a = \frac{1}{l} \int_0^l |Z(x)|\, dx, \tag{1}$$

$$R_z = R_p + R_d, \tag{2}$$

where: l is the length of the elementary segment (m); Z is the height of profile elements (m); R_p is the maximum profile peak height (m); R_d is the maximum profile valley depth (m).

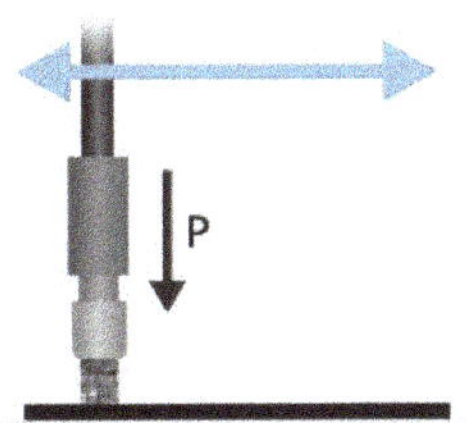

Figure 3. The Taber Linear Abraser model 5750.

2.3. The Resonance Frequency Estimation by the Analytical Approach

Each body or body system has its own natural frequency. Its value is influenced by both the body shape and the physical properties of the system, resulting, for example, from the material used. Knowledge of the natural frequencies allows avoiding the undesirable resonance phenomenon in the construction of machines and devices, which can lead to their damage. For this reason, the role of natural vibrations in the area of aviation is raised primarily in the context of safety aspects.

In [30], the impact of the emergency condition of the aircraft structure on the change of its natural vibrations was mentioned. As the author emphasizes, this is especially important in the case of military aircraft used for combat tasks, when there is a risk of local violations of the structure.

On the other hand, the authors of [31] refer to the necessity of conducting vibration tests before issuing the type for the aircraft being put into service. They emphasize that the obtained results of the experiment are not authoritative in the context of vibration tests—the element connected to the structure will behave differently than the element rigidly attached to the stand. However, they indicate that this type of research plays an important role in verifying Finite Element Method (FEM) model components based on measurements of fragmentary structures.

The authors refer to the dynamic analysis of thin-walled composite structures in [32], indicating numerous applications of these materials, including in the aviation industry. They indicate the need to determine the arrangement of reinforcement layers in the tested composites, which was confirmed by the obtained test results.

With regard to issues related to resonance frequencies, it was considered impossible to raise key messages related to their most important aspects. However, it has been done in the following section of this article. In this subsection, the natural frequency is estimated using the analytical approach by means of the Euler–Bernoulli theory.

The governing differential equation of motion was derived for the analytical estimation of the natural frequency by means of the Lagrange approach [33,34]:

$$\frac{d}{dt}\left(\frac{\partial T}{\partial \dot{v}}\right) - \frac{\partial T}{\partial v} + \frac{\partial P}{\partial v} = 0 \tag{3}$$

where T and Π are the kinetic and potential energies, respectively.

The geometrical relationships for the displacements, velocities, and curvature of the beam are expressed as a function of the deflection of the beam free end in form $v(x,t) = \psi(x)v(t)$, where $\psi(x)$ describes the beam deformation [35]:

$$\psi(x) = \cosh\left(\lambda\frac{x}{L}\right) - \cos\left(\lambda\frac{x}{L}\right) + \eta\left(\sinh\left(\lambda\frac{x}{L}\right) - \sin\left(\lambda\frac{x}{L}\right)\right) \text{ and } \eta = -\frac{\sinh\lambda - \sin\lambda}{\cosh\lambda - \cos\lambda} \tag{4}$$

where the eigenvalue λ is obtained from the transcendental equation:

$$1 + \cosh\lambda\cos\lambda = 0 \tag{5}$$

Introducing the corresponding derivatives of the kinetic T and potential Π energies to the Lagrange (Equation (3)) and neglecting the third- and higher-order terms, the differential equation of motion is in form of [35]:

$$\left[\varrho A N_1 + \frac{h^2}{12}\rho A N_4 + \left(\rho A N_3 + \frac{h^2}{12}\rho A N_9\right)v(t)^2\right]\ddot{v}(t) + \left(\rho A N_3 + \frac{h^2}{12}\rho A N_9\right)\dot{v}(t)^2 v(t) + EI N_6 v(t) = -\varrho A N_2 \ddot{q}(t) \tag{6}$$

where constants N_i in Equation (6) depend on the shape function and for the first mode shape are equal to:

$$N_1 = \int_0^l \psi(x)^2 dx = 0.25L \ [m], \qquad N_2 = \int_0^l \psi(x)ds = 0.38L \ [m],$$

$$N_3 = \int_0^l\left(\int_0^s (\psi'(x)^2 dx\right)^2 dx = 0.29/L \ [m^{-1}], \quad N_4 = \int_0^l \psi'(x)^2 ds = 1.16/L \ [m^{-1}], \tag{7}$$

$$N_6 = \int_0^l \psi''(x)^2 dx = 3.09/L^3 \ [m^{-3}], \qquad N_9 = \int_0^l \psi'(x)^4 ds = 1.80/L^3 \ [m^{-3}]$$

To find the natural frequency of the beam, the linearized form of the equation of motion for free response was taken into account, as follows:

$$\left[\varrho AN_1 + \frac{h^2}{12}\rho AN_4\right]\ddot{v}(t) + EIN_6 v(t) = 0 \tag{8}$$

Finally, the natural frequency for small vibrations from Equation (8) is given by the following:

$$f_n = \frac{1}{2\pi}\sqrt{\frac{EIN_6}{\varrho A\left(N_1 + \frac{h^2}{12}N_4\right)}} \tag{9}$$

Assuming that the analyzed composite structure has isotropic properties, one can apply its parameters (described in Table 1) for estimation of the natural frequency from Equation (9). In the case of the first mode shape, it is f_n = 40 Hz.

2.4. The Test Stand for Dynamical Behavior

For the experiment performed in the laboratory, the electromagnetic shaker system, TIRAvib 50101, was used. A schematic of the measuring armature is presented in Figure 4a. The data acquisition system was based on the multichannel dynamic LMS Scadias III controller, which was supported by Test.Lab software. The armature reproduced environmental conditions in a range of specified frequency bands for the beam structure. The excitation input signal was assumed as a sinusoidal at constant level that corresponds to 1 g, where g corresponds to the gravity acceleration and is taken as g = 9.81 m/s^2. The vibrations of the composite beam were measured in terms of the acceleration response by means of two acceleration sensors qualified for assumed frequency range (Figure 4b). The first sensor was installed on the shaker payload, which controlled the reference signal, while the second one located on the free end of the beam, and measured the on-time acceleration values of the composite structure [36].

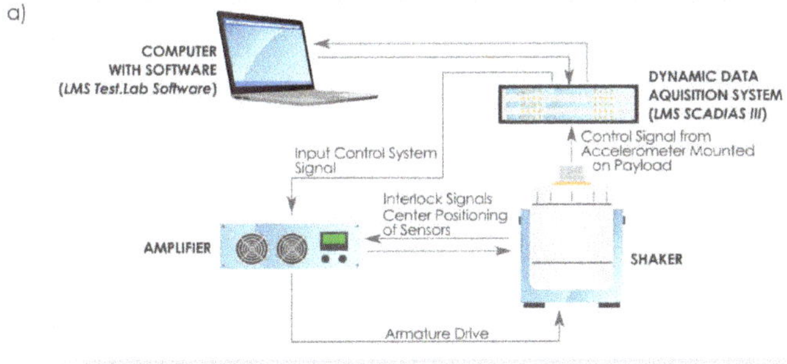

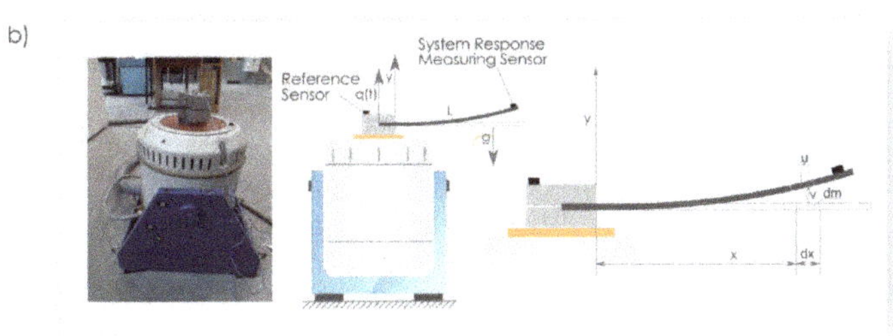

Figure 4. (**a**) The laboratory armature of the electro-dynamical vibration system. (**b**) The experimental set-up and the scheme of the vertically excited composite beam.

3. Results

The conducted research allowed us to determine the properties of a polymer composite dedicated to aviation solutions in the field of issues related to tribology and in the scope of the base resonance zone. Detailed results are presented in the following two subsections.

3.1. Selected Tribological Properties

The results of the measurements of roughness parameters and abrasive wear are presented in Figure 5. As the number of cycles increases, the expected increase in abrasive wear (higher weight loss) and the largest depression R_{max} occur. The increase of mass loss has a linear character and is described by a regression function $y = 9.3988 \times 10^{-6} + 0.0005$, with the coefficient of determination $R^2 = 0.9158$. Similarly, there is a noticeable increase in the average roughness profile R_a and the largest R_z profile. However, changes in parameters R_a, R_z, and R_{max} are not linear. In up to 600 cycles, there is a slight stabilization of the values of these parameters, followed by their almost doubled increase and a slight decrease again. Noticeable fluctuations in roughness values are associated with the destruction of subsequent layers of carbon composite reinforcement. Larger changes in roughness are associated with the exposure of carbon fibers, which, in contact with the counter-sample, are damaged and crumble, causing more and more changes in the surface profiles.

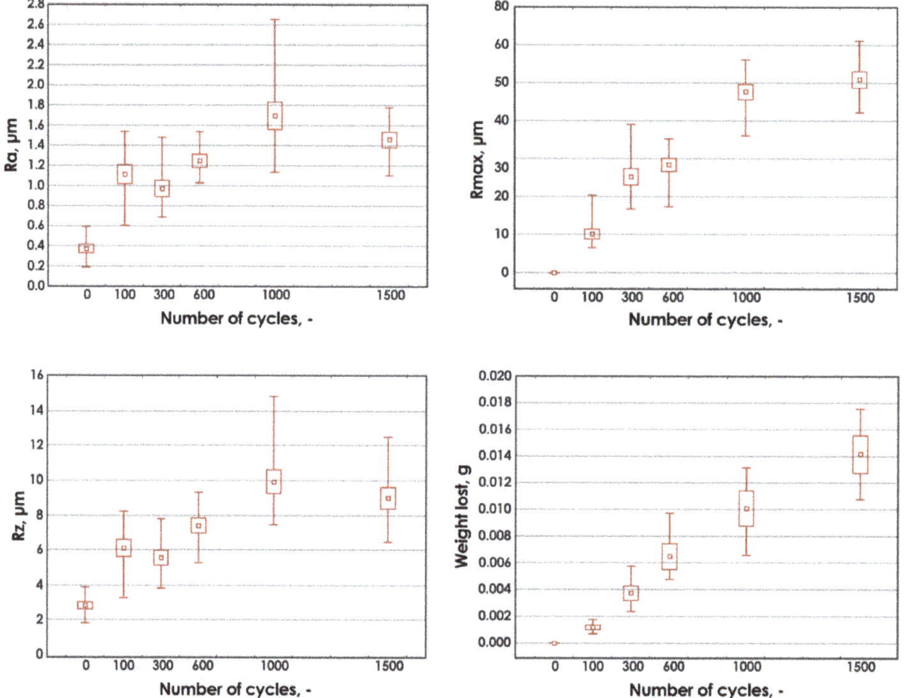

Figure 5. Dependencies of the arithmetical mean height (R_a), the maximum height of the profile (R_z), and maximum roughness depth (R_{max}) parameters and weight loss on the number of wear cycles.

The obtained tribological test results for the basic composite material constitute a starting point for further tests in the context of the dependencies between the modification of its composition and the observed abrasive wear. In addition, they are necessary to refer, in subsequent observations, to the impact of dependent variables (such as temperature, humidity, or UV radiation) on this material.

The wide range of planned composition modifications and seasoning conditions that the composite does not release from highlight the need to conduct research on the basic material, thanks to which it will be possible to conduct final reliability and economic analyses.

3.2. Base Resonance Zone

The second part of the measurements concerns the dynamical analysis, which was performed in a small series of the polymer composites of four samples for two sets of fabric orientation. For the six layers of the fabric configuration 0/22.5/45/0/22.5/45 in the x direction, the composite plate was cut out in perpendicular directions to each other, as shown in Figure 6, where samples 1–4 are cut along the x direction and samples 5–8 along the y direction. Assuming the cutting directions at a right angle of the x and y axes, the dynamical behavior of the beam was investigated for both fabric configuration sets: 0/22.5/45/0/22.5/45 and 90/67.5/45/90/67.5/45. For such cut and prepared composite beams, experimental measurements were performed. The results of the amplitude–excitation frequency responses provide answer about the dynamical behavior of the samples, while the beam was vibrating in the vicinity of the first resonance frequency.

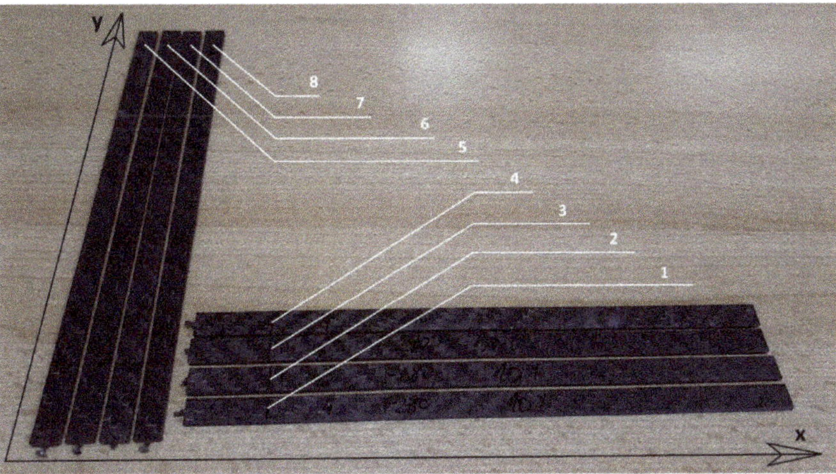

Figure 6. The sample cut along the x and perpendicular y directions.

Figure 7a,b presents the experimental results of the tests for both sets. Slight discrepancies are visible for each set of samples, which are caused by the beam thickness tolerance, as well as the clamping force on the shaker grip. It is clear that, for both sets of four samples, the cantilever beams reached the first mode shape at different excitation frequencies. The mean values of the resonance frequency for the first sample set (Figure 7a) is $f_e \approx 34$ Hz, but for the second set (Figure 7b) it equals $f_e \approx 40$ Hz, and this case is in accordance with the theoretically calculated excitation frequency from Equation (9). The dynamical analysis of the composite structure yielded a view about the stiffness of the material for one case of the fabric configuration, but was simultaneously excited in different (here, perpendicular) directions, as shown in Figure 6. The experiment revealed that the stiffness can be different. While the resonance frequency increased, the composite reached the progressive characteristic of the stiffness, and for decreasing resonance frequency, the material manifested the regressive characteristic. For the dimensions of the beams listed in Table 1, the measured response amplitudes of acceleration reached approximately $A_{out} \approx 70$ g at excitation force amplitude $A_{in} = 1$ g. This shows that this polymer composite structure is able to carry significant dynamical loads, which is an advantage of this structure. One can conclude by linking these results to the real measurements on a PZL SW-4 helicopter, provided by the research in [37]. The authors recorded a frequency spectrum taken from the vertical stabilizer, while the helicopter was flying horizontally at a constant velocity of 200 km/h on 1000 m altitude. The measurements reported the range of frequencies and amplitudes to enable identifying the important design variables. From this investigation, one can select the sensitive frequencies and set these values against the results presented in Figure 7. While helicopter is flying, the noticeable vibration levels occurred at frequencies are 22, 30 and 45 Hz [37]. Analyzing the composite material of both fabric configuration set, the natural frequencies are close to those above. This is crucial from an assembly point of view in the manufacturing process. Because the fabric configuration set from the same composite material gives an opportunity to reach a different natural frequency, it moves the resonance point of the assembled parts applied in the aircraft construction. Namely, the parts assembled of the material from samples 1–4 are more convenient for use as elements which will be exposed to vibrations at about 45 Hz, but the other parts, including the configurations of samples 5–8 should be saved for use in the construction of parts exposed to excitations at 22 or 30 Hz.

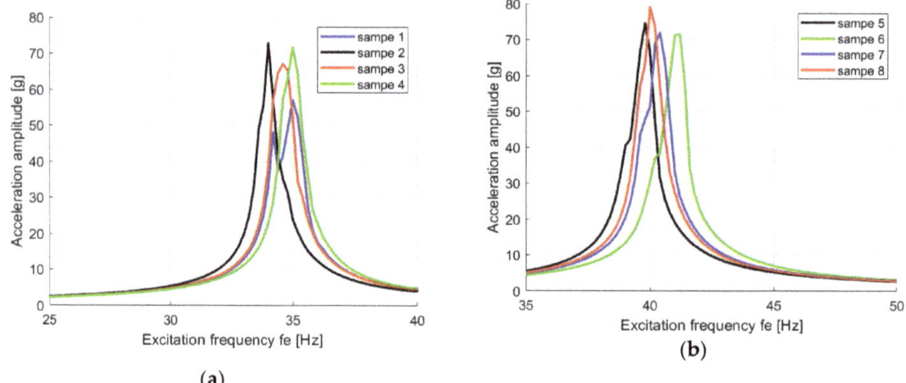

Figure 7. The response acceleration amplitude of the composite beams via excitation frequency at the first resonance zone of the samples cut along the x axis for the fabric configuration set 0/22.5/45/0/22.5/45 (**a**), and along the y axis for the fabric configuration set 90/67.5/45/90/67.5/45 (**b**).

4. Conclusions

The dynamical behavior of the polymer composite during vibrations in the vicinity of the base resonance zone reveals that the mechanical properties of the structure determined the considerable changes of the material amplitude–frequency response. This indicates the importance of fiber arrangement for applying the composites' plate components in the aircraft structure. For the vibrating parts in particular, it should be verified that the natural frequency is safe for the sake of the excitation sources. It is noticeable, based on the comparison the vibration frequency spectra of the PZL SW-4 helicopter and the composite dynamical behavior reported in amplitude responses. The significant frequencies of the analyzed composite materials are close to each other. Consequently, the fiber arrangement is essential for changing the dynamical characteristics of the material, which allows us to move the resonance zone points for safe aircraft construction.

The increase in abrasive wear resulted in the expected increase in surface profile depression. At the maximum average weight loss at level 0.01416 g, the maximum of the largest recess in the material after the friction process R_{max} was 50.93 μm. No similar correlation was observed with the roughness coefficients—the arithmetical mean height (R_a) and the maximum height of the profile (R_z). These coefficients increased by leaps and bounds. Nevertheless, after 300 and 1500 cycles, decreased values were observed compared to the values at 100 and 1000 cycles. The decrease in the average R_a was 14.5% and 16.3%, respectively, for R_z (in both cases) it was 10%. The observed decrease indicates that temporary stabilization of the arithmetical mean height and maximum height of the profile was observed, which was adequate to the process of damage (breaking and falling out of carbon fiber and resin) of the subsequent carbon fabric layers in the composite.

Due to the exposition of the polymer composites used in aviation to harmful vibrations and the related friction process, it is necessary to conduct further research to improve the tribological properties by applying additional physical modifiers (e.g., Al_2O_3, SiC) while maintaining high resistance to damage caused by long-term dynamic loads.

This research confirmed the assumed hypothesis related to the abrasive wear process for the newly developed material, as well as empirically obtaining the dependencies of the resonance zone from the fabric orientation sets.

Author Contributions: Conceptualization, A.K., M.B., E.K., and R.S.; methodology for dynamics analysis, M.B. and E.K.; methodology for tribology research, A.K. and R.S.; software of LSM system operation, M.B.; analytical approach of the beam models, M.B.; writing—original draft preparation, A.K., E.K., R.S., and M.B.; visualization, E.K.; funding acquisition, A.K. All authors have read and agreed to the published version of the manuscript.

Funding: The scientific work was carried out in the framework of the project "Analysis of the tribological properties of polymer composites used in aviation, exposed to surface layer wear," No. GB/5/2018/209/2018/DA, founded by the Ministry of National Defence in the years 2018–2022.

Conflicts of Interest: The authors declare no conflicts of interest.

References

1. Kosicka, E.; Gola, A.; Pawlak, J. Application-based support of machine maintenance. *IFAC-PapersOnLine* **2019**, *52*, 131–135. [CrossRef]

2. Greškovič, F.; Dulebová, L.; Duleba, B.; Krzyżak, A. Criteria of maintenance for assessing the suitability of aluminum alloys for the production of interchangeable parts injection mold. *Eksploat. Niezawodn.* **2013**, *15*, 434–440.

3. Nikoniuk, D.; Bednarska, K.; Sienkiewicz, M.; Krzesiński, G.; Olszyna, M.; Dähne, L.; Woliński, T.R.; Lesiak, P. Polymer fibers covered by soft multilayered films for sensing applications in composite materials. *Sensors* **2019**, *19*, 4052. [CrossRef] [PubMed]

4. Kuczmaszewski, J.; Zagórski, I. Methodological problems of temperature measurement in the cutting area during milling magnesium alloys. *Manag. Prod. Eng. Rev.* **2013**, *4*, 26–33. [CrossRef]

5. Szczepaniak, R.; Rolecki, K.; Krzyzak, A. The influence of the powder additive upon selected mechanical properties of a composite. *IOP Conf. Ser. Mater. Sci. Eng.* **2019**, *634*, 012007. [CrossRef]

6. Sławski, S.; Szymiczek, M.; Domin, J. Influence of the reinforcement on the destruction image of the composites panels after applying impact load. *AIP Conf. Proc.* **2019**, *2077*, 020050. [CrossRef]

7. Krzyżak, A.; Prażmo, J.; Kucharczyk, W. Effect of natural ageing on the physical properties of polypropylene composites. *Adv. Mater. Res.* **2014**, *1001*, 141–148. [CrossRef]

8. Komorek, A.; Przybylek, P.; Brzozowski, D. The influence of UV radiation upon the properties of fibre reinforced polymers. *Solid State Phenom.* **2015**, *223*, 27–34. [CrossRef]

9. Komorek, A.; Przybyłek, P.; Kucharczyk, W. Effect of sea water and natural ageing on residual strength of epoxy laminates, reinforced with glass and carbon woven fabrics. *Adv. Mater. Sci. Eng.* **2016**, *2016*, 3754912. [CrossRef]

10. Li, S.; Guo, Y.S.; Guo, W. Investigation on chaotic motion in hysteretic nonlinear suspension system with multi-frequency excitations. *Mech. Res. Commun.* **2004**, *31*, 229–236. [CrossRef]

11. Wang, W.; Cao, J.; Bowen, C.R.; Litak, G. Multiple solutions of asymmetric potential bistable energy harvesters: Numerical simulation and experimental validation. *Eur. Phys. J. B* **2018**, *92*, 254. [CrossRef]

12. Harris, P.; Bowen, C.R.; Kim, H.A.; Litak, G. Dynamics of a vibrational energy harvester with a bistable beam: Voltage response identification by multiscale entropy and "0–1" test. *Eur. Phys. J. Plus* **2016**, *136*, 109. [CrossRef]

13. Lucintel, Composites Market Report: Trends, Forecast and Competitive Analysis. January 2019. Available online: https://www.lucintel.com/composites-industry.aspx# (accessed on 12 February 2019).

14. Zhang, X.; Chen, Y.; Hu, J. Recent advances in the development of aerospace materials. *Prog. Aerosp. Sci.* **2018**, *97*, 22–34. [CrossRef]

15. Daliri, A.; Zhang, J.; Wang, C.H. Hybrid polymer composites for high strain rate applications. In *Lightweight Composite Structures in Transport*; Woodhead Publishing: Sawston, UK, 2016; pp. 121–163. [CrossRef]
16. Toensmeier, P.A. Advanced composites soar to new heights in Boeing 787. *Plast. Eng.* **2005**, *61*, 8–10.
17. Marsh, G. Boeing's 787: Trials, tribulations, and restoring the dream. *Reinf. Plast.* **2009**, *53*, 16–21. [CrossRef]
18. Merkisz, J.; Bajerlein, M. Materiały kompozytowe stosowane we współczesnych statkach powietrznych. *Logistyka* **2011**, *6*, 2829–2837.
19. Materials Used in the Boeing 787 Dreamliner Skin Structure. Available online: https://www.enterpriseproducts.com/images/area/inline_image_20180524124144.png (accessed on 12 February 2019).
20. Noistering, J.F. Carbon Fibre Composites as Stay Cables for Bridges. *Appl. Compos. Mater.* **2000**, *7*, 139–150. [CrossRef]
21. Kobets, L.P.; Deev, I.S. Carbon Fibres: Structure and Mechanical Properties. *Compos. Sci. Technol.* **1998**, *57*, 1571–1580. [CrossRef]
22. Shama Rao, N.; Simha, T.G.A.; Rao, K.P.; Ravi Kumar, G.V.V. Carbon Composites are becoming Competitive and Cost Effective. *Exter. Doc.* **2015**.
23. Krzyżak, A.; Mucha, M.; Pindych, D.; Racinowski, D. Analysis of abrasive wear of selected aircraft materials in various abrasion conditions. *J. KONES* **2018**, *25*, 2217–2222.
24. Policandriotes, T.; Filip, P. Effects of selected nanoadditives on the friction and wear performance of carbon–carbon aircraft brake composites. *Wear* **2011**, *271*, 2280–2289. [CrossRef]
25. Koutsomichalis, A.; Vaxevanidis, N.; Petropoulos, G.; Xatzaki, E.; Mourlas, A.; Antoniou, S. Tribological Coatings for Aerospace Applications and the Case of WC-Co Plasma Spray Coatings. *Tribol. Ind.* **2009**, *31*, 37–42.
26. Paszeczko, M.; Kindrachuk, M. *Tribologia*; Politechnika Lubelska: Lublin, Poland, 2017.
27. Cely Illera, L.; Cely Nino, J.; Cely Illera, C.V. Effects of corundum in the development of structural, mechanical and tribological properties of raw materials for the manufacture of structural products. *Cerâmica* **2018**, *64*, 352–358. [CrossRef]
28. Rathod, V.T.; Kumar, J.S.; Jain, A. Polymer and Ceramic Nanocomposites for Aerospace Applications. *Appl. Nanosci.* **2017**, *7*, 519–548. [CrossRef]
29. ISO. *PN-EN ISO 4287. Geometrical Product Specifications (GPS)—Surface Texture: Profile Method—Terms, Definitions and Surface Texture Parameters*; ISO: Geneva, Switzerland, 1999.
30. Błaszczyk, J. Analiza numeryczna drgań własnych samolotu z niesymetrycznym płatem nośnym. *Biul. Wojsk. Akad. Technol.* **2011**, *60*, 271–288.
31. Olejnik, A.; Rogólski, R.; Szcześniak, M. Pomiary drgań izolowanych fragmentów konstrukcji płatowcowych z zastosowaniem analizatora modalnego LMS SCADAS Lab. *Prob. Mech.* **2017**, *8*, 127–144.
32. Markiewicz, B.; Ziemiański, L. Analiza dynamiczna kompozytowych konstrukcji cienkościennych. *Czasopismo Inżynierii Lądowej Środowiska i Architektury* **2017**, *64*, 291–302. [CrossRef]
33. Friswell, M.I.; Ali, S.F.; Bilgen, O.; Adhikari, S.; Lees, A.W.; Litak, G. Non-linear piezoelectric vibration energy harvesting from a vertical cantilever beam with tip mass. *J. Intell. Mater. Syst. Struct.* **2012**, *23*, 1505–1521. [CrossRef]
34. Borowiec, M. Energy harvesting of cantilever beam system with linear and nonlinear piezoelectric model. *Eur. Phys. J. Spec. Top.* **2015**, *224*, 2771–2785. [CrossRef]
35. Borowiec, M.; Syta, A.; Litak, G. Energy Harvesting Optimizing with a Magnetostrictive Cantilever Beam System. *Int. J. Struct. Stab. Dyn.* **2019**, *19*, 1941002. [CrossRef]
36. Borowiec, M.; Bocheński, M.; Gawryluk, J.; Augustyniak, M. Analysis of the Macro Fiber Composite Characteristics for Energy Harvesting Efficiency. In *Dynamical Systems: Theoretical and Experimental Analysis. Springer Proceedings in Mathematics & Statistics*, 2nd ed.; Awrejcewicz, J., Ed.; Springer: Łódź, Poland, 2015; Volume 182, pp. 27–37. [CrossRef]
37. Stamos, M.; Nicoleau, C.; Toral, R.; Tudor, J.; Harris, N.R.; Niewiadomski, M.; Beeby, S.P. Screen-printed piezoelectric generator for helicopter health and usage monitoring systems. In Proceedings of the 8th International Workshop on Micro and Nanotechnology for Power Generation and Energy Conversion Applications (PowerMEMS 2008), Sendai, Japan, 9–12 November 2008.

Article

Numerical Simulation on In-plane Deformation Characteristics of Lightweight Aluminum Honeycomb under Direct and Indirect Explosion

Xiangcheng Li, Yuliang Lin * and Fangyun Lu

College of Liberal Arts and Science, National University of Defense Technology, Changsha 410073, China
* Correspondence: ansen_liang@163.com; Tel.: +86-138-0848-4146

Received: 6 June 2019; Accepted: 6 July 2019; Published: 10 July 2019

Abstract: Lightweight aluminum honeycomb is a buffering and energy-absorbed structure against dynamic impact and explosion. Direct and indirect explosions with different equivalent explosive masses are applied to investigate the in-plane deformation characteristics and energy-absorbing distribution of aluminum honeycombs. Two finite element models of honeycombs, i.e., rigid plate-honeycomb-rigid plate (RP-H-RP) and honeycomb-rigid plate (H-RP) are created. The models indicate that there are three deformation modes in the $X1$ direction for the RP-H-RP, which are the overall response mode at low equivalent explosive masses, transitional response mode at medium equivalent explosive masses, and local response mode at large equivalent explosive masses, respectively. Meanwhile, the honeycombs exhibit two deformation modes in the $X2$ direction, i.e., the expansion mode at low equivalent explosive masses and local inner concave mode at large equivalent explosive masses, respectively. Interestingly, a counter-intuitive phenomenon is observed on the loaded boundary of the H-RP. Besides, the energy distribution and buffering capacity of different parts on the honeycomb models are discussed. In a unit cell, most of the energy is absorbed by the edges with an edge thickness of 0.04 mm while little energy is absorbed by the other bilateral edges. For the buffering capacity, the honeycomb in the $X1$ direction behaves better than that in the $X2$ direction.

Keywords: aluminum honeycomb; deformation modes; shock wave; counter-intuitive behavior; energy distribution

---

1. Introduction

Aluminum honeycomb is a promising lightweight structure—a typical bionic structure [1–3]. Due to the homogeneous deformation mechanism [4], high strength to weight ratio [5], and large specific energy absorption [6,7], aluminum honeycombs have been frequently applied in many important engineering fields in recent years, such as aircraft, aerospace, transportation, building, and sporting equipment, which are involved in the dynamic impact or indirect explosive load. Hexagonal aluminum honeycomb, as one of the most widely used honeycombs, is orthotropic, i.e., one out-of-plane direction and two in-plane directions, respectively. Dynamic impact and indirect explosive load have been studied widely, while the compressive response of honeycomb structures subjected to direct explosive load is rarely researched. Hence, there is a pressing need to better understand the deformation behaviors of honeycombs subjected to direct explosion.

Numerical simulation is an effective way to study the deformation properties of honeycombs because it is difficult to observe the deformation modes of honeycombs experimentally. Therefore, there are some researchers who have conducted their work by numerical simulations [8–13]. For example, Ashab [8] investigated the mechanical behavior of aluminum hexagonal honeycombs subjected to dynamic indentation and compression loads only using ANSYS/LS-DYNA. Qiao [9] explored the in-plane uniaxial collapse response of a hierarchical honeycomb only by finite element simulations. Besides, there are some analogous deformation modes in references [8,9] to that studied in this work. Ruan et al. [10] observed three modes in one of the in-plane directions, X mode at low velocity, V mode at moderate velocity, and I mode at high velocity. Besides, V mode was also observed at low velocity and I mode at high velocity in the other direction. Hu et al. [11] obtained the same conclusion. Moreover, the cell-wall angle, the crushing velocity, and the honeycomb's relative density could determine the in-plane dynamic properties of hexagonal honeycomb at different cell-wall angles. Wang [14,15] studied the combined effects with a dynamic inclined load. Oblique loads [16] in both in-plane and out-of-plane directions produced deformation modes from X, V, I modes to Half-X mode, Half-X following I mode, V mode, and I mode. Sun et al. [17] concluded that the impacting velocities were related to the stress when all configuration parameters were kept constant. Numerical models could adequately approximate the study of the in-plane deformation properties of honeycombs.

In engineering applications, such as explosion-proof tank, weapon protection engineering, it is possible that honeycomb is subjected to explosive loading. Li et al. [18] investigated the blast resistance of a square sandwich including hexagonal aluminum honeycomb cores, with different cell side lengths of the core. They concluded that failure modes of face sheets were related to the scaled distance of the explosive and the deformation modes of honeycomb core included full densified, progressive buckling, shear deformation, and fragments. Zhu et al. [19] studied the dynamic properties of a sandwich structure with a honeycomb core by air blast experiments. They thought deformation modes were mainly related to face sheet thickness and honeycomb core density. Li et al. [20] investigated the pressure distribution and the structural response of a sandwich with graded honeycomb cores. Three deformation modes were classified: large inelastic bending deformation without a significantly localized core compression, large inelastic bending deformation with significantly localized core compression, and large inelastic bending deformation with localized penetration. Jin et al. [21] proposed an innovative sandwich structure with an auxetic re-entrant cell honeycomb core and investigated its blast resistance by simulation. They showed that both the graded honeycomb cores and the cross-arranged honeycomb cores were better than the ungraded honeycomb cores and regular-arranged cores in the resistance ability. More work with different configurations or various combined structures were also studied by explosive loading [22–25]. A common feature of these studies is the indirect interaction between explosive loading and honeycomb structures.

According to the literature above, deformation properties of honeycomb under dynamic impact or indirect explosive loading have mostly been analyzed, but it is rare to study in-plane deformation characteristics of aluminum honeycomb structure core impacted by direct blast load. There are two typical applications that led to this work. One is as a passive explosive wave energy sensor, which has the characteristics of a low-cost, no electricity measurement and shock wave energy measurement under bad environment, while the other is as a shock absorber in the spaceflight separator, which can effectively reduce the shock to protect the electronic equipment on the test platform. However, deformation characteristics and energy distribution of honeycomb structures subjected to direct explosive loads are unclear.

This work focuses on the direct interaction between the honeycomb structure and spherical blast wave. The layout of the rest of the paper is arranged as follows. In Section 2, a hexagonal honeycomb structure is proposed with a cell length of 2 mm and 0.04 mm thickness with four edges and 0.08 mm with two edges. Besides, double finite element models in Section 3, including honeycomb-rigid plate (H-RP) and rigid plate-honeycomb-rigid plate (RP-H-RP) in the two in-plane directions with a 50 cm fixed explosive height, are created. Then, in Section 4, deformation modes of the models are described, and more details are given about the deformation properties and mechanism. In particular, the energy distribution of different parts and contact stress comparisons during the whole process are discussed and analyzed in Section 5, followed by a conclusion in Section 6. It is noted that the study is only a preliminary numerical example, which provides a new idea to explore the in-plane deformation regularity of honeycombs under direct explosion by numerical simulation. This work enables the improvement of the energy absorption and buffering capacity of honeycomb used in an explosive environment.

2. Honeycomb Core and Loading Setup

2.1. Honeycomb Core

An orthotropic hexagonal honeycomb structure (Figure 1a) is constructed by some regular hexagonal cells (Figure 1b) using the stretch forming method. Briefly, the direction along the stretch forming of the honeycomb structure is the $X1$ direction. The vertical direction of $X1$ in the in-plane is referred to as the $X2$ direction, and $X3$ is along the out-of-plane direction. The in-plane dimension of the aluminum honeycomb sample ($X1$-$X2$) is 47.6 mm × 46.0 mm with a height of 4 mm, based on an experimental honeycomb sample, and the number of repeated unit cells in the $X1$ and $X2$ directions is 14 and 15, respectively. The sketch of a unit cell is shown in Figure 1b. The edge length of the unit is 2 mm and the angle between each adjacent edge is 120°. Due to the processing method, the cell wall thickness in the $X1$ direction, which is 0.08 mm, is twice that of the other directions.

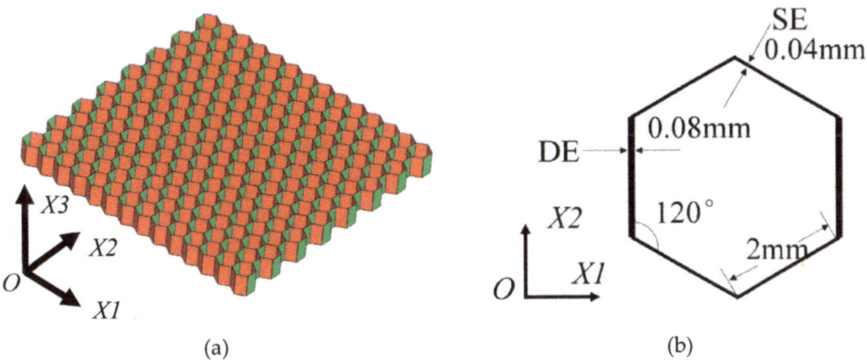

Figure 1. (**a**) A regular hexagonal honeycomb and (**b**) the sketch of a single unit cell.

2.2. Loading Setup

The present work creates two types of models under air blast, which are the honeycomb-rigid panel (H-RP) and the rigid panel-honeycomb-rigid (RP-H-RP), respectively, as sketched in Figure 2. The H-RP is used to analyze the mechanical and energy-absorbing properties of honeycomb structures subjected to the direct shock wave. For comparison, RP-H-RP is subjected to the indirect shock wave, where deformation of the rigid plate is ignored. In both models, the bottom rigid plate is fixed. Besides, a spherical explosive wave is realized by an air explosion. The honeycomb is loaded in both the $X1$ direction and the $X2$ direction, respectively, for each model. Besides, the height of the burst in this work is fixed as 50 cm, which describes the distance between the explosive and the ceiling of the model.

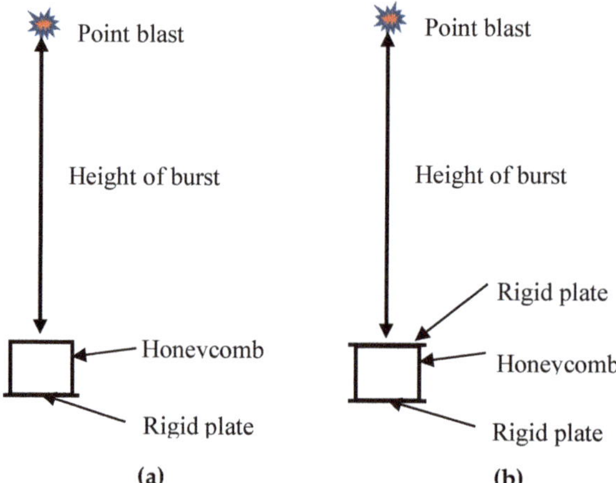

Figure 2. Two types of sketch models under air blast, (**a**) honeycomb-rigid plate and (**b**) rigid plate-honeycomb-rigid plate.

3. Numerical Simulation

3.1. Finite Element Models

Numerical simulation is conducted by ANSYS/LS-DYNA (Version 971), a popular software to simulate the dynamic impact or detonation. Based on two types of models in Section 2.2, related finite element in-plane models are created, as shown in Figure 3. Three parts are included in the H-RP (Figure 3a), in which Part 1 is the cell's double edges with 0.08 mm thickness (marked as DE), Part 2 is the cell's single edges with 0.04 mm thickness (marked as SE), and Part 3 is the bottom rigid plate which is constrained in all directions. The ceiling of the honeycomb is a loaded surface. In addition, there are four parts in the RP-H-RP (Figure 3b). In this model, the former parts are the same as those in the H-PR, while Part 4 is the ceiling rigid plate with 0.08 mm thickness. The plate, as the loaded surface in the RP-H-RP, is only translated in the load direction. Besides, the size of all the grids is set as 0.125 mm to enable steady calculation. Both models are created from the loading directions of *X1* and *X2*, respectively, to research the effect of different loading surface shapes on the deformation characteristics.

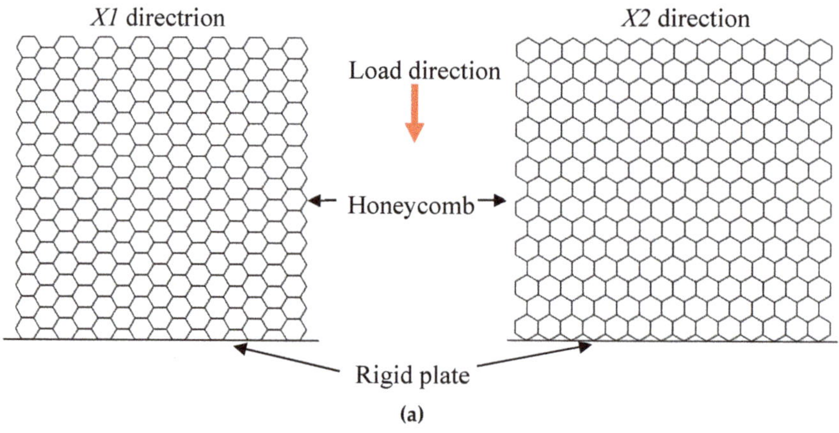

Figure 3. *Cont.*

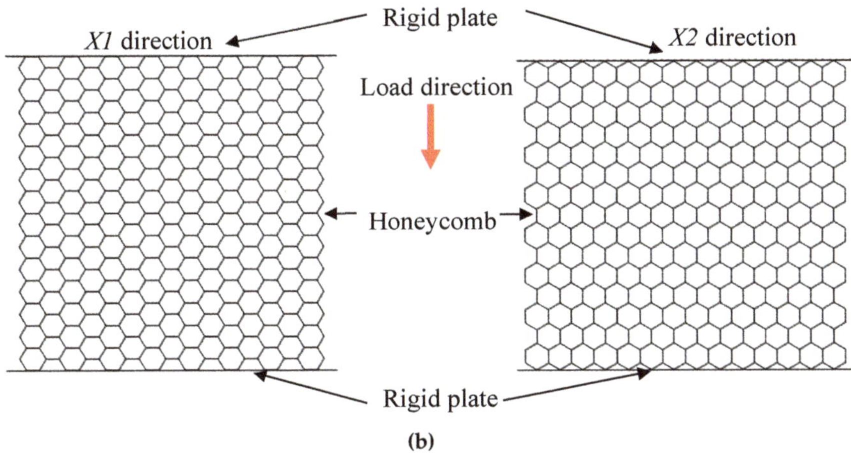

Figure 3. Equivalent finite element models, (**a**) honeycomb-rigid plate (H-RP) in the *X1* and *X2* direction; (**b**) rigid plate-honeycomb-rigid plate (RP-H-RP) in the *X1* and *X2* direction.

3.2. Material Properties and Control Setup

Here, the keyword LOAD_BLAST is used to define an air blast function for the application of pressure loads to explosives in conventional weapons, in which trinitrotoluene (TNT) is used as the explosive to simulate the detonation in the keyword file. Related parameters of TNT are taken from reference [26]. Besides, MAT_PLASTIC_KINEMATIC is a material model in the ANSYS/LS-DYNA (Version 971) software to describe the honeycomb structure of the parent material aluminum alloy 3003H18. MAT_RIGID is a rigid material to describe the rigid plate in the bottom/ceiling, where steel material parameters are used to replace parameters of the rigid body. The specific material parameters are shown in Table 1.

Table 1. Material parameters.

Parts	Density/(g·cm^{-3})	Elastic Modulus/(GPa)	Poisson Ratio	Ultimate Strength/(MPa)
Honeycomb	2.7	69	0.33	289.6
Steel plate	7.83	210	0.3	-

In these models, all finite elements are quadrilateral 2D grids, where the thickness of the SE is 0.04 mm, the thickness of the DE is 0.08 mm, and the type of these elements is the Belytschko-Tsay Shell 163 with five through-thickness integration points to provide simulation accuracy. All nodes limit the freedom of translation in the X3 direction. It is noted that additional components should be installed to prevent the out-of-plane deformation in actual applications. As for contacts, the automatic double-sided face is employed between parts of the models. Self-contact is used for the bucking contact of the same part. Hourglass energy and energy dissipation are computed and included in the energy balance to make the simulation accuracy. The scale factor for the computed time step is 0.9 and the output time step of energy and force is 5 μs. The shell elements between the different parts of the honeycomb structure are connected by common nodes. The main variable in the work is the equivalent mass of the TNT controlled by LOAD_BLAST. Three typical cases with equivalent explosive masses of 10 g, 20 g, and 40 g are employed in the work.

In addition, the effect of different grid densities is analyzed. As a result, the elements with 0.125 mm are created in the following discussion. Besides, the mesh size also fits well with the numerical simulation results in reference [8]. However, if the mesh is further refined, it will have less and less influence on the deformation result, but it will take more time to calculate the model. Therefore, it could not only guarantee the efficiency of calculation but also make simulation results consistent.

In order to analyze the effect of grid density, three sizes of grids are selected, 0.5 mm, 0.25 mm, and 0.125 mm, respectively. Taking the 10 g TNT explosive as an example, the deformation characteristics of the different mesh sizes at four typical moments are obtained, as shown in Figure 4. At the moment of 1700 µs, the difference in both the overall deformation and the local deformation is obvious. Concretely, the overall deformation increased from 0.26 to 0.32 with the size of the elements from 0.5 mm to 0.125 mm. Besides, the concentrated deformation zone in the local position was more prominent in the model with a mesh size of 0.125 mm. It should be noted that the compressive deformation strain is the ratio of the displacement of the upper panel to the length of the honeycomb compression direction, marked as a symbol, ε.

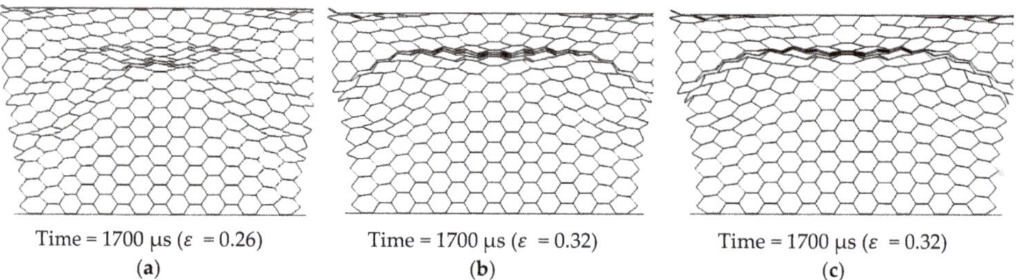

Time = 1700 µs (ε = 0.26)   Time = 1700 µs (ε = 0.32)   Time = 1700 µs (ε = 0.32)

(**a**)   (**b**)   (**c**)

Figure 4. The deformation characteristics of three sizes of grids, (**a**) mesh size = 0.5 mm, (**b**) mesh size = 0.25 mm, and (**c**) mesh size = 0.125 mm.

4. Deformation Modes

4.1. The X1-RP-H-RP

With the increase of equivalent explosive masses (10 g, 20 g, and 40 g), there are three deformation modes in the *X1* direction, the overall response mode in the small equivalent explosive masses case, the transitional response mode in the medium equivalent explosive masses case, and the local response mode in the large equivalent explosive masses case, respectively, as shown in Figure 5. In this figure, the upper figures describe the deformation of the whole honeycomb structure, and the bottom figures record the deformation of the SE in order to observe the deformation characteristics clearly. In the bottom figures, the red lines are used to describe the shape of the local deformation belt. Besides, the subsequent description of the deformation characteristics is also presented in this way. The specific meanings of the three deformation modes are as follows, in which the deformation diagrams on display are selected based on whether the local deformation is clearly visible.

The overall response mode means all parts in the model take part in the process of deformation. In the deformation mode, the honeycomb sample is compressed by the upper panel, but the compact state of the honeycomb is not achieved. In detail, at 550 µs, the shock wave begins to propagate into the honeycomb structure. At the moment of 1000 µs, the compressive strain of the honeycomb structure is up to 0.14. A "C" shape folding interface appears near the impact section. Then, an "X" folding interface (or double "C" folding interface back to back) occurs near the upper plate at 1360 µs and a "C" interface appears near the bottom plate. At the moment, the honeycomb structure is compressed to the maximum deformation which is 0.18. After that, the upper panel starts to rebound with a constant speed of 5 m/s at 1800 µs because of the elastic potential energy stored in the honeycomb structure and the compression strain decreases to 0.15. Besides, an interesting phenomenon occurs that some cells near the central axis keep the regular hexagonal configuration in the whole process, and it seems that the cells never received any load.

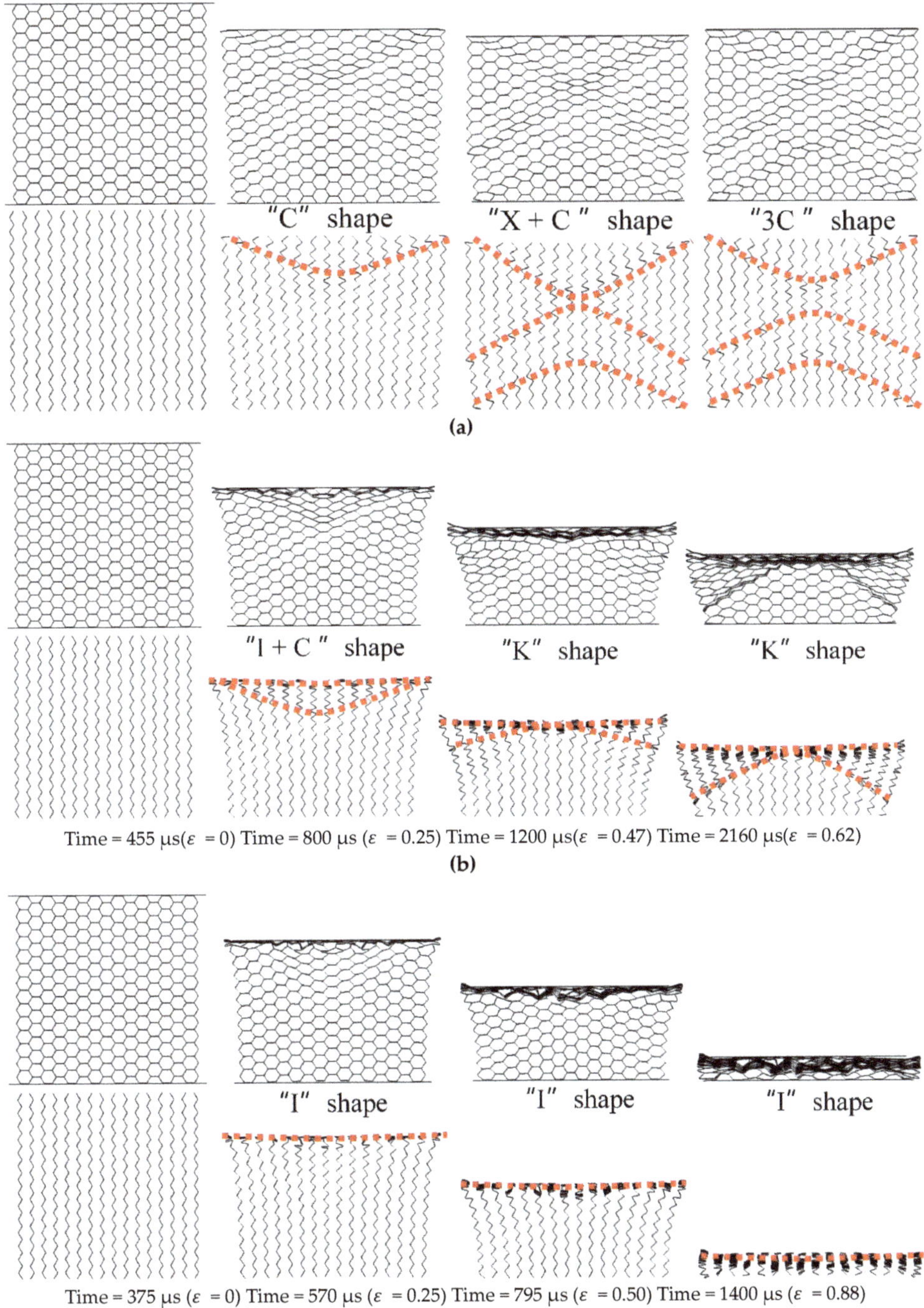

Figure 5. Deformation characteristics in the *X1*-RP-H-RP. (**a**) Overall response mode, (**b**) Transitional response mode, and (**c**) Local response mode.

In the transitional response mode, the honeycomb sample can be exactly compact. The blast wave reaches the honeycomb structure at 455 µs. At 800 µs, the compression strain of the whole honeycomb structure is 0.25, which has exceeded the maximum deformation in the overall response mode. At this moment, the clear deformation interface in the honeycomb structure is a mixed buckling region of "I" and "C" shape, while the area near the upper panel shows the local compaction state and presents the "I" shape. Then, the shape of "C" becomes plain gradually and becomes the "K" mode at the moment of 1200 µs. Once the folding degree of the cells close to the upper panel at the end of the impact section increased to 0.62 at 1800 µs, a larger span of "K" shape occurs. At last, the upper panel bounces back at 6.7 m/s. In addition, the "regular cells" mentioned in the overall response mode disappear in the transitional response mode.

The local response mode is similar to the high-speed loading of deformation proposed by Hu [8]. At 375 µs, the honeycomb structure begins to be loaded by a shock wave. On the upper panel, a local "I" shape interface appears in the honeycomb structure at 570 µs, while honeycomb cells near the bottom rigid plate are not deformed and then at 795 µs, the "I" shape interface is clearer. The maximum compression strain is 0.88 at 1400 µs. The sample reaches the compact state in the mode. The upper panel rebounds back eventually at 7.4 m/s.

4.2. The X2-RP-H-RP

The *X2* direction is perpendicular to the *X1* direction in the in-plane region. In order to compare the deformation difference with the *X1*-RP-H-RP, the explosion load is also applied in the *X2*-RP-H-RP. Two types of deformation modes occur, expansion mode in the small equivalent explosive masses case and local inner concave mode in the large equivalent explosive masses case, respectively, as illustrated in Figure 6.

Taking the 10 g TNT as an example of the small equivalent mass cases, the honeycomb sample is expanded across the whole compression process in the expansion mode, but it fails to reach the compact state. In this model, the time for the explosive wave to reach the loading surface is the same as that in *X1*-RP-H-RP, both of which are 550 µs, because they have the same explosive yield and blast height. At 890 µs ($\varepsilon = 0.14$), there is an obvious lateral expansion by the way of stretching transversely of the SE. The mechanism makes the cell shape change from the convex hexagon to the square or even concave hexagon. At the moment of 1260 µs, the deformation reaches the maximum, i.e., $\varepsilon = 0.18$. The impacted region forms "V" shape area, a series of "I" interfaces. At last, the overall elasticity properties make the ceiling plate rebound with 5.7 m/s. The case of 20 g is similar to the case of 10 g. Local inner concave mode occurs at large equivalent explosive masses (for example, 40 g TNT in Figure 6b). The explosion shock wave begins to propagate in the honeycomb sample at 375 µs. At the moment of 560 µs, the strain of the honeycomb sample is compressed to be 0.25. An "I" interface appears near the impact region. Then, a "bow" shape forms at 790 µs ($\varepsilon = 0.50$) and an "M" interface becomes contour of honeycomb structure at 1570 µs ($\varepsilon = 0.88$).

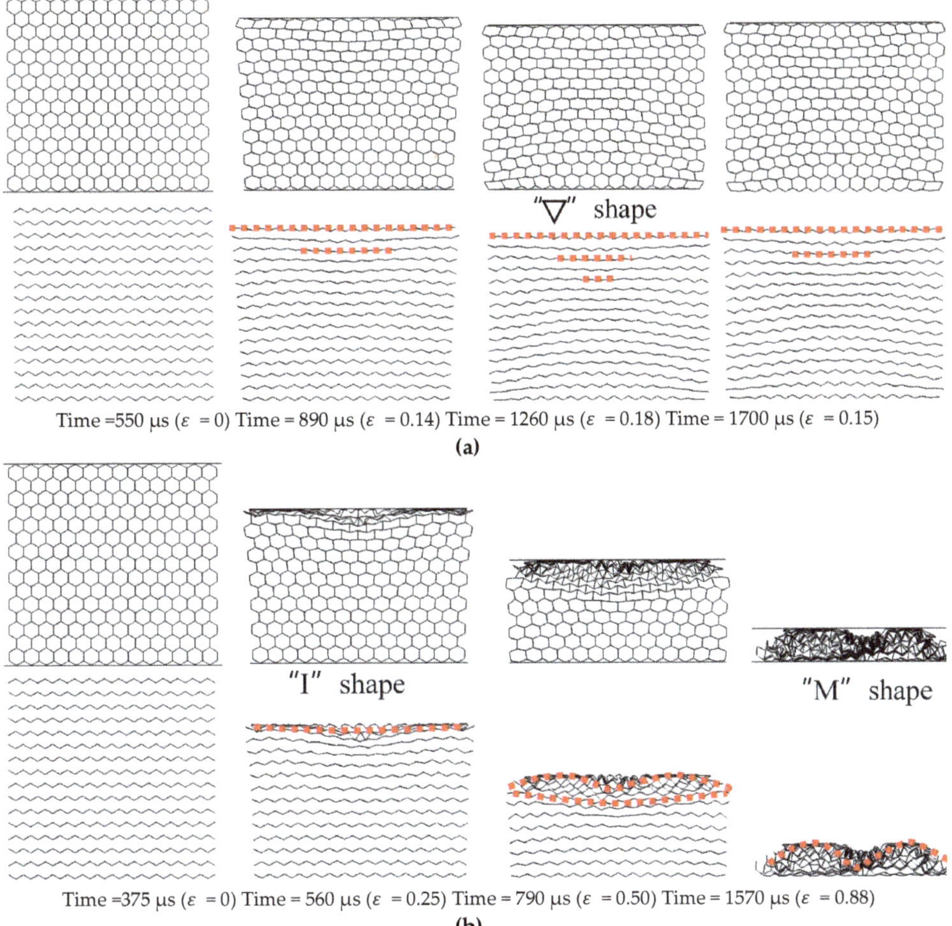

Time =550 μs (ε = 0) Time = 890 μs (ε = 0.14) Time = 1260 μs (ε = 0.18) Time = 1700 μs (ε = 0.15)

(a)

Time =375 μs (ε = 0) Time = 560 μs (ε = 0.25) Time = 790 μs (ε = 0.50) Time = 1570 μs (ε = 0.88)

(b)

Figure 6. Deformation modes in the X2-RP-H-RP, (**a**) Expansion mode and (**b**) Local inner concave mode.

4.3. H-RP

In-plane deformation characteristics of the honeycomb sandwich structures (rigid plate-honeycomb-rigid plate) are obtained with the above two models. In order to obtain the effect of the direct explosive load on the honeycomb and analyze the influence of the shape of the explosive load surface on the deformation characteristics of honeycomb, H-RP is studied in this work. Taking 10 g *X1*-H-RP and 10 g *X2*-H-RP as examples, as illustrated in Figure 7, the same moments are used to compare with the previous two models. Similarity and unique deformation characteristics in the H-RP are shown as follows.

The similarity is shown in the internal deformation characteristics of the honeycomb structure. In *X1*-H-RP in Figure 7a, a "C" shape folding interface occurs near the impact section as well at 1000 μs. The "C" interface and the upper interface form a lip-shaped interface. Then, an "X" interface also occurs near the impacted section and a "C" interface also occurs at the opposite from the blast at 1360 μs. At 1800 μs, the honeycomb structure is also rebounded because of the elastic properties. In the X2-H-RP in Figure 7b, the honeycomb sample is also expanded, but the honeycomb structure is not compacted. This expansion changes the shape of the cellular structure of the honeycomb from a convex hexagon to a quadrilateral as well at 890 µs. At 1260 µs, the impacted region also forms a "V" shape area, consisting of a series of horizontal lines. Besides, the deformation similarity is also verified in the cases of medium (20 g) and large explosives (40 g).

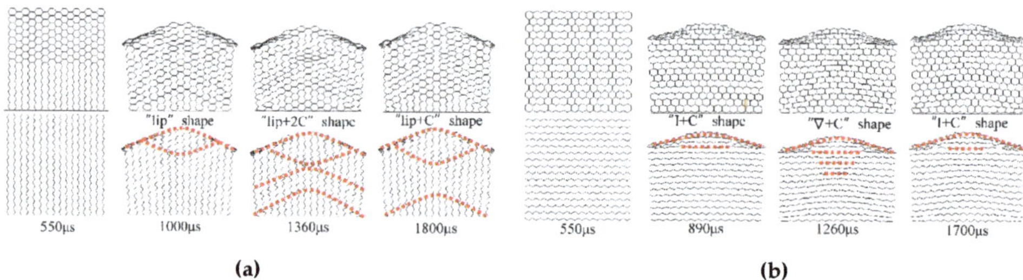

Figure 7. Deformation modes in the H-RP. (**a**) 10 g *X1*-H-RP and (**b**) 10 g *X2*-H-RP.

Different from the deformation characteristics of the RP-H-RP, a counter-intuitive deformation phenomenon appears near the explosive load surface in the H-RP. In this phenomenon, the shape of the loading wave is a concave face while the deformation surface of the honeycomb structure is a convex face in in-plane directions, as shown in Figure 7. It's worth mentioning that the counter-intuitive phenomenon always exists, with the increase of equivalent explosive masses.

4.4. Discussion on Deformation Mechanism

Due to the low density of the honeycomb structure, the variation of the honeycomb structure deformation characteristics is essentially caused by the influence of the upper and lower panels on the propagation of the blast wave in these models.

The propagation process of the shock wave in the honeycomb structure is shown in Figure 8a for RP-H-PR and Figure 8b for H-RP. In Figure 8, the shock wave just reaches the upper interface of the honeycomb at the moment T1. The wave is a concave interface. Then, the shock wave is reflected at the moment T2, whose shape is a convex interface because of the fixed bottom panel. T3 represents the moment when the reflected shock wave reaches the upper interface and reflects again. For the RP-H-RP with an upper panel, the concave shock wave is reflected again because of the reflection of the upper rigid panel. However, for the H-RP without the upper panel, the reflected wave is a concave sparse wave due to the upper interface, which is a free surface. Therefore, the "C"-shaped interface in the honeycomb is actually the propagation of the interface of the shock wave in the honeycomb structure. The shape of the ceiling surface of the honeycomb structure is the same as that of the convex shock wave, rather than the concave one. This is the reason for the counter-intuitive deformation phenomenon.

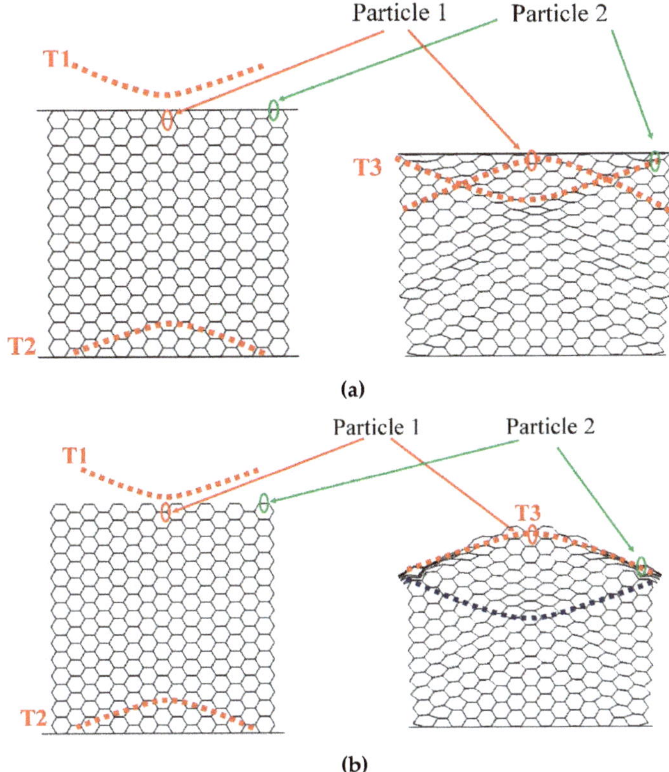

Figure 8. Wave propagation at different moments. (**a**) RP-H-RP and (**b**) H-RP.

Two typical locations are selected from the honeycomb to analyze the honeycomb deformation distribution. Particle 1 is a particle in the middle of the upper interface of the honeycomb, indicated by a red circle. Particle 2 is a particle in the right area of the interface on the honeycomb, represented by a green circle, as shown in Figure 8. At the moment T3, the motion state of the two particles is shown in Figure 9. The particle velocity is the same as the wave direction in the shock wave, while the particle velocity is opposite in the sparse wave. In Figure 9, $v1$ is the velocity after the action of the upward convex shock wave, $v2$ is the velocity after the action of the reflected wave at the upper interface, and $v0$ is the velocity of the upper panel. The directions of these speeds is shown in Figure 9. For RP-H-RP in Figure 9a, the direction of $v1$ and $v2$ is opposite, so both particle 1 and particle 2 have moved downward at the velocity of $v0$, which is consistent with the deformation state in Figure 8. However, For H-RP in Figure 9b, the $v1$ and $v2$ are moving in the same direction, so Particle 1 is moving up at $(v1 + v2)$, and Particle 2 is moving up at $v4$, which is the vertical component of $(v1 + v2)$. Therefore, Particle 1 is above particle 2 because $v4$ is less than $(v1 + v2)$. In addition, $v3$, the horizontal component of $(v1 + v2)$, causes Particle 2 to shift in the horizontal direction, so the upper interface will expand to both sides.

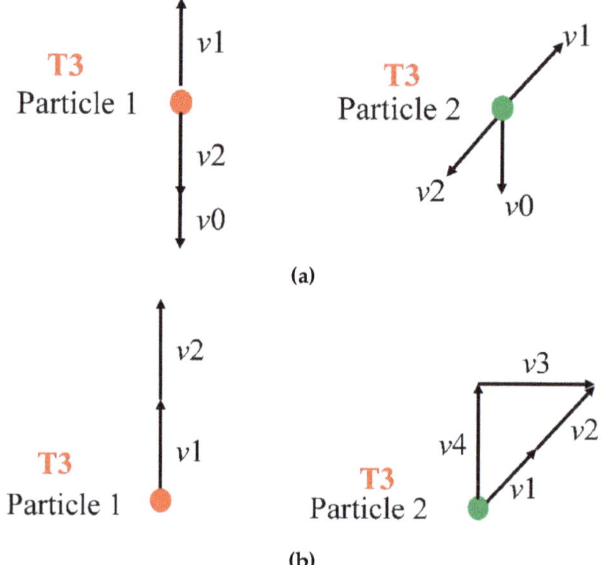

Figure 9. Schematic diagram of particles' motion at the moment T3, (**a**) the particles near the top with a rigid plate and (**b**) the particles near the top without a rigid plate.

5. Analysis of Energy and Stress Properties

5.1. Energy Distribution

For RP-H-RP, the total energy is divided into five parts at different moments, internal energy of SE, internal energy of DE, kinetic energy of rigid plate (marked as RP), kinetic energy of SE, and kinetic energy of DE, respectively. In order to compare the energy distribution of each part during the compression process, the energy ratio ζ is defined as follows,

$$\zeta = \frac{E_{part}}{E_{total}} \times 100\% \tag{1}$$

where E_{part} represents one or more of the five components mentioned above, E_{total} is the total energy of the model.

Once the shock wave reaches the rigid plate in *X1*-RP-H-RP (Figure 10) or *X2*-RP-H-RP (Figure 11), most of the energy is converted immediately to the kinetic energy of the rigid plate (76.5% in overall response mode, 77.5% in transitional response mode, 59% in local response mode, 87.5% in expansion mode, and 62.0% in the local inner concave mode), and a little becomes the internal energy of honeycomb structure. Then, a partition of the kinetic energy of RP and the elastic energy of the honeycomb are transferred into the kinetic energy of the honeycomb. Before bouncing back of the rigid plate, the kinetic energy of all the parts decreases. However, the internal energy of the SE increases greatly and the internal energy of the DE increases slightly. With the increase of the explosive yield, the effect becomes more obvious. In addition, the kinetic energy of the SE and DE remains basically the same in the process.

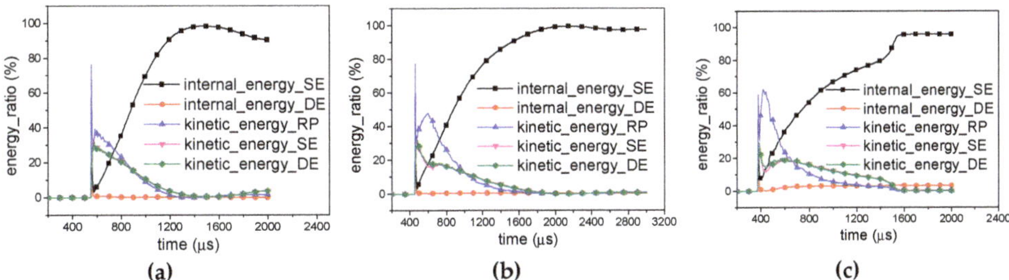

Figure 10. Energy distribution in *X1*-RP-H-RP. (**a**) Overall response mode; (**b**) Transitional response mode; (**c**) Local response mode.

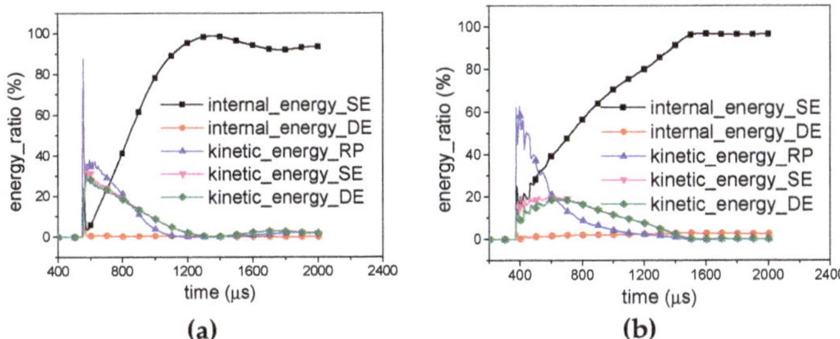

Figure 11. Energy distribution in *X2*-RP-H-RP. (**a**) Expansion mode and (**b**) Local inner concave mode.

According to Figures 10 and 11, the deformation mode of the honeycomb can be judged from the kinetic energy and internal energy of DE. If the kinetic energy of DE is less than the internal energy of DE, the deformation mode of the honeycomb presents as an overall response or expansion mode. Otherwise, the deformation mode of the honeycomb is the local response mode.

For the H-RP, the total energy is divided into four parts, the kinetic energy and internal energy of SE and the kinetic energy and internal energy of DE, respectively, as illustrated in Figure 12. Once the shock wave reaches the ceiling of the honeycomb, 79.6% of the total energy is distributed to the kinetic energy of the DE for *X1*-H-RP and 76.3% of the total energy is distributed to kinetic energy of SE for *X2*-H-RP. In the subsequent process, the kinetic energies of SE and DE are basically equal. Finally, the total energy is converted to internal energy of SE.

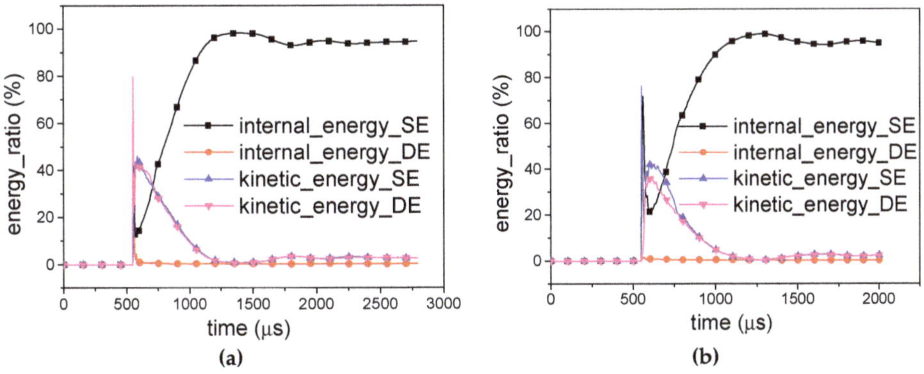

Figure 12. Energy distribution of H-RP. (**a**) *X1*-H-RP-10g and (**b**) *X2*-H-RP-10g.

It is noted that SE produces plastic deformation and converts total energy into deformation energy, while DE merely transfers energy in the form of kinetic energy. Therefore, in order to improve the energy absorption effect of a honeycomb structure, DE can be involved in the plastic deformation process through the optimization design. This provides a good way to enhance energy absorption.

Although the RP-H-RP and H-RP differ only by the ceiling rigid plate, the plate has an obvious effect on the energy absorption of the honeycomb structure. The internal energy of the SE in the -RP and RP-H-RP is compared in Figure 13. Once the shock wave reaches, 41.4% is absorbed for H-RP while 19.9% for RP-H-RP in the *X1* direction and 71.6% is absorbed for H-RP while 23.8% for RP-H-RP in the *X2* direction, respectively. Besides, the H-RP has more proportion of the energy, including elastic energy and plastic deformation energy of SE. However, the plate does not affect the energy absorption rate of the honeycomb structure.

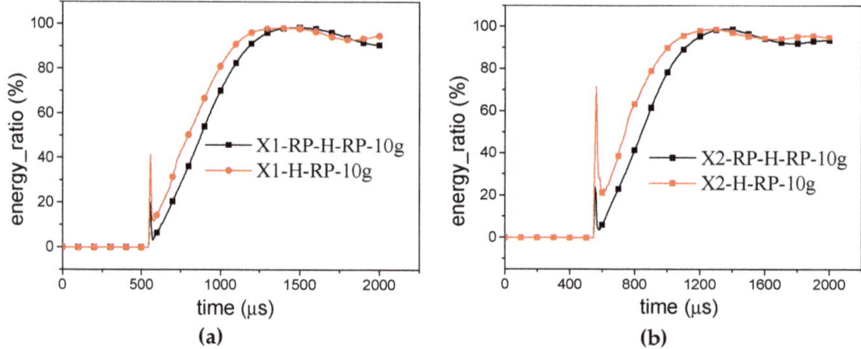

Figure 13. Comparison of energy distribution between RP-H-RP-10g and H-RP-10g (**a**) in the X1 direction and (**b**) in the X2 direction.

5.2. Stress Analysis

For a buffering material, the peak stress is an important parameter to access the buffering ability of honeycombs. In order to describe the stress attenuation effect of the honeycomb structure, the buffering coefficient η is defined as follows,

$$\eta = \frac{\sigma_{cp} - \sigma_{bp}}{\sigma_{cp}} \times 100\% \tag{2}$$

where σ_{cp} is the initial contact peak stress between the honeycomb and the ceiling rigid plate and σ_{bp} is the initial contact peak force between honeycomb and the bottom rigid plate. The large buffering coefficient is beneficial to the structure buffering effect.

For 10 g RP-H-RP, as shown in Figures 14a and 15a, the ceiling rigid plate gets the initial peak stress and about 15 µs later, the peak stress occurs in the bottom rigid plate, but its amplitude is decreased. Then, the stress in the ceiling rigid plate decreases rapidly and the stress in the bottom rigid plate increases. They are equal at about 800 µs. In the next stage, the stress response of the bottom plate is a platform. It is larger than the stress in the ceiling plate. It has a similar characteristic for 20 g RP-H-RP, as shown in Figure 14b. However, the bottom plate has a longer platform stress response. The stress in the bottom rigid plate increases rapidly at the final state for 40 g RP-H-RP in Figures 14c and 15b. At this moment, the honeycomb specimen has been crushed. Besides, in different deformation modes, η is calculated as shown in Table 2. It can be seen that η is not correlated with the equivalent explosive masses, which is 73.06% in the *X1* direction and 65.72% in the *X2* direction, though a small rise with the masses. Therefore, the buffering effect in the *X1* direction is better than that in the *X2* direction.

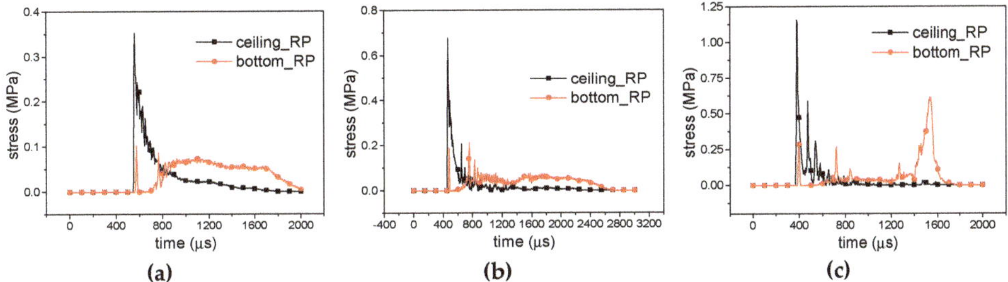

Figure 14. Stress-time curves of *X1*-RP-H-RP; (**a**) Overall response mode; (**b**) Transitional response mode and (**c**) Local response mode.

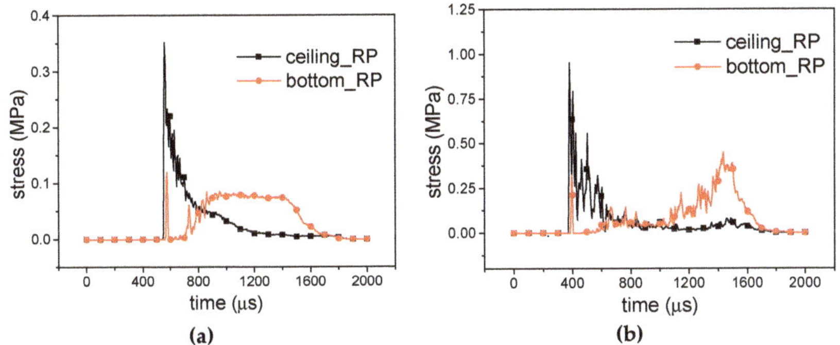

Figure 15. Stress-time curves of *X2*-RP-H-RP. (**a**) Expansion mode and (**b**) Local inner concave mode.

Table 2. The buffering effect under different equivalent explosive masses.

Cases	Deformation Modes	σ_{cp}/MPa	σ_{bp}/MPa	η/%
10 g *X1*-RP-H-RP	Overall	0.35	0.10	70.86
20 g *X1*-RP-H-RP	Transitional	0.68	0.19	72.24
40 g *X1*-RP-H-RP	Local	1.16	0.28	76.07
Average		–	–	73.06
10 g *X2*-RP-H-RP	Expansion	0.35	0.12	65.79
40 g *X2*-RP-H-RP	Local inner concave	0.95	0.32	66.08
Average		–	–	65.72

6. Conclusions

Lightweight aluminum honeycomb is a nice bionic material to absorb energy under explosion. Two finite element models of honeycombs whose parent material is lightweight aluminum alloy 3003H18, including rigid plate-honeycomb-rigid plate (RP-H-RP) and honeycomb-rigid plate (H-RP) are created to investigate the in-plane deformation properties of honeycombs subjected to air blast. Conclusions can be drawn as follows,

(1) At the different equivalent explosive masses, there are three deformation modes in the *X1* direction, i.e., the overall response mode in the small equivalent explosive masses case in which there is a "C" or multi "C" folding shape, transitional response mode in the middle equivalent explosive masses case, in which there is an "I" and "C" folding shape, and the local response mode in the large equivalent explosive masses case in which there is only an "I" folding interface. There are mainly two deformation modes in the *X2* direction, expansion mode and local inner concave mode. In the expansion mode, the shape of the convex cells is changed into concave and some "I" shape folding interfaces form "V"

shape area. In the local inner concave mode, "I" and "bow" interfaces appear successively, and finally the "M" interface is formed.

(2) The mechanism of deformation characteristics under different explosive equivalent conditions is given by wave propagation, and the counterintuitive phenomenon is explained due to the reflection of circle shock wave by the bottom plate but there is no secondary reflection on the ceiling plate in the H-RP.

(3) The honeycomb structure can absorb most of the energy carried by a shock wave. In the energy absorption process, it is mainly the deformation energy of the cell single edges with 0.04 mm thickness (marked as SE) that plays a major role. However, there is a low proportion of the internal energy of the cell double edges with 0.08 mm thickness (marked as DE). It provides a new way to improve the energy-absorbed capacity of honeycombs.

(4) The honeycomb structure can effectively attenuate the stress peak of a shock wave. Besides, the buffering effect in the $X1$ direction is better than that in the $X2$ direction.

The model presented is not yet validated experimentally; therefore, it represents a preliminary numerical approach of the deformation behavior of honeycomb structures under direct and indirect explosion.

Author Contributions: Conceptualization, X.L.; methodology, Y.L.; software, X.L.; validation, Y.L. and F.L.; formal analysis, Y.L.; investigation, X.L.; resources, Y.L.; data curation, X.L. and Y.L.; writing-original draft preparation, X.L.; writing-review and editing, X.L., Y.L. and F.L.; visualization, X.L. and Y.L.; supervision, Y.L. and F.L.; project administration, Y.L.; funding acquisition, Y.L. and F.L.

Funding: This research was funded by the National Natural Science Foundation of China (grant numbers 11672329 and 11872376).

Acknowledgments: All authors thank Zhang Yuwu for helping to check the manuscript before submission.

Conflicts of Interest: The authors declare no conflict of interest.

References

1. Wierzbicki, T.; Abramowicz, W. On the Crushing Mechanics of Thin-Walled Structures. *J. Appl. Mech.* **1983**, *50*, 727–734. [CrossRef]

2. Baker, W.E.; Togami, T.C.; Weydert, J.C. Static and dynamic properties of high-density metal honeycombs. *Int. J. Impact Eng.* **1998**, *21*, 149–163. [CrossRef]

3. Qin, H.X.; Yang, D.Q.; Ren, C.H. Design method of lightweight metamaterials with arbitrary poisson's ratio. *Materials* **2018**, *11*, 1574. [CrossRef] [PubMed]

4. Gibson, L.J.; Ashby, M.F. *Cellular Solids, Structural and Properties*, 2nd ed.; Cambridge University Press: Cambridge, UK, 1999.

5. Miller, W.; Smith, C.W.; Evans, K.E. Honeycomb cores with enhanced buckling strength. *Compos. Struct.* **2011**, *93*, 1072–1077. [CrossRef]

6. Zhang, X.; Zhang, H.; Wen, Z. Experimental and numerical studies on the crush resistance of aluminum honeycombs with various cell configurations. *Int. J. Impact Eng.* **2014**, *66*, 48–59. [CrossRef]

7. Li, X.C.; Li, K.; Lin, Y.L.; Chen, R.; Lu, F.Y. Inserting stress analysis of combined hexagonal aluminum honeycombs. *Shock Vib.* **2016**, *2016*. [CrossRef]

8. Ashab, A.; Ruan, D.; Lu, G.; Bhuiyan, A.A. Finite element analysis of aluminum honeycombs subjected to dynamic indentation and compression loads. *Materials* **2016**, *9*, 162. [CrossRef]

9. Qiao, J.; Chen, C. In-plane crushing of a hierarchical honeycomb. *Int. J. Solids Struct.* **2016**, *85–86*, 57–66. [CrossRef]

10. Ruan, D.; Lu, G.X.; Wang, B.; Yu, T.X. In-plane dynamic crushing of honeycombs-a finite element study. *Int. J. Impact Eng.* **2003**, *28*, 161–182. [CrossRef]

11. Hu, L.L.; You, F.F.; Yu, T.X. Effect of cell-wall angle on the in-plane crushing behaviour of hexagonal honeycombs. *Mater. Des.* **2013**, *46*, 511–523. [CrossRef]

12. Papka, S.D.; Kyriakides, S. Experiments and full-scale numerical simulations of in-plane crushing of a honeycomb. *Acta Mater.* **1998**, *46*, 2765–2776. [CrossRef]

13. Arash, F.; Christian, B.; Ralf, S. Numercial fracture analysis and model validation for disbonded honeycomb core sandwich composites. *Compos. Struct.* **2019**, *210*, 231–238.

14. Wang, Z.G.; Lu, Z.J.; Yao, S.; Zhang, Y.B.; Hui, D.; Feo, L. Deformation mode evolutional mechanism of honeycomb structure when undergoing a shallow inclined load. *Compos. Struct.* **2016**, *147*, 211–219. [CrossRef]

15. Wang, Z.G.; Liu, J.F.; Hui, D. Mechanical behaviors of inclined cell honeycomb structure subjected to compression. *Compos. Part B Eng.* **2017**, *110*, 307–314. [CrossRef]

16. Tounsi, R.; Markiewicz, E.; Zouari, B.; Chaari, F.; Haugou, G. Numerical investigation, experimental validation and macroscopic yield criterion of Al5056 honeycombs under mixed shear-compression loading. *Int. J. Impact Eng.* **2017**, *108*, 348–360. [CrossRef]

17. Sun, D.Q.; Zhang, W.H.; Wei, Y.B. Mean out-of-plane dynamic plateau stresses of hexagonal honeycomb cores under impact loadings. *Compos. Struct.* **2010**, *92*, 2609–2621.

18. Li, X.; Zhang, P.W.; Wang, Z.H.; Wu, G.Y.; Zhao, L.M. Dynamic behavior of aluminum honeycomb sandwich panels under air blast: Experiment and numerical analysis. *Compos. Struct.* **2014**, *108*, 1001–1008. [CrossRef]

19. Zhu, F.; Zhao, L.M.; Lu, G.X. Deformation and failure of blast loaded metallic sandwich panels—Experimental investigations. *Int. J. Impact Eng.* **2008**, *35*, 937–951. [CrossRef]

20. Li, S.Q.; Li, X.; Wang, Z.H.; Wu, G.Y.; Lu, G.X.; Zhao, L.M. Sandwich panels with layered graded aluminum honeycomb cores under blast loading. *Compos. Struct.* **2017**, *173*, 242–254. [CrossRef]

21. Jin, X.C.; Wang, Z.H.; Ning, J.G.; Xiao, G.S.; Liu, E.Q.; Shu, X.F. Dynamic response of sandwich structures with graded auxetic honeycomb cores under blast loading. *Compos. Part B Eng.* **2016**, *106*, 206–217. [CrossRef]

22. Dharmasena Kumar, P.; Wadley, H.N.; Xue, Z.Y.; Hutchinson John, W. Mechanical response of metallic honeycomb sandwich panel structures to high-intensity dynamic loading. *Int. J. Impact Eng.* **2008**, *35*, 1063–1074. [CrossRef]

23. Karagiozova, D.; Nurick, G.N.; Langdon, G.S. Behaviour of sandwich panels subject to intense air blasts—Part 2: Numerical simulation. *Compos. Struct.* **2009**, *91*, 442–445. [CrossRef]

24. Cui, X.; Zhao, L.; Wang, Z.; Zhao, H.; Fang, D. Dynamic response of metallic lattice sandwich structures to impulsive loading. *Int. J. Impact Eng.* **2012**, *43*, 1–5. [CrossRef]

25. Li, S.; Lu, G.; Wang, Z.; Zhao, L.; Wu, G. Finite element simulation of metallic cylindrical sandwich shells with graded aluminum tubular cores subjected to internal blast loading. *Int. J. Mech. Sci.* **2015**, *96–97*, 1–12. [CrossRef]

26. Edri, I.E.; Grisaro, H.Y.; Yankelevsky, D.Z. TNT equivalency in an internal explosion event. *J. Hazard. Mater.* **2019**, *374*, 248–257. [CrossRef] [PubMed]

Review

From Local Structure to Overall Performance: An Overview on the Design of an Acoustic Coating

Hongbai Bai [1,2], Zhiqiang Zhan [1,2], Jinchun Liu [1,2,*] and Zhiying Ren [1,2,*]

1   College of Mechanical Engineering and Automation, Fuzhou University, Fuzhou 350116, China
2   Engineering Research Center for Metal Ruber, Fuzhou University, Fuzhou 350116, China
*   Correspondence: www333coca@sina.com (J.L.); renzyrose@126.com (Z.R.)

Received: 19 June 2019; Accepted: 5 August 2019; Published: 7 August 2019

Abstract: Based on the requirements of underwater acoustic stealth, the classification and research background of acoustic coatings are introduced herein. The research significance of acoustic coatings is expounded from the perspective of both the military and civilian use. A brief overview of the conventional design process of acoustic coatings is presented, which describes the substrates used in different countries. Aimed at the local design of acoustic coatings, research progress on passive and semi-active/active sound absorption structure is summarized. Focused on the passive acoustic coatings; acoustic cavity design and optimization, acoustic performance of acoustic coatings with rigid inclusions or scatterers, and acoustic coatings with a hybrid structure are discussed. Moreover, an overview of the overall design of acoustic coatings based on the sound field characteristics of the submarine is also presented. Finally, the shortcomings of the research are discussed, breakthroughs in acoustic coating design research are forecast, and the key technical issues to be solved are highlighted.

Keywords: acoustic stealth; acoustic coating; passive sound absorption; active sound absorption; acoustic characteristics of a submarine; finite element method (FEM)

---

1. Introduction

As acoustic stealth equipment widely used in submarines, the acoustic coating can absorb the sound waves emitted by the active sonar. It can also suppress the vibration of the hull and isolate the noise inside the boat. Therefore, the acoustic coating becomes the only key component on the submarine that can effectively counter active enemy sonar and passive sonar detection. Based on the sound field characteristics of the submarine, namely, radiation noise characteristics, self-noise characteristics, and target strength characteristics [1], acoustic coatings can be roughly divided into two types according to function: sound insulation decoupling tiles and anechoic tiles, as shown in Figure 1. The main function of decoupling tiles is to reduce the radiation noise and self-noise of the submarine. In contrast, anechoic tiles reduce the target strength characteristics of the submarine and the reflection of active sonar sound waves. Moreover, some acoustic coatings actually have both the functions mentioned above.

In practice, properties of acoustic materials can change under hydrostatic pressure conditions (such as the shape of the cavity, modulus of materials, etc.), which affects the sound absorption performance of the materials. With the development of low-frequency sonar, high-efficiency sound absorption of underwater acoustic stealth materials with finite thickness under low frequency (generally considered to be below 1 kHz) and wide frequency conditions has become a hot topic in current research [2]. In the military field, the acoustic coating is one of the effective means to improve the sound stealth performance of submarine. The "quiet submarine" with an acoustic coating has become an important direction for the development of modern submarines, which has practical significance for the development of naval equipment and the construction of national defense military equipment.

In the civilian field, acoustic coating generally refers to various functional materials used underwater. In civil water-sound systems, involving seabed resource exploration [3], seabed mapping [4], fish stock information detection [5], and remote sensing [6], acoustic coatings can absorb unnecessary sound waves in the marine environment to improve the accuracy of positioning, searching, and the communication capability of the equipment. In addition, acoustic coatings play an important role in improving the marine environment and reducing noise pollution [7]. Moreover, the acoustic coatings should be used to cover the muffler pool during the measurement or calibration of underwater acoustic equipment and other hydroacoustic tests [8].

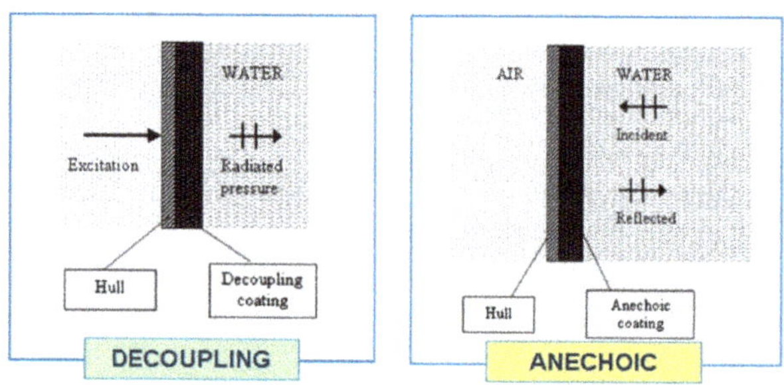

Figure 1. Two main types of acoustic coating [9].

To acquire a better understanding of the design of acoustic coatings, in this study a brief overview of the conventional design process for acoustic coatings is provided first, and then the local structure design of acoustic coatings was introduced (especially for the design of passive acoustic coatings). An overview of the overall performance of acoustic coatings was also outlined. Finally, the shortcomings of the research were discussed, followed by the development of a new prospect.

2. Conventional Design Process for Acoustic Coatings

The design process of the acoustic coating was developed around the performance requirements presented in Figure 2, starting from small-scale design to large-scale design, as shown in Figure 3. The conventional acoustic coating design process begins with polymer and filler design while carrying out the local acoustic structure design, and finally, the overall performance design (including the hull). The formula design is the most important step in the process, which mainly involves viscoelastic polymer materials, such as butyl rubber, styrene-butadiene rubber, polyurethane, polysulfide rubber, polybutylene rubber, neoprene, fiber-reinforced polymer, polyurethane/epoxy copolymer, silicone rubber, etc., as substrates. Different countries use different substrate materials in acoustic coatings, as presented in Table 1.

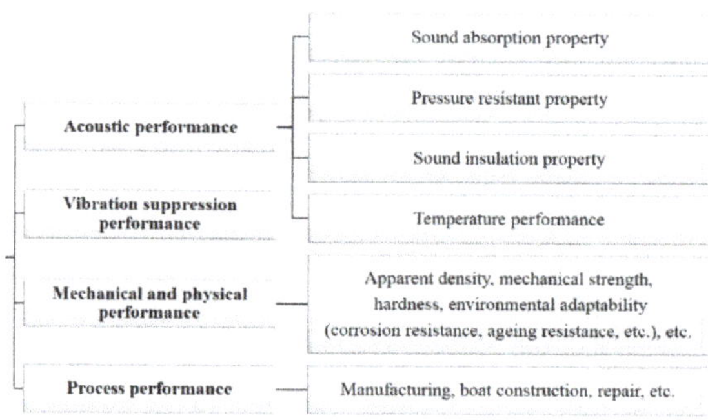

Figure 2. Performance requirements for an acoustic coating [1].

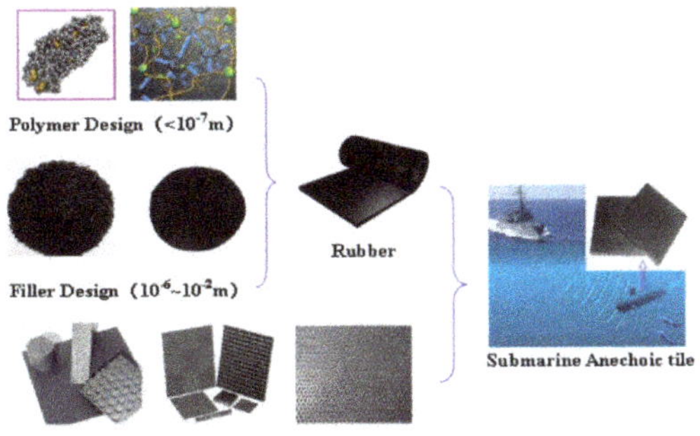

Figure 3. Conventional design process of acoustic coatings [10].

Table 1. Substrate materials of acoustic coatings [11].

Country	Acoustic Coating Substrates
Germany	Composite rubber
United States	Polyurethane, glass fiber, butyl rubber
Russia	Styrene-butadiene rubber, polybutadiene rubber, rubber ceramics
France	Polyurethane, polysulfide rubber
United Kingdom	Polyurethane
Japan	Neoprene
China	Styrene-butadiene rubber, polyurethane

3. Local Design of Acoustic Coatings

Local design of acoustic coatings mainly refers to designing of local acoustic structure. Acoustic coatings are usually laid on the hull piece by piece; therefore, design of a single-piece acoustic coating (or unit structure) can also be defined as local design. Local design consists of passive and semi-active/active acoustic coatings' design (including research on pressure resistance, sound insulation, etc.). Passive sound absorption design involves optimization of acoustic cavities, acoustic performance of acoustic coatings with rigid inclusions or scatterers, and hybrid structure acoustic coatings. In contrast, semi-active or active sound absorption is achieved by semi-active or active control of sound waves.

3.1. Acoustic Cavity Design and Optimization on Passive Acoustic Coatings

The origin of the world's first passive acoustic structure dates back to the end of World War II when the concept of German "Alberich"-type acoustic coating was introduced. The historical synthetic rubber coatings installed on German submarines were designed with lattices of resonant cavities, which had sound-absorbing properties [12]. With the increasing complexity of acoustic coatings, analytical methods are difficult to use for calculations of complex structures. Since the 1990s, Hennion [13] and Easwaran [14] began to study acoustic coatings by the finite element method (FEM). Nowadays, with the development of numerical simulation software (for example, ANSYS, COMSOL, etc.), the FEM can not only optimize the traditional structure but also facilitate the research of new material structure.

The study of oblique incidence provides a more practical and comprehensive significance for characterizing the acoustic performance of the Alberich-type acoustic coating; therefore, sound absorption performance of the steel-backed acoustic coating was studied when the acoustic wave was obliquely incident. By using numerical calculation methods, the sound absorption coefficient of the Alberich-type acoustic coating was calculated, and the mechanism of sound absorption was studied in terms of the structural displacement vector and deformation, as presented in Figure 4a. Notably, there is a wide and strong silencer region at larger angles of incidence and lower frequency. However, at higher frequencies, the sound absorption coefficient decreases with the increase in the angle of incidence [15].

Most of the current researches focus on acoustic coatings laid on the surface of rigid steel plates; however, different acoustic coatings laid on the surface of water-immersed steel sheets have rarely been investigated. Therefore, differential evolution algorithms combined with FEM were used for optimizing two types of Alberich acoustic coatings. The two resonance modes of acoustic coatings were different; therefore, we could obtain good absorption properties through acoustic coupling between two acoustic coatings and the cooperation with the damping rubber [16], as shown in Figure 4b.

Acoustic and physical characteristics of acoustic coatings can be modified by adding various components. By redesigning materials and structures of the Alberich-type acoustic coating, the tapered bore structure led to the enhancement in its low-frequency performance in the use of underwater testing and calibration cells [17].

In the aspect of the establishment and calculation of an analysis model for acoustic coatings, an analytical model was established based on the homogenization theory, which has higher computational efficiency. The results shown in Figure 4c indicate that strong coupling of cavity resonance leads to broadband attenuation of the sound [18]. According to the accurate requirements of viscoelastic dynamic parameters, such as complex elastic modulus and Poisson's ratio, a new parameter identification method was proposed based on the measured reflection coefficient. Acoustic properties of acoustic coating with horizontally distributed cylindrical holes were also studied, as shown in Figure 4d. The results showed that the smaller the horizontal spacing between two adjacent holes, the better the low-frequency sound absorption effect [19].

Most literatures have studied the optimization of structure and material on single-layer acoustic coatings; however, acoustic coatings with a multilayer structure have not been extensively studied. A program based on finite element simulation was developed in ANSYS. Through the comparison between a cone frustum hole and a cylindrical hole, the frustum hole (as shown in Figure 4e) was found to have a large transmission coefficient [20]. A finite element model with the frustum cavity (as shown in Figure 4f) was constructed by using COMSOL software. The results showed the superior absorption properties of the multilayered structure, and the sound absorption performance of the frustum cavity was better than that of the cylindrical and ellipsoidal cavities [21].

The development of finite element simulation technology is of great significance, which aids in the design of complex acoustic cavities. To improve the decoupling ability of acoustic coating, a complex shape of the cavity was designed by optimization, as shown in Figure 4g. Furthermore, an equivalent fluid model was established for processing acoustic coatings with periodically distributed and axisymmetric cavities [22]. The sound insulation performance of two types of honeycomb acoustic coatings was studied, and it was concluded that the negative Poisson's one has better sound insulation performance at a certain frequency [23], as shown in Figure 4h. The effectiveness of the use of small air Helmholtz resonator structure was studied, and the echo suppression results showed that the resonator has a broad application prospect in achieving underwater stealth [24].

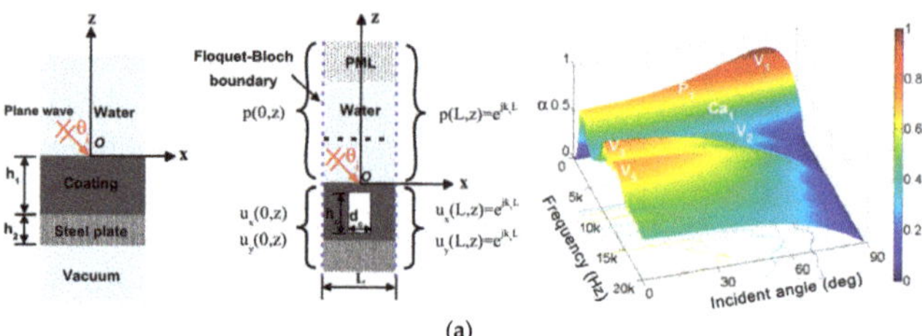

(a)

Figure 4. *Cont.*

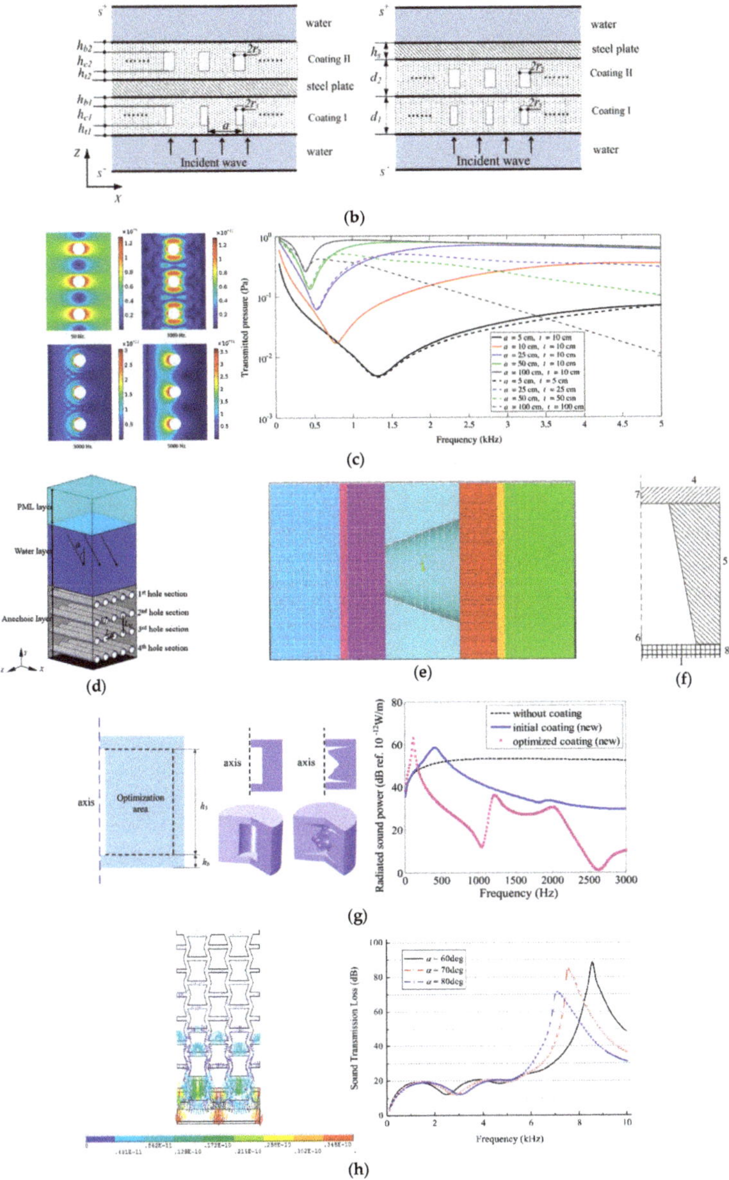

Figure 4. Design and optimization of the acoustic cavity. (**a**) Two-dimensional analytical model, periodic numerical model, and sound absorption coefficients of the Alberich anechoic coating [15]. (**b**) Coatings on both surfaces and the outer surface of a steel plate and their absorption coefficients [16]. (**c**) Deformation of the polydimethylsiloxane (PDMS) medium and transmitted pressure obtained analytically [18]. (**d**) Schematic of the finite element method (FEM) model [19]. (**e**) Cross-section of the absorbent coating, including a cone frustum hole [20]. (**f**) Two-dimensional axisymmetric model [21]. (**g**) Sketch of the optimization area for the cavity, initial cavity, optimized cavity, and radiated sound power associated with different coatings attached to an infinite plate [22]. (**h**) Sound transmission loss and displacement contour of the negative Poisson's ratio honeycomb-hole coatings [23].

With the deepening of research in this area, pressure resistance performance of acoustic coatings has also received increasing attention. Under hydrostatic pressure, acoustic properties of acoustic coatings change. On the one hand, dynamic mechanical properties of materials change with pressure, which leads to the change in parameters of acoustic coating materials. On the other hand, when the acoustic coating is subjected to hydrostatic pressure, the cavity is deformed such that the structural parameters of acoustic coatings change.

In terms of the influence of hydrostatic pressure on sound absorption performance, a spherical cavity acoustic coating unit model was built by using COMSOL software, which was based on the two-dimensional (2D) theory of sound waves under normal incidence. A coupled model of acoustic performance calculation based on the compression deformation of the cavity was obtained, and the influence of the internal pressure of the cavity was considered. With increasing hydrostatic pressure, the peak frequency was toward the high frequency and became wider, absorption effect became more significant for overall degradation, and internal air pressure reduced the cavity deformation and improved the low-frequency sound absorption effect [25].

For the problem of the sound insulation effect associated with sound insulation decoupling material under hydrostatic pressure, the influence of hydrostatic pressure on the insulation performance of acoustic coating was analyzed, as shown in Figure 5. With the increase in the hydrostatic pressure, the equivalent sound velocity of the material increased, and the sound insulation performance of acoustic coating decreased [26]. Therefore, deformation resistance of the material under pressure can be improved by adding a rigid confinement structural unit inside the acoustic cavity of the sound insulation decoupling materials [27].

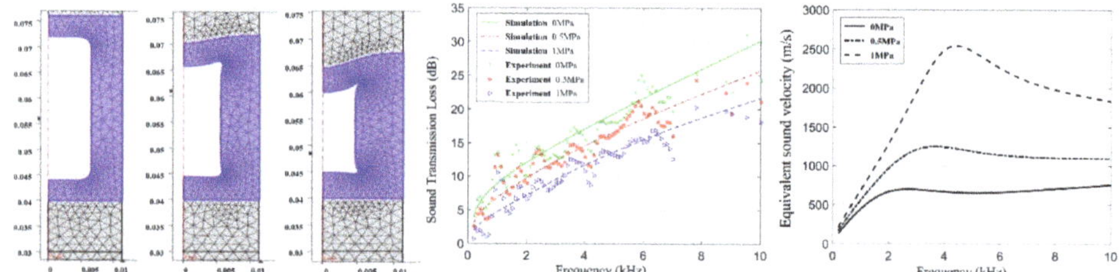

Figure 5. Acoustic properties of acoustic coatings under hydrostatic pressure. Deformed shape of the unit cell, transmission loss, and the equivalent sound velocity of the sound insulation layer under different static pressure [26].

3.2. Acoustic Properties of Passive Acoustic Coatings with Rigid Inclusions or Scatterers

Structural design of the acoustic cavity better solves the problem of low-frequency sound absorption, but it also needs to consider its pressure resistance performance. A United States Patent discloses a composite material comprising a spherical shell inclusion for a subsea platform, which exhibited a strong static stiffness [28]. This provides an idea for the study of the relationship between acoustic and pressure properties of acoustic coatings, i.e., enhancement of pressure resistance by adding rigid inclusions or scatterers. Traditional underwater acoustic stealth materials are limited by the mass density law. Larger sized of materials are required to increase the acoustic wave propagation path for improving the energy dissipation of low-frequency sound waves. Of note, local resonant acoustic metamaterials can achieve long-wavelength sound waves' control at a small scale, which provides ideas for solving low-frequency sound absorption problems [8].

In 2000, Liu et al. proposed the concept of locally resonant sonic metamaterials. A lead ball coated with a layer of soft silicone rubber was embedded in the hard epoxy resin matrix to form a three-dimensional lattice structure. The acoustic band gap was generated by resonance, thereby achieving the purpose of absorbing sound waves and dissipating energy [29]. Comparison of sound absorption properties between spherical and cylindrical local resonance structures indicated that a cylindrical one can further improve the low-frequency sound absorption performance [30]. Use of a locally resonant phononic woodpile structure (as shown in Figure 6a) provided a new approach to solve the associated weight loss problem [31]. Figure 6b shows a phononic glass with an interpenetrating network structure, exhibiting dual characteristics of high mechanical strength and excellent underwater sound absorption capacity [32].

For a more realistic understanding, the influence of frequency-dependent material parameters on the performance of local resonance acoustic metamaterials (LRAMs), a generalized Maxwell model was used to simulate the behavior of multi-polymer materials. Studies have shown that the parameters of the rubber coating material vary with the frequency and influence the wave attenuation effect in the frequency range around the band gap [33]. The acoustic performance of locally resonant phononic crystals consisting of periodic steel cylinders in a viscoelastic rubber medium was studied analytically and numerically, and the approach may be useful for customized designs of acoustic metamaterials, such as rapid prototyping and trend analysis, in underwater applications [34].

Current research focuses on symmetric resonant cavities, such as spherical resonators or coaxial cylindrical resonant cavities. However, characteristics of non-symmetric resonators, such as non-concentric resonators or non-coaxial resonators, have been less studied. Thus, FEM was used to study the absorption characteristics of a viscoelastic plate periodically embedded in an infinitely long non-coaxial cylindrical local scatterer, as shown in Figure 6c. The results indicated the existence of two typical resonance modes: One is the overall resonance mode caused by the steel back, in which the core position had little effect on it. The other is the core resonance mode caused by the local resonance scatterer. Furthermore, with the increase in the core eccentricity, the core resonance mode moved toward the high-frequency direction [35].

To overcome the shortcomings of the narrow range of the effective sound-absorption frequency of local resonance phononic crystals, a novel multilayer local resonance acoustic metamaterial was proposed, in which a multilayer local resonance scatterer was embedded into the matrix structure, as shown in Figure 6d. Coupling resonance expanded the band gap and enhanced the sound absorption performance of metamaterials [36].

Since the existing technology cannot realize effective absorption of subwavelength structure, an underwater element structure was proposed to achieve low-frequency broadband sound absorption below 1000 Hz, which was based on the structural combination design of common viscoelastic damping materials and metal materials, as shown in Figure 6e. Physical mechanisms of acoustic coating included waveform transformation, multiple scattering, etc. Different geometries and materials of the helical structure aided in the adjustment of the sound absorption properties of materials [37].

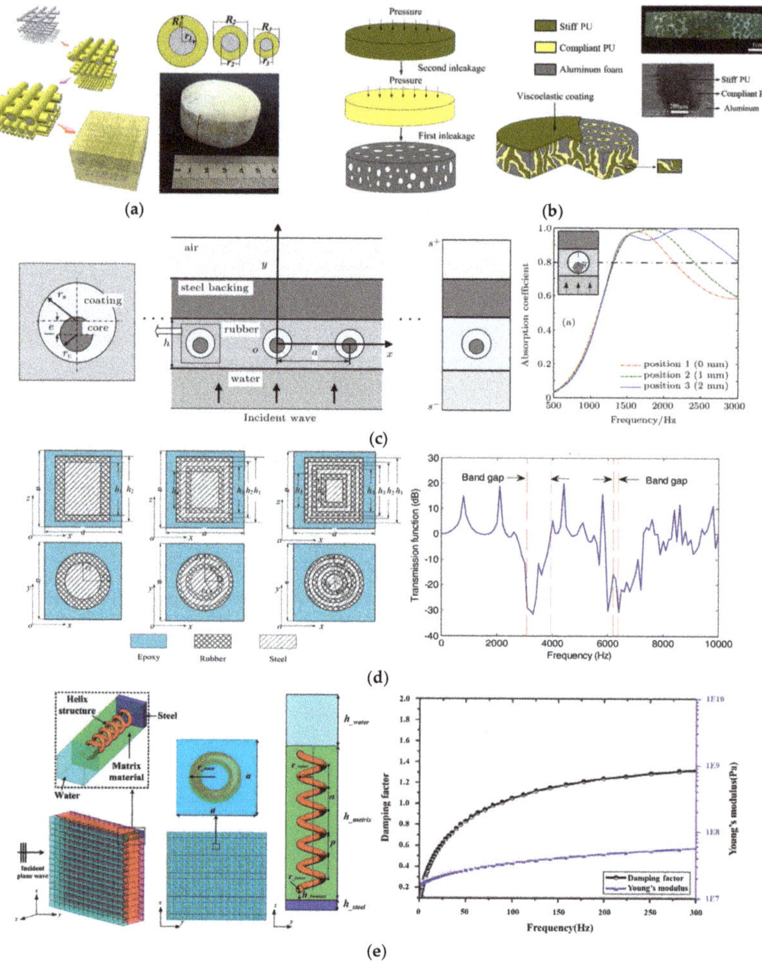

Figure 6. Acoustic properties of acoustic coatings with rigid inclusions or scatterers. (**a**) Schematic of synthesis route and optical image of locally resonant phononic woodpile [31]. (**b**) Schematic of synthesis, the structure of phononic glass, optical and SEM images of a typical phononic glass sample [32]. (**c**) Cross-section of one locally resonant scatterer unit, absorption structure with steel backing, one Bloch unit cell, and comparison among the absorptances for three core positions [35]. (**d**) Conventional local resonance acoustic metamaterials (LRAMs), double-layered cylindrical scatterers, three-layered cylindrical scatterers, and band diagram of the LRAMs with three-layered locally resonant scatterers [36]. (**e**) Three-dimensional geometric dimensions and experimental results of dynamic mechanical parameters of viscoelastic damping rubber [37].

3.3. Passive Acoustic Coatings with Hybrid Structure

An acoustic coating with a hybrid structure takes sound absorption and pressure resistance performance into account, which incorporates other rigid materials while retaining the acoustic cavity. Sound absorption performance of a composite structure combining rubber material containing a cylindrical cavity and porous metal material was studied. The results revealed that the addition of porous metal material could improve the low-frequency sound absorption capability [38]. For acoustic coating containing both cavities and hard scatterers, the equivalent medium theory was used to analyze and numerically calculate underwater sound absorption characteristics of phononic crystals in soft rubber media. A finite element model for the periodic distribution of cavities and hard inclusions in the rubber media was established, as shown in Figure 7. It has been found that acoustic coating consisting of a layer of hard inclusions plus a small cavity in the direction of sound propagation exhibits a high sound absorption capacity over a wide frequency range [39].

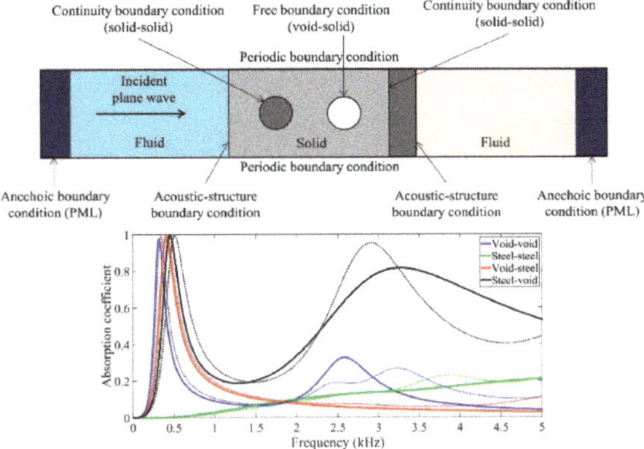

Figure 7. Acoustic coatings with hybrid structure. Schematic diagram of the numerical model and effect of steel backing plate on the sound absorption coefficient of a phononic crystal [39].

3.4. Design on Semi-Active/Active Acoustic Coatings

Passive sound absorption technology relies on the structural design of acoustic coatings and material modification; however, its sound absorption mechanism is limited, and control of low frequency and wide frequency sound waves is not good. Therefore, a semi-active/active control method for sound waves was proposed.

Piezoelectric composites have been widely used in semi-active acoustic coatings. Through internal loss, part of the acoustic energy can be directly converted into thermal energy; another part of the acoustic energy can be converted into electrical energy due to the piezoelectric effect and then converted into thermal energy by piezoelectric passive control. The effect of rubber damping materials in the piezoelectric parallel sound absorption layer was investigated, and a design for acoustic coating with a multilayer structure (damping rubber, piezoelectric composite and lead zirconate titanate (PZT) ceramic) was proposed, which effectively improved the low-frequency sound absorption coefficient [40]. Combined with parallel impedance under the constraint of the underwater finite frequency sound absorption coefficient, a practical multi-layer piezoelectric acoustic coating sound absorption control method was proposed [41], as shown in Figure 8a. Another design of semi-active compound acoustic coating exhibited periodic sub-wavelength piezoelectric array, which broadened the bandwidth of the first absorption peak [42], as presented in Figure 8b.

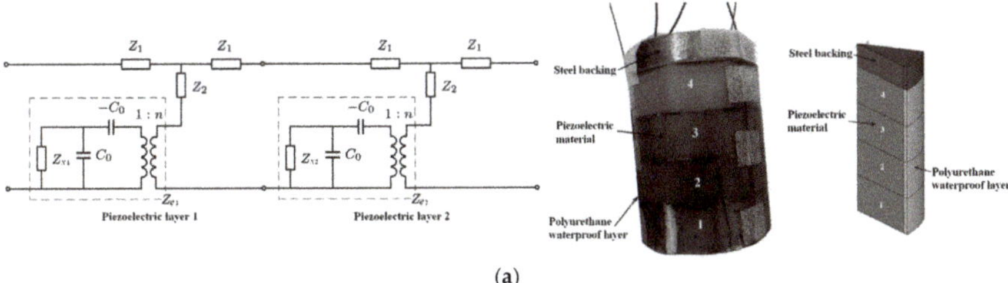

(**a**)

Figure 8. *Cont.*

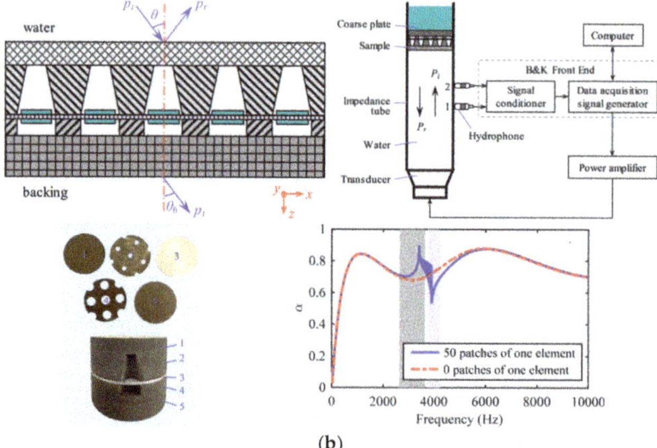

Figure 8. Design on semi-active acoustic coatings. (**a**) Mason equivalent circuit of two piezoelectric layers, experimental sample and finite element model with polyurethane waterproof layer [41]. (**b**) Schematic diagram of the composite coating, test schematic of a hydroacoustic impedance tube, test sample, and comparison between theoretical predictions of the coating [42].

Active sound absorption controls the secondary sound source to generate sound waves that are opposite to the primary sound source through the principle of coherent cancellation, thus achieving the objective of active sound absorption [11]. A low-frequency echo suppression technique based on active impedance matching was proposed, which employed a tile projector designed to cover a wide area, such as the surface of a ship. Moreover, its low-frequency echo reduction performance was tested [43]. A low-frequency miniaturized active control unit with a thickness of less than 50 mm using giant magnetostrictive material was designed, which was able to achieve significant noise reduction in less than a one-second interval. Even under high pressure, its sound absorption coefficient was far more than 0.8 [44], as shown in Figure 9.

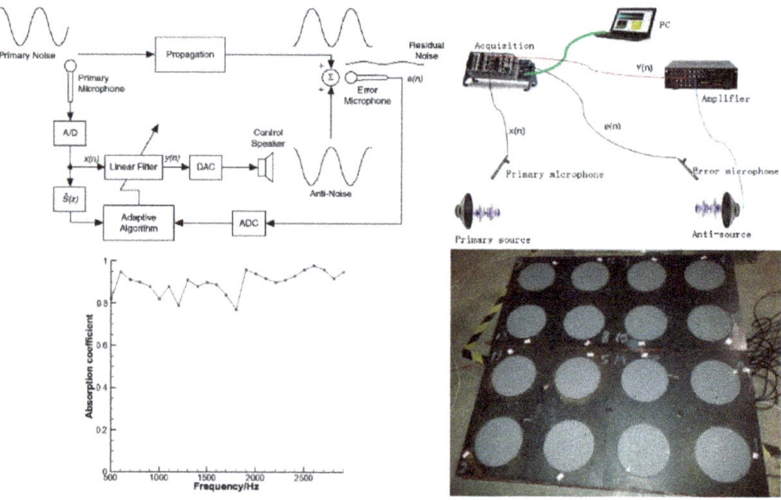

Figure 9. Design on active acoustic coatings. Active control procedure, absorption coefficients at the stable pressure value of 0.96 MPa and an active large plate sample [44].

4. Overall Design on Acoustic Coatings

In recent years, overall design on acoustic coatings based on the characteristics of the submarine sound field has progressively received more attention, with a focus on acoustic radiation of underwater cylindrical shells, including finite/infinite long cylindrical shells, elastic or viscoelastic single/double shells, stiffened cylindrical shells, etc. The finite element and boundary element model of the submarine section were built, and sound radiation calculations revealed that functionally graded materials exhibited better vibration and noise reduction performance [45]. By matching the velocity method, a method for estimating the equivalent material parameters of acoustic coatings, was proposed, which simplified the problem associated with complex acoustic radiation prediction to the problem of the equivalent homogeneous viscoelastic acoustic coating structure [46]. An equivalent modulus method based on impedance transfer formula was proposed, which reduced the number of calculating nodes. The study indicated that an increase in the mass of backing caused absorption peak to move to low frequency [47], as shown in Figure 10.

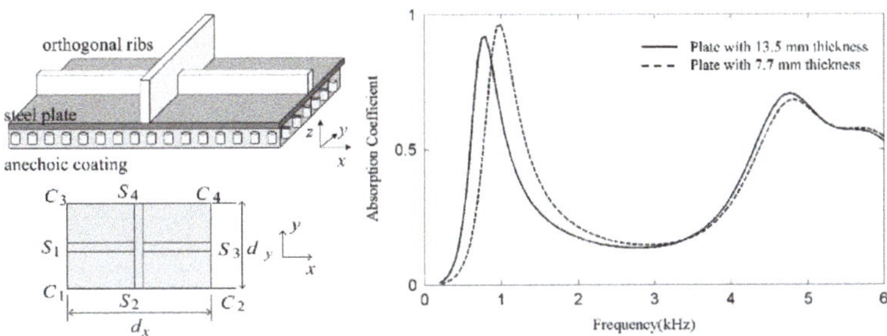

Figure 10. Acoustic radiation of underwater cylindrical shells. Calculating model of a unit cell, its top view, and absorption coefficients of the Alberich anechoic coating backed with different steel plates [47].

Other studies were on the target strength characteristics of the submarine. A high-frequency analysis program based on Kirchhoff approximation was developed for the prediction of submarine target strength (TS) [48], as presented in Figure 11a. Figure 11b exhibits the analysis of underwater vehicle model based on the established near-field equation, and the influence of parameters, such as distance and frequency, on the model was researched [49]. Based on Monte Carlo–Plate Element Method (MC-PEM), the relationship between target strength and shedding rate of the cylindrical shell was quantitatively analyzed. The results showed that the inspection can be carried out to hold strength at a relatively low level within the range of 0.3t to 0.5t (t represents the average life of anechoic tile) [50], as shown in Figure 11c.

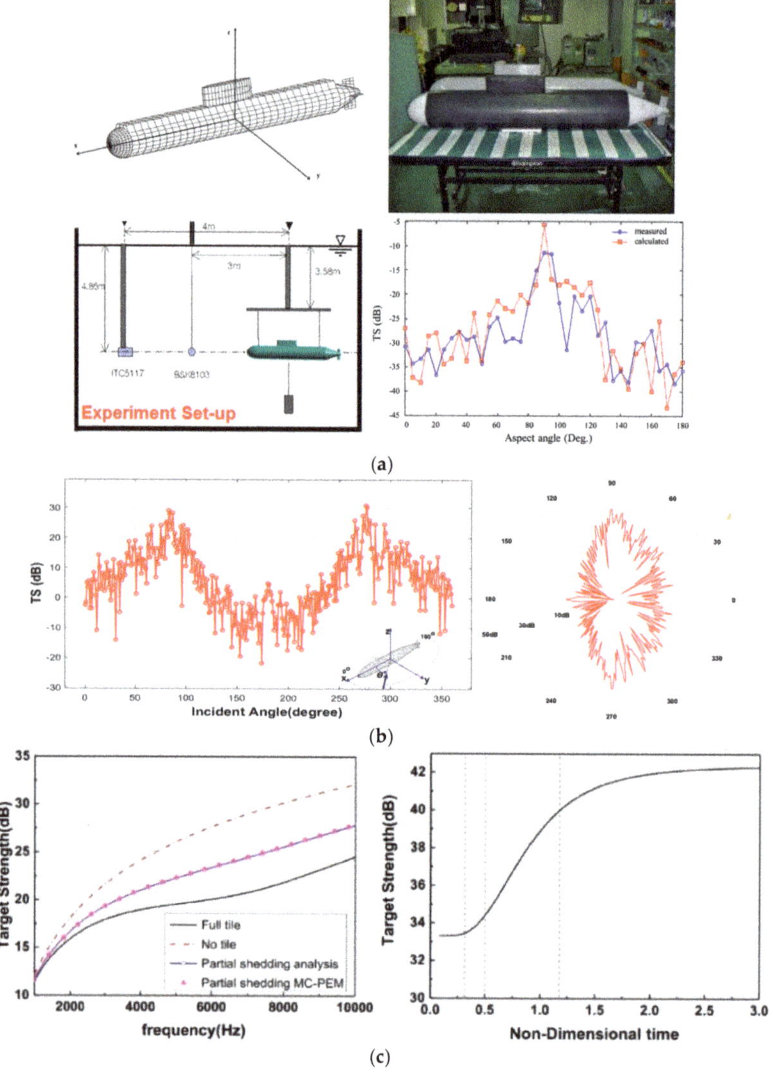

Figure 11. Research on target strength characteristics of the submarine. (**a**) Mesh model, test model, experiment equipment, and comparison of experimental and numerical solutions for target strength (TS) [48]. (**b**) TS of the rectangular and polar coordinate of the submarine model at a frequency of 5 kHz and a distance of 500 m [49]. (**c**) Comparison of the analytical method and Monte Carlo–Plate Element Method (MC-PEM) and the relationship between TS and non-dimensioned time [50].

5. Summary and Outlook

5.1. Summary

Of note, material parameters and structural forms of acoustic coatings are the main factors affecting the acoustic stealth performance. Overall, material properties are affected by its structural form. From local structure to overall performance, design of acoustic coatings needs to grasp the key research content of each stage. Wang et al. classified the underwater sound absorption materials, introduced the corresponding sound absorption mechanism, and compared their advantages and disadvantages, as summarized in Table 2.

Table 2. Classification of underwater sound absorption materials, main sound absorption mechanism, advantages and disadvantages [51].

Classification	Main Sound Absorption Mechanism	Advantages and Disadvantages
pure polymer material	viscous dissipation, heat transfer absorption, molecular relaxation absorption	excellent physical and chemical properties, easy to vulcanization molding, high viscoelasticity and excellent damping properties; large size, generally low strength, not resistant to hydrostatic pressure
particle-filled material	acoustic scattering, waveform transformation	improving sound absorption performance, improving the overall strength; large size, random distribution of the particle, its own characteristics are diluted by the matrix material
impedance-grading type material	viscoelastic internal friction, elastic relaxation	simple structure, good process molding, good sound absorption effect; large size, energy consumption mechanism is relatively simple
porous (foam) material	fluid vibration and friction in the hole	lightweight, high strength, resistant to hydrostatic pressure; bad low-frequency sound absorption performance, not resistant to seawater corrosion
cavity resonance type material	cavity resonance, waveform transformation, intrinsic properties of polymer material	better solving the problem of low-frequency sound absorption; low material strength, not resistant to hydrostatic pressure
phononic crystal	Bragg scattering, local resonance sound absorption	realizing control of long-wavelength acoustic waves by small-scale materials; regulation band is narrow, polymer matrix will cause bad sound absorption performance with the increase of hydrostatic pressure

After a literature investigation, we find that current research still has some shortcomings:

(1)　Although the local-structure design based on traditional acoustic theory is clear, its preparation is difficult, or the parameters of the prepared materials are contradictory to the setting parameters, which make it hard to achieve the expected performance [10].

(2)　Some theoretical studies are limited in terms of preparation techniques, and they lack experimental verification. Examples are specified in Table A1 of Appendix A. Owing to its complicated structure, preparation of acoustic coating is difficult, which cannot be verified by experiments, thus making it more difficult for practical applications.

(3)　Few researches focus on other performance requirements for acoustic coatings during design, such as temperature performance [52], inspection [53]. In different seas, depths, and seasons, seawater temperature changes significantly [1]. For more realistic underwater conditions, it is suggested to investigate mechanical and acoustic performances (such as complex elastic modulus) of elastomers in different temperatures and pressures [52].

(4)　Researches on low-frequency sound absorption of acoustic coatings are few, especially the study of underwater acoustic materials below 1000 Hz.

5.2. Outlook

(1)　Compared to passive control, the active control of sound waves is more adaptable and has broad application prospects. Accordingly, the design of active acoustic coatings will be a major direction for future development.

(2)　Overall modeling and prediction of acoustic coatings under the coupling state of the shell structure require further study. When acoustic coatings are fabricated on the shell, acoustic characteristics in the coupling state of the structure and fluid are different from that in the natural state [54].

(3)   Applications of other acoustic metamaterials in underwater environments are also worthy of further research, such as thin-film acoustic metamaterials [55,56], Helmholtz resonator type acoustic metamaterials [57], double-negative (negative effective density and negative bulk modulus) acoustic metamaterials [58], acoustic cloak [59], and so on.

(4)   Theoretical breakthroughs have been brought by the electricity–mechanics–acoustics analogy [60] and mutual learning between acoustic methods and optical methods; structural design breakthrough brought by bionics [61]. Modern acoustics has a strong crossover and extensibility. This is especially true on the design of acoustic coatings. It involves various fields, such as mechanical engineering, material, chemical industry, etc., which requires communication and cooperation between scholars in different fields, and the strengthening of our own study in these fields.

However, it must be emphasized that there are still some key technical issues to be solved [45] as follows:

(1)   The realization of low-frequency sound absorption performance of acoustic coatings without increasing thickness and weight of materials.

(2)   The guarantee of strong sound absorption effect of acoustic coatings in low frequency and broadband range.

(3)   The maintenance of sound absorption performance of acoustic coatings under deep-sea hydrostatic pressure.

Author Contributions: Conceptualization and methodology, H.B., J.L., and Z.R.; validation, Z.Z. and Z.R.; investigation, H.B., Z.Z., J.L. and Z.R.; writing—original draft preparation, H.B. and Z.Z.; writing—review and editing, H.B., Z.Z., J.L., and Z.R.; supervision, Z.R.; project administration, H.B. and J.L.

Funding: This research was funded by National Natural Science Foundation of China (Grant No. 51805086), the Fujian Provincial Natural Science Foundation (2019J01210) and High end bearing tribology technology and application, National Joint Engineering Laboratory Open Fund Project, Henan University of Science and Technology (201802).

Conflicts of Interest: The authors declare no conflict of interest.

Appendix A

The appendix is to specify for some references the nature of the work.

Table A1. The nature of the work for some references.

References	Theoretical/ Numerical Work	Experimental Work	Year of Publication	References	Theoretical/ Numerical Work	Experimental Work	Year of Publication
15	√		2018	34	√		2019
16	√		2018	35	√		2015
17	√	√	2010	36	√		2019
18	√		2017	37	√		2019
19	√		2019	38	√		2017
20	√		2018	39	√		2019
21	√		2016	40	√		2017
22	√		2018	41	√	√	2015
23	√		2018	42	√	√	2019
24	√		2019	43	√	√	2014
25	√		2017	44	√	√	2017
26	√	√	2016	45	√		2013
27	√	√	2012	46	√		2014
30	√		2012	47	√		2016
31	√	√	2009	48	√	√	2017
32	√	√	2012	49	√	√	2018
33	√		2017	50	√		2018

References

1. Zhu, B.L.; Huang, X.C. *Key Technologies for Submarine Stealth Design of Acoustic Coating*; Shanghai Jiao Tong University Press: Shanghai, China, 2012. (In Chinese)
2. Sun, W.H.; Gao, K.J.; Liu, B. Research of underwater vehicles acoustic stealthy materials. *J. Qingdao Univ. Sci. Technol. Nat. Sci. Ed.* **2017**, *38*, 67–74. (In Chinese)
3. Yoo, C.M.; Joo, J.; Lee, S.H.; Ko, Y.; Chi, S.B.; Kim, H.J.; Seo, I.; Hyeong, K. Resource assessment of polymetallic nodules using acoustic backscatter intensity data from the Korean exploration area, northeastern equatorial pacific. *Ocean Sci. J.* **2018**, *53*, 381–394. [CrossRef]
4. Boswarva, K.; Butters, A.; Fox, C.J.; Howe, J.A.; Narayanaswamy, B. Improving marine habitat mapping using high-resolution acoustic data; a predictive habitat map for the Firth of Lorn, Scotland. *Cont. Shelf Res.* **2018**, *168*, 39–47. [CrossRef]
5. Bacheler, N.M.; Shertzer, K.W.; Buckel, J.A.; Rudershausen, P.J.; Runde, B.J. Behavior of gray triggerfish Balistes capriscus around baited fish traps determined from fine-scale acoustic tracking. *Mar. Ecol. Prog. Ser.* **2018**, *606*, 133–150. [CrossRef]
6. Brown, M.G.; Godin, O.A.; Zang, X.; Ball, J.S.; Zabotin, N.A.; Zabotina, L.Y.; Williams, N.J. Ocean acoustic remote sensing using ambient noise: Results from the Florida Straits. *Geophys. J. Int.* **2016**, *206*, 574–589. [CrossRef]
7. Simmonds, M.P.; Dolman, S.J.; Jasny, M.; Parsons, E.C.M.; Weilgart, L.; Wright, A.J.; Leaper, R. Marine noise pollution–increasing recognition but need for more practical action. *J. Mar. Technol. Soc.* **2014**, *9*, 71–90.
8. Chen, M. Investigation of Broadband Absorption Mechanisms of Underwater Sound-Absrobing Materials Based on Coupled Resonators. Ph.D. Thesis, University of Chinese Academy of Sciences, Beijing, China, 2015. (In Chinese)
9. Méresse, P.; Audoly, C.; Croënne, C.; Hladky-Hennion, A.C. Acoustic coatings for maritime systems applications using resonant phenomena. *C. R. Mec.* **2015**, *343*, 645–655. [CrossRef]
10. Qian, C.; Li, Y. Review on Multi-scale Structural Design of Submarine Stealth Composite. In Proceedings of the 2017 2nd International Conference on Architectural Engineering and New Materials, Guangzhou, China, 25–26 February 2017; DEStech Publications: Lancaster, PA, USA, 2017.
11. Zhou, Q. Research on Active Control of Acoustic Wave Based on Metamaterial. Master's Thesis, Harbin Institute of Technology, Harbin, China, 2016. (In Chinese)
12. Audoly, C. Perspectives of Metamaterials for Acoustic Performances of Submerged Platforms and Systems. In Proceedings of the Undersea Defence Technology Conference 2016, Oslo, Norway, 1–3 June 2016.
13. Hladky-Hennion, A.C.; Decarpigny, J.N. Analysis of the scattering of a plane acoustic wave by a doubly periodic structure using the finite element method: Application to Alberich anechoic coatings. *J. Acoust. Soc. Am.* **1991**, *90*, 3356–3367. [CrossRef]
14. Easwaran, V.; Munjal, M.L. Analysis of reflection characteristics of a normal incidence plane wave on resonant sound absorbers: A finite element approach. *J. Acoust. Soc. Am.* **1993**, *93*, 1308–1318. [CrossRef]
15. Zhou, F.; Fan, J.; Wang, B.; Peng, Z. Absorption Performance of an Anechoic Layer with a Steel Plate Backing at Oblique Incidence. *Acoust. Aust.* **2018**, *46*, 317–327. [CrossRef]
16. Zhao, D.; Zhao, H.; Yang, H.; Wen, J. Optimization and mechanism of acoustic absorption of Alberich coatings on a steel plate in water. *Appl. Acoust.* **2018**, *140*, 183–187. [CrossRef]
17. Zeqiri, B.; Gélat, P.; Norris, M.; Robinson, S.; Beamiss, G.; Hayman, G. Design and Testing of a Novel Alberich Anechoic Acoustic Tile. In Proceedings of the 39th International Congress on Noise Control Engineering 2010, Lisbon, Portugal, 13–15 June 2010.
18. Sharma, G.S.; Skvortsov, A.; MacGillivray, I.; Kessissoglou, N. Sound transmission through a periodically voided soft elastic medium submerged in water. *Wave Motion* **2017**, *70*, 101–112. [CrossRef]
19. Tao, M.; Ye, H.; Zhao, X. Acoustic performance prediction of anechoic layer using identified viscoelastic parameters. *J. Vib. Control* **2019**, *25*, 1164–1178. [CrossRef]
20. Sohrabi, S.H.; Ketabdari, M.J. Numerical simulation of a viscoelastic sound absorbent coating with a doubly periodic array of cavities. *Cogent Eng.* **2018**, *5*, 1529721. [CrossRef]
21. Liu, G.Q.; Lou, J.J.; Li, H.T. Absorption Characteristics of Multi-layered Material Anechoic Coating on Frustum-of-a-cone Cavities. In *MATEC Web of Conferences*; EDP Sciences: Les Ulis, France, 2016.

22. Huang, L.; Xiao, Y.; Wen, J.; Zhang, H.; Wen, X. Optimization of decoupling performance of underwater acoustic coating with cavities via equivalent fluid model. *J. Sound Vib.* **2018**, *426*, 244–257. [CrossRef]

23. Ye, H.F.; Tao, M.; Zhang, W.Z. Modeling and Sound Insulation Performance Analysis of Two Honeycomb-hole Coatings. In *Journal of Physics: Conference Series*; IOP Publishing: Bristol, UK, 2018.

24. Park, S.J.; Kim, J. Simulation of underwater echo reduction using miniaturized Helmholtz resonators. *J. Acoust. Soc. Korea* **2019**, *38*, 67–72. (In Korean)

25. Zhang, C.; He, S.P.; Yi, S.Q. Model and absorption performance of anechoic coating embedding sphere cavities. *J. Ship Mech.* **2017**, *21*, 99–106. (In Chinese)

26. Zhou, J.L.; Yin, J.F.; Wen, J.H.; Yu, D.L. Underwater Insulation Properties and Mechanisms of Alberich Anechoic Coating. In Proceedings of the 2016 national conference on acoustics, Hubei, China, 11–13 February 2016. (In Chinese)

27. Sun, Y.F.; Zhang, Y.B.; Wang, X.W.; Wang, B. Research on the Influence of Restraint Structure on Acoustic Performance of Sound Insulation and Decoupling Materials. In Proceedings of the 2012 academic conference on ship materials and engineering applications, Gansu, China, 16–17 December 2012. (In Chinese)

28. Spero, A.C.; Godoy, C.M.; Harari, A.; Teague, J.M. Pressure Resistant Anechoic Coating for Undersea Platforms. U.S. Patent 7,205,043, 2007.

29. Liu, Z.; Zhang, X.; Mao, Y.; Zhu, Y.Y.; Yang, Z.; Chan, C.T.; Sheng, P. Locally resonant sonic materials. *Science* **2000**, *289*, 1734–1736. [CrossRef] [PubMed]

30. Ivansson, S.M. Anechoic coatings obtained from two-and three-dimensional monopole resonance diffraction gratings. *J. Acoust. Soc. Am.* **2012**, *131*, 2622–2637. [CrossRef] [PubMed]

31. Jiang, H.; Wang, Y.; Zhang, M.; Hu, Y.; Lan, D.; Zhang, Y.; Wei, B. Locally resonant phononic woodpile: A wide band anomalous underwater acoustic absorbing material. *Appl. Phys. Lett.* **2009**, *95*, 104101. [CrossRef]

32. Jiang, H.; Wang, Y. Phononic glass: A robust acoustic-absorption material. *J. Acoust. Soc. Am.* **2012**, *132*, 694–699. [CrossRef]

33. Lewińska, M.A.; Kouznetsova, V.G.; Van Dommelen, J.A.W.; Krushynska, A.O.; Geers, M.G.D. The attenuation performance of locally resonant acoustic metamaterials based on generalised viscoelastic modelling. *Int. J. Solids Struct.* **2017**, *126*, 163–174. [CrossRef]

34. Sharma, G.S.; Skvortsov, A.; MacGillivray, I.; Kessissoglou, N. Acoustic performance of periodic steel cylinders embedded in a viscoelastic medium. *J. Sound Vibr.* **2019**, *443*, 652–665. [CrossRef]

35. Zhong, J.; Wen, J.H.; Zhao, H.G.; Yin, J.F.; Yang, H.B. Effects of core position of locally resonant scatterers on low-frequency acoustic absorption in viscoelastic panel. *Chin. Phys. B* **2015**, *24*, 084301. [CrossRef]

36. Shi, K.; Jin, G.; Liu, R.; Ye, T.; Xue, Y. Underwater sound absorption performance of acoustic metamaterials with multilayered locally resonant scatterers. *Results Phys.* **2019**, *12*, 132–142. [CrossRef]

37. Gao, N.; Zhang, Y. A low frequency underwater metastructure composed by helix metal and viscoelastic damping rubber. *J. Vib. Control* **2019**, *25*, 538–548. [CrossRef]

38. Yu, Y.L. Optimization of Sound Absorption Performance of Underwater Composite Structure. Master's Thesis, Huazhong University of Science Technology, Wuhan, China, 2017.

39. Sharma, G.S.; Skvortsov, A.; MacGillivray, I.; Kessissoglou, N. Sound absorption by rubber coatings with periodic voids and hard inclusions. *Appl. Acoust.* **2019**, *143*, 200–210. [CrossRef]

40. Feng, L.; Li, S.P.; Hu, Q. The Research of Damped Absorbing Layer Based on Piezoelectric Shunt. In Proceedings of the 2017 Symposium on Piezoelectricity, Acoustic Waves, and Device Applications (SPAWDA), Chengdu, China, 27–30 October 2017.

41. Sun, Y.; Li, Z.; Huang, A.; Li, Q. Semi-active control of piezoelectric coating's underwater sound absorption by combining design of the shunt impedances. *J. Sound Vib.* **2015**, *355*, 19–38. [CrossRef]

42. Zhang, Z.; Huang, Y.; Huang, Q. Low-frequency broadband absorption of underwater composite anechoic coating with periodic subwavelength arrays of shunted piezoelectric patches. *Compos. Struct.* **2019**, *216*, 449–463. [CrossRef]

43. Lee, J.W.; Woo, S.; Jang, H.; Lee, K.; Kim, W.G.; Kang, H.S.; Ohm, W.S.; Park, Y.; Yoon, S.W.; Seo, Y. Low-frequency active echo reduction using a tile projector. *J. Acoust. Soc. Korea* **2014**, *33*, 366–374. (In Korean) [CrossRef]

44. Wang, W.; Thomas, P.J. Low-frequency active noise control of an underwater large-scale structure with distributed giant magnetostrictive actuators. *Sens. Actuator A Phys.* **2017**, *263*, 113–121. [CrossRef]

45. Zhang, Y.B.; Ren, C.Y.; Zhu, X. *Research on Vibration and Sound Radiation from Submarine Functionally Graded Material Nonpressure Cylindrical Shell*; Advanced Materials Research; Trans Tech Publications: Stafa-Zurich, Switzerland, 2013.

46. Zhang, C.; Shang, D.; Xiao, Y. Prediction of Anechoic Coating's Equivalent Material Parameters by Matched-velocity. In Proceedings of the 21st International Congress on Sound and Vibration 2014, Beijing, China, 13–17 July 2014.

47. Jin, Z.; Yin, Y.; Liu, B. Equivalent modulus method for finite element simulation of the sound absorption of anechoic coating backed with orthogonally rib-stiffened plate. *J. Sound Vibr.* **2016**, *366*, 357–371. [CrossRef]

48. Kwon, H.W.; Hong, S.Y.; Song, J.H. A study for acoustic target strength characteristics of submarines using kirchhoff approximation. *Mar. Technol. Soc. J.* **2017**, *51*, 52–58. [CrossRef]

49. Kim, J.Y.; Hong, S.Y.; Cho, B.G.; Song, J.H.; Kwon, H.W. Development of Acoustic Target Strength Near-Field Equation for Underwater Vehicles. Available online: https://journals.sagepub.com/doi/abs/10.1177/1475090218779292. (accessed on 11 March 2019).

50. Bo, H. Research of Statistical Characteristics of Target Strength of a Single-layer Cylindrical Shell with Random Coating. In *IOP Conference Series: Earth and Environmental Science*; IOP Publishing: Bristol, UK, 2018.

51. Wang, Y.R.; Miao, X.H.; Jiang, H.; Chen, M.; Liu, Y.; Xu, W.S.; Meng, D. Review on underwater sound absorption materials and mechanisms. *Adv. Mech.* **2017**, *47*, 92–121. (In Chinese)

52. Huang, Y.; Hou, H.; Oterkus, S.; Wei, Z.; Zhang, S. Mechanical and acoustic performance prediction model for elastomers in different environmental conditions. *J. Acoust. Soc. Am.* **2018**, *144*, 2269–2280. [CrossRef]

53. Kilcullen, P.; Shegelski, M.; Na, M.; Purschke, D.; Hegmann, F.; Reid, M. Terahertz spectroscopy and brewster angle reflection imaging of acoustic tiles. *J. Spectrosc.* **2017**, *2017*, 2134868. [CrossRef]

54. Zhang, C. *Modeling and Calculation of Underwater Viscoelastic Composite Cylindrical Shells Coupled with Acoustic Vibration*; Harbin Engineering University Press: Harbin, China, 2017; pp. 2–4. (In Chinese)

55. Mei, J.; Ma, G.; Yang, M.; Yang, Z.; Wen, W.; Sheng, P. Dark acoustic metamaterials as super absorbers for low-frequency sound. *Nat. Commun.* **2012**, *3*, 756. [CrossRef]

56. Leroy, V.; Strybulevych, A.; Lanoy, M.; Lemoult, F.; Tourin, A.; Page, J.H. Superabsorption of acoustic waves with bubble metascreens. *Phys. Rev. B* **2015**, *91*, 020301. [CrossRef]

57. Fang, N.; Xi, D.; Xu, J.; Ambati, M.; Srituravanich, W.; Sun, C.; Zhang, X. Ultrasonic metamaterials with negative modulus. *Nat. Mater.* **2006**, *5*, 452. [CrossRef]

58. Li, J.; Chan, C.T. Double-negative acoustic metamaterial. *Phys. Rev. E* **2004**, *70*, 055602. [CrossRef]

59. Cummer, S.A.; Schurig, D. One path to acoustic cloaking. *New J. Phys.* **2007**, *9*, 45. [CrossRef]

60. Skvortsov, A.; MacGillivray, I.; Sharma, G.S.; Kessissoglou, N. Sound scattering by a lattice of resonant inclusions in a soft medium. *Phys. Rev. E* **2019**, *99*, 063006. [CrossRef]

61. Wang, Y.; Zhang, C.; Ren, L.; Ichchou, M.; Galland, M.A.; Bareille, O. Sound absorption of a new bionic multi-layer absorber. *Compos. Struct.* **2014**, *108*, 400–408. [CrossRef]

Article

Extension of Solid Solubility and Structural Evolution in Nano-Structured Cu-Cr Solid Solution Induced by High-Energy Milling

Liyuan Shan [1], Xueliang Wang [2] and Yaping Wang [1,3,*

[1] MOE Key Laboratory for Nonequilibrium Synthesis and Modulation of Condensed Matter, Xi'an Jiaotong University, Xi'an 710049, China; shanliyuan7981@163.com
[2] MOE Key Laboratory of Thermo-Fluid Science and Engineering, Xi'an Jiaotong University, Xi'an 710049, China; xlwang082@mail.xjtu.edu.cn
[3] State Key Laboratory for Mechanical Behavior of Materials, Xi'an Jiaotong University, Xi'an 710049, China
* Correspondence: ypwang@mail.xjtu.edu.cn

Received: 30 October 2020; Accepted: 2 December 2020; Published: 4 December 2020

Abstract: In Cu-Cr alloys, the strengthening effects of Cr are severely limited due to the relatively low Cr solid solubility in Cu matrix. In addition, apart from the dissolved Cr, it should be noted that high proportion of Cr in Cu matrix work as the second phase dispersion strengthening. Therefore, it is of great significance to extend the Cr solid solubility and decrease the size of the undissolved Cr phase to nano-structure. In this work, the nano-sized Cu-5 wt.% Cr solid solution was achieved through high energy ball milling (HEBM) only for 12 h. The Cr solubility of ~1.15 at.% was quantitatively calculated based on XRD patterns, which means supersaturated solid solution was realized. Except for the dissolved Cr, the undissolved Cr phase was with nano-sized work as the second phase. Upon milling of the Cu-Cr powders with coarse grains, the crystallite sizes and grain sizes are found to decrease with the milling time, and remain almost unchanged at a steady-state with continued milling. In addition, it was found that the stored energy induced by dislocation density increment and grain size refinement would be high enough to overcome the thermodynamic barrier for the formation of solid solution.

Keywords: Cu-Cr system; mechanical alloying; solid solubility extension; structural evolution; thermodynamic

---

1. Introduction

Cu-Cr alloys are widely used in the electrical industry, such as contact materials and lead frames in integrated circuits, due to the well mechanical properties and high electrical conductivity [1–3]. The excellent mechanical properties are mainly produced by the formation of fine tiny Cr precipitates in Cu matrix through aging treatment, known as age or precipitation hardening [4]. However, the effect of precipitating strengthening of Cr particles is limited due to the limited maximum solubility at eutectic temperature [5]. In addition, for Cu-Cr alloys with high Cr content (>5 wt.%), the undissolved Cr phase possesses the majority proportion compared with dissolved Cr, and the coarse undissolved Cr particles may deteriorate the mechanical properties [6,7]. Therefore, it is of great significance to extend the Cr solid solubility in Cu matrix and achieve nano-sized undissolved Cr particles in Cu matrix, so that the precipitation hardening and dispersion strengthening can be combined.

In recent years, to extend the Cr solubility in Cu matrix, technology methods such as rapid solidification (RS) [8], severe plastic deformation (SPD) [9], high energy ball milling (HEBM) have been introduced and improved [10]. The common feature of the aforementioned techniques is the possibility to drastically extend the solid solubility level of nearly immiscible elements in alloys by non-equilibrium growth conditions [11,12], offering a promising route to attain high-performance composites. However, in the case of the undissolved Cr phase, RS method can only achieve micron-sized (0.1 μm) residual Cr particles, which cannot strengthen Cu matrix effectively. Mechanical alloying (MA) using high energy ball mill is not only a promising method to extend the solid solubility, but also a practical approach to achieve the homogeneous distribution of nano-structural materials from starting blended powders [10]. In addition, during the milling process, a large number of structural defects occur, such as vacancies, dislocations and stacking faults. Additionally, high energy can transfer easily to the powders due to the increase in the grain boundary volume. Under these conditions, the diffusivity of the solute atoms increase, and solid solubility extension can be achieved, namely supersaturated solid solution.

Sheibani and Fang et al. have prepared Cu-Cr alloying powders by MA treatment for 50 h or a longer time and supersaturated Cu-based solid solution (<1 at.% in case of Cu–Cr) has been formed [13–15]. Although the results showed that the MA method can increase Cr solubility in Cu matrix, and Cu-Cr bulk fabricated from the powders with MA treatment possessed high mechanical properties (1.6 Gpa) [16], there was no quantitative calculation of Cr solubility as a function of milling time and the strengthening effect of residual nano-sized Cr phase. A quantitative understanding of the formation process of supersaturated solid solution could help to design high-strength and high-conductivity nano-structural Cu-Cr alloy. More importantly, it should be mentioned that amount of impurities may get into the milled powder and contaminate it during MA treatment for such a long time (>50 h) [11,17]. In some cases, contaminations, such as Fe element, cannot be ruled out in MA materials and might significantly influence the measured properties, especially when the container is produced by stainless-steel [10].

In the present study, Cu-5 wt.% Cr blended powders were MA treated for 12 h, which is shorter than that of other studies. XRD patterns are employed to study the effect of milling time on the microstructure evolution, microstructural parameters such as the crystallize size, lattice parameter and dislocation density. The Cr solid solubility of ~1.15 at.% was quantitatively calculated based on XRD patterns, which means supersaturated solid solution was realized. Except for the dissolved Cr, the undissolved Cr phase was with nano-sized work as the second phase. In addition, thermodynamic analysis of the driving force and mechanism for the formation of supersaturated solid solution in the Cu-Cr system was conducted. The morphology and microstructure were characterized by scanning electron microscopy (SEM) and transmission electron microscopy (TEM) equipment.

2. Experimental Procedure

This work aims to extend the solid solubility of Cr and achieve nano-sized undissolved Cr particles in Cu matrix using the HEBM method. Figure 1 shows the schematic illustration of morphology changes of Cu-Cr powders during MA process, which mainly includes the fracture and welding stage. Commercial pure Cu powders and Cr powders (purity >99.5% and particle size smaller than 75 μm) were used as the raw materials. A total of 10 g of blend compositions with 5 wt.% Cr were subjected to high-energy ball milling for 12 h in stainless-steel grinding media by a Spex-8000 mill. The ball to powder weight ratio was set as 20:1, and the diameter of the milling ball was in the range of 5–8 mm. During the milling process, 2 wt.% ethanol was added as a process control agent to alleviate the aggregation of powders and realize a homogeneous supersaturated solid solution [13]. The effects of HEBM on the microstructure evolution and solid solubility extension of Cu-Cr alloy powders were investigated through different milling times (2, 4, 8, 10 and 12 h). To further evaluate the effect of the milling process on the dissolution ability of Cr in Cu matrix, Cu-5 wt.% Cr powders with MA treatment for 12 h was aging treated at 550 °C for 3 h.

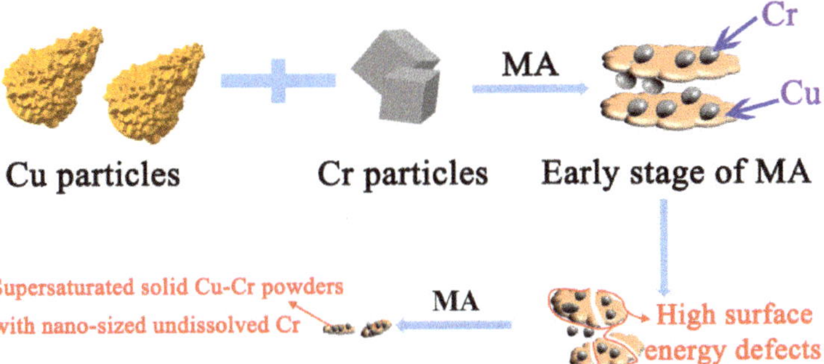

Figure 1. Schematic presentation of morphology changes during mechanical alloying (MA) process average particle size.

X-ray diffraction (XRD, Bruker D8 ADVANCE, Leipzig, Germany) with Cu Kα radiation ($\lambda = 0.15406$ nm) was utilized to determine the structural evolution of Cu-Cr powders. The patterns were recorded in a 2θ scan ranging from 20° to 100°. The crystallite size (d) powder was determined according to the Williamson–Hall plot [18]. The line broadening due to the instrument was calculated from Warren's methods [19,20]. The internal strain (ε) was calculated according to Equation (1) [21,22],

$$\varepsilon = \frac{\beta}{4 \tan \theta} \tag{1}$$

In this work, the dislocation density (ρ) can be estimated through Equation (2) based on the XRD data [1,21–23],

$$\rho = \frac{k\varepsilon}{b^2} \tag{2}$$

where k is a geometrical constant with a value of 16.1 for face-centered cubic materials, b is the Burgers vector of Cu with the value of 0.256 nm [21,24]. Meanwhile, the lattice parameter of Cu was also evaluated based on the XRD patterns.

The morphology and microstructure of Cu-Cr alloying powders were investigated using SEM (JEOL JSM-7000F, Tokyo, Japan) technique. TEM images of the samples were observed with a JEOL JEM-2100 TEM instrument operated at 120 keV with the samples ultrasonicated in ethanol for 30 min and dispersed on carbon film supported on copper grids (200 mesh). The average particle size (D) was estimated from the SEM images (the average value is obtained from more than 100 measurements).

3. Thermodynamic Theory

The thermodynamic theory was utilized to predict the stability of the Cu-Cr system. In the standard state, the Gibbs free energy (ΔG) for forming ideal A(B) solid solution from elemental powders of pure A and B can be defined as [25]:

$$\Delta G = \Delta H - T \times \Delta S \tag{3}$$

where ΔH and ΔS are the formations of enthalpy and entropy of solid solution, respectively. T is the absolute temperature at which solid solution formation is realized. Considering the rise in temperature during the milling process, the temperature of 473 K is adopted to calculate ΔG. Specifically, ΔS can be presented as follows:

$$\Delta S = -R(x_A \times lnx_A + x_B \times lnx_B) \tag{4}$$

where R is the universal gas constant (8.31 J/(mol·K)), x_A and x_B are the mole fractions of elementals A and B, respectively, $x_A + x_B = 1$.

ΔH consists of three terms based on the semi-empirical model, which can be evaluated by the following equation [25]:

$$\Delta H = \Delta H_C + \Delta H_E + \Delta H_S \tag{5}$$

where ΔH_C is the enthalpy change resulting from chemical contribution, ΔH_E represents elastic mismatch energy associating with size mismatch between solid solution and starting materials. ΔH_S is the structural contribution enthalpy due to the crystal structure difference in the mixture atoms. For the Cu-Cr system, the effect of ΔH_S is so small that it can be neglected in the calculation process [26].

Specifically, ΔH_C for binary alloy can be characterized by Equation (6) [27]:

$$\Delta H_C = 2P \times \frac{x_A \times V_A^{\frac{2}{3}} \times V_B^{\frac{2}{3}} / (x_A \times V_A^{\frac{2}{3}} + x_B \times V_B^{\frac{2}{3}})}{n_A^{\frac{-1}{3}} + n_B^{\frac{-1}{3}}} \times [-\varphi^2 + \frac{Q}{P}(n_A^{\frac{1}{3}} - n_B^{\frac{1}{3}})^2] \tag{6}$$

where P and Q are the empirical parameters related to elementals A and B. In numerical calculation, P is adopted as 14.1, and Q/P is determined as 9.4 for the Cu-Cr system [27]. V, φ and n are the mole volume, electron chemical potential and electronic density, respectively [28,29].

ΔH_E arises from elastic contribution would be given by Equation (7):

$$\Delta H_E = x_A \times x_B \left(x_A \times \Delta H_A^{\ B} + x_B \times \Delta H_B^{\ A} \right) \tag{7}$$

where $\Delta H_A^{\ B}$ and $\Delta H_B^{\ A}$ are the elastic energy contributed by A dissolving in B and B dissolving in A, respectively. In addition, $\Delta H_A^{\ B}$ and $\Delta H_B^{\ A}$ would be estimated as follows:

$$\Delta H_A^{\ B} = \frac{2K_A \times G_B \ (V_A - V_B)^2}{3K_A \times V_A + 4G_B \times V_A} \tag{8}$$

$$\Delta H_B^{\ A} = \frac{2K_B \times G_A \ (V_B - V_A)^2}{3K_B \times V_B + 4G_A \times V_B} \tag{9}$$

where K and G denote the bulk modulus and shear modulus, respectively.

Table 1 gives the input parameters in Equations (3)–(9). Accordingly, the formation enthalpy, entropy and Gibbs free energy of the Cu-Cr system at the temperature of 473 K can be calculated, as shown in Figure 2. It is easily deduced that the contribution of ΔH_C is main to form Cu-Cr solid solution. This may be due to the obvious difference of bonding energy between solid solution and the starting mixture. However, the contribution of elastic energy can be neglected due to the near zero value of ΔH_E, which can be confirmed by the small difference in molar volume between Cu and Cr, as shown in Table 1. Moreover, it should be noted that the value of Gibbs free energy is positive in alloy materials, which means that external energy should be introduced to overcome the energy barrier to achieve stable Cu-Cr supersaturated solid solution.

Table 1. Parameters to calculate enthalpy, entropy and Gibbs free energy for the Cu-Cr system.

	G (GPa)	V (10^{-6} m^3/mol)	K (GPa)	Φ (V)	$n^{1/3}$ (m^{-1})
Cu	48	7.1	137	4.45	147
Cr	115.3	7.12	160.2	4.65	173

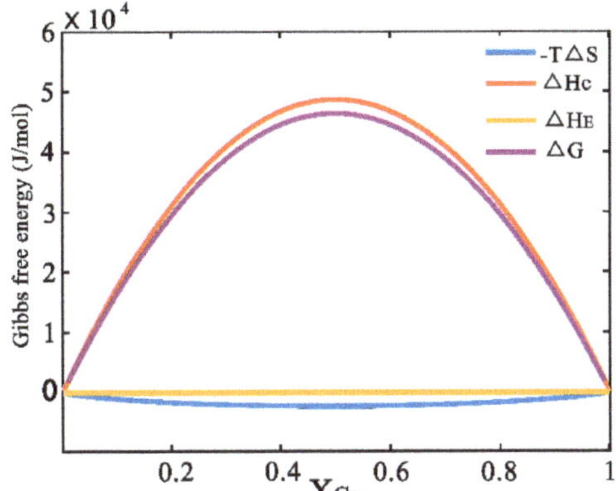

Figure 2. Enthalpy, entropy and Gibbs free energy of the Cu-Cr system at the temperature of 473 K.

4. Results and Discussion

4.1. Microstructure Evolution

XRD analysis was performed to evaluate the effect of milling time on the structural evolution of Cu-Cr powder, the pattern of raw Cu-5 wt.% Cr mixtures are presented as a reference. In the XRD patterns of all samples, the strongest face-centered cubic (fcc) diffraction peak of Cu (111) appears at $2\theta = 43.297°$. In addition, the strongest body-centered cubic (bcc) Cr (110) peak in the standard pattern appears at $2\theta = 44.392°$. According to the studies of Bachmaier and Zhao, in which the lattice parameter changes for Cu, calculated from the (200, 220, 311) and (111, 222) peaks, are quite similar [27,28]. Therefore, the specific XRD patterns are analyzed based on Cu (111) peak within the range of $2\theta = \{41.5, 45.5\}$ in this work, as shown in Figure 3. No significant signs of Cr peaks can be observed in the XRD pattern, this can be attributed to the lower volume of Cr phase.

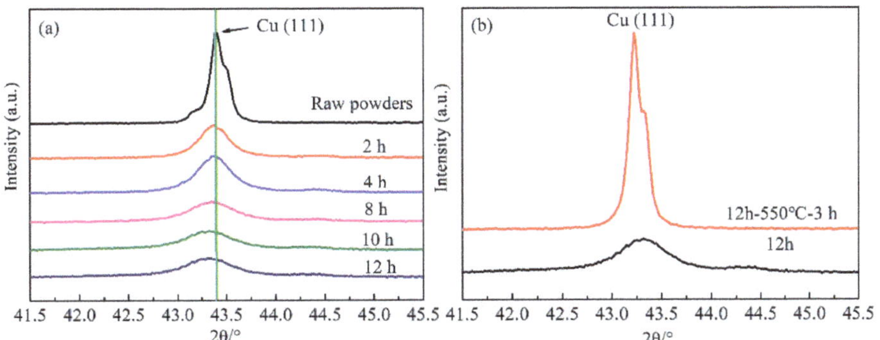

Figure 3. XRD patterns of Cu-Cr powders: (a) pattern changes as a function of milling time; (b) comparison of Cu (111) peak in the final samples before and after aging treatment.

In Figure 3a, it can be clearly visible that the fcc Cu (111) peaks become wider with the increase of milling time. The Cu peaks are broadened by increased milling time since the continuous deformation of powder particles during the milling process results in crystallite refinement and an increase in lattice microstrain, which is consistent with the research of Sheibani [13,30]. In addition, the peak intensity reduction and shift of diffraction angle can be observed, which display the typical behaviors of MA treatment powders. Interestingly, the relative intensity and width of Cu (111) peaks exhibit no significant change with the milling time of 8, 10 and 12 h, indicating that the diffusivity of Cr in Cu substrate becomes slow after MA treatment for 8 h.

Figure 3b shows the XRD patterns of MA treatment for 12 h to Cu-5 wt.% Cr powders and then followed by aging treatment. In comparison, the Cu (111) peak of the sample with aging treatment becomes sharper, as well as the intensity is higher. It seems like the alloying powders after aging treatment transformed into the mixture of the Cu phase and Cr phase. The difference in XRD patterns results from the dissolved Cr atoms precipitated out of Cu substrate during the subsequent heat treatment.

To further evaluate the effect of milling time on Cu-Cr alloyed powders, the crystallize size, lattice parameter and dislocation density of Cu in all samples were analyzed based on the XRD data. As shown in Table 2, the crystallize size decrease gradually with the increase of milling time. In addition, Cu lattice parameter has a different rising due to the dissolubility of Cr atoms. These results are in excellent agreement with the previous research, in which Sheibani believed that the only evidence of solid solubility extension was lattice parameter changes [13]. Deserved to be mentioned, up to 8 h of milling time, the rapid changes of crystallize size and lattice parameter can be noticed, after that a slower change trend is observed, which could be also confirmed with the small change of diffraction peak intensity and peak width as previously mentioned. Therefore, it could be concluded that in the Cu-Cr powders with coarse grains, the crystallite sizes are found to decrease gradually with the milling time, and remain small changes with continued milling. This phenomenon can be explained by the following reasons:

Table 2. Variations in the crystallize size, lattice parameter and dislocation density.

Milling Time (h)	0	2	4	8	10	12
Crystallize size (nm)	598.35 ± 4.6	235.35 ± 5.6	208.25 ± 3.2	156.45 ± 4.9	146.89 ± 2.1	133.54 ± 4.9
Lattice parameter (nm)	0.36156	0.36175	0.36178	0.36183	0.36184	0.36186
Dislocation density ($\times 10^{16}$ m/m^3)	-	6.46	8.60	10.6	12.0	13.0

At the beginning of MA process, the ductile Cu matrix was subjected to the ball collisions, leading to the improvement of surface activity. The brittle Cr phase fractured and size decreased more quickly, being easy to adhere onto the soft Cu surface and transfer into Cu lattice. However, under repeated collisions, the surface layer of Cu particles got hardened due to the dissolution of solute atoms, so that the diffusion of Cr and variation of crystallize size are going to be difficult. The steady-state value was proposed and needs to be verified further.

4.2. Morphology of Cu-Cr Alloying Powders

SEM morphologies concerning milling time are shown in Figure 4. As marked in Figure 4a, raw Cu powders have the typical dendritic appearance, coarse Cr powders are used for the strengthening phase. The mixed powders transform to a layer-shaped structure due to the excellent ductility of Cu in the initial stage of MA process, tending to be easily cold welding and appear as large layer [14]. After that, the fracture occurs more frequently than cold welding, so that the layered structure gradually refined, as presented in Figure 4c. In addition, the increase of surface energy and the defect can make the refined Cr adhere easily onto Cu particles. By comparing Figure 4e,f, there is no big difference in particle size and morphology. However, the coarsening of powders cannot be avoided even milling with 12 h due to the existence of cold welding, as displayed in Figure 4f. During the long-time of milling, the harder Cr particles not only act as the abrasive particles but also can dissolve in the softer Cu particles due to the high density of structural defects.

Figure 4. Morphology changes of Cu-Cr powders with the milling time. (**a**) 0 h, (**b**) 2 h, (**c**) 4 h, (**d**) 8 h, (**e**) 10 h, (**f**) 12 h.

The average particle size changes as a function of milling time are shown in Figure 5. It can be observed that the average particle size of sample milled with 2 h is a little larger than that of raw materials. In addition, MA treatment for 12 h to sample possess the minimum particle size. This result is consistent with that of Figure 4, in which the particle size becomes coarse firstly and then decreases tending to remain unchanged with continued milling. The mentioned phenomena also indicate that the solid solubility increase trend becomes slower after 8 h of milling, which will be estimated in the next part.

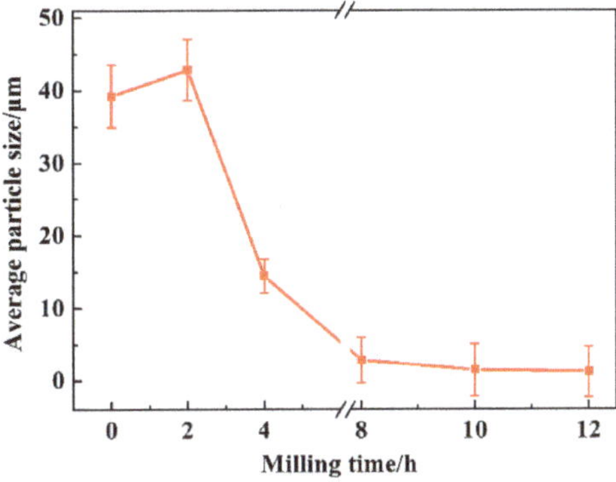

Figure 5. The average particle size of Cu-Cr alloying powders as a function of milling time.

4.3. Effect of Milling Time on the Solid Solubility

The solid solubility level is related to the lattice parameter changes of the matrix obtained either for equilibrium or nonequilibrium processing techniques [10]. More importantly, the structure factors for alloyed phases produced by method of rapidly solidified, irradiated, sputtered and MA have been found to be very similar [31]. As mentioned previously, Sheiban believed that the only evidence of solid solubility extension was lattice parameter changes [12]. In this regard, the solid solubility of Cr in Cu matrix obtained by MA treatment can be determined by Equation (10) [4]:

$$c = \delta a / 0.00026 \tag{10}$$

where c is the Cr content in MA treated Cu-Cr alloy, at.%, δa denotes the lattice parameter variation due to the dissolubility of Cr atoms, nm. The value 0.00026 is the factor, nm/at.%, which was established for the extended solid solubility obtained by chill block melt spinning [4,32].

Figure 6 displays the changes in Cr solubility and Cu internal strain concerning milling time. With the increase of milling time, Cr solubility and the internal strain of Cu both increase rapidly, especially at the 4 h of milling time for Cr solubility. The increase of Cu internal strain is mainly due to the refinement of crystallize size, as displayed in Table 2. Crystalize size decreased rapidly up to 4 h of milling time, and slowly decreased from there. However, the lattice parameter increased in a steady-stage during the MA process. In the case of the internal strain, the increase is slow for the entire time of milling, while the Cr solubility increased sharply within the initial 4 h of milling. The slower increase rate after 4 h would be attributed to the diffusion of Cr becoming difficult. In addition, it can be noted that the Cr solubility (1.07 at.%, 1.15 at.%) reach to the usual saturation (0.89 at.%) with milling time more than 8 h. The supersaturated Cu-Cr solid solution powder is prepared successfully with MA treatment for more than 8 h in this work. According to the trace of Cr solubility curve, the Cr solubility tends to increase very slowly or reach a constant with the increase in milling time, as well as the crystallize size.

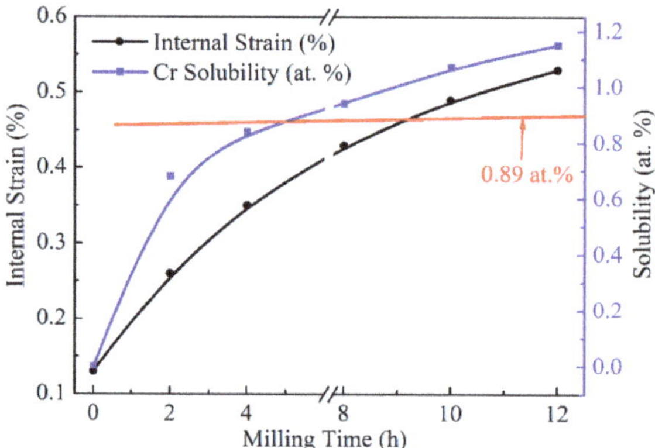

Figure 6. Effect of milling time on Cu internal strain and Cr solid solubility.

4.4. Gibbs Free Energy Changes of Cu-Cr Alloying Powders

The Gibbs free energy in the Cu-Cr system will change due to the formation of solid solution. As mentioned in Table 2, with the increase of milling time, the typical crystallite size refinement and dislocation density increase can be observed. On one hand, the increment of dislocation density during the milling process would contribute to the total Gibbs free energy (ΔG_t). The energy change due to the dislocation density increment, ΔG_d, can be calculated by Equation (11) [13]:

$$\Delta G_d = \zeta \rho V_m \tag{11}$$

where ρ and V_m are the dislocation density and molar volume, respectively. The molar volume of Cu is 7.1 cm³/mol [13]. ζ, the dislocation elastic energy per unit length of dislocation lines, can be calculated using the following equation [33]:

$$\zeta = \frac{Gb^2}{4\pi} \times ln\left(\frac{R_e}{b}\right) \tag{12}$$

where $G = 4.8 \times 10^{10}$ N/m² is the shear modulus for Cu [22], b is the burgers vector. R_e is the outer cutoff radius of dislocation, which can be taken as crystallite size in nanocrystalline materials [13,32].

On the other hand, the Gibbs free energy in the Cu-Cr system also increases with the refinement of crystallite size, which determined by Equation (13):

$$\Delta G_c = \gamma\left(\frac{A}{V}\right)V_m \tag{13}$$

where γ is the grain boundary energy with the value of 625 mJ/m^2 for Cu, A/V and V_m are the surface volume ratio and molar volume, respectively. To determine ΔG_c, spherical crystallite was considered [34].

Figure 7 displays the Gibbs free energy increment due to the variation of crystallite size and dislocation density. It can be seen that the increment of Gibbs free energy resulting from crystallite size and dislocation density changes are ~1.45 and 0.99 kJ/mol with MA treatment for 12 h, respectively. It can be visible that the contribution of dislocation density is higher than that of crystallite size. The total Gibbs free energy, ΔG_t, also was calculated based on ΔG_c and ΔG_d. Therefore, the maximum ΔG_t with milling time for 12 h would be 2.44 kJ/mol. By comparing 2.44 kJ/mol with the value in ΔG trace in Figure 1, it can be concluded that the energy deriving from MA process would be high enough to form the Cu-Cr supersaturated solid solution. In other words, the stored energy obtained by the variation of crystallite size and dislocation density can extend the Cr solid solubility in Cu matrix.

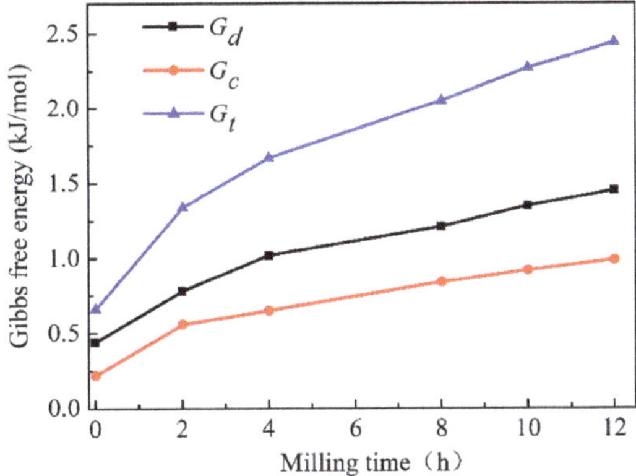

Figure 7. Effect of Cu crystallite size and dislocation density on Gibbs free energy.

However, the thermodynamic barrier is about 2.62 kJ/mol in Figure 1 while forming the same content Cr (1.15 at.%, MA treatment for 12 h) supersaturated solid solution, which is a little higher than 2.44 kJ/mol. The small difference in these two values may be due to only mean crystallite size was employed to estimate the energy. The smaller size crystallite was neglected which presents a lognormal distribution [34]. In addition, although a large number of structural defects occur during MA process, such as vacancies, dislocations and stacking faults, only dislocations were considered to estimate the energy increase.

4.5. HRTEM Characterization

In addition to the XRD technique, TEM equipment was also employed to study the microstructure of the sample subjected to milling for 12 h. Figure 8a shows the typical bright-field image, numerous nanograins (~100 nm) can be visible in the bright-field image, which is agreed well with that estimated by the Debye Scherrer equation. The corresponding selected area electron diffraction (SAED) analysis is inserted in the bottom-right corner, in which Debye–Scherrer rings exhibit two kinds of diffraction rings. As marked, (200), (220) and (311) diffraction rings belong to the fcc Cu phase, and an additional (110) ring pattern can be attributed to the bcc Cr phase. The existing Cr ring pattern demonstrates that there are residual nano Cr particles or Cr-rich phase except for the dissolved Cr in Cu matrix, which cannot be observed in XRD patterns due to the small amount. It should be mentioned that the residual nano-sized Cr phase also can act as secondary strengthening particles by dispersing uniform in Cu matrix. This phenomenon is comparable with the calculated solid solubility result, in which the solid solubility value is ~1.15 at.% after MA treatment for 12 h. More importantly, the lattice parameter of fcc Cu (200), Cu (220) and Cu (311) ring patterns in Figure 8a are 0.36199, 0.36212 and 0.36215 nm, respectively, which are larger than that of pure Cu. The increase in lattice parameter can be attributed to Cr atoms dissolve in Cu matrix, which has been estimated before in the analysis of XRD patterns. There is a little difference in the result of lattice parameters determined from XRD and TEM since both methods present different sensitivities concerning small misorientations of the crystal lattice [17]. Moreover, from the SEM and TEM images, it can be observed that samples subjected to milling for 12 h have good homogeneity.

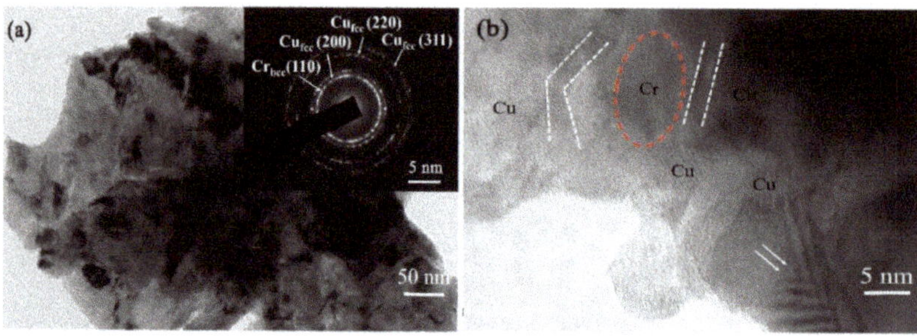

Figure 8. The microstructure of Cu-Cr alloying powders subjected to MA for 12 h: (**a**) bright-field image (the inset bottom-right is SAED pattern); (**b**) typical HRTEM image.

To further understand the atomic structure of supersaturated solid solution Cu-Cr powders, the high-resolution TEM (HRTEM) image was taken, as presented in Figure 8b. It can be observed that several Cu crystallites are distributed around a residual nano-sized Cr crystallite based on the difference of lattice fringe. In addition, Cu grains have evident twins and deformed zone (as indicated by double arrows), which is different from Cr grains. The interface of Cu and Cr (labeled with white dashed line) is ~2–3 nm, which is visible clearly. However, the interface at other regions is ambiguous, this may be caused by Cr atoms dissolving in Cu lattice or by severe deformation after MA treatment, so that two or more grains overlapping.

5. Conclusions

Supersaturated Cu-Cr solid solution has been formed by the MA method. This could be noticed from the XRD data and the variation in the lattice parameter of Cu. The changes of lattice parameter are believed the only evidence of solubility extension, and the quantitative calculated results show that the Cr solubility is ~1.15 at.% for the 12 h milling powders. Morphology and particle size have been changed during MA, in the early stage processing, especially up to 4 h, the average particle size decreased rapidly with increasing milling and the internal strain accumulation. In the case of thermodynamic changes, the Gibbs free energy induced by crystallite size decrease and dislocation density increment in the Cu-Cr alloy system would be high enough to increase the Cr solubility.

Author Contributions: Conceptualization, L.S. and Y.W.; data curation, L.S.; investigation, L.S. and Y.W.; methodology, L.S.; writing—original draft preparation, L.S.; writing—review and editing, X.W.; supervision, Y.W.; validation, X.W. All authors have read and agreed to the published version of the manuscript.

Funding: This work was supported by the National Natural Science Foundation of China-State Grid Joint Fund for Smart Grid [U1866203], the National Natural Science Foundation of China [51906189], Key Scientific and Technological Innovation Team of Shaanxi province [2020TD-001].

Conflicts of Interest: The authors declare no conflict of interest.

References

1. Peng, L.M.; Mao, X.M.; Xu, K.D.; Ding, W.J. Property and thermal stability of in situ composite Cu–Cr alloy contact cable. *J. Mater. Process. Technol.* **2005**, *166*, 193–198. [CrossRef]
2. Su, J.H.; Dong, Q.M.; Liu, P.; Li, H.J.; Kang, B.X. Research on aging precipitation in a Cu–Cr–Zr–Mg alloy. *Mater. Sci. Eng. A.* **2005**, *392*, 422–426. [CrossRef]
3. Sun, Z.B.; Zhang, C.Y.; Zhu, Y.M.; Yang, Z.M.; Ding, B.J.; Song, X.P. Microstructures of melt-spun Cu100−x–Crx (x=3.4–25) ribbons. *J. Alloys Compd.* **2003**, *361*, 165–168. [CrossRef]
4. Correia, J.B.; Davies, H.A.; Sellars, C.M. Strengthening in rapidly solidified age hardened Cu-Cr and Cu-Cr-Zr alloys. *Acta Mater.* **1997**, *45*, 177–190. [CrossRef]
5. Massalski, T.B. Binary alloy phase diagrams. *Am. Soc. Met.* **1986**, *1*, 819–838.
6. Jin, Y.; Adachi, K.; Takeuchi, T.; Suzuki, G. Ageing Characteristics of Cu-Cr in-situ Cu-Cr composite. *J. Mater. Sci.* **1998**, *33*, 1333–1341. [CrossRef]
7. He, W.; Wang, E.; Hu, L.; Yu, Y.; Sun, H. Effect of extrusion on microstructure and properties of a submicron crystalline Cu–5wt.%Cr alloy. *J. Mater. Process. Technol.* **2008**, *208*, 205–210. [CrossRef]
8. Christiansen, L.; Kirchlechner, I.; Breitbach, B.; Liebscher, C.H.; Manjon, A.; Springer, H.; Dehm, G. Synthesis, microstructure, and hardness of rapidly solidified Cu-Cr alloys. *J. Alloys Compd.* **2019**, *794*, 203–209.
9. Kapoor, G.; Kvackaj, T.; Heczel, A.; Bidulska, J.; Kocisko, R.; Fogarassy, Z.; Simcak, D.; Gubicza, J. The Influence of Severe Plastic Deformation and Subsequent Annealing on the Microstructure and Hardness of a Cu-Cr-Zr Alloy. *Materials* **2020**, *13*, 2241. [CrossRef]
10. Suryanarayana, C. Mechanical alloying and milling. *Prog. Mater Sci.* **2001**, *46*, 1–184. [CrossRef]
11. Bachmaier, A.; Pfaff, M.; Stolpe, M.; Aboulfadl, H.; Motz, C. Phase separation of a supersaturated nanocrystalline Cu–Co alloy and its influence on thermal stability. *Acta Mater.* **2015**, *96*, 269–283. [CrossRef]
12. Raghavan, R.; Harzer, T.P.; Chawla, V.; Djaziri, S.; Phillipi, B.; Wehrs, J.; Wheeler, J.M.; Michler, J.; Dehm, G. Comparing small scale plasticity of copper-chromium nanolayered and alloyed thin films at elevated temperatures. *Acta Mater.* **2015**, *93*, 175–186. [CrossRef]
13. Sheibani, S.; Heshmati-Manesh, S.; Ataie, A. Structural investigation on nano-crystalline Cu–Cr supersaturated solid solution prepared by mechanical alloying. *J. Alloys Compd.* **2010**, *495*, 59–62. [CrossRef]
14. Sahani, P.; Mula, S.; Roy, P.K.; Kang, P.C.; Koch, C.C. Structural investigation of vacuum sintered Cu–Cr and Cu–Cr–4% SiC nanocomposites prepared by mechanical alloying. *Mater. Sci. Eng. A* **2011**, *528*, 7781–7789. [CrossRef]
15. Fang, Q.; Kang, Z.X. An investigation on morphology and structure of Cu–Cr alloy powders prepared by mechanical milling and alloying. *Powder Technol.* **2015**, *270*, 104–111. [CrossRef]
16. Fang, Q.; Kang, Z.; Gan, Y.; Long, Y. Microstructures and mechanical properties of spark plasma sintered Cu–Cr composites prepared by mechanical milling and alloying. *Mater. Des.* **2015**, *88*, 8–15. [CrossRef]
17. Wang, Z.B.; Lu, K.; Wilde, G.; Divinski, S.V. Interfacial diffusion in Cu with a gradient nanostructured surface layer. *Acta Mater.* **2010**, *58*, 2376–2386. [CrossRef]
18. Williamson, G.K.; Hall, W.H. X-ray line broadening from filed aluminium and wolfram. *Acta Metall.* **1953**, *1*, 22–31. [CrossRef]
19. Cullity, B.D.; Stock, S.R. *Elements of X-ray Diffraction*, 3rd ed.; Prentice Hall: Upper Saddle River, NJ, USA, 2001.
20. Lipson, H.; Steeple, H. *Interpretation of X-ray Powder Diffraction Patterns*; Macmillan: London, UK, 1970.
21. Chander, S.; Dhaka, M.S. Influence of thickness on physical properties of vacuum evaporated polycrystalline CdTe thin films for solar cell applications. *Physica E* **2016**, *76*, 52–59. [CrossRef]
22. Chander, S.; Dhaka, M.S. Impact of thermal annealing on physical properties of vacuum evaporated polycrystalline CdTe thin films for solar cell applications. *Physica E* **2016**, *80*, 62–68. [CrossRef]
23. Guo, X.; Xiao, Z.; Qiu, W.; Li, Z.; Zhao, Z.; Wang, X.; Jiang, Y. Microstructure and properties of Cu-Cr-Nb alloy with high strength, high electrical conductivity and good softening resistance performance at elevated temperature. *Mater. Sci. Eng. A* **2019**, *749*, 281–290. [CrossRef]
24. Shumaker, C. *Introduction to Metallurgical Thermodynamics*, 2nd ed.; McGraw-Hill: New York, NY, USA, 1981.
25. Miedema, A.R.; DeBoer, F.R.; Boom, R. Predicting heat effects in alloys. *Physica B* **1981**, *103*, 67–81. [CrossRef]

26. Bakker, H.; Zhou, G.G.; Yang, H. Mechanically driven disorder and phase transformations in alloys. *Prog. Mater. Sci.* 1995 39, 159–241. [CrossRef]

27. Niessen, A.K.; Miedema, A.R.; Andries, R. The Enthalpy Effect on Forming Diluted Solid Solutions of Two 4d and 5d Transition Metals. *Phys. Chem. Chem. Phys.* **1983**, *87*, 717–725. [CrossRef]

28. Niessen, A.K.; Miedema, A.R.; Boer, F.R.; Boom, R. Enthalpies of formation of liquid and solid binary alloys based on 3d metals. *Alloys Cobalt Phys. B+C* **1988**, *151*, 401–432. [CrossRef]

29. Bachmaier, A.; Rathmayr, G.B.; Bartosik, M.; Apel, D.; Zhang, Z.; Pippan, R. New insights on the formation of supersaturated solid solutions in the Cu–Cr system deformed by high-pressure torsion. *Acta Mater.* **2014**, *69*, 301–313. [CrossRef]

30. Sheibani, S.; Heshmati-Manesh, S.; Ataie, A. Influence of Al_2O_3 nanoparticles on solubility extension of Cr in Cu by mechanical alloying. *Acta Mater.* **2011**, *58*, 6828–6834. [CrossRef]

31. Wagner, C.N.; Boldrick, M.S. The structure of amorphous binary metal-metal alloys prepared by mechanical alloying. *J. Alloys Compd.* **1993**, *194*, 295–302. [CrossRef]

32. Tenwick, M.J.; Davies, H.A. Enhanced Strength in High Conductivity Copper Alloys. *Mater. Sci. Eng. A* **1998**, *98*, 543–546. [CrossRef]

33. Williamson, G.K.; Smallman, R.E. Dislocation densities in some annealed and cold-worked metals from measurements on the X-ray debye-scherrer spectrum. *Philos. Mag.* **1956**, *1*, 34–46. [CrossRef]

34. Aguilar, C.; Martinez, V.d.P.; Palacios, J.M.; Ordoñez, S.; Pavez, O. A thermodynamic approach to energy storage on mechanical alloying of the Cu–Cr system. *Scr. Mater.* **2007**, *57*, 213–216. [CrossRef]

Publisher's Note: MDPI stays neutral with regard to jurisdictional claims in published maps and institutional affiliations.

 *materials*

Letter

In Situ Formation of Ti47Cu38Zr7.5Fe2.5Sn2Si1Nb2 Amorphous Coating by Laser Surface Remelting

Peizhen Li, Lingtao Meng, Shenghai Wang *, Kunlun Wang, Qingxuan Sui, Lingyu Liu, Yuying Zhang, Xiaotian Yin, Qingxia Zhang and Li Wang *

School of Mechanical, Electrical & Information Engineering, Shandong University (Weihai), Weihai 264209, China; peizhenl@yeah.net (P.L.); mlt142536@163.com (L.M.); wkl@sdu.edu.cn (K.W.); qingxuansui@163.com (Q.S.); l1024lly@163.com (L.L.); zyy741790710@163.com (Y.Z.); 15262019723@163.com (X.Y.); 17862703363@163.com (Q.Z.)

* Correspondence: shenghaiw@163.com (S.W.); wanglihxf@sdu.edu.cn (L.W.);
   Tel.: +86-631-568-8224 (S.W. & L.W.)

Received: 21 October 2019; Accepted: 4 November 2019; Published: 7 November 2019

Abstract: In previous studies, Ti-based bulk metallic glasses (BMGs) free from Ni and Be were developed as promising biomaterials. Corresponding amorphous coatings might have low elastic modulus, remarkable wear resistance, good corrosion resistance, and biocompatibility. However, the amorphous coatings obtained by the common methods (high velocity oxygen fuel, laser cladding, etc.) have cracks, micro-pores, and unfused particles. In this work, a Ti-based Ti47Cu38Zr7.5Fe2.5Sn2Si1Nb2 amorphous coating with a maximum thickness of about 100 μm was obtained by laser surface remelting (LSR). The in-situ formation makes the coating dense and strongly bonded. It exhibited better corrosion resistance than the matrix and its corrosion mechanism was discussed. The effects of LSR on the microstructural evolution of Ti-based prefabricated alloy sheets were investigated. The nano-hardness in the heat affected zone (HAZ) was markedly increased by 51%, meanwhile the elastic modulus of the amorphous coating was decreased by 18%. This demonstrated that LSR could be an effective method to manufacture the high-quality amorphous coating. The in-situ amorphous coating free from Ni and Be had a low modulus, which might be a potential corrosion-resistant biomaterial.

Keywords: in situ amorphous coating; laser surface remelting; Ti-based alloy

---

1. Introduction

Ti-based alloys are widely used in dental implants and orthopedic prostheses because of their high corrosion resistance and good biocompatibility. However, the elastic modulus of the alloy implant is greater compared with that of bone, causing stress shielding and bone resorption, which ultimately leads to implant failure [1]. To meet the requirements of replacing bones, developing metallic materials with high strength, low elastic modulus, high wear and corrosion resistance, and good biocompatibility is desired [2–4]. Ti-based bulk metallic glasses (BMGs) have potential applications in biomedical fields due to their low density, high specific strength, good corrosion resistance, and excellent biocompatibility, etc. [5,6]. However, the application of BMGs is very difficult, for one reason, the preparation of amorphous alloy requires cooling the molten metal liquid at an extremely fast quenching rate ($\sim 10^5$ K/s), where the larger the workpiece, the harder to achieve. Otherwise the material will crystallize and lose the remarkable properties of the amorphous. In the Ti-based BMGs, a high content of Be (a toxic element) and Pd (an expensive metal element) are often added to maximize the glass forming ability (GFA), which confines the application of Ti-based BMGs as a biomaterial [6,7]. In this situation, if an amorphous coating can be prepared on the surface of a bulk material, it is possible to obtain significant surface corrosion resistance, wear resistance, and biological adhesion. We no longer consider whether the substrate is completely amorphous, because only the surface properties of the material are considered. Common methods for preparing amorphous coatings are high velocity oxygen fuel (HVOF) [8], laser cladding [9], ion bombardment [10], surface mechanical attrition treatment (SMAT) [11], etc. The coatings obtained by ion bombardment and SMAT are nanoscale, which is easily broken down by abrasion and usually has partial crystallization. The coatings prepared by HVOF and laser cladding are completely amorphous, but have a large number of cracks, micro-pores, and unfused particles, which greatly reduces their immersion corrosion resistance [12]. The in situ formation method might form a coating without these defects. Laser surface remelting (LSR) is accompanied by an instantaneous cooling rate on the order of 10^5–10^8 K/s, which is considered to be a more effective method for the surface modification of titanium alloys [13,14]. To date, studies of laser surface modification have usually focused on refining grains and obtaining crystal phases [14,15]. We believe that LSR may be a potential method for preparing amorphous coatings due to its fast cooling rate, ease of changing process parameters, and low cost.

In previous studies, it was found that Ti-based BMGs free from Ni and Be can be formed in a Ti–Cu–Zr–Fe–Sn–Si alloy system. This is considered to be promising for biomedical applications [16,17]. We added a small amount of Nb element to this system to improve the corrosion resistance of the alloy [18]. A high Nb element content will reduce the GFA of the system. Finally, we chose a nominal composition of Ti47Cu38Zr7.5Fe2.5Sn2Si1Nb2 (at. %) as the subject. We intended to develop an amorphous coating with notable corrosion resistance and potential bio-application prospects.

In the present work, LSR was conducted for a prefabricated Ti-based alloy sheet with the laser-induced microstructures and phase changes carefully characterized by back-scattered electron (BSE) imaging and x-ray diffraction (XRD) techniques. In addition, hardness variation across the laser modified zones was examined, and corrosion resistance was also analyzed by anodic polarization experiments.

2. Materials and Methods

Alloy ingots with a nominal composition of Ti47Cu38Zr7.5Fe2.5Sn2Si1Nb2 (at. %) were prepared by arc-melting the mixture of the pure elements under a high-purity argon atmosphere. The intermediate alloy was melted five times to ensure the uniformity of ingredients. Plate samples with thicknesses of 2 mm, 1.5 mm, and 1 mm were fabricated by suction casting with water cooled copper molds. The surface of the specimens was polished to 3000-grit by silicon carbide paper and cleaned by absolute ethanol. The laser surface remelting (LSR) experiment was carried out under the protection of a flowing high-purity argon atmosphere, accomplished by a commercial selective laser melting machine (SLM Solutions 125HL, Lubeck, Germany) equipped with a 1067 nm wavelength fiber laser of up to 400 W laser power with a beam diameter of about 80 μm. The surface of the melt receives the effect of spheroidization, causing uneven surface topography (Figure 1a). Figure 1b shows the scheme of the LSR experiment.

The linear energy density (LED), as the single line effective energy input during LSR, can be calculated by P/v (J/mm) [19,20]. P represents the laser power and v represents the scanning speed. In this paper, scanning speed (v) was 2000 mm/s, hatch distance (h) was 140 μm, and laser powers (P) were 0 W, 140 W, 160 W, 180 W, 200 W, 220 W, 240 W, 260 W, 280 W, 300 W, and 320 W. All combinations of different thicknesses of samples and laser powers were processed three times to ensure repeatability. The structures of the treated samples were conducted via x-ray diffraction (XRD, Ultima IV, Rigaku, Tokyo, Japan) with Cu Kα radiation. The nanoindentation characteristics of the sample cross sections were acquired by a Nano Indenter NHT2 (Anton Paar, Austria). The maximum load was 15 mN, and the loading speed was 30 mN/min. Microstructural characteristics under laser treatment were investigated by scanning electron microscopy (SEM, Nova Nano SEM450, FEI Sirion, Hillsboro, OR, USA). To evaluate the corrosion resistance, anodic polarization experiments were implemented in 3.5 wt. % NaCl aqueous solution at room temperature.

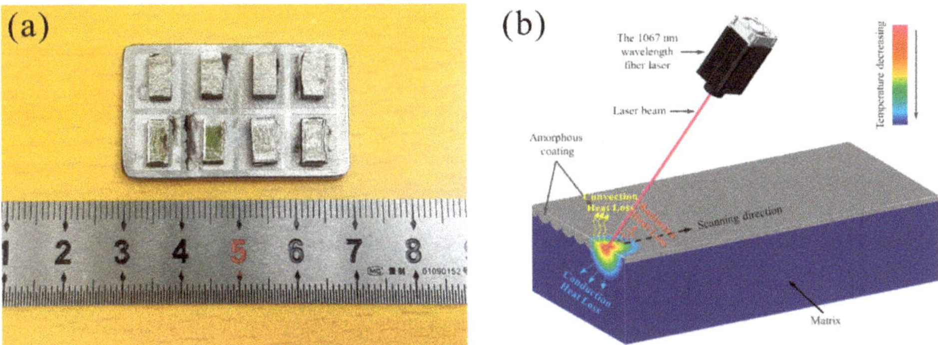

Figure 1. (**a**) The plate samples disposed by different laser powers; (**b**) Scheme of the laser surface remelting.

3. Results and Discussion

XRD patterns of the Ti-based alloy plates treated with different laser powers are shown in Figure 2. All diffraction peaks of the untreated samples with a thickness of 2 mm matched $CuTi_2$, CuTi, and Cu_4Ti_3 phases. Similar to the treatment with 200 W (LED = 0.10 J/mm), the 240 W (LED = 0.12 J/mm) exhibited a main halo without any Bragg peaks in the XRD pattern, indicating the amorphous structure. The tight combination of the melt and the solid matrix causes rapid heat extraction during solidification, allowing the melt to form a metallic glass under very high cooling rates on the order of 10^5–10^8 K/s [14]. For the 280 W, LED = 0.14 J/mm, sharp peaks corresponding to intermetallic compounds $CuTi_2$ and CuTi superimposed on the broad peak of amorphous phase were observed (Figure 2a).

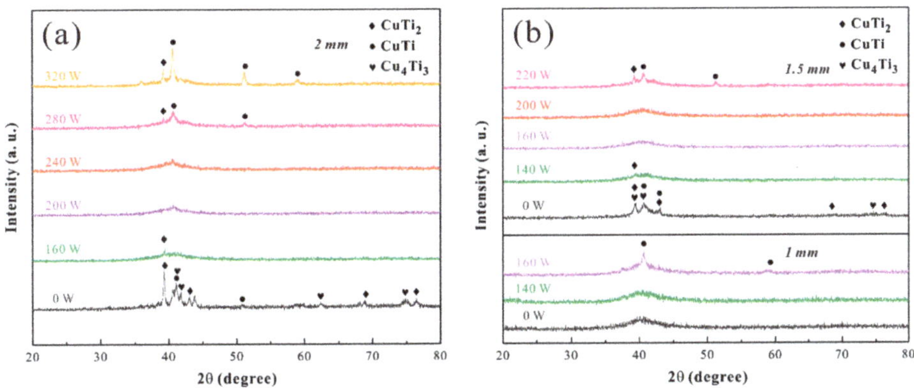

Figure 2. X-ray diffraction (XRD) patterns of samples with (**a**) 2 mm treated under different laser powers; (**b**) 1.5 mm and 1 mm treated under different powers.

The raw samples with a thickness of 1.5 mm had more nanocrystalline structure inside than the 2 mm samples. The nanocrystalline structure was closer to the long-range disorder than the dendrites of the matrix, which had a higher entropy value [21]. From a thermodynamic point of view, the high-conflict atom stacking arrangement corresponded to the "high-entropy" property, and makes the Gibbs free energy drop faster with decreasing temperature, which is conducive to the formation of amorphous structures during solidification. In addition, the laser heat melts the titanium plate during LSR. The generated heat can be dissipated by convection, radiation, and the material itself in a conduction way during remelting [15], while conduction heat loss in a dominant manner (see the blue arrow in Figure 1b). Thick critical dimension heats up more slowly when subjected to the same amount of conduction heat, resulting in a larger temperature gradient, which contributes to achieving a higher cooling rate. Therefore, the crystal in the sample of 1 mm was induced at a lower critical power (160 W, Figure 2b) than the 2 mm sample (280 W, Figure 2a) and the 1.5 mm sample (220 W, Figure 2b).

Figure 3a is the SEM image of the untreated alloy in backscattered electron (BSE) mode, showing two regions with different morphologies. The surface of the sample was obtained at a faster cooling rate to form a uniform nanocrystal, and the black dendrites were continuously grown due to the slower cooling rate inside. In Figure 3b, the material in the bright region has undergone laser remelting and re-solidification to form an amorphous coating (AC) with a maximum thickness of about 100 μm. There was another modification zone labeled as the heat affected zone (HAZ) beneath the AC. It can be seen that the microstructure of the heat-affected zone is similar to the surface structure in Figure 3a, consisting of uniformly fine nanocrystals. This illustrates the proximity of the HAZ cooling rate to the water-cooled copper mold. After the LSR treatment, the fine and uniform nanocrystalline structure begins to grow under the thermal effect, and grows radially as the dominant form of dark nanocrystals (Figure 3c, inset). Since the nanocrystals have a large atomic diffusion activation energy, the atomic co-diffusion ability is reduced. With the influence of laser heat, the atomic diffusion coefficient increases, and solute redistribution occurs further between dendrites. Ti atoms are enriched in the dark dendritic region, and Zr elements appear in a large number of light regions (Figure 3f). The nanocrystalline structure is gradually consumed as the dendrite grows. Randomly occurring radial dendrites are interconnected (Figure 3c inset), forming directional large-sized dendrite crystals (Figure 3d). The dendrites at the junction grow perpendicular to the interface, consistent with the direction in which the temperature gradient increases (see the red arrow in Figure 3a,b). There were more nanocrystals and few dendrite crystals in the initial state of the 1.5 mm sample (Figure 3e, inset). The temperature gradient of 160 W laser treatment was significantly lower than that of the 200 W sample. The energy density was also relatively low, so the radial dendrites had just been induced to form and cannot continue to grow, which is why it does not have a distinct HAZ region (Figure 3e). It can be seen that the dark dendritic crystal is randomly distributed in the matrix (Figure 3d), with the characteristic size of 100 to 300 μm. In combination with the results of XRD and Energy-dispersive X-ray spectroscopy (EDX), the dark phase was characterized to be primary Cu_4Ti_3, and the white regions were Zr-rich, Sn-rich intermetallic compounds with a low melting point.

Figure 4a shows the measured load–displacement curves for the matrix, HAZ, and AC regions in the sample with a thickness of 1.5 mm under 200 W. The corresponding experimental data are given in Table 1. The indentation hardness (HIT) of the matrix, AC, and HAZ are 7995.1, 8514.4, and 12,083 MPa, respectively. The effective elastic modulus in the different regions were obtained based on the obtained indenter's elastic modulus and Poisson's ratio, A and S values (Table 1). The effective elastic modulus was 151.78 GPa in the matrix region, 125.55 GPa in the AC, and 153.59 GPa in the HAZ. The intrinsic microstructural origin for the plasticity of amorphous alloys is related to the features of the flow units. In amorphous alloys, the nanoscale liquid-like region (flow units) exhibits a lower atomic bulk density, higher energy states, easier shear deformation, and easier flow than the surrounding area [22–24]. The existence of large free volume in amorphous alloys increases the interatomic distance, weakening the atomic bonding energy and atomic migration barrier, resulting in a decrease in hardness and elastic modulus [25]. The load–displacement curve of the AC shows a prominent trait of the intermittent plastic deformation, termed serration or serrated flow [26]. It is generally accepted that the deformation mechanism for amorphous alloys is the nucleation and expansion of shear bands. In our load-controlled tests, the operation of a shear band gives rise to a burst of displacement [27]. Since a serrated flow seems to correspond to each shear band, we expect that larger plastic deformation originates in the accumulation of single shear displacements [27,28]. The distribution of effective elastic modulus from the AC to the matrix region in a sample with thickness of 2 mm under 200 W is illustrated in Figure 4b. Nanoindentation experiments were performed three times for the samples to ensure repeatability. The distribution of the effective elastic modulus in different regions showed the same trend. Furthermore, the effective elastic modulus in the AC zone (123.22 ± 4.64 GPa) was clearly lower than that in the matrix region (149.03 ± 5.76 GPa) and HAZ (153.22 ± 3.96 GPa). The nano-hardness of the 2 mm samples treated by 200 W in the HAZ (10.035 ± 1.45 GPa) was higher than that in the matrix region (8.381 ± 0.279 GPa) and AC zone (8.082 ± 0.396 GPa). Microscopic examination revealed that nanocrystalline dispersion would provide effective precipitation strengthening. Since the structure of nanocrystalline has a short average interatomic distance and strong atomic bonding due to the annihilation of free volume, it has the highest hardness and modulus of elasticity [29].

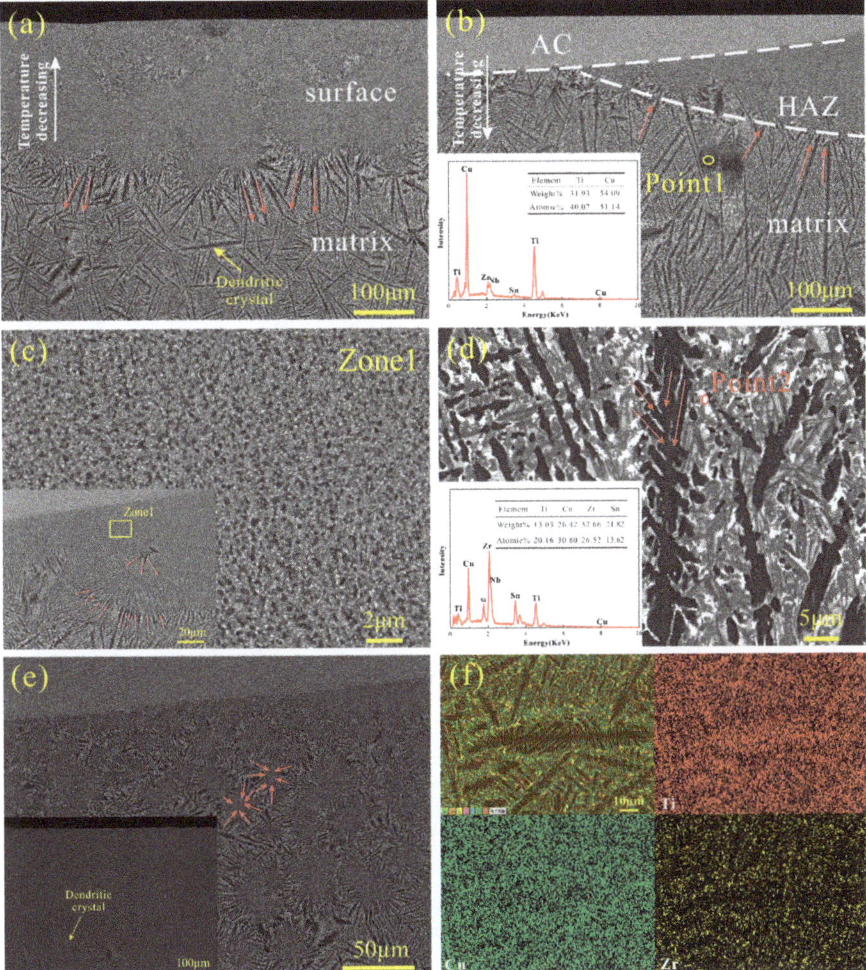

Figure 3. Back-scattered scanning electron microscopy (SEM) micrographs of the (**a**) cross-section of the 2 mm Ti alloy untreated; (**b**) the 2 mm sample under 200 W and the Energy-dispersive X-ray spectroscopy (EDX) corresponding to point 1; (**c**) Heat affected zone (HAZ); and (**d**) matrix regions in the sample with 2 mm under 200 W at high magnification and the EDX of point 2; (**e**) the 1.5 mm sample under 160 W with the inset showing untreated; (**f**) SEM-EDX mapping of dendritic crystals in sample with 2 mm under 200 W.

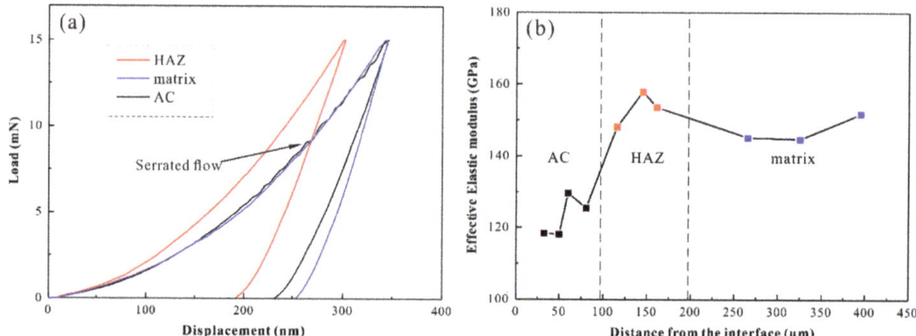

Figure 4. (**a**) Load–displacement curve in different regions of the 1.5 mm sample by 200 W; (**b**) Effective elastic modulus in the different regions near the surface of the 2 mm sample by 200 W.

Table 1. Nano-hardness and elastic modulus in the different zones of the 1.5 mm sample by 200 W.

Test Parameters	HAZ	Matrix	AC
Maximum load (P_{max}, mN)	15	15	15
Indenter's Poisson's ratio	0.07	0.07	0.07
Indenter's elastic modulus (GPa)	1141	1141	1141
Contact depth (h_c, nm)	231.93	288.88	279.59
Contact area (A, nm²)	1,244,136.48	1,878,274.12	1,763,841.78
Contact stiffness (S, mN/nm)	0.1763	0.2143	0.1753
Nano-hardness (H, GPa)	12.083	7.9951	8.5144
Effective Elastic modulus (E*, GPa)	153.59	151.78	125.55

The anodic polarization curve for the in situ Ti-based amorphous coating in a 3.5% NaCl solution at room temperature in open air, and the curve for specimens deposited under different laser powers are also presented for comparison, as shown in Figure 5. All the anodic portions of the polarization curves exhibit atypical passivation behavior. The corrosion current density (I_{corr}) values of all the specimens are shown in the inset image of Figure 5. The I_{corr} value of 240 W was 1.78×10^{-6} A/cm², which was about 15% lower than that of the untreated sample (2.12×10^{-6} A/cm²). Furthermore, the I_{corr} value of 200 W was 1.80×10^{-6} A/cm², which was about 27% lower than that of the 320 W (2.45×10^{-6} A/cm²). According to previous studies, the remarkable corrosion resistance of metallic glasses is contributed by the composition homogeneity with low residual stresses and single-phase nature without grain boundaries and second-phase particles [30–33]. Ti and Zr are considered to play an important role in corrosion resistance. Zr, Ti, and Nb elements have strong affinity to oxygen, thus, they easily form ZrO_2, TiO_2, and Nb_2O_5 on the surface of samples after polarization, preventing the direct contact of the corrosive solution with the alloy and thus reducing dissolution of the alloy elements (especially Cu) [34]. Therefore, the existence of a large amount of Ti and Zr in the material leads to a decrease in corrosion current density. However, the Cu element in the Cu-rich region reacts with chloride ions to form CuCl, which subsequently hydrolyzes to Cu_2O. It provides a galvanic coupling effect for the rapid local dissolution of the Zr-rich and Ti-rich nanometer active regions on the surface of the sample. These active metals also react with the solution, further inducing the formation of local galvanic cells involving Cu [35]. A similar galvanic coupling effect has also been mentioned in another study [18]. In that paper, the addition of the Nb element induced a dual structure containing Ti/Zr/Nb-rich crystalline dendrites and amorphous matrix phases. Regarding the preferential corrosion of the amorphous matrix derived from the formation of a micro galvanic cell between the amorphous matrix and dendrite phases, the results were attributed to the difference in elemental content. In this work, however, the crystalline structure with elemental segregation formed an amorphous coating under laser induction. The amorphous coating has no elemental segregation zone, which is immune to the local dissolution and typically does not form undesirable heterogeneous galvanic cells. Additionally, the addition of Nb promotes the formation of more oxides on the surface, preventing further corrosion of the material.

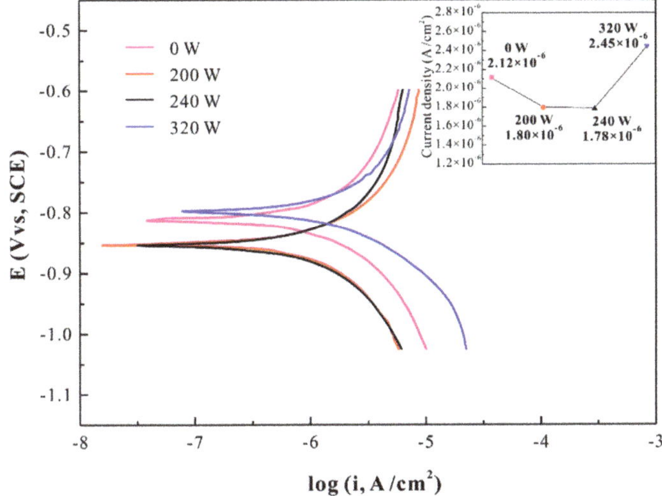

Figure 5. Anodic polarization curves and corrosion current density for the samples with a thickness of 2 mm deposited under different laser powers in a 3.5% NaCl solution.

4. Conclusions

In this paper, the prefabricated alloy plate (Ti47Cu38Zr7.5Fe2.5Sn2Si1Nb2) was processed by laser surface remelting (LSR) technology to obtain an in situ, strongly binding amorphous coating. The in situ formation method makes the coating free from cracks, micro-pores, and unfused particles. The completely amorphous coatings can be obtained with the line energy density (LED) of 0.1 to 0.12 J/mm in 2 mm samples. The corresponding LED was 0.08 to 0.1 J/mm in 1.5 mm samples. By performing the LSR with P = 200 W, v = 2000 mm/s, two distinct modification zones were obtained. The superior hardness promotion (by 51%) was obtained in the heat affect zone (HAZ), and a marked decrease (by 18%) of the elastic modulus in the amorphous coating zone (AC) was present. After the polarization, the addition of the Nb element to form an oxide prevented further corrosion of the material. The amorphous structure without element segregation leads to the failure of micro galvanic cells formation. In addition, the corrosion current density (I_{corr}) of the amorphous coating was about 15% lower than that of the untreated specimen (2.12×10^{-6} A/cm²). Therefore, the in situ amorphous coating obtained by LSR has relatively remarkable corrosion resistance. The composition free from Ni and Be ensures non-toxicity for biological applications. A low modulus of the coating can be better matched to the bone. This might be helpful for the development of corrosion resistant biomaterials.

Author Contributions: Conceptualization, S.W.; Methodology, P.L., X.Y., S.W., and K.W.; Software, P.L. and Q.S.; Validation, P.L., S.W., and L.W.; Formal analysis, P.L., L.M., X.Y., L.L., Y.Z., and Q.Z.; Investigation, P.L. and K.W.; Resources, S.W. and L.W.; Data curation, L.M., P.L., K.W., Y.Z., and Q.Z.; Writing—original draft preparation, P.L.; Writing—review and editing, P.L., L.M. and S.W.; Visualization, P.L. and Q.S.; Supervision, S.W.; Project administration, S.W. and L.W.; Funding acquisition, S.W. and L.W.; all authors gave their final approval.

Funding: This work was financially supported by the Natural Science Foundation of Shandong Province (no. ZR2019MEM040).

Acknowledgments: The authors are grateful to Jun Mi (Experimentalist), Qiuhong Huo (Experimentalist), Yanqing Xin (Experimentalist), and Kunlun Wang (Experimentalist) for their technical assistance. The authors are also grateful for the Physical-Chemical Test and Analysis Center of Shandong University at Weihai.

Conflicts of Interest: The authors declare no conflicts of interest.

References

1. Geetha, M.; Singh, A.K.; Asokamani, R.; Gogia, A.K. Ti based biomaterials, the ultimate choice for orthopaedic implants—A review. *Prog. Mater. Sci.* **2009**, *54*, 397–425. [CrossRef]
2. Inoue, A. Stabilization of metallic supercooled liquid and bulk amorphous alloys. *Acta Mater.* **2000**, *48*, 279–306. [CrossRef]
3. Chen, M. A brief overview of bulk metallic glasses. *Npg Asia Mater.* **2011**, *3*, 82–90. [CrossRef]
4. WWang, H.; Dong, C.; Shek, C.H. Bulk metallic glasses. *Mater. Sci. Eng. R* **2004**, *44*, 45–89. [CrossRef]
5. Chen, Q.; Thouas, G.A. Metallic implant biomaterials. *Mater. Sci. Eng. R* **2015**, *87*, 1–57. [CrossRef]
6. Hanawa, T. Research and development of metals for medical devices based on clinical needs. *Sci. Technol. Adv. Mater.* **2012**, *13*, 064102. [CrossRef]
7. Wang, T.; Wu, Y.D.; Si, J.J.; Cai, Y.H.; Chen, X.H.; Hui, X.D. Novel Ti-based bulk metallic glasses with superior plastic yielding strength and corrosion resistance. *Mater. Sci. Eng. A* **2015**, *642*, 297–303. [CrossRef]
8. Zhang, J.; Liu, M.; Song, J.; Deng, C.; Deng, C. Microstructure and corrosion behavior of Fe-based amorphous coating prepared by HVOF. *J. Alloy. Compd.* **2017**, *721*, 506–511. [CrossRef]
9. Huang, G.; Qu, L.; Lu, Y.; Wang, Y.; Li, H.; Qin, Z.; Lu, X. Corrosion resistance improvement of 45 steel by Fe-based amorphous coating. *Vacuum* **2018**, *153*, 39–42. [CrossRef]
10. Wan, Q.; Yang, B.; Chen, Y.M.; Liu, H.D.; Ren, F. Grain size dependence of the radiation tolerances of nano-amorphous Ti-Si-N composite coatings. *Appl. Surf. Sci.* **2019**, *466*, 179–184. [CrossRef]
11. Chen, K.; Yang, H.; Gao, J.; Yang, Y. Microstructure detection of Mg-Zi-Ti-Si amorphous composite coating prepared by SMAT. *Mater. Charact.* **2018**, *135*, 265–269. [CrossRef]
12. Bai, Y.; Li, X.; Xing, L.; Wang, Z.; Li, Y. A novel non-skid composite coating with higher corrosion resistance. *Ceram. Int.* **2017**, *43*, 15095–15106. [CrossRef]
13. Chai, L.; Wu, H.; Zheng, Z.; Guan, H.; Pan, H.; Guo, N.; Song, B. Microstructural characterization and hardness variation of pure Ti surface-treated by pulsed laser. *J. Alloy. Compd.* **2018**, *741*, 116–122. [CrossRef]
14. Yao, Y.; Li, X.; Wang, Y.Y.; Zhao, W.; Li, G.; Liu, R.P. Microstructural evolution and mechanical properties of Ti–Zr beta titanium alloy after laser surface remelting. *J. Alloy. Compd.* **2014**, *583*, 43–47. [CrossRef]
15. Qin, Y.S.; Han, X.L.; Song, K.K.; Tian, Y.H.; Peng, C.X.; Wang, L.; Sun, B.A.; Wang, G.; Kaban, I.; Eckert, J. Local melting to design strong and plastically deformable bulk metallic glass composites. *Sci. Rep.* **2017**, *7*, 42518. [CrossRef]
16. Liu, Y.; Pang, S.; Li, H.; Hu, Q.; Chen, B.; Zhang, T. Formation and properties of Ti-based Ti–Zr–Cu–Fe–Sn–Si bulk metallic glasses with different (Ti + Zr)/Cu ratios for biomedical application. *Intermetallics* **2016**, *72*, 36–43. [CrossRef]
17. Pang, S.; Liu, Y.; Li, H.; Sun, L.; Li, Y.; Zhang, T. New Ti-based Ti–Cu–Zr–Fe–Sn–Si–Ag bulk metallic glass for biomedical applications. *J. Alloy. Compd.* **2015**, *625*, 323–327. [CrossRef]
18. Yang, Y.J.; Fan, X.D.; Wang, F.L.; Qi, H.N.; Yue, Y.; Ma, M.Z.; Zhang, X.Y.; Li, G.; Liu, R.P. Effect of Nb content on corrosion behavior of Ti-based bulk metallic glass composites in different solutions. *Appl. Surf. Sci.* **2019**, *471*, 108–117. [CrossRef]
19. Weingarten, C.; Buchbinder, D.; Pirch, N.; Meiners, W.; Wissenbach, K.; Poprawe, R. Formation and reduction of hydrogen porosity during selective laser melting of AlSi10Mg. *J. Mater. Process. Technol.* **2015**, *221*, 112–120. [CrossRef]
20. Guo, M.; Gu, D.D.; Xi, L.X.; Du, L.; Zhang, H.M.; Zhang, J.Y. Formation of scanning tracks during Selective Laser Melting (SLM) of pure tungsten powder: Morphology, geometric features and forming mechanisms. *Int. J. Refract. Met. Hard Mater.* **2019**, *79*, 37–46. [CrossRef]
21. Yeh, J.; Chen, S.; Lin, S.; Gan, J.; Chin, T.; Shun, T.; Tsau, C.; Chang, S. Nanostructured High-Entropy Alloys with Multiple Principal Elements: Novel Alloy Design Concepts and Outcomes. *Adv. Eng. Mater.* **2004**, *6*, 299–303. [CrossRef]
22. Huo, L.S.; Zeng, J.F.; Wang, W.H.; Liu, C.T.; Yang, Y. The dependence of shear modulus on dynamic relaxation and evolution of local structural heterogeneity in a metallic glass. *Acta Mater.* **2013**, *61*, 4329–4338. [CrossRef]
23. Wang, W.H. The elastic properties, elastic models and elastic perspectives of metallic glasses. *Prog. Mater. Sci.* **2012**, *57*, 487–656. [CrossRef]

24. Lu, Z.; Jiao, W.; Wang, W.H.; Bai, H.Y. Flow unit perspective on room temperature homogeneous plastic deformation in metallic glasses. *Phys. Rev. Lett.* **2014**, *113*, 045501. [CrossRef] [PubMed]

25. Turnbull, D.; Cohen, M.H. On the Free-Volume Model of the Liquid-Glass Transition. *J. Chem. Phys.* **1970**, *52*, 3038–3041. [CrossRef]

26. Song, S.X.; Nieh, T.G. Direct measurements of shear band propagation in metallic glasses—An overview. *Intermetallics* **2011**, *19*, 1968–1977. [CrossRef]

27. Schuh, C.A.; Nieh, T.G. A nanoindentation study of serrated flow in bulk metallic glasses. *Acta Mater.* **2003**, *51*, 87–99. [CrossRef]

28. Drozdz, D.; Kulik, T.; Fecht, H.J. Nanoindentation studies of Zr-based bulk metallic glasses. *J. Alloy. Compd.* **2007**, *441*, 62–65. [CrossRef]

29. Wang, W.H.; Wang, R.J.; Yang, W.T.; Wei, B.C.; Wen, P.; Zhao, D.Q.; Pan, M.X. Stability of ZrTiCuNiBe bulk metallic glass upon isothermal annealing near the glass transition temperature. *J. Mater. Res.* **2002**, *17*, 1385–1389. [CrossRef]

30. Sun, Y.; Huang, Y.; Fan, H.; Wang, Y.; Ning, Z.; Liu, F.; Feng, D.; Jin, X.; Shen, J.; Sun, J.; et al. In vitro and in vivo biocompatibility of an Ag-bearing Zr-based bulk metallic glass for potential medical use. *J. Non-Cryst. Solids* **2015**, *419*, 82–91. [CrossRef]

31. Huang, Y.; Guo, Y.; Fan, H.; Shen, J. Synthesis of Fe–Cr–Mo–C–B amorphous coating with high corrosion resistance. *Mater. Lett.* **2012**, *89*, 229–232. [CrossRef]

32. Pang, S.J.; Shek, C.H.; Asami, K.; Inoue, A.; Zhang, T. Formation and corrosion behavior of glassy Ni–Nb–Ti–Zr–Co(–Cu) alloys. *J. Alloy. Compd.* **2007**, *434–435*, 240–243. [CrossRef]

33. Raicheff, R.; Zaprianova, V.; Gattef, E. Effect of structural relaxation on electrochemical corrosion behaviour of amorphous alloys. *J. Mater. Sci. Lett.* **1997**, *16*, 1701–1704. [CrossRef]

34. Gu, J.-L.; Shao, Y.; Zhao, S.-F.; Lu, S.-Y.; Yang, G.-N.; Chen, S.-Q.; Yao, K.-F. Effects of Cu addition on the glass forming ability and corrosion resistance of Ti-Zr-Be-Ni alloys. *J. Alloy. Compd.* **2017**, *725*, 573–579. [CrossRef]

35. Yang, Y.J.; Zhang, Z.P.; Jin, Z.S.; Sun, W.C.; Xia, C.Q.; Ma, M.Z.; Zhang, X.Y.; Li, G.; Liu, R.P. A study on the corrosion behavior of the in-situ Ti-based bulk metallic glass matrix composites in acid solutions. *J. Alloy. Compd.* **2019**, *782*, 927–935. [CrossRef]

Article

Influence of Effective Grain Size on Low Temperature Toughness of High-Strength Pipeline Steel

Yanlong Niu [1], Shujun Jia [1,*], Qingyou Liu [1], Shuai Tong [1], Ba Li [1], Yi Ren [2] and Bing Wang [1]

[1] Department of Structural Steel, Central Iron and Steel Research Institute, Beijing 100083, China; n18254101846@163.com (Y.N.); 13501064911@sina.com (Q.L.); e200912137011@163.com (S.T.); 1867648003@163.com (B.L.); 13867541062@163.com (B.W.)

[2] State Key Laboratory of Metal Materials for Marine Equipment and Applications of Iron & Steel Research Institutes of Ansteel Group Corporation, Anshan 114009, China; 138540058154@sina.com

* Correspondence: jiashujun101@163.com

Received: 20 September 2019; Accepted: 28 October 2019; Published: 7 November 2019

Abstract: In this study, the series temperature Charpy impact and drop-weight tear test (DWTT) were investigated, the misorientation angles among structural boundaries where the cleavage crack propagated were identified, and angles of {100} cleavage planes between adjacent grains along the cleavage crack propagated path were calculated in five directions (0°, 30°, 45°, 60°, and 90° to the rolling direction) of high-grade pipeline steel. Furthermore, the effective grain size (grain with misorientation angles greater than 15°) was redefined, and the quantitative influences of the redefined effective grain size on Charpy impact and DWTT is also discussed synthetically. The results showed that the microstructure presented a typical acicular ferrite characteristic with some polygonal ferrite and M-A islands (composed of martensite and retained austenite), and the distribution of the high-angle grain boundaries were mainly distributed in the range of 45°–65° in different directions. The Charpy impact energy and percent shear area of DWTT in the five directions increased with refinement of the redefined effective grain size, composed of grains with {100} cleavage planes less than 35° between grain boundaries. The ductile-to-brittle transition temperature also decreased with the refining of the redefined effective grain size. The redefined effective grain boundaries can strongly hinder fracture propagation through electron backscattered diffraction analysis of the cleavage crack path, and thus redefined effective grain can act as the effective microstructure unit for cleavage.

Keywords: pipeline steel; toughness; cleavage unit; crack propagation; misorientation angles

1. Introduction

Pipeline steels are widely used for transporting crude oil and natural gas over long distances because of their excellent combination of high strength and toughness [1–4]. At present, with the increasing global demand for energy, the exploitation of oil and gas has already extended to remote areas or oceans, which requires pipeline steel that can be used at low temperatures, as well as across more complex geological landforms. Therefore, research on the strength and toughness of pipeline steels in low temperature environments has become a hot topic [5,6]. Some investigations showed that a fully-refined acicular ferrite (AF) microstructure with some polygonal ferrite (PF) can ensure that the material has an excellent combination of strength and toughness in a low temperature environment [7–9].

It is known that the strength and toughness of most steels at a low temperature can be improved by grain refinement during the controlled rolling and cooling process in steel production. However, many studies have indicated that the controlled rolling and cooling technology can also result in anisotropy of properties, especially low toughness, which can restrict the optimum design of the materials [6,10,11]. Therefore, investigation into toughness anisotropy in high-strength pipeline steels is very important. However, most studies on high-strength pipeline steels are concerned with the microstructure characteristics and mechanical properties [12–16], while the relationship between grain size and fracture toughness in different directions is still far from being understood due to its complicated structure. The current view from some investigators is that the structure units affecting toughness are grains with grain boundary misorientation angles greater than 15° in acicular ferrite pipeline steels. In other words, the high-angle grain boundaries (HAGB) with grain boundary misorientation angles greater than 15° are the effective grains, and cracks will be arrested and then deflected by a large angle when encountering HAGB in the process of crack propagation [12,17,18]. However, a large number of experiments show that cracks can also pass straight through HAGB between grains (as shown in Figure 1). Therefore, the effective grain defined previously is limited in explaining this phenomenon in acicular ferrite steels. It is necessary to redefine the effective grain and study the relationship between the effective grain size and fracture toughness in acicular ferrite steel.

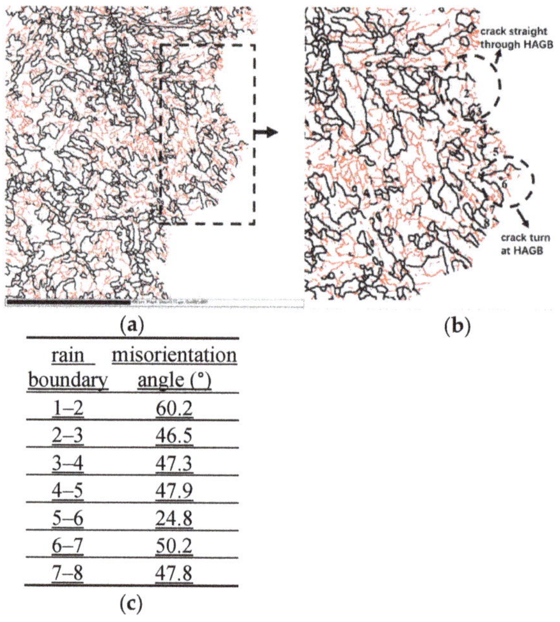

rain boundary	misorientation angle (°)
1–2	60.2
2–3	46.5
3–4	47.3
4–5	47.9
5–6	24.8
6–7	50.2
7–8	47.8

(c)

Figure 1. (a) Boundary distribution along the crack propagation path; (b) enlarged image of the dotted wire frame in Figure 1a; (c) misorientation angle between the grain boundaries.

For bcc metals, the cleavage crack always propagates along {100} cleavage planes. The angles between adjacent grains will influence the ability of changing the crack propagation direction. Some investigations have studied the correlations between toughness and structure of martensitic steels by means of calculating the {100} cleavage plane angles. For example, Deng Can-Ming et al. and Shen Jun-clang et al. [19,20] found that blocks are the microstructure units controlling the cleavage fracture and ductile-to-brittle transition temperature (DBTT) of lath martensite by means of analyzing {100} cleavage plane angles. However, the relationship between toughness and structure in different directions of acicular ferrite pipeline steels by means of analyzing {100} cleavage plane angles has not yet been reported.

In this study, the complex microstructure of acicular ferrite pipeline steel was studied by means of optical microscopy (OM), scanning electron microscopy (SEM), and transmission electron microscope (TEM). The toughness in the five typical directions (0°, 30°, 45°, 60°, and 90° to the rolling direction) of acicular pipeline steel was estimated through the Charpy impact and drop weight tear test (DWTT).

The misorientation angles among grain boundaries where the cleavage crack propagated were identified by electron backscattered diffraction (EBSD), and the angles of the {100} cleavage planes of adjacent grains were statistically counted. Based on the experiments above, the final purpose was to obtain the relationship between low temperature toughness and microstructure in different directions of acicular ferrite pipeline steels, and to understand the role of the boundaries during cleavage crack.

2. Experimental Procedure

2.1. Material and Chemical Composition

A commercial high-strength low alloy pipeline steel coil (used in the investigation) with a critical thickness of 21.4 mm was industrially processed on a hot strip mill equipped with an ultra-fast cooling system (for use after the rolling mill). The chemical composition of the strip is presented in Table 1.

Table 1. Chemical composition of the studied steel (wt %).

Element	C	Si	Mn	S	P	Nb	Mo	V	Ti	Ni	Cr	Cu
Content	0.06	0.2	1.75	0.001	0.007	0.075	0.23	0.022	0.014	0.21	0.25	0.01

2.2. Mechanical Property Tests

The tensile, Charpy, and DWTT specimens were prepared in five directions ($0°, 30°, 45°, 60°$, and $90°$ to the rolling direction), as shown in Figure 2. Tensile specimens were machined with a diameter of 5 mm and a gauge length of 25 mm. The tensile tests were performed on a CMT5105-SANS machine in accordance with SASTM A370 (Chinese standard) at room temperature with a draw speed of 1 mm/min. The Charpy specimens were in a standard V-notched geometry with a dimension of $10 \times 10 \times 55$ mm. The Charpy tests were carried out at 20 °C, 0 °C, −20 °C, −40 °C, −60 °C, −80 °C, −100 °C, −120 °C, and −196 °C with a K-type thermocouple attached to the specimen to control the test temperature. The DWTT ($305 \times 75 \times 21.4$ mm) specimens were carried out at 20 °C, 0 °C, −15 °C, −30 °C, −45 °C, and −60 °C in accordance with the SY/T 6476 Chinese standard.

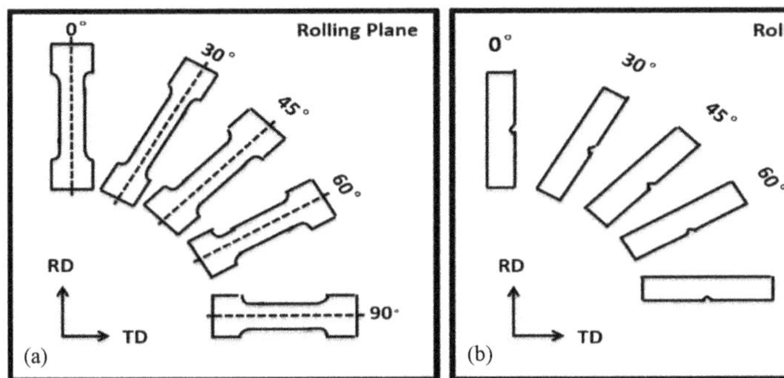

Figure 2. Orientation of the mechanical test specimens relative to the steel coil, where RD and TD stand for the rolling and transverse directions respectively: (**a**) directions of tensile test specimens and (**b**) different orientation of the Charpy impact and DWTT specimens.

2.3. Microstructural Characterization and Fracture Propagation Observation

The specimens for microstructural studies were polished and etched with 4% Nital, and then the microstructures were observed using a Leica DMIRM optical microscope (OM) and a Zeiss Ultra-55 field emission scanning electron microscope (SEM, Beijing, China) equipped with an electron back-scattered diffraction (EBSD, Beijing, China) system. The specimens for EBSD were electropolished with an electrolyte containing 100 mL glacial acetic acid and 100 ml distilled water at 0.1 A for 8–10s at room temperature (about 20 °C). The scanned area for EBSD analysis was 50×50 μm^2 with a step size of 0.15 μ at a magnification of 2000×, or 100×100 μm^2 with a step size of 0.2 μ at a magnification of 1000×. Selected specimens were evaluated by a FEI Tecnai G2 F20 transmission electron microscope (TEM, Beijing, China) at an accelerating voltage of 200 kV. The thin foils for TEM were ground mechanically to 30–50 μm thickness and then made into 3 mm diameter discs, which were electropolished in a two-jet machine at −20 °C. Figure 3 shows the observation plane positions of the fracture propagation of Charpy impact and DWTT samples. The crack propagation behavior of Charpy impact (−196 °C) and DWTT samples (−60 °C) were also studied by OM, SEM, and EBSD after preserving the fracture surface by electroplating with nickel, and angles of the {100} cleavage plane of the grains in different directions were measured.

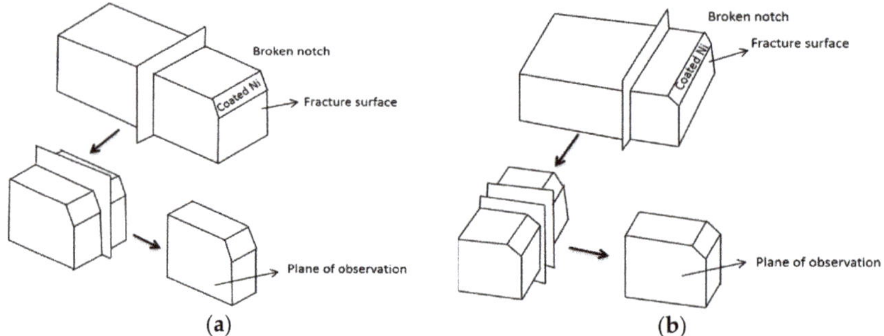

Figure 3. Observation plane positions of fracture propagation for Charpy impact and DWTT samples: (a) fracture propagation plane of Charpy impact, (b) fracture propagation plane of DWTT.

In order to calculate the angle of {100} cleavage planes of adjacent grains, two adjacent grains were assumed to be grain 1 and grain 2, the orientation matrix of grain 1 obtained from the EBSD test was regarded as G1, and the orientation matrix {100} cleavage plane was regarded as P. The normal vector (a_1, b_1, c_1) of the {100} cleavage surface of grain 1 can be calculated by G1 × P, as demonstrated in Equation (1). Similarly, the normal vector (a_2, b_2, c_2) of the {100} cleavage plane of grain 2 was deduced. The angles $(\varphi_1, \varphi_2, \varphi_3, \varphi_4, \varphi_5, \varphi_6, \varphi_7, \varphi_8, \varphi_9)$ between vectors $a_1, b_1, c_1, a_2, b_2, c_2$ was calculated by the vector formula, and the minimum angle among $\varphi_1, \varphi_2, \varphi_3, \varphi_4, \varphi_5, \varphi_6, \varphi_7, \varphi_8,$ and φ_9 was the {100} cleavage planes of grain 1 and grain 2.

$$G_1 = \begin{pmatrix} u & r & h \\ v & s & k \\ w & t & l \end{pmatrix}, P = \begin{pmatrix} 0 & 0 & 1 \\ 0 & 1 & 0 \\ 0 & 0 & 1 \end{pmatrix}, G_1 \times P = \begin{pmatrix} u & r & h \\ v & s & k \\ w & t & l \end{pmatrix}\begin{pmatrix} 0 & 0 & 1 \\ 0 & 1 & 0 \\ 0 & 0 & 1 \end{pmatrix}, \tag{1}$$

3. Results

3.1. Microstructures

Similar elongated grain structures related to the large austenite deformation rolled below the non-recrystallization temperature (Tnr) were observed, as shown in Figure 4. This can also be seen from the similar microstructure of the five directions, which mainly consisted of AF and PF with a small number of M-A Islands. AF was composed of approximately parallel slab ferrite, which combined into bundles with each other. The width of ferrite lath generally ranged from 200 to 500 nm (indicated by the arrow in Figure 5a), which would be beneficial for the strength and low temperature toughness of the material [12,21,22]. In contrast, PF is an essentially equiaxed grain structure, which concentrates at the flatted austenite grain boundary. A M-A island was formed along with the formation of acicular ferrite, which was a basic feature of acicular ferrite. The uniformly distributed M-A islands can pin grain boundaries and improve the toughness of the material [23] (indicated by the arrow in Figure 5b).

The microstructure of the five directions was refined by the thermo-mechanical control process and their grain size, shown in Figure 4, ranged from 3.9 μm to 4.1 μm. Figure 4 shows that the difference of microstructure morphologies, distribution, and grain size from samples with five directions were not obvious in the steel. Hence, the effect of microstructure in the different directions on anisotropic behavior can be reasonably ignored in the following study.

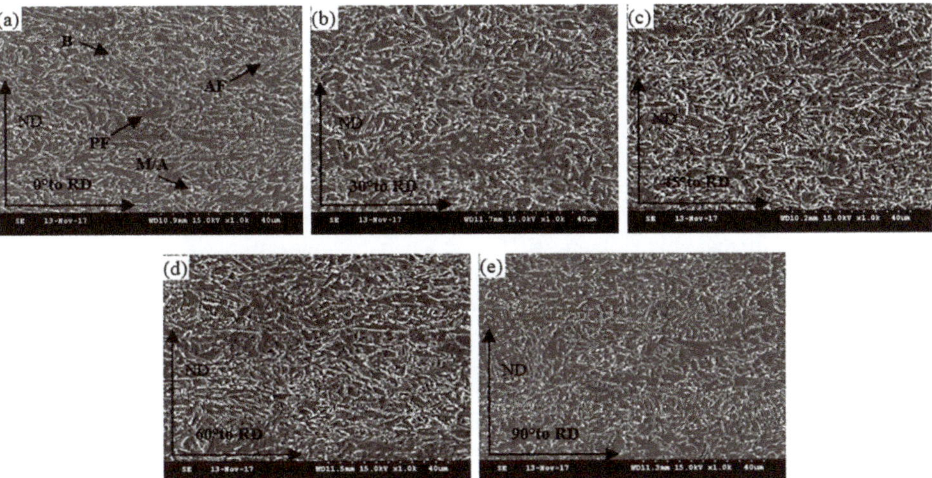

Figure 4. Scanning electron micrographs of samples in five directions: (**a**) 0° to RD, (**b**) 30° to RD, (**c**) 45° to RD, (**d**) 60° to RD, and (**e**) 90° to RD.

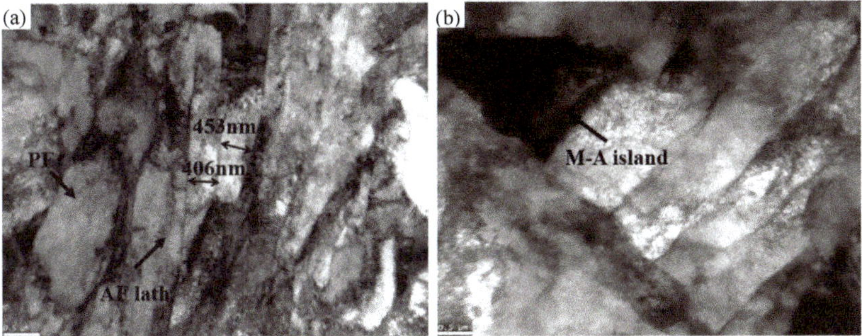

Figure 5. Transmission electron micrographs: (**a**) AF and PF, (**b**) M-A islands.

3.2. Mechanical Properties

The typical yield strength (YS) and tensile strength (TS) of high-strength low alloy pipeline steels in different directions were tested, and the results are illustrated in Table 2. The tensile tested data show that the yield strengths with different directions all exceeded 550 MPa, while their tensile strengths ranged from 640 MPa to 695 MPa, and the elongations of all the specimens were greater than 17%. The YS/TS ratios of all the specimens lie in the range of 0.83–0.87. It can also be seen from Table 2 that YS and TS are basically similar in most directions, except for the transverse direction (90° to RD). In brief, the strengths vary little in different directions.

Table 2. Mechanical proprieties of high-strength pipeline steel in different directions.

	Directions	YS (R$_{p0.2}$) MPa	TS MPa	YS/TS (%)	A%	Section Shrinkage (%)
	0°	560	675	0.83	27.5	81
	30°	555	640	0.87	31	80
21.4 mm	45°	565	655	0.86	27	85
	60°	560	670	0.84	26.5	84
	90°	600	695	0.86	23.5	80

The Charpy impact energy and percent shear area (SA%) to DWTT, varying with different directions and different temperatures, are presented in Figures 6 and 7. As expected, the Charpy impact energy and SA% to DWTT all decreased with the decrease of tested temperature. The Charpy impact energy and SA% to DWTT were obviously better along the longitudinal (0° to RD) direction than other directions. The Charpy impact energy in different directions were similar, and all were above 300 J when the temperature was higher than −40 °C; and when the temperature was lower than −40 °C, the impact energy at 45° to RD and 60° to RD decreased rapidly with the decrease of temperature. Compared to the other directions, the impact toughness of the sample at 45° to RD was the worst; in other words, the decreasing trend of the impact energy of the sample at 45° to RD was the most significant one with the decrease of temperature. When the temperature was −80 °C or −100 °C, the impact energy dropped to 112 J and 25 J, respectively; while the impact energy of samples at 0°, 30°, and 90° to RD still remained 200 J or more at −80 °C, and remained 150 J or more at −100 °C. The Charpy impact energy of the sample at 60° to RD was slightly better than that at 45° to RD. Figure 6b showed that the favorable upper shelf region, the 100% ductile range in the steel, can still be attained along the direction of 30°, 60°, and 90° with the test temperature ranging from 20 °C to −60 °C. However, the favorable upper shelf region of samples at 0° to RD and 45° to RD were attained at temperatures ranging from 20 °C to −80 °C and 20 °C to −40 °C, respectively. Compared with the results in the impact energy, Figure 7 shows that the SA% to DWTT in different directions were different, except at 0 °C. Even when the temperature decreased to −15 °C, the SA% decreased significantly at 60° and 90° to RD compared to other directions. Furthermore, a brittle fracture character began to appear in the fracture morphology when the temperature dropped to −30 °C, and the SA% to DWTT at 60° to RD and 90° to RD were 48% and 42%, respectively, which illustrated that the fracture mode was brittle fracture. By contrast, the SA% to DWTT at 0°, 30°, and 45° to RD remained above 80%. When the temperature dropped to −60 °C, dimples were dominant in the fracture surface in all directions. The tendency of SA% to DWTT to change with temperature was different from that of impact energy changing with temperature because there are some complex factors that affect the DWTT, such as full-scale thickness and multi-dimensional stress [24,25].

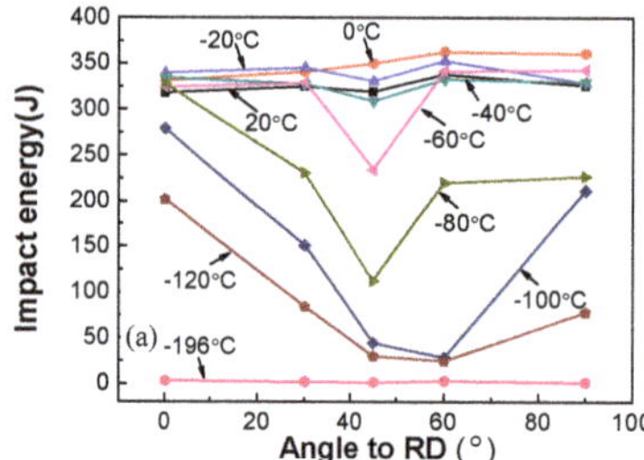

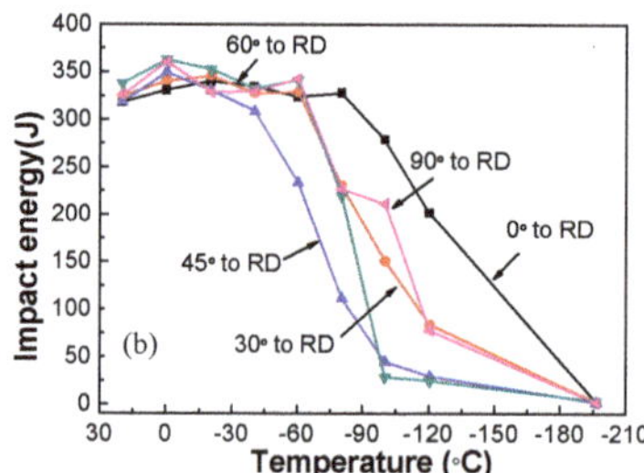

Figure 6. Charpy impact test result: (**a**) curve between impact energy and directions, (**b**) curve between impact energy and temperature.

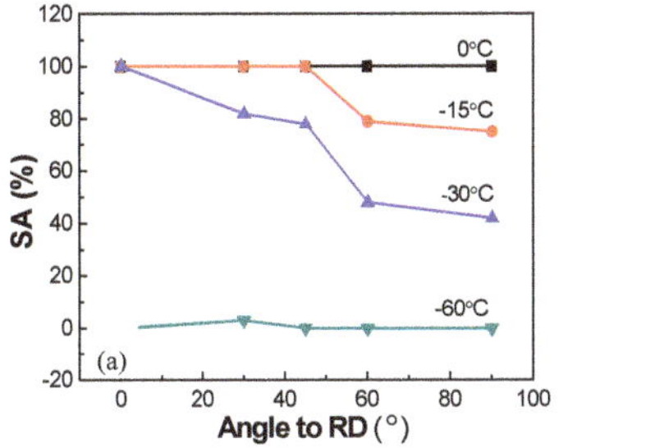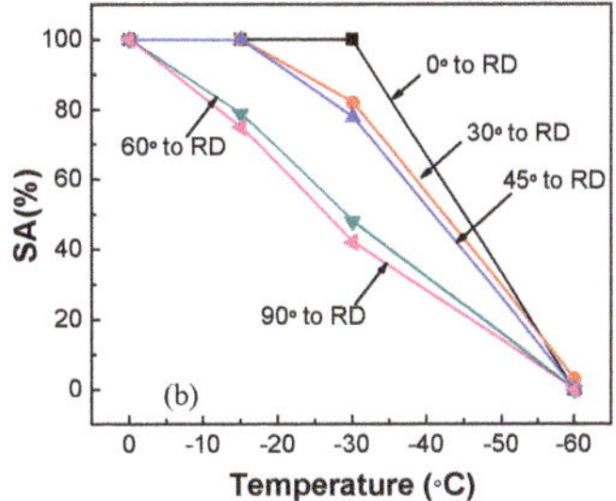

Figure 7. DWTT results: (**a**) curve between SA% (percent shear area) and directions, (**b**) curve between SA% and temperature.

Figure 8 shows the change in fracture mode for impact specimens in three directions (0° to RD, 45° to RD, and 90° to RD) at three temperatures (−40 °C, −60 °C, and −80 °C). It can be seen that the fractography of the samples at 45° to RD were mainly dimples at −40 °C, which indicated that the sample fractured in a ductile manner. However, the cleavage plane could be observed in the radial zone of the specimens when the temperature dropped to −60 °C and it almost covered all of the fracture surface at −80 °C. The cleavage can be clearly seen in Figure 8h, which indicates that the sample fractured in a brittle way. However, Figure 8a,c,d,f also show that the dimples of the samples at 0° to RD and 90° to RD were still dominant in the fracture surface when the test temperature was above −80 °C. This is consistent with the results obtained from Figure 6.

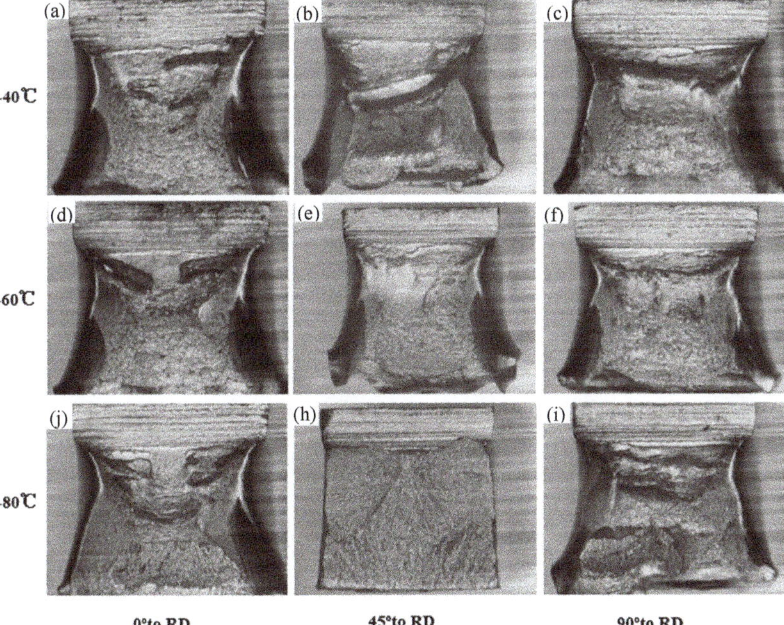

Figure 8. Selected fracture surfaces of Charpy specimens in three orientations (0° to RD, 45° to RD and 90° to RD) at three temperatures (−40 °C, −60 °C, and −80 °C): (**a**)−40 °C and 0° to RD, (**b**) −40 °C and 45° to RD, (**c**) −40°C and 90° to RD, (**d**) −60°C and 0° to RD, (**e**)−60 °C and 45° to RD, (**f**) −60°C and 90° to RD, (**j**) −80 °C and 0° to RD, (**h**) −80°C and 45° to RD, (**i**) −80 °C and 90° to RD.

Figure 9 shows the change in fracture mode for DWTT specimens in three directions (0° to RD, 45° to RD, and 90° to RD) at three temperatures (0 °C, −30 °C, and −60 °C). It was expected that the ductile deformation stage would reduce, while the brittle propagation stage would extend, resulting in deterioration of the toughness with the decrease of the test temperature from 0 °C to −60 °C. The typical ductile fracture was observed at 0 °C in three directions. When the temperature dropped to −30 °C, the dominant cleavage fracture was observed at 90° to RD, large cleavage surfaces existed in it, and a small amount of ductile fracture existed. However, the ductile fracture was still the main fracture mode in 0° to RD and 45° to RD with some ductile fractures observed on the fracture surface, as shown in Figure 9d–f. The dominant cleavage fracture occurred at −60 °C in three directions.

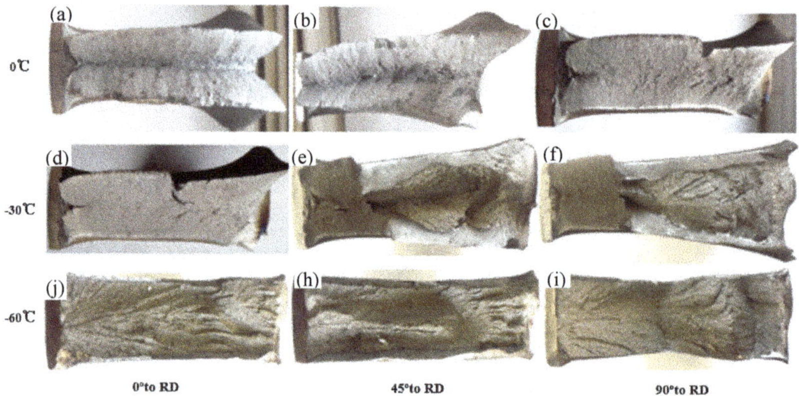

Figure 9. Selected fracture surfaces of the three directions (0° to RD, 45° to RD, and 90° to RD); DWTT specimens at three temperatures (0 °C, −30 °C, and −60 °C): (**a**) 0 °C and 0° to RD, (**b**) 0 °C and 45° to RD, (**c**) 0 °C and 90° to RD, (**d**) −30 °C and 0° to RD, (**e**) −30 °C and 45° to RD, (**f**) −30 °C and 90° to RD, (**j**) −60 °C and 0° to RD, (**h**) −60 °C and 45° to RD, (**i**) −60 °C and 90° to RD.

The results discussed above show that AF and PF were obtained in all directions of the tested steels, and that the microstructure types and sizes of the test steels were similar. However, the toughness (Charpy impact energy and SA% to DWTT) of the steels and fracture behaviors in different directions were different, which indicates that the effective grain sizes of the steels affecting the low temperature toughness and fine structure in different directions were significantly different. Therefore, it was necessary to systematically research the crystal orientation relationship between the fine substructures and effective grain sizes in different directions.

3.3. Crystal Orientation Relation and Effective Grain Size

The distribution map of the HAGB of the high strength pipeline steel in different directions was obtained (Figure 10). It can be seen that the tendency of the frequency distribution curves of the HAGB were similar to each other and the frequencies were slightly different in different directions. The HAGB in all directions were mainly distributed in the range 45°–65°, and the frequency distribution of 55° was the strongest. More than 65% of the HAGB ranged from 45° to 65° in the five directions. This is in agreement with recent investigations on acicular pipeline steels, and the distribution of HAGB showed no significant difference in the five direction samples [26].

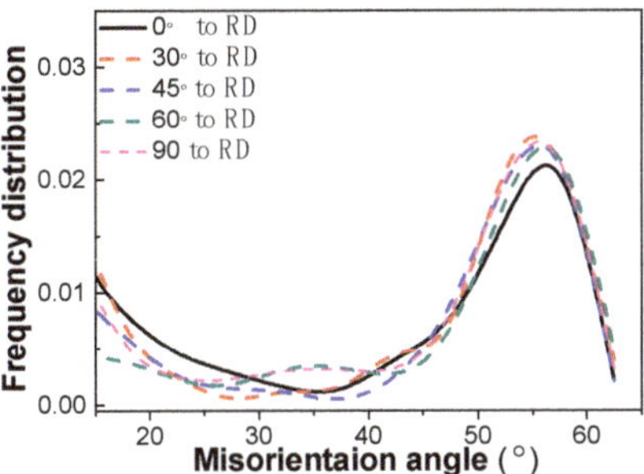

Figure 10. Angle boundaries distribution map of the pipeline steel in different directions.

The acicular ferrite had a body-centered cubic (bcc) structure. Therefore, it was most likely to slip on the {110} plane [12,26,27] and the {100} cleavage angles between adjacent grains will influence the ability to change the crack propagation direction, thereby affecting the fracture toughness of the materials. Therefore, it is very important to analyze the influence of the {100} cleavage angles on the low temperature toughness of high strength pipeline steel.

Figure 11 and Table 3 are diagrams of the fracture propagation and statistics, respectively, of the crystal orientation relationship between Charpy impact and DWTT samples, which show the distribution of misorientation angles in the samples. The thin black lines marked in Figure 11 and Table 3 represent HAGB, the thick black lines represent cleavage unit boundaries, and the thin red lines represent low angle grain boundaries (LAGB) with misorientation angles lower than 15°. It can also clearly be seen that there are a number of unordered grains distributed randomly in the metallographic structure, most of which are HAGB. In addition, there are substructures in most grains, and most of the substructures are LAGB.

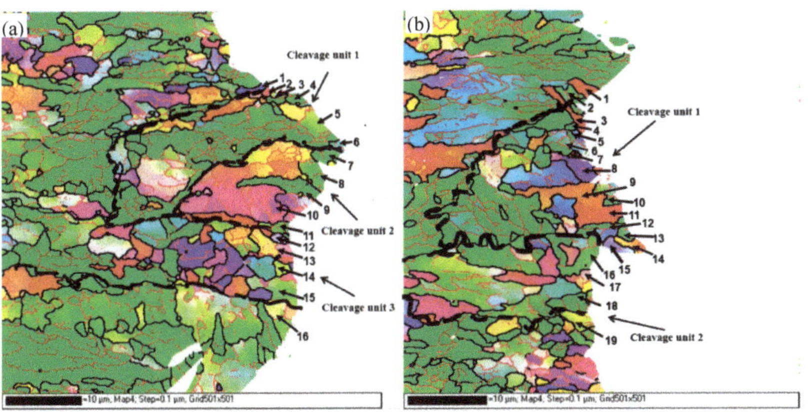

Figure 11. Schematic diagram of the fracture propagation of Charpy impact and DWTT samples: (**a**) Charpy impact, (**b**) DWTT.

Table 3. Statistics of the crystal orientation relationship between Charpy impact and DWTT samples: Charpy impact corresponding to Figure 11a,b DWTT corresponding to Figure 11b.

Charpy Impact				DWTT			
Position	Grain	Misorientation Angle	{100} Planes Angle	Position	Grain	Misorientation Angle	{100} Planes Angle
cleavage unit 1and adjacent grain	1–2	48.9	40.23	cleavage unit 1 and adjacent grain	1–2	55.26	20.12
cleavage unit 1	2–3	47.5	12.25		2–3	54.12	41.23
	3–4	46.3	14.62		3–4	58.31	20.43
	4–5	54.1	17.59		4–5	58.22	30.23
	5–6	48.7	12.34		5–6	52.33	25.32
cleavage unit2	6–7	56.3	35.62	cleavage unit 1	6–7	37.22	17.24
	7–8	52.1	21.68		7–8	52.56	38
	8–9	51.6	22.17		8–9	49.37	10.34
	9–10	54.1	20.96		9–10	37.45	15.24
	10–11	52.5	42.24		10–11	25.08	20.18
cleavage unit 3	11–12	58.4	12.11		11–12	52.12	19.15
	12–13	54.1	35.24		12–13	55.15	21.55
	13–14	55.2	32.17		13–14	36.78	18.45
	14–15	54.3	32.48		14–15	60.25	40.4
cleavage unit 3 and adjacent grain	15–16	52.6	45.19	cleavage unit 2	15–16	62.12	44.87
					16–17	55.31	25.12
					17–18	54.32	37.25
				cleavage unit 2 and adjacent grain	18–19	49.22	40.21

The orientation relationship of grains and angles between the {100} cleavage plane is obtained by measuring more than 350 fracture crack EBSD images. The average grain sizes of flattened austenite and grain size were measured to be about 200 grains. The average cleavage unit size was calibrated by measuring about 300 grains for each direction. The quantitative result of the subunit size of the microstructure in different directions is shown in Table 4. It can be seen that the flatted austenite grain size (about 16 μm) and grain size (about 4 μm) are basically the same, produced in the same process. However, the impact energy and SA% to DWTT in different directions are clearly different when tested at a low temperature. Therefore, the size of flattened austenite and grain size are not the effective size, which determines the low temperature toughness of materials.

Table 4. Statistical diagram of microstructure unit size and toughness.

Project	0° to RD	30° to RD	45° to RD	60° to RD	90° to RD
Flatted austenite grain size (μm)	15.8	16.1	15.8	15.9	16.0
Cleavage unit size of impact sample (μm)	14.1	14.5	16.7	15.1	13.9
Cleavage unit size of DWTT sample (μm)	15	15.1	16.5	16.8	16.6
Grain size (μm)	4.0	4.1	3.9	4.2	4.1
50% FATT (K)- impact sample	405	377	348	361	385
50% FATT (K)-DWTT sample	317	319	311	301	302
−60 °C impact energy (J)	324	329	234	342	343
−80 °C Charpy energy (J)	327	230	112	220	226
−15 °C SA (%) to DWTT	100	100	100	78	75
−30 °C SA (%) to DWTT	100	77	80	45	42

3.4. Quantitative Analysis of Subunit Size and Its Influence on Performance

Table 4 and Figure 12 show that the Charpy impact at −60 °C and −80 °C, and the SA% to DWTT at −15 °C and −30 °C in different directions, all generally increase with the refining of the cleavage unit size (in the same cleavage unit, the angles of {100} cleavage planes of adjacent grains are less than 35°). In detail, the Charpy impact increased from 234 J to 343 J at −60 °C, and from 112 J to 327 J at −80 °C, and the SA% to DWTT increased from 75% to 100% at −15 °C, and 42% to 100% at −30 °C. While the cleavage unit size of Charpy impact and DWTT were refined from 16.6 μm to 13.9 μm, and 16.2 μm to 15.2 μm, respectively, indicating that the refinement of cleavage unit size makes a large contribution to the toughness of the acicular ferritic pipeline steel. Figure 13 shows the relationship between the 50% FATT (fracture appearance transition temperature) and the cleavage unit size. The 50% FATT to Charpy impact and DWTT were all strongly dependent on the cleavage unit size, with a linear relationship existing between the $d^{-1/2}$ of the cleavage unit size and the corresponding 50% FATT to Charpy impact and DWTT. This is consistent with the Cottrell–Petch ductile–brittle fracture formula. The ductile brittle transition temperature depends on the cleavage unit size. The cleavage fracture occurs when the cleavage fracture and yield strength of the material are equal, which also fully demonstrates that cleavage unit size plays a decisive role in low temperature toughness. The cleavage unit size was thus identified as being the unit crack path for cleavage fracture. However, in previous studies, it was shown that the structural unit controlling the toughness of acicular ferrite pipeline steels is the grain size. Obviously, this point was not supported by the present study. Because Table 4 showed that the grain size in different directions was similar and had no effect on the low temperature toughness. Moreover, the size of the cleavage unit was similar to that of flattened austenite, which may indicate that the size of the cleavage unit can be refined by refining the original austenite.

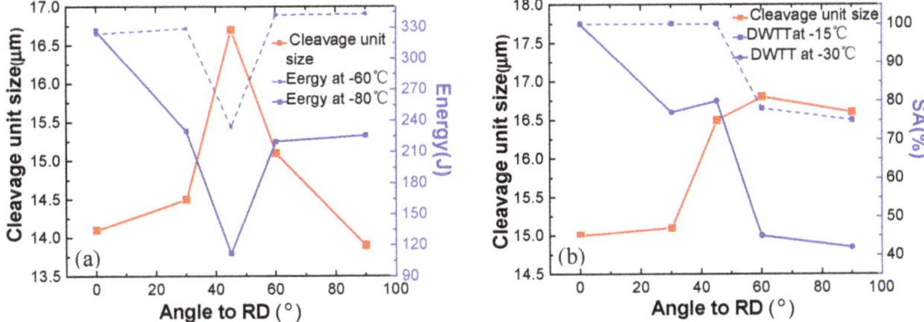

Figure 12. Relationship between toughness and cleavage unit size in different directions: (**a**) curve between impact energy and cleavage unit size, (**b**) curve between SA% and cleavage unit size.

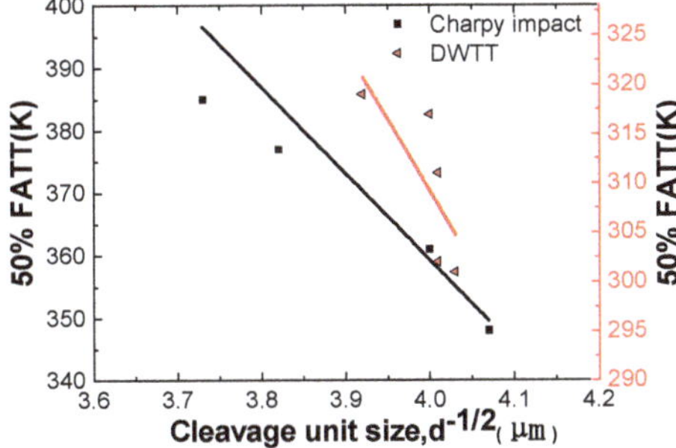

Figure 13. Relationship between cleavage unit size and 50% DBTT (Charpy impact and DWTT).

Table 4 and Figure 13 show that the cleavage unit size of the impact sample was refined from 16.7 μm to 13.9 μm, and the corresponding ductile to brittle transition temperature (DBTT) to Charpy impact decreased from 405 K to 361 K. The cleavage unit size of the DWTT sample was refined from 16.7 μm to 13.9 μm, and the corresponding DBTT to Charpy impact decreased from 405 K to 361 K. The refinement of the cleavage unit size had a significant influence on decreasing the DBTT of the material, indicating that the refinement of the cleavage unit contributes more to the cleavage fracture strength than to the yield strength. Therefore, the refinement of cleavage unit size resulted in a decrease in DBTT. According to Griffith theory, the critical fracture stress is expressed as Equation (2) [28].

$$\delta_f = \left[\frac{\pi E \gamma_P}{(1 - v^2)d} \right]^{1/2} \tag{2}$$

where δ_f is critical fracture stress, E is the Young's modulus, γ_p is the plastic deformation energy, v is Poisson's ratio, and d is the effective grain size, which can also be regarded as the cleavage unit size in this study as the cleavage fracture unit is determined as being the cleavage unit.

From the mechanical aspect, the fracture appearance transition behavior can be interpreted as a result of the competition between the yield strength and fracture strength. The effect of temperature on them is different, which leads to a fracture appearance transition of low alloy steel with a body-centered cubic. Generally, the yield strength increases rapidly with a decrease in temperature, while the fracture stress is independent of temperature. At the temperature ranges in which the fracture stress is lower than the yield strength, as the ductility is caused by the slip motion of dislocation, the resistance of dislocation motion increases with a decrease in temperature. Therefore, the yield strength increases with a decrease in temperature, while the cleavage fracture strength is less affected by temperature. When the temperature is lower than the fracture appearance transition temperature and the yield strength is higher than the cleavage fracture strength, the fracture mode changes from a microporous aggregated ductile fracture to a cleavage brittle fracture, and the fracture surfaces will reveal cleavage because cleavage fractures occur before plastic deformation. Obviously, its low temperature toughness is poor.

The toughness of the materials is influenced by the crack initiation and crack propagation path. Fracture theory suggests that the initiation of a crack mainly relies on the degree of stress concentration. In the process of crack propagation of the Charpy impact and DWTT samples, dislocation piles up at the secondary phases, such as M-A islands, coarse cementite, or nonmetallic inclusions, inducing the stress to concentrate at second phases, or the interface between second phases and matrix. As non-metallic inclusions are well controlled in the pipeline strip, cracks do not deflect sharply when they encounter non-metallic inclusions generally. The critical fracture stress is related to the fracture path and frequency of crack propagation in a sharp deflection. Moreover, it is known that the smaller the cleavage unit size is, the higher the frequency of sharp crack deflection is. In addition, the crack propagation path can be effectively hindered by deflecting the crack propagation path during the Charpy impact and DWTT test as the size of the cleavage unit decreases.

3.5. Cleavage Crack Propagation Path

Figures 14 and 15 are the typical actual crack propagation paths of Charpy impact and DWTT specimens obtained by EBSD analysis. The zone surrounded by grains with {100} cleavage planes lower than 35° represents a cleavage unit. It showed that the cleavage cracks crossed HAGB directly (Figure 14, 2–11, in Charpy samples, and Figure 15, 1–9, in DWTT samples) or only occurred as frequent small deflections at HAGB in the same cleavage unit. Only when the crack extended to the boundary of two cleavage units (Figure 14, 6–7 and Figure 15, 5–6), a great deflection occurred, and more energy was needed for the cleavage crack to cross the cleavage unit boundary. The experimental data was consistent with the result discussed in Figure 11. Obviously, grains 6 and 7 in Figure 14 and grains 5 and 6 in Figure 15 belonged to different cleavage units. The angles of {100} cleavage planes between adjacent cleavage units were all above 35°. When the cleavage crack encountered the grain boundary of 6 and 7 in Figure 14 or 5 and 6 in Figure 15, it was arrested and then deflected by a large angle, and a sharp deflection occurred along the propagation direction of the crack, which effectively retards the propagation speed of the crack. Because the positive bonding strength of the {100} crystal plane of body-centered cubic metals was lower than that of other crystal planes, when the stress exceeded the cleavage fracture strength, the cleavage fracture crack first propagated along the {100} crystal plane rapidly. When the orientation of different {100} crystal planes changed, the cleavage crack propagation was hindered. As a result, more energy was used for the cleavage crack to cross the grain boundaries with {100} cleavage plane angles of adjacent grains greater than 35°, compared to those less than 35°.

The cleavage unit was thus further identified as the minimum effective structural unit for affecting the low temperature toughness of materials.

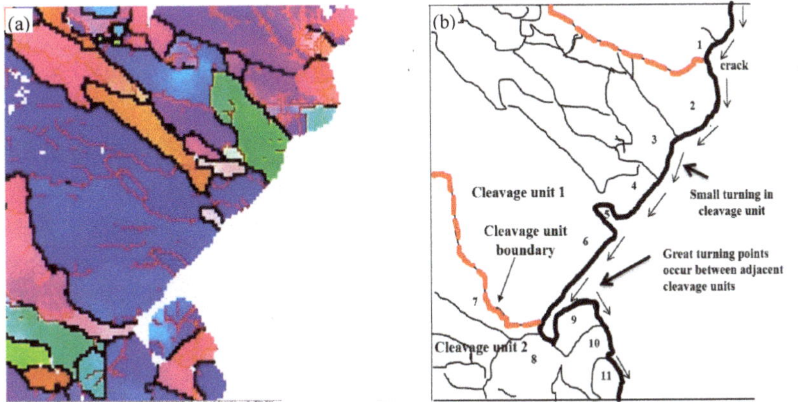

Figure 14. Orientation and schematic map of Charpy impact crack propagation: (**a**)orientation map of Charpy impact crack propagation, (**b**) schematic map of Charpy impact crack propagation according to Figure 14a.

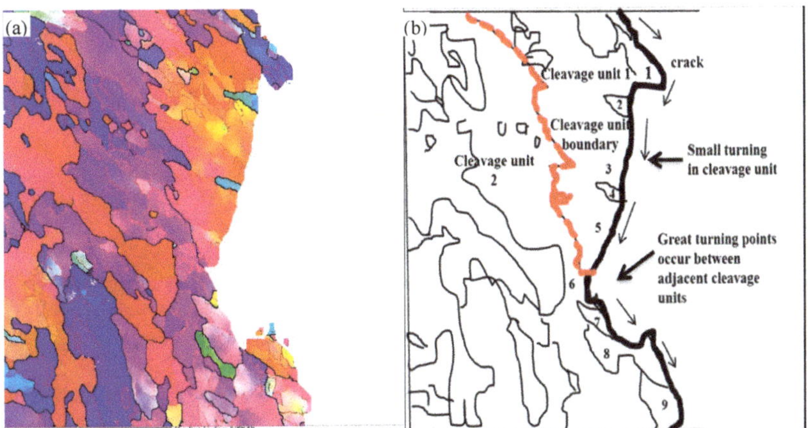

Figure 15. Orientation and schematic map of DWTT crack propagation: (**a**)orientation map of DWTT crack propagation, (**b**) schematic map of DWTT crack propagation according to Figure 15a.

4. Conclusions

The toughness, including the Charpy impact, drop-weight tear test (DWTT), and crack propagation behaviors in five different directions were investigated in detail in this study, and the different crystallographic relationships and angles between the {100} planes were obtained. Scanning electron microscopy (SEM), transmission electron microscopy (TEM), and electron backscattered diffraction (EBSD) results and fracture behaviors were applied to analyze the relationships among direction, the cleavage unit, and the low toughness of Charpy impact and DWTT properties. The following conclusions are drawn from the research.

1. The microstructure in different directions all consists of refined acicular ferrite (AF) and polygonal ferrite (PF). The distribution of high angle grain boundaries (HAGB) was similar and mainly distributed in the range of 45°–65°, which can ensure acicular ferrite has excellent

toughness. However, they have no significant influence on the anisotropy of toughness in acicular pipeline steel.

2. The impact energy and SA% to DWTT in the five directions increased when the cleavage unit was composed of grains with a {100} cleavage plane less than 35° between grain boundaries, and the ductile to brittle transition temperature decreased. The cleavage unit composed of grains with a {100} cleavage plane less than 35° between grain boundaries was identified as the effective grain for cleavage fracture.

3. In the actual process of crack propagation, when the cleavage crack encounters another cleavage unit (refined redefined effective grain), it will be arrested and then largely change its propagation direction, while when it encounters HAGB, it is possible to cross the HAGB directly. Therefore, it is important to refine the redefined effective grain size to improve the low temperature toughness of materials.

Author Contributions: Y.N. mainly participated in scientific research projects or experiments, collected data, made experiments and analyzed results, and wrote the first draft of the paper, which was the main author of the paper. S.J. and Q.L. were the decision makers and main organizers of the research. S.T., B.L., Y.R., and B.W. mainly assisted with trials, calculations, supervision and review of this articles.

Funding: This research was funded by National Key Research and Development Program of China (Grant No.2018YFC0310302).

Conflicts of Interest: The submitted dissertation is the result of my independent research under the guidance of my tutor except for the quoted contents, this paper does not include any other works or achievements that have been published or written by individuals or groups. Individuals and collectives who have made important contributions to the study of this paper have been clearly identified in this paper. The legal result of this statement shall be borne by me.

References

1. Shukla, R.; Ghosh, S.K.; Chakrabarti, D.; Chatterjee, S. Microstructure, texture, property relationship in thermo-mechanically processed ultra-low carbon micro-alloyed steel for pipeline application. *Mater. Sci. Eng. A* **2013**, *587*, 201–208. [CrossRef]

2. Hwang, B.; Lee, C.G.; Kim, S.J. Low temperature toughening mechanism in thermomechanical processed high strength Low-Alloy Steels. *Metal. Mater. Trans. A* **2011**, *42*, 717–728. [CrossRef]

3. Venkatsurya, P.K.C.; Misra, R.D.K.; Mulholland, M.D.; Manohar, M.; Hartmann, J.E. The impact of microstructure on yield strength anisotropy in pipeline steels. *Mater. Sci. Eng. A* **2014**, *556*, 2335–2342.

4. Zong, C.; Zhu, G.; Mao, W. Effect of crystallographic texture on the anisotropy of Charpy impact behavior in pipeline steel. *Mater. Sci. Eng. A* **2013**, *563*, 1–7. [CrossRef]

5. Zhang, Y.Q.; Guo, A.M.; Shang, C.J.; Liu, Q.Y.; Gray, M.J.; Frank, B. Latest development and application of high strength and heavy gauge pipeline steel in China. *J. Mech. Eng. Auto.* **2016**, *6*, 19–24. [CrossRef]

6. Yang, X.L.; Xu, Y.B.; Tan, X.D.; Wu, D. Influences of crystallography and delamination on anisotropy of Charpy impact toughness in API X100 pipeline steel. *Mater. Sci. Eng. A* **2014**, *607*, 53–62. [CrossRef]

7. Zhao, M.C.; Hanamura, T.; Qiu, H.; Nagai, K. Difference in the role of non-quench aging on mechanical properties between acicular ferrite and ferrite-pearlite pipeline steels. *ISIJ Int.* **2005**, *45*, 116–120. [CrossRef]

8. Xiao, F.R.; Liao, B.; Shan, Y.Y.; Qiao, G.Y.; Zhong, Y.; Zhang, L.C. Challenge of mechanical properties of an acicular ferrite pipeline steel. *Mater. Sci. Eng. A* **2006**, *431*, 41–52. [CrossRef]

9. Zhao, M.C.; Hanamura, T.; Qiu, H.; Yang, K. Lath boundary thin-film martensite in acicular ferrite ultralow carbon pipeline steels. *Mater. Sci. Eng. A* **2005**, *395*, 327–332. [CrossRef]

10. Joo, M.S.; Suh, D.-W.; Bae, J.H.; Bhadeshia, H.K.D.H. Toughness anisotropy in X70 and X80 pipeline steels. *Mater. Sci. Eng. A* **2012**, *556*, 601–606. [CrossRef]

11. Jabr, H.M.A.; Speer, J.G.; Matlock, D.K.; Zhang, P.; Cho, S.H. Anisotropy of mechanical properties of API X70 spiral welded pipe steels. *Mater. Sci. Forum* **2013**, *753*, 538–541.

12. Xiao, F.R.; Liao, B.; Ren, D.L.; Shan, Y.Y.; Yang, K. Acicular ferritic microstructure of a low-carbon Mn-Mo-Nb micro-alloyed pipeline steel. *Mater. Charact.* **2005**, *54*, 305–314. [CrossRef]

13. Sanchez, N.; Petrov, R.; Bae, J.H.; Kim, K. Texture dependent mechanical anisotropy of X80 pipeline steel. *Adv. Eng. Mater.* **2010**, *12*, 973–980.
14. Byun, J.S.; Shim, J.H.; Suh, J.Y.; Oh, Y.J.; Cho, Y.W.; Shim, J.D.; Lee, D.N. Inoculated acicular ferrite microstructure and mechanical properties. *Mater. Sci. Eng. A* **2001**, *319–321*, 326–331. [CrossRef]
15. Massa, F.; Piques, R.; Laurent, A. Rapid crack propagation in polyethylene pipe: Combined effect of strain rate and temperature on fracture toughness. *J. Mater. Sci.* **1997**, *32*, 6583–6587. [CrossRef]
16. Cheng, S.; Zhang, X.; Zhang, J.; Feng, Y.; Ma, J.; Gao, H. Effect of coiling temperature on microstructure and properties of X100 pipeline steel. *Mater. Sci. Eng. A* **2016**, *666*, 156–164. [CrossRef]
17. Duan, Q.; Yan, J.; Zhu, G.H.; Cai, Q.W. Effects of grain size and misorientation on anisotropy of X80 pipeline steel. *Hot Working Tech.* **2013**, *42*, 107–109.
18. Masoumi, M.; Silva, C.C.; Abreu, H.F.G.D. Effect of rolling in the recrystallization temperature region associated with a post-heat treatment on the microstructure, crystal orientation, and mechanical properties of API 5L X70 pipeline steel. *J. Mater. Eng. Perfor.* **2018**, *27*, 1694–1705. [CrossRef]
19. Deng, C.M.; Li, Z.D.; Sun, X.J.; Zhou, Y.; Yong, Q.L. Influence mechanism of high angle boundary on propagation of cleavage cracks in low carbon lath martensite steel. *Mater. Mech. Eng.* **2014**, *38*, 20–24.
20. Shen, J.C.; Luo, Z.J.; Yang, C.F.; Zhang, Y.Q. Effective grain size affecting low temperature toughness in lath structure of HSLA steel. *J. Iron Steel Res. Int.* **2014**, *26*, 70–76.
21. Yang, X.L.; Xu, Y.B.; Tan, X.D.; Wu, D. Relationships among crystallographic texture, fracture behavior and Charpy impact toughness API X100 pipeline steel. *Mater. Sci. Eng. A* **2015**, *641*, 96–106. [CrossRef]
22. Zhao, J.; Hu, W.; Wang, X.; Wang, J.; Yuan, G.; Di, H.; Misra, R.D.K. Effect of microstructure on the crack propagation behavior of micro-alloyed 560MPa(X80) strip during ultra-fast cooling. *Mater. Sci. Eng. A* **2016**, *666*, 214–224. [CrossRef]
23. Cui, G.B.; Ju, X.H.; Peng, G.; Yin, L.X. EBSD characterization on the structure of M-A islands of X80 pipeline steel. *J. Elec. Micro. Soc.* **2014**, *33*, 423–428.
24. Hwang, B.; Lee, S.; Kim, Y.M.; Kim, N.J.; Yoo, J.Y.; Woo, C.S. Analysis of abnormal fracture occurring during drop-weight tear test of high-toughness line-pipe steel. *Mater. Sci. Eng. A* **2004**, *368*, 18–27. [CrossRef]
25. Zhou, M.; Du, L.X.; Yi, H.L.; Liu, X.H. Factors affecting DWTT property of X80 pipeline steel. *J. Iron Steel Res. Int.* **2009**, *21*, 33–36.
26. Zhang, X.L.; Feng, Y.R.; Zhuang, C.J.; Zhao, W.Z.; Huo, C.Y. Study on effective particle size of high-grade pipeline steels and relationship between CVN. *Mater. Mech. Eng.* **2007**, *31*, 4–8.
27. Fuentes, M.D.; Mendia, A.; Gutiérrez, I. Analysis of different acicular ferrite microstructures in low-carbon steels by electron backscattered diffraction. *Metal. Mater. Trans. A* **2003**, *34*, 2505–2516. [CrossRef]
28. Wang, C.F.; Wang, M.Q.; Shi, J.; Hui, W.J.; Dong, H. Effect of microstructural refinement on the toughness of low carbon martensitic steel. *Script. Mater.* **2008**, *58*, 492–495. [CrossRef]

Article

Fracture Toughness, Breakthrough Morphology, Microstructural Analysis of the T2 Copper-45 Steel Welded Joints

Hao Ding [1], Qi Huang [1], Peng Liu [1], Yumei Bao [2] and Guozhong Chai [1,*

[1] Key Laboratory of E&M, Ministry of Education & Zhejiang Province, Zhejiang University of Technology, Hangzhou 310014, China; dinghao@zjut.edu.cn (H.D.); huangqijob@sina.com (Q.H.); liu_394962328@163.com (P.L.)

[2] School of Mechanical Engineering, Zhijiang College of Zhejiang University of Technology, Shaoxing 312030, China; baoym@zjut.edu.cn

* Correspondence: chaigz@zjut.edu.cn

Received: 21 December 2019; Accepted: 16 January 2020; Published: 20 January 2020

Abstract: The performance and flaws of welded joints are important features that characteristics of the welding material influence. There is significant research activity on the performance and characteristics of welding joint materials. However, the properties of dissimilar welding materials and the cracking problem have not been thoroughly investigated. This investigation focuses on the evaluation and analysis of fracture mechanics, including fracture toughness, microstructural analysis, and crack initiation of T2 copper-45 steel dissimilar welding materials. Standard tensile and three-point bending experiments were performed to calculate the ultimate strength, yield strength, and elastic modulus for fracture toughness. The macro/micro-fracture morphology for tensile fracture and three-point bending fracture were analysed. Based on these investigations, it was concluded that the fracture types were quasi-cleavage and an intergranular brittle fracture mixed model. The deflection of the crack path was discussed and it was determined that the crack was extended along the weld area and tilted towards the T2 copper. Finally, the crack propagation and deflecting direction after the three-point bending test could provide the basis for improvement in the performance of welded joints based on experimental testing parameters and ABAQUS finite element analysis.

Keywords: dissimilar welding materials; electron-beam welding; fracture morphology; fracture toughness; crack deflection; three-point bending test

---

1. Introduction

The welding of dissimilar metals has been an area of active investigations for many years. This objective reflects an overall industrial need of increasing importance that is predicated on the technical and economic potential of the process [1]. Dissimilar metals are welded to achieve physical flexibility, but this practice often results in problems that negatively affect the performance of the weld [2]. Many researchers have investigated the effects of the welding method used for different materials that are characterized by different electrochemical [3], thermal [4], optical [5], and mechanical properties [6–9], especially dissimilar metals [9–21]. In general, for conventional joints with two dissimilar metals, the primary concern is the potential effect of the unique properties of the materials on the fusing process and further determines the mechanical behavior of the joint [22]. It has been determined that welding defects are highly related to mechanical properties. In the case of keyhole pores, the formation is controlled by the temperature gradient and surface tensions of the liquid/solid interface [23], when the selective laser melting (SLM) defects quantity increase to a certain proportion, the tensile strength, fatigue life, and hardness of the dissimilar joint are dramatically affected [24].

Generally, the physical and chemical properties of copper and carbon steel are quite different. The thermal conductivity of copper is 7–11 times that of steel and the melting point is 400–500 degrees Celsius lower than steel. However, at high temperatures, the atomic radius, lattice types, and lattice constants of Fe and Cu are very close. These similarities are beneficial in the welding of copper and steel dissimilar materials [15]. T2 copper is a commonly used metal material in industry. It is often used as the material of large container structure because of its high chemical stability and good corrosion resistance in a calcium salt environment [25]. However, T2 copper has low strength and a large specific gravity, which results in its limitation in lightweight design [11]. In contrast, high strength and easy cutting behavior characterize 45 Steel. These contrasting properties imply that the resulting joints due to these two metals would have broad application prospects. The fusion zone (FZ) microstructures in the electron beam welding of copper-stainless steel were investigated and the results indicated the existence of some defects, such as porosity and micro-fissures, which are mainly influenced by the process and geometry parameters [2]. More interestingly, by appropriately adjusting the welding parameters during electron beam welding, the porosity and micro-cracks can be effectively reduced in the heat-affected zone (HAZ) and FZ; this was strongly controlled by a high-temperature gradient [25]. The high-temperature gradient in the electron beam welding process compensated for the influence of the temperature difference between copper and steel on the solid-liquid interface and FZ [3].

Many researchers have investigated the mechanical behavior of dissimilar welding joints; however, most of the studies have focused on the effect of welding defects on the tensile strength of dissimilar welding materials [1]. Analysis of the fracture characteristics using finite element simulation has seldom been performed, while the control of microstructure during various welding processes has been well investigated [26]. The effect of an intermetallic compound on mechanical properties has also been well-studied, but the available information on fracture performance is still limited [27]. The formation of intermetallic phases greatly affects the interfacial strength of dissimilar welding materials as a result of the different melting temperatures, particularly for copper-steel dissimilar welding materials [2]. The crack propagation mechanism of dissimilar welding materials has generated general interest, and some reports have demonstrated the mechanism of surface crack propagation of these materials by combining microstructures e.g., ferrite and austenite [27,28]. Moreover, many investigators have analysed the effect of crack position on the fracture behaviour that is based on the three-point bending tests [29]. A few researchers have summarized the effect of some regular patterns on crack propagation [30]. Crack ductility fracture occurs in low-strength materials and the distance between the crack initiation point and interface affect the fracture behavior [31]. The numerical simulation of crack propagation was consistent with experimental results. The extended finite element method (XFEM) [32,33] is frequently utilized to simulate crack propagation [34]. XFEM has also been applied to simulate the crack propagation of contact fatigue [35], which was analysed based on two-dimensional and three-dimensional contact fatigue tests [36,37]. Various models of crack propagation have been established to be consistent with actual situations [38].

At present, research on the welding materials of T2 copper and 45 steel mainly focuses on the influence of the welding solder on the interfacial strength. However, correlative research on electron beam welding is limited. As such, it is important to study the joint property and fracture behaviour of the electron beam welding specimens.

In this work, a copper/steel dissimilar welding specimen was prepared by electron beam welding (without filler wire) to examine the weld properties of these joints. The microstructure of the copper/steel dissimilar welding materials of the weld area was analysed via the combination of micro-topography and macroscopic appearance. Accordingly, the relationship between crack deflection was demonstrated based on the material properties. Moreover, the fracture mechanics test parameter was examined on the basis of tensile properties and bending experiments.

2. Materials and Methods

2.1. Parameter Test Method for Joint Property

The test materials were prepared while using commercial welding processes and electron beam welding equipment. The model of electron beam welding machine was SEBW and the manufacturer was Guilin Shichuang vacuum CNC Equipment Co., Ltd. (Guilin, China). After investigation, Table 1 shows the parameter cases of some scholars in the electron beam welding of copper-steel.

Table 1. Several cases of electron beam welding parameters of copper-steel.

Case	Thickness /mm	Acceleration Voltage/kV	Electron Beam/mA	Welding Speed/mm·min^{-1}
Kar, J. [25,39]	3	60	65, 73, 80	1000
Guo, S. [40]	5	60	43–70	600
Zhang, B.G. [15,41]	2.7	60	25, 30, 35	100, 200, 300
Chen, G. [42]	5	60	15	400
Tomashchuk, I. [43]	2	20–40	20–40	200–900

We preferentially adjust the parameters with reference to Table 1. After actual testing, the electron beam welding parameters of T2 Copper/45 steel dissimilar welding materials were as follows: acceleration voltage 80 kV, electron beam 100 mA, vacuum degree 5×10^{-2} torr, and welding speed 300 mm/min. The surface of the welding sample and the HAZ had a visible dividing line with the weld, which is clearly shown in Figure 1. Table 2 shows the chemical composition of T2 copper and 45 steel.

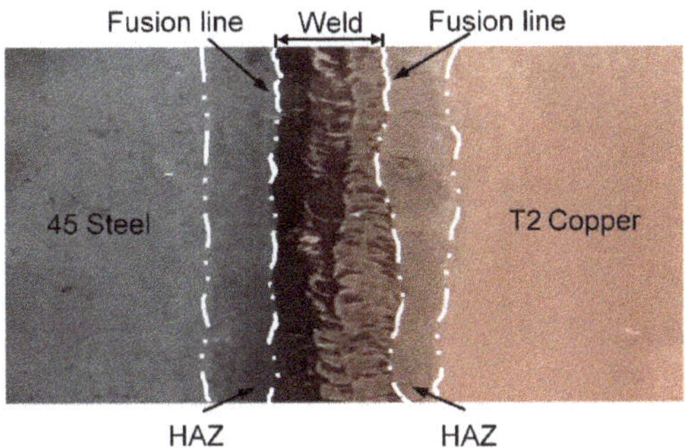

Figure 1. Welded sample.

Table 2. Chemical composition of T2 copper [44] and 45 steel [45].

Sample	Cu + Ag (Minimum Value)	Bi	Sb	As	Fe	Pb	S
T2 copper	99.90	0.001	0.002	0.002	0.005	0.005	0.005

Sample	C	Si	Mn	P	S	Cr	Ni	Cu
45 steel	0.42–0.50	0.17–0.37	0.50–0.80	0.035	0.035	0.25	0.30	0.25

A $4 \times 8 \times 10$ mm square sample was taken from the welded sample by Electrical Discharge Machining (EDM), as shown in Figure 2a, and Figure 2b–e show the microscopic topography of welded area via scanning electron microscopy (SEM) after polishing. The magnifications were 50 and 500 times, respectively. Figure 2b displays the iron and copper ends of the weld area. In Figure 2c–e, some pores, microcracks, and the insufficient welding area can be seen. These defects are the important factors that affect the welded bond quality of T2 copper-45 steel.

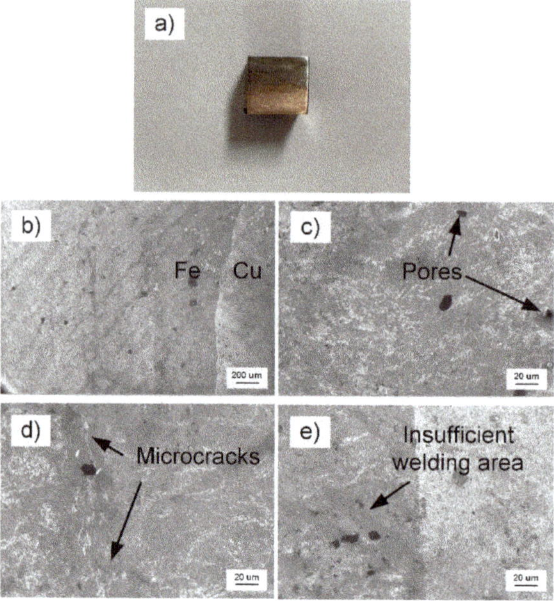

Figure 2. Scanning electron microscopy (SEM) micro-morphology of welded area (**a**) square sample by Electrical Discharge Machining (EDM); (**b**) magnification is 50 times; and, (**c,d,e**) magnification is 500 times.

The tensile specimen was obtained by wire cutting according to GB/T228.1-2010 [46] (Metallic Materials-Tensile Testing-Part 1: Method of testing at room temperature). The ultimate strength, yield strength, and elastic modulus were determined while using the INSRON-8801 Servohydraulic Fatigue Testing System (Instron, Darmstadt, Germany) with a loading rate of 1 mm/min. Figure 3 shows the tensile specimen.

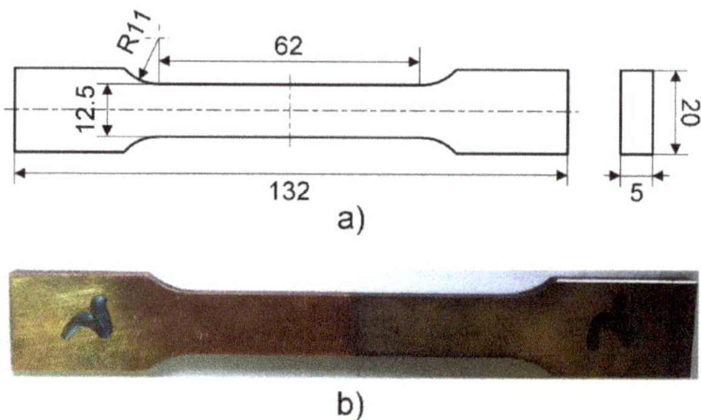

Figure 3. Tensile specimen (**a**) dimension (mm); and, (**b**) actual sample.

The processing of the three-point bending specimen was based on GB/T21143-2014 [30] (unified method of test for determination of quasi-static fracture toughness) and GB/T 28896-2012 [47] (metallic materials-method of test for the determination of quasi-static fracture toughness of welds). The sampling orientation of the fracture surface of the fracture toughness specimen in the weld zone was NQ, as shown in Figure 4a, the maximum fatigue preformed twill force was set according to the smaller value of Equations (1) and (2), the maximum fatigue crack stress was calculated as F_f is 1344.7169 N at the last 1.3 mm or 50% fatigue precracking propagation, and the stress ratio r is 0.5.

$$F_f = 0.8 \times \frac{B(W - a_0)^2}{S} \times R_{p0.2} \tag{1}$$

$$F_f = \xi \cdot E \left[\frac{(W \cdot B \cdot B_N)^{0.5}}{g_1\left(\frac{a_0}{W}\right)} \right] \cdot \left(\frac{W}{S}\right) \tag{2}$$

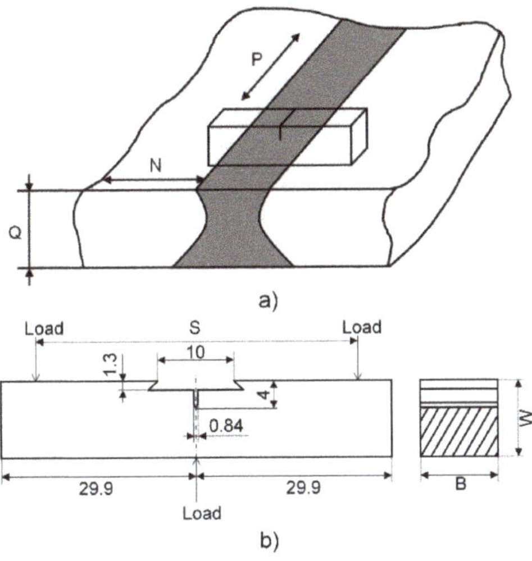

Figure 4. Three-point bending (**a**) specimen sampling orientation; and, (**b**) dimension (mm).

In the preceding Equations (1) and (2), the dimensional coefficient ξ is 1.6×10^{-4} m$^{1/2}$, B is the sample thickness that is shown in Figure 4b, W is the width of the specimen, B_N is the net thickness of the specimen and B, B_N, W are 13 mm; the span S is 52 mm, the initial crack length a_0 is 6 mm, the stress intensity factor coefficient $g(a_0/W)$ is 2.29; E is the elastic modulus; and, $R_{p0.2}$ is the specified plastic elongation strength of the material in the vertical crack plane 0.2% at the test temperature.

The fatigue crack was prepared while using the constant load method. After this process, the fatigue precracking of the three specimens was: 2.02, 1.96, and 2.04 mm. Figure 5 shows the final specimen of three-point bending.

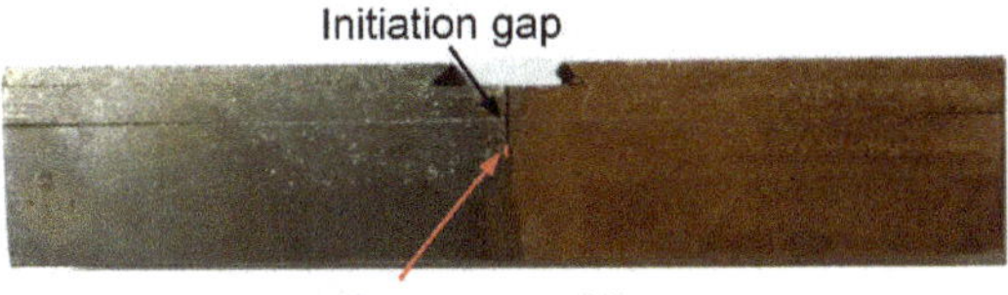

Figure 5. Final specimen of three-point bending.

2.2. Characterization Results of Joint Property Parameters

Table 3 shows the performance parameters were obtained by standard tensile tests and the results. The displacement-force curve (P-V curve) of the notch opening was obtained based on the three-point bending test that is shown in Figure 6.

Table 3. Tensile test results.

No.	Ultimate Strength	Yield Strength	Elastic Modulus
	σ_b/MPa	σ_s/MPa	E/GPa
1	102.45	81.02	88.37
2	84.67	78.64	110.53
3	94.06	66.45	127.68
Average value	93.73	75.37	108.86

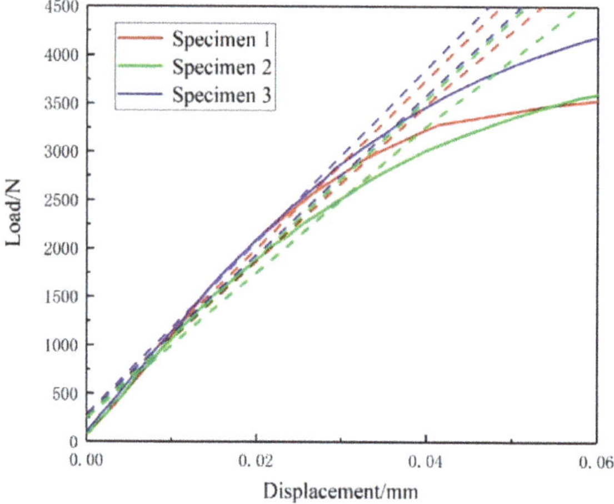

Figure 6. Displacement-force (P-V) curve of three-point bending.

After the P-V curve of the three-point bending specimen was shifted, the value F_Q of the three specimens was 3286.569, 2727.193, and 3581.864 N.

The judgment basis is as follows.

$$\frac{F_{max}}{F_Q} \geq 1.1 \tag{3}$$

where F_Q is the maximum force and F_{max} is the maximum force that the specimen can withstand.

Given that F_{max}/F_{Q1}, F_{max}/F_{Q2}, and F_{max}/F_{Q3} are greater than 1.1, K_{max} (conditional value of K_{IC}) was calculated while using Equation (4).

$$K_{max} = K_Q = \left[\left(\frac{S}{W} \right) \frac{F_Q}{(B \cdot B_N \cdot W)^{0.5}} \right] \cdot \left[g_1 \left(\frac{a_0}{W} \right) \right] \tag{4}$$

The judgment on plane strain fracture toughness K_{IC} is represented, as follows.

$$a_0 = 2.5 \left(\frac{K_Q}{R_{p0.2}} \right)^2 \tag{5}$$

$$(W - a_0) = 2.5 \left(\frac{K_Q}{R_{p0.2}} \right)^2 \tag{6}$$

$$B = 2.5 \left(\frac{K_Q}{R_{p0.2}} \right) \tag{7}$$

$$K_f = 0.6 K_Q \left(\frac{(R_{p0.2})_p}{(R_{p0.2})_e} \right) \tag{8}$$

where K_Q can be acquired from the three-point bending test, $(R_{p0.2})_e$ is the plastic extension strength corresponding to the bias 0.2% at the test temperature, and $(R_{p0.2})_p$ is the plastic elongation corresponding to the fatigue precracking offset 0.2%.

After the aforementioned judgement, a_0 is the initial crack length, $W - a_0$ is the difference between the sample width and the initial crack, and K_f is the maximum value of the stress intensity factor in the final stage of the prepared fatigue crack. Table 4 presents these parameters.

$$K_{IC} = \frac{K_{Q1} + K_{Q2} + K_{Q3}}{3} = 6.027 \text{MPa} \cdot \text{m}^{1/2} \tag{9}$$

Table 4. K_{IC} data from calculation.

No.	K_Q/MPa·m$^{1/2}$	a_0/mm	B/mm	$(W - a_0)$/mm	K_f/MPa·m$^{1/2}$
1	5.827	10.17	12.35	0.33	31.426
2	6.072	10.21	12.47	0.35	33.325
3	6.181	11.08	12.61	0.36	35.217

According to results from the data that are contained in Table 4 and Equation (9), the fracture toughness of the welded joint calculated in the three-point bending test is 6.027 MP·m$^{1/2}$.

3. Results and Discussion

3.1. Fracture Analysis of Tensile Test

After tensile testing, the macroscopic fracture area of the specimen is shown in Figures 7 and 8. The specimen breaks in the weld zone and crack propagation was biased towards the copper interface. The copper can be observed on the fracture surface, which was partially attached to the ends.

Figure 7. Macroscopic tensile fracture (**a**) for specimen 1; (**b**) for specimen 2; and, (**c**) for specimen 3; steel on the left and copper on the right.

The microscopic topography of the tensile fracture of the T2 copper/45 steel dissimilar welding materials is shown in Figure 9 via SEM at a magnification of 1000 times for each specimen two-fracture end face, according to the order of the macroscopic fracture morphology in the immediately preceding figures.

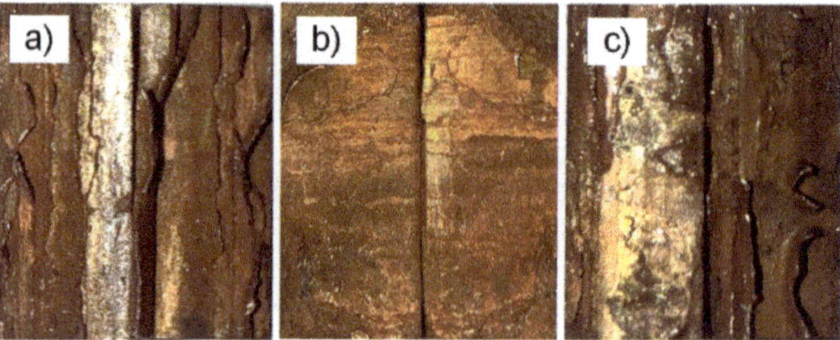

Figure 8. Macro-morphology fracture surface of tensile specimen (**a**) for specimen 1; (**b**) for specimen 2; and, (**c**) for specimen 3; steel on the left and copper on the right.

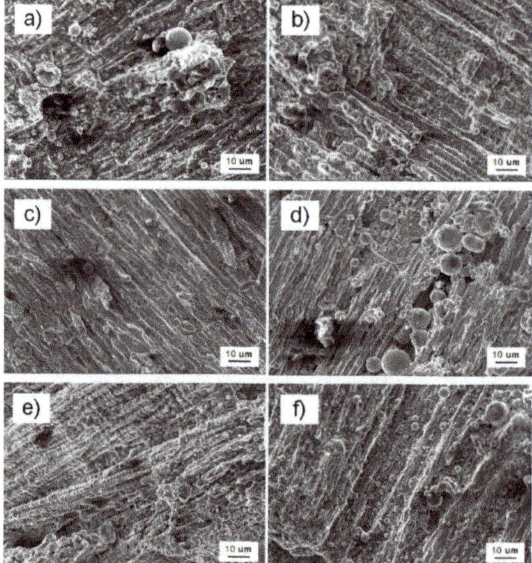

Figure 9. SEM micro-morphology fracture surface of tensile specimen (**a,b**) for specimen 1; (**c,d**) for specimen 2; and, (**e,f**) for specimen 3; steel on the left and copper on the right.

The macroscopic shape of specimen 1 had obvious gloss and irregular geometry, and the shiny surface of the fracture was almost perpendicular to the normal stress, which is associated with the brittle fracture characteristic; and, significant grain-brittle fracture characteristics at the microscopic level. There was a network structure after fracture due to an external force, which was a relatively obvious network brittle phase, as shown in Figure 9b. The reason for this fracture was the brittle precipitation phase on the grain boundary, which results in the formation of a continuous carbide network by allotropes of iron during electron beam welding, which led to a thin layer of brittle fracture splitting.

Brittle fracture also characterized the macroscopic fracture of specimen 2, which appeared as herringbone and radial patterns at the fracture with a shiny surface. There were fluvial, blocky, and spherical structures in the microscopic topography with cleavage steps and tearing ribs, which exhibited the microscopic features of crystal brittleness and cleavage fracture [41,48].

A few flaky smooth surfaces existed in the macroscopic fracture of specimen 3 and the entire fracture surface was relatively flat. The cleavage characteristics of trapezoidal and river patterns also appeared in the microscopic morphology, with tiny cleavage steps and tearing ribs that are associated with the cleavage fracture. The defects in the weld area and the impurities of the welding material caused this microscopic appearance.

3.2. Fracture Analysis of Three-Point Bending Test

The macroscopic fracture surface is shown in Figure 10 after the three-point bending test and the prepared breach prepared fatigue crack and crack extension zone can be observed.

Figure 10. Macro-morphology fracture surface of three-point bending specimen (**a**) for specimen 1; (**b**) for specimen 2; and (**c**) for specimen 3; steel on the left and copper on the right.

The purplish-red hue gradually deepens from the top to the bottom in the crack extension zone and the copper attached to the fracture surface gradually increased. It was known that the crack gradually deflected along the copper thereby tearing copper that was attached to the surface, as shown in Figure 11.

The three-point bending fracture was observed at 1000 times magnification while using SEM. As shown in Figure 12, these fractures in the macroscopic image have an obvious shiny surface and irregular geometry. The fluvial, blocky, and spherical structures were readily apparent in the microscopic topography with the cleavage steps and tearing ribs distributed, therefore, it was a typically mixed mode of brittle intergranular and quasi-cleavage fracture.

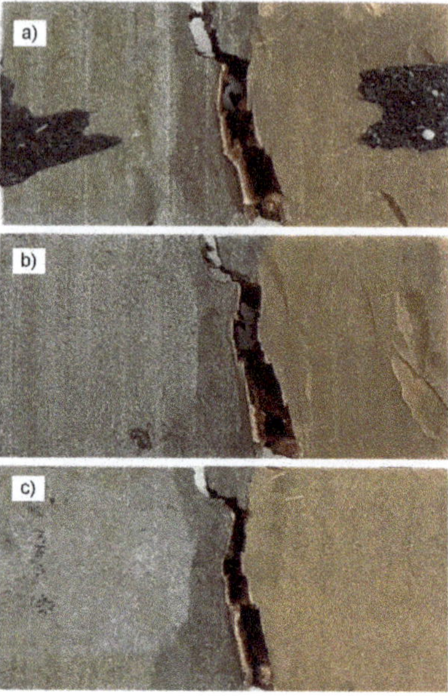

Figure 11. Macroscopic crack propagation of three-point bending specimen (**a**) for specimen 1; (**b**) for specimen 2; and (**c**) for specimen 3; steel on the left and copper on the right.

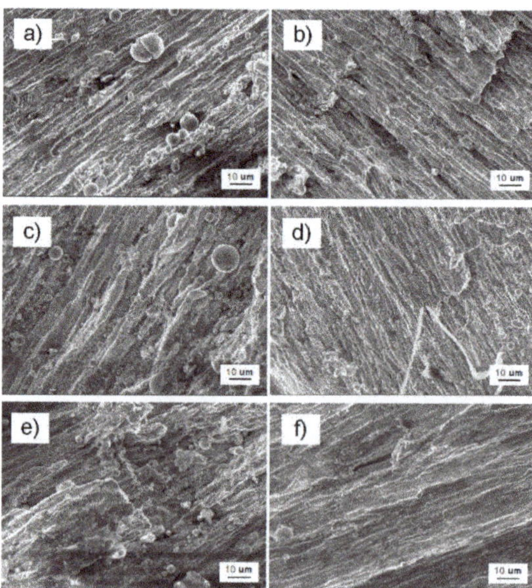

Figure 12. SEM micro-morphology fracture surface of three-point bending specimen (**a**,**b**) for specimen 1; (**c**,**d**) for specimen 2; and, (**e**,**f**) for specimen 3; steel on the left and copper on the right.

3.3. Analysis of Crack Propagation Direction

The crack propagation path deflection of the dissimilar metal welding materials always deflects to the low strength material region [19]. The attached copper on the fracture area was caused by the deflection tear of the fracture path based on the aforementioned tensile test fracture morphology. Figure 13a shows a schematic diagram of the crack deflection of the standard tensile specimen, which was similar to the three-point bending test of T2 copper/45 steel dissimilar welding materials deflection path. This resulted in the phenomenon of stepwise reduction of the resistance to fracture due to the difference in the toughness between the weld area, HAZ and base metal in the electron beam welding process. Based on the three-point bending test of T2 copper/45 steel dissimilar welding materials cracking failure, the crack path was deflected due to the difference in the toughness, subject to factors, such as pores, micro-cracks in the weld area, crack deflection to T2 copper, as shown in the schematic diagram in Figure 13b.

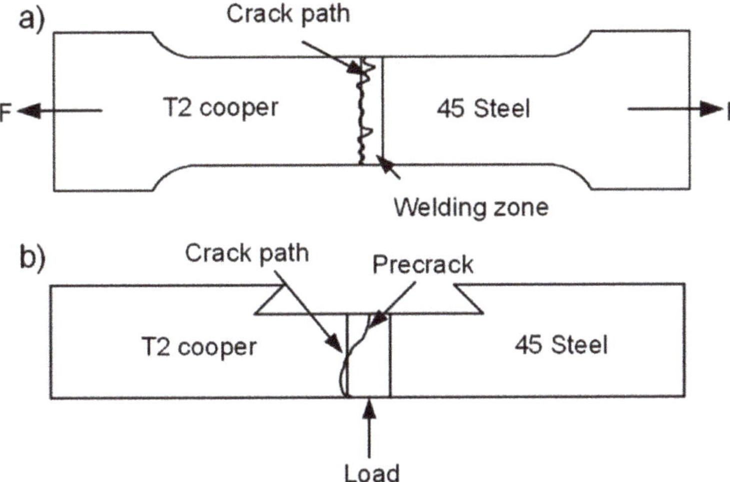

Figure 13. Crack deflection diagram (**a**) Standard tensile test; and (**b**) Three-point bending test.

Therefore, the crack propagation path of T2 copper/45 steel dissimilar welding materials always deflected to the low strength side of the T2 copper because the strength mismatch between three regions was comparatively large and the toughness decreases from the weld area to HAZ and then the base metal.

According to the test parameters and conditions, the simulation of crack propagation of the three-point bending test was performed by ABAQUS, and the results are shown in Figure 14. With the increase of the expansion step, the crack expanded along the weld seam position and it was initially biased toward the T2 copper. The crack expanded along the junction until the specimen broke when the crack extended to the junction of the weld seam area and T2 copper. It is clear that the ABAQUS simulation results are consistent with these observations, as represented in Figure 11.

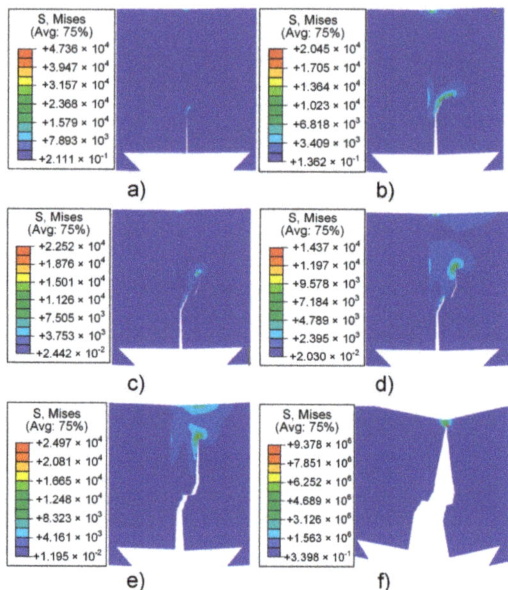

Figure 14. Three-point bending crack propagation simulation (**a**) step 1; (**b**) step 2; (**c**) step 3; (**d**) step 4; (**e**) step 5; (**f**) step 6.

4. Conclusions

For the T2 copper-45 steel dissimilar welding materials that were made by electron beam welding, the joint strength, microstructural analysis, and crack initiation were explored. Based on the standard tensile test, the ultimate strength of T2 copper/45 steel dissimilar welding materials were determined to be 93.73 MPa, the yield strength was 75.37 MPa, and the elastic modulus was 108.86 GPa. It can be seen that the mechanical properties of the weld area are significantly different from those of copper and steel, which causes the strength mismatch between three regions. Through the three-point bending test, the fracture toughness was determined to be 6.027 MPa·m$^{1/2}$, which was lower than that of pure copper (approximately 8 MPa·m$^{1/2}$–10 MPa·m$^{1/2}$) [49]. This is due to welding defects in the weld area. Some pores and microcracks were found in SEM micro-morphology of the welded area, which directly leads to the reduction of the mechanical properties. Weld defects indicate that, in practical application, the electron beam welding process needs to be optimized, or more suitable welding methods need to be found.

The SEM micro-morphology fracture surface of three-point bending specimen shows that the fracture type was a mixed mode of brittle intergranular and quasi-cleavage fracture. The observation results of macroscopic crack propagation of three-point bending specimen were consistent with the theoretical and ABAQUS analysis, it was concluded that the cracking path was extended along the weld area and biased towards the T2 copper. Moreover, the strength of mismatch and toughness reduction controlled the deflection.

Author Contributions: Conceptualization, H.D.; Data curation, P.L.; Formal analysis, Q.H. and P.L.; Funding acquisition, H.D., Y.B. and G.C.; Project administration, G.C.; Supervision, Y.B. and G.C.; Writing—original draft, H.D. and P.L.; Writing—review & editing, Q.H. All authors have read and agreed to the published version of the manuscript.

Funding: This research was funded by National Natural Science Foundation of China, grant number (51575489) and Natural Science Foundation of Zhejiang Province, grant number (LSY19H180004, LQY18E050001 and LY20A020007).

Conflicts of Interest: The authors declare no conflict of interest.

References

1. Verma, J.; Taiwade, R.V. Effect of welding processes and conditions on the microstructure, mechanical properties and corrosion resistance of duplex stainless steel weldments—A review. *J. Manuf. Process.* **2017**, *25*, 134–152. [CrossRef]
2. Magnabosco, I.; Ferro, P.; Bonollo, F.; Arnberg, L. An investigation of fusion zone microstructures in electron beam welding of copper–stainless steel. *Mater. Sci. Eng. A* **2006**, *424*, 163–173. [CrossRef]
3. Chung, F.K.; Wei, P.S. Mass, Momentum, and Energy Transport in a Molten Pool When Welding Dissimilar Metals. *J. Heat Transf.* **1999**, *121*, 451–461. [CrossRef]
4. Sun, Z.; Karppi, R. The application of electron beam welding for the joining of dissimilar metals: An overview. *J. Mater. Process. Technol.* **1996**, *59*, 257–267. [CrossRef]
5. Zumelzu, E.; Cabezas, C. Study on welding such dissimilar materials as AISI 304 stainless steel and DHP copper in a sea-water environment. Influence of weld metals on corrosion. *J. Mater. Process. Technol.* **1996**, *57*, 249–252. [CrossRef]
6. Mai, T.A.; Spowage, A.C. Characterisation of dissimilar joints in laser welding of steel-kovar, copper-steel and copper-aluminium. *Mater. Sci. Eng. A* **2004**, *374*, 224–233. [CrossRef]
7. Srinivasan, P.B.; Muthupandi, V.; Dietzel, W.; Sivan, V. An assessment of impact strength and corrosion behaviour of shielded metal arc welded dissimilar weldments between UNS 31803 and IS 2062 steels. *Mater. Des.* **2006**, *27*, 182–191. [CrossRef]
8. Wei, P.S.; Kuo, Y.K.; Ku, J.S. Fusion Zone Shapes in Electron-Beam Welding Dissimilar Metals. *J. Heat Transf.* **2000**, *122*, 626–631. [CrossRef]
9. Tosto, S.; Nenci, F.; Jiandong, H. Microstructure of copper-AISI type 304L electron beam welded alloy. *Mater. Sci. Technol.* **2003**, *19*, 519–522. [CrossRef]
10. Weigl, M.; Schmidt, M. Modulated laser spot welding of dissimilar copper-aluminium connections. In Proceedings of the 6th International Conference on Multi-Material Micro Manufacture, Forschungszentrum Karlsruhe, Germany, 23–25 September 2009.
11. Velu, M.; Bhat, S. Metallurgical and mechanical examinations of steel-copper joints arc welded using bronze and nickel-base superalloy filler materials. *Mater. Des.* **2013**, *47*, 793–809. [CrossRef]
12. Wu, M.F.; Si, N.C.; Chen, J. Contact reactive brazing of Al alloy/Cu/stainless steel joints and dissolution behaviors of interlayer. *Trans. Nonferrous Met. Soc. China* **2011**, *21*, 1035–1039. [CrossRef]
13. Yaghi, A.H.; Hyde, T.H.; Becker, A.A.; Sun, W. Finite element simulation of residual stresses induced by the dissimilar welding of a P92 steel pipe with weld metal IN625. *Int. J. Press. Vessel. Pip.* **2013**, *111*, 173–186. [CrossRef]
14. Yao, C.W.; Xu, B.S.; Zhang, X.C.; Huang, J.; Fu, J.; Wu, Y.X. Interface microstructure and mechanical properties of laser welding copper-steel dissimilar joint. *Opt. Lasers Eng.* **2009**, *47*, 807–814. [CrossRef]
15. Zhang, B.G.; Zhao, J.; Li, X.P.; Chen, G.Q. Effects of filler wire on residual stress in electron beam welded QCr0.8 copper alloy to 304 stainless steel joints. *Appl. Therm. Eng.* **2015**, *80*, 261–268. [CrossRef]
16. Liu, S.; Liu, F.; Xu, C.; Zhang, H. Experimental investigation on arc characteristic and droplet transfer in CO_2 laser–metal arc gas (MAG) hybrid welding. *Int. J. Heat Mass Transf.* **2013**, *62*, 604–611. [CrossRef]
17. Liu, F.; Zhang, Z.; Liu, L. Microstructure evolution of Al/Mg butt joints welded by gas tungsten arc with Zn filler metal. *Mater. Character.* **2012**, *69*, 84–89. [CrossRef]
18. Dong, H.; Hu, W.; Duan, Y.; Wang, X.; Dong, C. Dissimilar metal joining of aluminum alloy to galvanized steel with Al–Si, Al–Cu, Al–Si–Cu and Zn–Al filler wires. *J. Mater. Process. Technol.* **2012**, *212*, 458–464. [CrossRef]
19. Chen, S.H.; Li, L.Q.; Chen, Y.B.; Liu, D.J. Si diffusion behavior during laser welding-brazing of Al alloy and Ti alloy with Al-12Si filler wire. *Trans. Nonferrous Met. Soc. China* **2010**, *20*, 64–70. [CrossRef]
20. Li, H.M.; Sun, D.Q.; Cai, X.L.; Dong, P.; Wang, W.Q. Laser welding of TiNi shape memory alloy and stainless steel using Ni interlayer. *Mater. Des.* **2012**, *39*, 285–293. [CrossRef]
21. Miles, M.; Kohkonen, K.; Weickum, B.; Feng, Z. Friction Bit Joining of Dissimilar Material Combinations of High Strength Steel DP 980 and Al Alloy AA 5754. *SAE Tech. Pap.* **2009**. [CrossRef]
22. Curtis, T.; Widener, C.; West, M.; Jasthi, B.; Hovanski, Y.; Carlson, B.; Szymanski, R.; Bane, W. Friction Stir Scribe Welding of Dissimilar Aluminum to Steel Lap Joints. In *Friction Stir Welding and Processing VIII*; Mishra, R.S., Mahoney, M.W., Sato, Y., Hovanski, Y., Eds.; Springer: Cham, Switzerland, 2015.

23. Semak, V.; Matsunawa, A. The role of recoil pressure in energy balance during laser materials processing. *J. Phys. D Appl. Phys.* **1999**, *30*, 2541. [CrossRef]

24. IMAM. How Do SLM Process Defects Impact Ti64 Mechanical Properties? Available online: http://www.insidemetaladditivemanufacturing.com/blog/how-doslm-process-defects-impact-ti64-mechanical-properties (accessed on 1 May 2019).

25. Kar, J.; Roy, S.K.; Roy, G.G. Effect of beam oscillation on electron beam welding of copper with AISI-304 stainless steel. *J. Mater. Process. Technol.* **2016**, *233*, 174–185. [CrossRef]

26. Wei, P.S.; Chung, F.K. Unsteady Marangoni Flow in a Molten Pool When Welding Dissimilar Metals. *Metall. Mater. Trans. B* **2000**, *1*, 1387–1403. [CrossRef]

27. Blouin, A.; Chapuliot, S.; Marie, S.; Niclaeys, C.; Bergheau, J.M. Brittle fracture analysis of Dissimilar Metal Welds. *Eng. Fract. Mech.* **2014**, *131*, 58–73. [CrossRef]

28. Gilles, P.; Brosse, A.; Pignol, M. Simulation of Ductile Tearing in a Dissimilar Material Weld up to Pipe Wall Break-Through. In Proceedings of the Asme Pressure Vessels & Piping Division/k-pvp Conference, New York, NY, USA, 18–22 July 2010.

29. Samal, M.K.; Seidenfuss, M.; Roos, E.; Balani, K. Investigation of failure behavior of ferritic–austenitic type of dissimilar steel welded joints. *Eng. Fail. Anal.* **2011**, *18*, 999–1008. [CrossRef]

30. *GB/T 21143-2014 Metallic Materials-Unified Method of Test for Determination of Quasistatic Fracture Toughness*; Standards Press of China: Beijing, China, 2014.

31. Faidy, C. Structural Integrity of Bi-Metallic Welds in Piping Fracture Testing and Analysis. In Proceedings of the Asme Pressure Vessels & Piping Conference, New York, NY, USA, 18–22 July 2010.

32. Ashari, S.E.; Mohammadi, S. Delamination analysis of composites by new orthotropic bimaterial extended finite element method. *Int. J. Numer. Methods Eng.* **2011**, *86*, 1507–1543. [CrossRef]

33. Belytschko, T.Y.; Black, T. Elastic Crack Growth in Finite Elements with Minimal Remeshing. *Int. J. Numer. Methods Eng.* **2015**, *45*, 601–620. [CrossRef]

34. Nicak, T.; Schendzielorz, H.; Keim, E.; Meier, G. STYLE: Study on Transferability of Fracture Material Properties from Small Scale Specimens to a Real Component. In Proceedings of the Asme Pressure Vessels & Piping Conference, New York, NY, USA, 18–22 July 2010.

35. Motamedi, D.; Mohammadi, S. Dynamic crack propagation analysis of orthotropic media by the extended finite element method. *Int. J. Fract.* **2009**, *161*, 21–39. [CrossRef]

36. Zhang, Z.; Ma, W.L.; Wu, H.L.; Wu, H.P.; Jiang, S.F.; Chai, G.Z. A rigid thick Miura-Ori structure driven by bistable carbon fibre-reinforced polymer cylindrical shell. *Compos. Sci. Technol.* **2018**, *167*, 411–420. [CrossRef]

37. Zhang, Z.; Li, Y.; Wu, H.L.; Chen, D.D.; Yang, J.; Wu, H.P.; Jiang, S.F.; Chai, G.Z. Viscoelastic bistable behaviour of antisymmetric laminated composite shells with time-temperature dependent properties. *Thin-Walled Struct.* **2018**, *122*, 403–415. [CrossRef]

38. Rivalin, F.; Besson, J.; Pineau, A.; Fant, M.D. Ductile tearing of pipeline-steel wide plates: II. Modeling of in-plane crack propagation. *Eng. Fract. Mech.* **2001**, *68*, 347–364. [CrossRef]

39. Kar, J.; Dinda, S.K.; Roy, G.G.; Roy, S.K.; Srirangam, P. X-ray tomography study on porosity in electron beam welded dissimilar copper–304SS joints. *Vacuum* **2018**, *149*, 200–206. [CrossRef]

40. Guo, S.; Zhou, Q.; Kong, J.; Peng, Y.; Xiang, Y.; Luo, T.; Wang, K.; Zhu, J. Effect of beam offset on the characteristics of copper/304stainless steel electron beam welding. *Vacuum* **2016**, *128*, 205–212. [CrossRef]

41. Zhang, B.G.; Zhao, J.; Xiao-Peng, L.I.; Feng, J.C. Electron beam welding of 304 stainless steel to QCr0.8 copper alloy with copper filler wire. *Trans. Nonferrous Met. Soc. China* **2014**, *24*, 4059–4066. [CrossRef]

42. Chen, G.; Shu, X.; Liu, J.; Zhang, B.; Feng, J. Crystallographic texture and mechanical properties by electron beam freeform fabrication of copper/steel gradient composite materials. *Vacuum* **2020**, *171*, 109009. [CrossRef]

43. Tomashchuk, I.; Sallamand, P.; Jouvard, J.M.; Grevey, D. The simulation of morphology of dissimilar copper–steel electron beam welds using level set method. *Comput. Mater. Sci.* **2010**, *48*, 827–836. [CrossRef]

44. *GB/T 5231-2012 Designation and Chemical Composition of Wrought Copper and Copper Alloys*; Standards Press of China: Beijing, China, 2012.

45. *GB/T 699-2015 Quality Carbon Structure Steels*; Standards Press of China: Beijing, China, 2015.

46. *GB/T 228.1-2010 Metallic Materials-Tensile Testing-Part 1: Method of Test at Room Temperature*; Standards Press of China: Beijing, China, 2010.

47. *GB/T 28896-2012 Metallic Materials-Method of Test for the Determination of Quasistatic Fracture Toughness of Welds*; Standards Press of China: Beijing, China, 2012.
48. Turichin, G.A.; Klimova, O.G.; Babkin, K.D.; Pevzner, Y.B. Effect of Thermal and Diffusion Processes on Formation of the Structure of Weld Metal in Laser Welding of Dissimilar Materials. *Met. Sci. Heat Treat.* **2014**, *55*, 569–574. [CrossRef]
49. Qin, E.W.; Lu, L.; Tao, N.R.; Tan, J.; Lu, K. Enhanced fracture toughness and strength in bulk nanocrystalline Cu with nanoscale twin bundles. *Acta Mater.* **2009**, *57*, 6215–6225. [CrossRef]

Article

The Efficiency of Utilisation of High-strength Steel in Tubular Profiles

Ieva Misiūnaitė, Viktor Gribniak *, Arvydas Rimkus and Ronaldas Jakubovskis

Laboratory of Innovative Building Structures, Vilnius Gediminas Technical University, Sauletekio av. 11, LT-10223 Vilnius, Lithuania; Ieva.Misiunaite@vgtu.lt (I.M.); Arvydas.Rimkus@vgtu.lt (A.R.); Ronaldas.Jakubovskis@vgtu.lt (R.J.)
* Correspondence: Viktor.Gribniak@vgtu.lt; Tel.: +370-6134-6759

Received: 7 February 2020; Accepted: 28 February 2020; Published: 6 March 2020

Abstract: The use of high-strength steel (HSS) is a current trend of the construction industry. Tubular profiles are widely used in various structural applications because of their high stiffness-to-weight ratio, exceptional resistance to torsion, and aesthetic appearance. However, the increase of the strength for the same elastic modulus of the material and geometry of tubular profiles is often not proportional to the rise of the load-bearing capacity of the structural element. The obtained experimental results support the above inference. The study was based on the flexural test results of two groups of HSS and normal-strength steel (NSS) tubular specimens with a 100 × 100 × 4 mm (height × width × thickness) cross-section. Numerical (finite element) simulation results demonstrated that the shape of the cross-section influenced the efficiency of utilisation of HSS. The relationship between the relative increase of the load-bearing capacity of the beam specimen and the corresponding change of the steel strength determined the utilisation efficiency.

Keywords: cold-formed profiles; high-strength steel; local deformations; bending test; load-bearing capacity

1. Introduction

The use of advanced materials such as high-strength steel (HSS) [1–6] is a current trend of the construction industry. The increase of the strength of the materials enables the reduction of the self-weight of the structural elements. Tubular profiles are widely used in various structural applications because of their high stiffness-to-weight ratio, exceptional resistance to torsion, and aesthetic appearance. Hot-rolling and cold-forming processes are predominant in the production of the steels. The application result of the latter technology—a square hollow section (SHS) profile—is the research object of this study.

Production technology has a substantial effect on altering mechanical properties of the steel [7]. Characteristics of the hot-rolled material remain practically unchanged. However, the cold-forming process causes a strength enhancement of the steel, smoothing the shape of the resultant stress-strain diagram; at the same time, the process reduces ultimate deformations, increasing the brittleness of the steel [8–11]. Substantial plastic deformations appear in the corner regions of SHS that causes uneven distribution of material properties in the section [8,11].

The failure of tubular profiles is a consequence of the excessively increased deformations (the stiffness condition) or fracture of the steel (the strength term). The efficient utilisation of the material requires an equilibrium between the above limitations. On the other hand, an increase of the strength under the same elastic modulus and geometry of the profile causes inefficient utilisation of the material. Misiūnaitė et al. [6] revealed that local effects cause the failure of the SHS specimens made from both normal-strength steel (NSS) and HSS. However, the increase of the load-bearing capacity of the tubular profiles was not proportional to the rise of the strength of the steel. Several studies investigated the adequacy of slenderness limits, identifying the extent to which the local buckling controls the resistance of the section [12–14]. An alternative approach is based on the identification of a correlation between the cross-section resistance and deformation capacity [15]. However, none of the works mentioned above define conditions ensuring efficient utilisation of HSS in tubular profiles that can be nominally estimated as the ratio between the alteration of the load-bearing capacity of the element and the increase of the steel strength.

The effect of an increase of the material strength on the enhancement of the mechanical resistance of the profiles is the focus of this research. Two groups of HSS and NSS tubular beam specimens with $100 \times 100 \times 4$ mm (height × width × thickness) cross-section were tested until failure, estimating the efficiency mentioned above. The experimental technique [6] was used to assess the local bearing capacity of the specimens. Numerical simulations illustrated the proposed efficiency concept.

2. Test Program

Square tubular profiles are not susceptible to torsional deformations. The stiff closed shape of the cross-section makes the elements resistant to lateral buckling. These aspects emphasise the structural efficiency of such profiles. At the same time, strengthening of tubular profiles, using supplemental materials, is a more complex problem than that which is characteristic of the open cross-sections [16]. Thus, the efficiency of the utilisation of HSS, reflecting the increase of the load-carrying capacity as the interaction between the local bearing mechanism and global bending resistance of the cold-formed tubular profiles, is the research object of this study.

2.1. Characterisation of the Materials

Flat coupons, extracted from the square hollow section (SHS) profiles, were used to determine material properties. Three tensile specimens were produced for each profile. One coupon was cut from the flange opposite to the welding seam; the other two coupons were extracted from the web. The 100 mm gauge length of the coupons had the thickness of the profile and the 20 mm width. The geometry corresponded to the standard EN ISO 6892-1 requirements [17]. This study employed two steel grades with the stress values of 355 MPa (NSS) and 900 MPa (HSS). Table 1 shows the chemical composition of the steel specified in the mill certificates.

Table 1. Chemical composition of the steel (%).

Material	C	Si	Mn	P	S	Al	Cr	Ni	Mo	Cu	Nb	V	Ti
HSS	0.100	0.250	1.30	0.020	0.010	0.015	–	–	–	–	0.050	0.050	0.070
NSS	0.140	0.180	1.13	0.006	0.014	0.030	0.040	0.020	–	0.040	0.020	–	0.003

The universal electromechanical testing machine LFM 100 (walter + bay ag, Löhningen, Switzerland) of 100 kN capacity with system controller and data acquisition software DION STAT (v.7, w + b Software, Löhningen, Switzerland) (the displacement control error was less than 0.1%) was utilised for the tension test (Figure 1a). Figure 1b shows the deformed shape of the coupons extracted from a nominally straight profile. Figure 1c–f demonstrate the selected coupon samples before and after testing. Figure 2 shows the stress–strain relationships of both considered steel grades. It also includes the graphs of each coupon (NSS-1 … NSS-3 and HSS-1 … HSS-3) and the averaged diagrams, which give the mean stresses corresponding to particular strain value. The outputs of the tension tests (Figure 1a) were collected every 0.05 seconds. Thus, more than 2000 records were composed in the stress-strain database. This made it possible to use the linear interpolation to harmonise the measurements of different test samples (coupons).

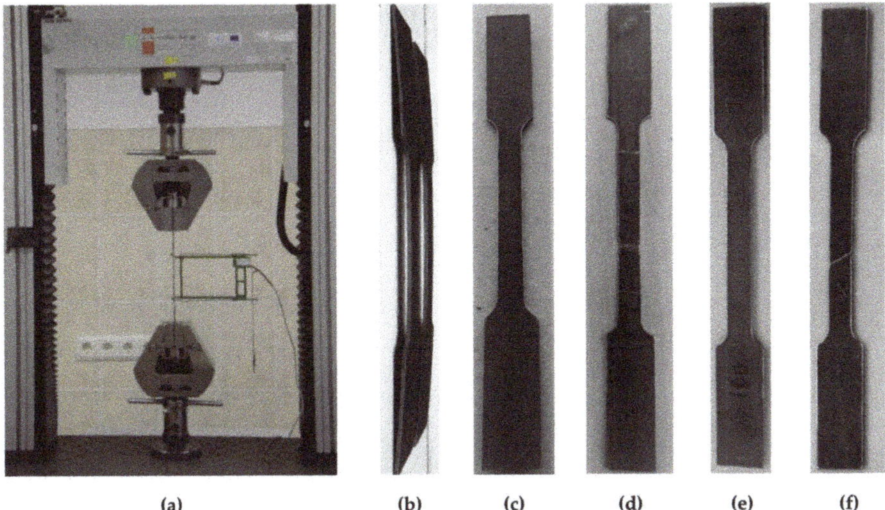

Figure 1. Tensile tests of the steel coupons: (**a**) test setup; (**b**) a deformed shape normal-strength steel (NSS) coupons cut from a straight square hollow section (SHS) profile; (**c**,**d**) NSS coupon before and after testing, respectively; (**e**,**f**) HSS coupon before and after testing, respectively.

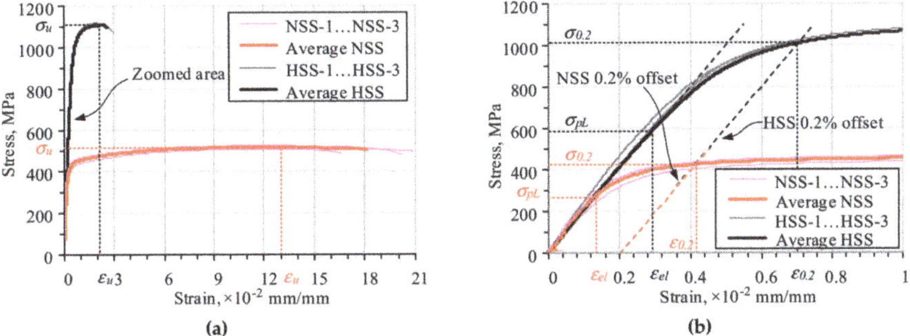

Figure 2. Stress–strain relations of flat tensile coupons of NSS and HSS: (**a**) actual graphs; (**b**) initial part of the diagrams

The averaged diagrams (Figure 2b) had a rounded shape with a moderate strain hardening that reflected typical material behaviour of carbon steel after the cold-forming process. Figure 2 demonstrates that the coupons made from HSS had a substantially increased strength, but decreased ductility comparing to the NSS coupons. The diagrams of both steel grades defined almost identical elastic moduli, which predominantly characterised the deformational behaviour of the profiles.

An adequate assessment of the elastic modulus is vital for materials having a smoothed stress-strain diagram (e.g., Figure 2) because it determines the equivalent yield stress ($\sigma_{0.2}$) defining the theoretical strength of the material. The existing specifications [18–20] recommend two methods to estimate the elastic modulus. The approach [20] was based on the linear approximation of two characteristic points of the stress–strain diagram expressing the elastic modulus as

$$E = \frac{\sigma_2 - \sigma_1}{\varepsilon_2 - \varepsilon_1} \tag{1}$$

where stresses σ_1 and σ_2 correspond to the strains $\varepsilon_1 = 0.0005$ and $\varepsilon_2 = 0.0025$, respectively. Such a simplifying assumption increases the sensitivity of the resultant estimation to non-linear effects caused by production peculiarities of the profiles. The cold-forming process induces residual stresses in the material, of which relaxation distorts the planar shape of the tensile coupons. Figure 1b shows the initial deformed shape of the coupons cut from a straight NSS profile. The stretching of such samples (coupons) caused a discrepancy of the "elastic" part of the stress–strain diagram from a linear form. Therefore, Equation (1) was not used to approximate the graphs shown in Figure 2b.

Another method to determine the elastic modulus employs a regression technique to describe the linear part of the stress-strain curve [17–19]—the slope of the regression line is the analysis parameter. No strict requirements exist to determine the limits of the linear portion of the stress–strain curve except the ISO [17] that defines the boundaries for the cold-formed carbon steel as 10% and 50% of the nominal value of the stress $\sigma_{0.2}$.

In this study, the limits of the linear portion of the stress–strain relationship were determined by maximising two parameters: the determination coefficient value (that should be not less than 0.99) and the length of the analysed part of the diagram. Figure 2b shows the defined regression lines. Table 2 gives the corresponding material parameters. In the table, E is the elastic modulus; $\sigma_{0.2}$ is the equivalent yielding stress; σ_u is the ultimate stress; ε_{el} and $\varepsilon_{0.2}$ are the strains corresponding to the proportional stress σ_{pL} and the equivalent yielding stress $\sigma_{0.2}$, respectively (Figure 2b); and A_{gt} is the elongation percentage at the maximum force. The elastic moduli of NSS and HSS were respectively estimated at the 30%–60% and 10%–50% ranges of the nominal stresses $\sigma_{0.2}$. The assessment of the HSS agreed well to the range recommended by ISO [17], whereas the elastic modulus determination range of NSS was modified due to the initial non-linearity of the shape of the coupon samples (Figure 1b). The lower boundary of the elastic modulus range was arbitrarily increased up to 30% to minimise the straightening effect of the coupon.

Table 2. Averaged mechanical properties of the steel.

Steel Grade	E (GPa)	$\sigma_{0.2}$ (MPa)	σ_u (MPa)	ε_{el} ($\times 10^{-3}$)	$\varepsilon_{0.2}$ ($\times 10^{-3}$)	$\sigma_u/\sigma_{0.2}$ (–)	A_{gt} (%)
S900 (HSS)	200.6	1013	1112	2.94	7.05	1.1	2.07
S355 (NSS)	200.0	430	515	1.35	4.15	1.2	13.03

2.2. Beam Tests

The previous investigation [6] revealed that local effects could cause the failure of the SHS specimens. The same procedure, based on the application of a digital image correlation (DIC) system, was utilised for monitoring the local deformations in this study. The four-point bending layout was used to induce a uniform distribution of stresses in the pure bending zone localising the ultimate strains close to the load application point. Linear variable displacement transducers (LVDT, Novotechnik, Southborough, Mass) monitored vertical displacements. The specimens were tested until failure.

The experiments were carried out in two groups comprising HSS and NSS beam specimens; each group consisted of two tubular beams. Table 3 describes the geometry parameters of the flexural specimens. The notation letter of the specimens designates the normal-strength ("N") or high-strength ("H") steel, whereas the numeral corresponds to the specimen number. In Table 3, B and H are the width and height of the cross-section, t is the thickness of the profile, and r is the outer radius of the cross-section corner (the cross-section in Figure 3 defines the notations). The theoretical section moduli (elastic W_{el} and plastic W_{pl}) are also given in the table.

Figure 3 shows the loading scheme and the dimensions of the specimens. Steel rollers and bearing steel plates were used, ensuring the simple support conditions. Wooden bricks were put inside the profile to avoid the local failure at the supports. The loading points of the beams were left unstiffened to investigate the interaction of the local and global failure mechanisms of the profiles. The beams were loaded under displacement control using a servohydraulic testing machine of 5000 kN capacity; the loading rate was 0.1 mm/min. A load cell measured the reaction to the applied load.

Table 3. Geometry parameters of the beam specimens.

Specimen	B (mm)	H (mm)	t (mm)	r (mm)	W_{el} ($\times 10^3$ mm^3)	W_{pl} ($\times 10^3$ mm^3)
N-1	100.33	100.43	3.75	7.49	43.26	51.51
N-2	100.62	99.80	3.72	7.45	42.70	51.22
H-1	101.74	101.60	3.99	7.98	46.81	56.33
H-2	101.65	101.68	4.00	8.00	46.93	56.53

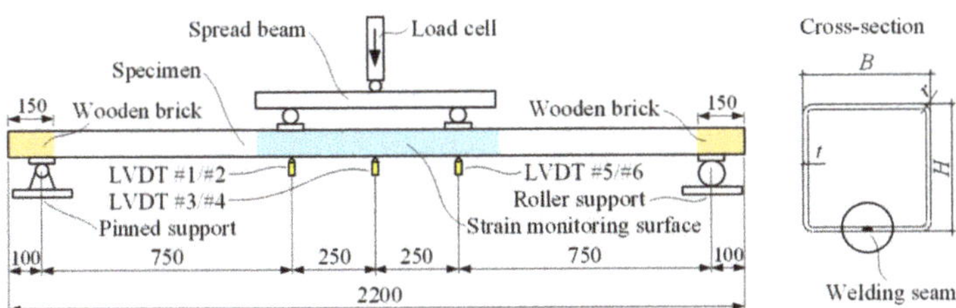

Figure 3. Loading scheme and cross-section of the beam specimen.

As shown in Figure 3, paired LVDT were used to monitor vertical displacements in the pure bending zone—two 100 mm LVDT devices (#3 and #4) were located at the mid-span and two sets of the 50 mm LVDT (#1, #2, and #5, #6) were placed at the boundaries of the bending zone. A data logger ALMEMO 2890-9 recorded the reading of the load cell and all LVDT devices every second.

The DIC system monitored deformation of the surface (coloured in Figure 3) at the rate of one frame per kN. The monitored surfaces of the specimens were prepared (white painting with contrast pattern) to facilitate tracking the displacement tensors. The DAVIS 8.1.6 software package by LA VISION (Göttingen, Germany) was used to process the collected data. This system enables the monitoring of movements of any points arbitrarily set on the monitored surface after the physical tests [6].

3. Test Results

As expected, the failure of all beam specimens was a consequence of the strain localisation close to the load application point (Figure 4a). Figure 4b demonstrates deformed shapes of the tested beams. The LVDT devices were unable to identify the local deformation mechanism. Therefore, these devices were used to assess the curvature of the pure bending zone and the global deformation of the profiles expressed in terms of the vertical displacements.

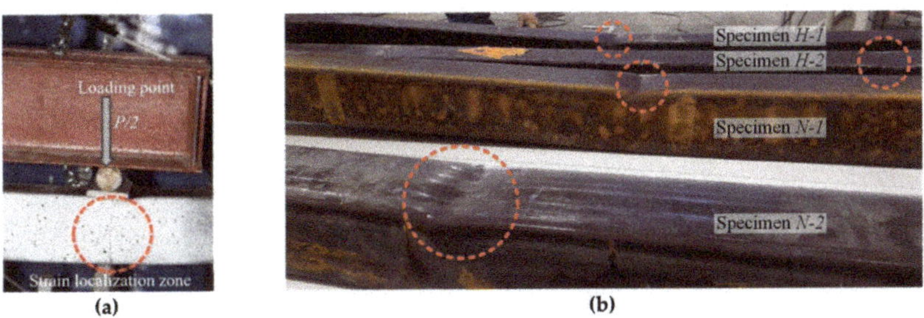

Figure 4. Failure localisation of the beams: (**a**) a typical failure; (**b**) the tested specimens.

The local deformation behaviour of the beam specimens was the focus of this study. The DIC system was used to monitor the strain localisation process. Figure 5a shows the stereoscopic layout of digital cameras. Figure 5b,c demonstrate the typical test results of NSS and HSS profiles.

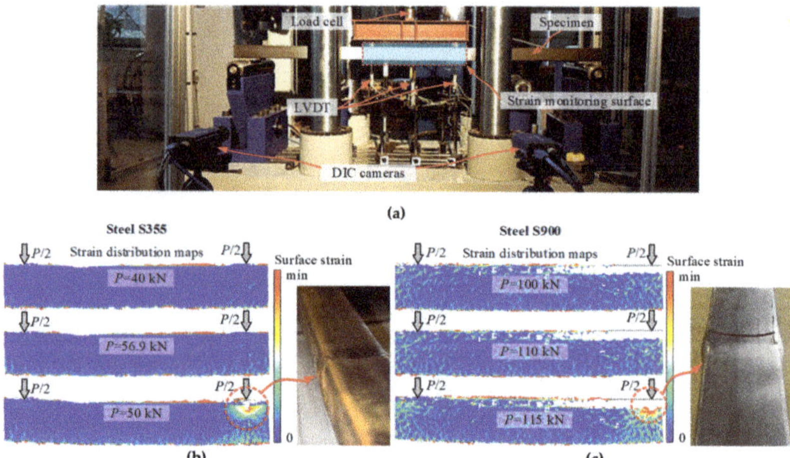

Figure 5. Deformation analysis of the SHS bending profiles: (**a**) loading sheme; (**b**,**c**) strain distribution maps identified by the digital image correlation (DIC) system and local failure of the NSS and HSS beam specimens.

Although the deformed shape of the NSS and HSS profiles shown in Figures 4 and 5b,c look similar, the corresponding deformation mechanisms were different. The respective compressive strain distribution maps (Figure 5b,c) identified by the DIC system illustrated the evolution of local deformations with the load increase. By comparing these maps, it can be observed that the strain localisation in the NSS profile corresponded to the descending branch of the loading diagram. Figure 6a shows the respective moment-curvature diagram. On the contrary, the HSS element demonstrated continuously increased deformations until the ultimate loading stage (see Figure 6b for reference).

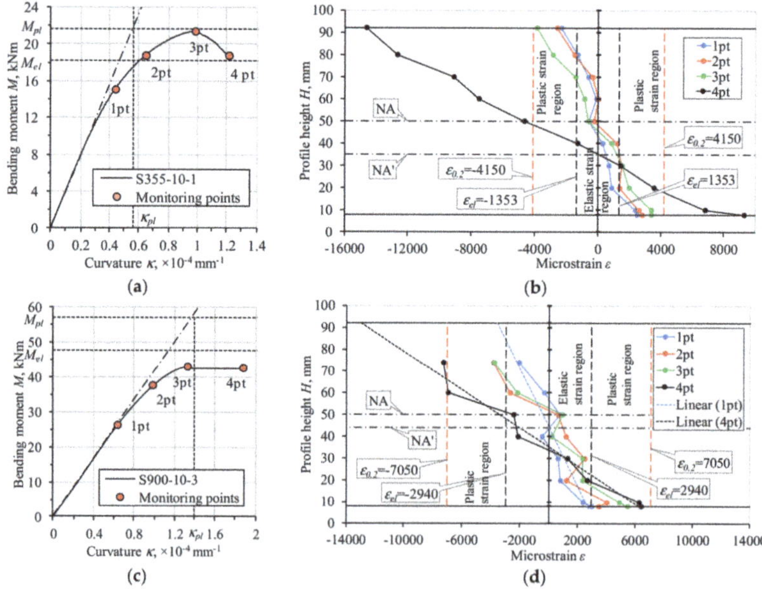

Figure 6. Moment-curvature diagrams: (**a**) NSS, (**b**) HSS; and strain distribution: (**c**) NSS, (**d**) HSS.

Table 4 summarises the test results, where M_u is the ultimate bending moment corresponding to the maximum load, and M_{el} and M_{pl} are the theoretical values of the elastic and plastic moments, respectively. Both theoretical bending moments were calculated using the yield strength $\sigma_{0.2}$ from Table 2 and the corresponding section modulus from Table 3. The following equation determines the curvature of the pure bending zone κ_u, corresponding to the maximum bending moment M_u:

$$\kappa_u = \frac{24\delta_{L/2}}{3L^2 - 4L_s^2} \tag{2}$$

Table 4. Results of the bending tests.

Specimen	M_u (kNm)	$\kappa_u \times 10^{-4}$ (mm^{-1})	M_{el} (kNm)	M_{pl} (kNm)	M_u/M_{el} (–)	M_u/M_{pl} (–)
N-1	21.34	1.00	18.17	21.63	1.17	0.99
N-2	21.41	1.01	17.93	21.38	1.19	1.00
H-1	43.05	1.35	47.44	57.08	0.91	0.75
H-2	43.13	1.26	47.56	57.29	0.91	0.75

In Equation (2), $\delta_{L/2}$ is the vertical displacement at the mid-span defined as the average outcome of the LVDT devices #3 and #4 (Figure 3), and L and L_s are the total span and the shear span of the beam, respectively ($L = 2$ m, $L_s = 0.75$ m).

In parallel to the results shown in Figure 5b,c, Table 4 illustrates the difference between the deformation mechanisms, which are characteristic of the NSS and HSS profiles subjected to flexure. In this table, M_u and κ_u are the experimental ultimate bending moment and the corresponding curvature of the pure bending zone, and M_{el} and M_{pl} are the theoretical elastic and plastic moments, respectively. The actual bending resistance of the NSS specimens described by the moment M_u is comparable to the theoretical plastic moment M_{pl}—the tubular specimens were capable of resisting the equivalent yielding stress ($\sigma_{0.2}$). In other words, the strength term controlled the load-bearing capacity of the NSS beams. Alternatively, the bending resistance of the HSS profiles was well below the theoretical moment M_{el}. This means that the stiffness condition governed the failure behaviour of the HSS specimens.

A detailed analysis of the strain distribution maps generated by the DIC system (Figure 5b,c) was carried out to determine the failure mechanisms of the beam specimens. The virtual strain gauges created by the DaVis 8.1.6 software were applied to monitor the deformation localisation process near the loading point. Figure 7 shows the distribution of the 20 mm virtual gauges. Figure 6c,d show the deformation diagrams of the flat part of the web identified using the virtual devices. The grey-coloured areas define the elastic strain regions limited by the average value of the strain ε_{el} given in Table 2. The red-filled zones correspond to the plastic strain regions limited by the average strain $\varepsilon_{0.2}$ (Table 2). The strain distribution diagrams correspond to the loading points highlighted in the moment-curvature graphs shown in Figure 6a,b.

Figure 6c shows the strain distribution of the beam specimen N-1. It can be observed that the compressive strain, corresponding to the first point ("1pt") of the diagram given in Figure 6a, exceeded the proportional limit ε_{el}. The deformations did not transcend the yield limit $\varepsilon_{0.2}$ until the load reached the third point ("3pt"). The shift of NA to the position NA' at that moment caused the increase of the local deformations, but the specimen N-1 was still capable of sustaining partial yielding before failure occurred.

Although DIC system can help to monitor the failure mechanisms, even this technology is not always capable of capturing local deformations. Two HSS beam specimens were tested (Table 4), but only the specimen H-1 demonstrated the evident deformation localisation process. There are two possible explanations for this outcome: (1) only one surface was exposed to the DIC, or (2) the deformation localisation process was uniformly distributed between the loading points.

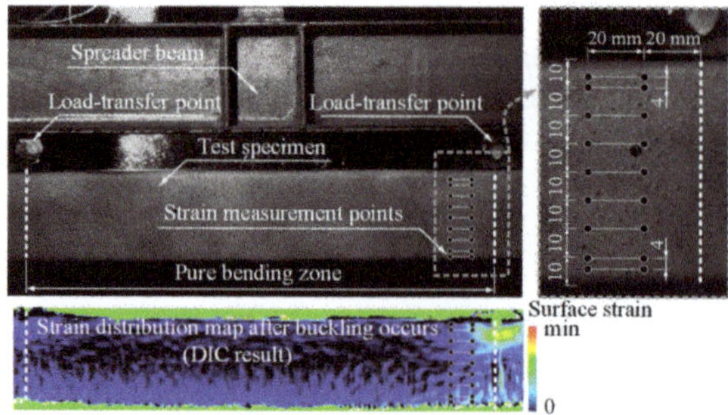

Figure 7. Virtual strain gauges of the DIC system.

Further analysis is based on the DIC results of the specimen *H-1*. Figure 6d shows the respective strain distribution. Unfortunately, the results of the virtual gauges placed in the compressive zone were lost due to an accidental light reflection. Therefore, linear trends were used to extrapolate the strain diagrams. The shift of NA to the position NA′ indicated a premature nonlinear deformation of the HSS bending profile. This process can be observed at the loading level corresponding to the first point ("1pt") of the diagram shown in Figure 6b. It also demonstrated that the deformations were localised in the outer layers of the section up to the third load point ("3pt"). The failure of the beam specimen was a consequence of the strain concentration realised before the yielding limit that reveals a predominant role of the local buckling effects on the deformation behaviour of the HSS profile.

4. Discussion of the Results

The relationship between the enhancement of the resistance of the SHS profiles to the mechanical load and the increase of the strength of the material is the focus of this research. The test results showed that the increase of the material strength for the same elastic modulus and geometry of SHS was not proportional to the rise of the load-bearing capacity of the tubular element subjected to bending. Comparative analysis of the local deformational behaviour of the SHS profiles identified the differences of failure mechanisms of the NSS and HSS specimens

- NSS: The beam specimens were capable of attaining the load-bearing resistance comparable to the theoretical plastic moment M_{pl}. The considered geometry of SHS enabled the efficient utilisation of the material. The M_u/M_{el} ratio given in Table 4 defined the average increase of the theoretical load-bearing capacity by 18%.
- HSS: A premature local buckling limited the load-bearing capacity of the tubular specimen. The considered geometry and structural performance of the profile disabled the ability to utilise HSS efficiently due to local bearing failure. The M_u/M_{el} ratio (Table 4) demonstrated the 9% deficit of the ultimate load, considering the theoretical elastic moment as the reference.

An appropriate alteration of the cross-section geometry is necessary to ensure efficient structural utilisation of advanced materials. The existing design codes employ a slenderness limit approach to identify the extent to which the local buckling resistance controls the load-bearing capacity of a tubular profile. The equivalent slenderness of the flange or web of the cross-section can be calculated using Equation (3), as follows:

$$\lambda = \frac{b}{t} \sqrt{\frac{f_y}{E}} \tag{3}$$

where *b* is the width of the flat part of the web or flange, *t* is the thickness of the corresponding part of the cross-section, f_y is the yield strength of steel, and *E* is the elastic modulus.

Several investigations [12–14] were carried out to verify the applicability of the slenderness limit for the structural design using HSS tubular profiles. Misiūnaitė et al. [6] proved the adequacy of this approach for the analysis of HSS profiles considered in this study. Equation (3) indicates that the slenderness increases together with the strength (f_y) if all other input parameters are remaining constant. Such estimation agrees well with the results of this study.

A finite element (FE) model developed and verified in the referred work [6] is used to illustrate the efficiency concept. For simplicity, the four-point bending test with the symmetry condition at the mid-section of the profile is the simulation object. The loading scheme is the same as it was used in the physical experiment (Figure 3). Figure 8 shows the FE models and boundary conditions. Two cross-sections are modelled. One SHS ($100 \times 100 \times 4$ mm) is the same as was used for the physical tests (Section 2.2), whereas another profile ($60 \times 60 \times 4$ mm) is considered as the alternative for demonstrating the utilisation efficiency of HSS. The FE size (= 10 mm) is identical in both models.

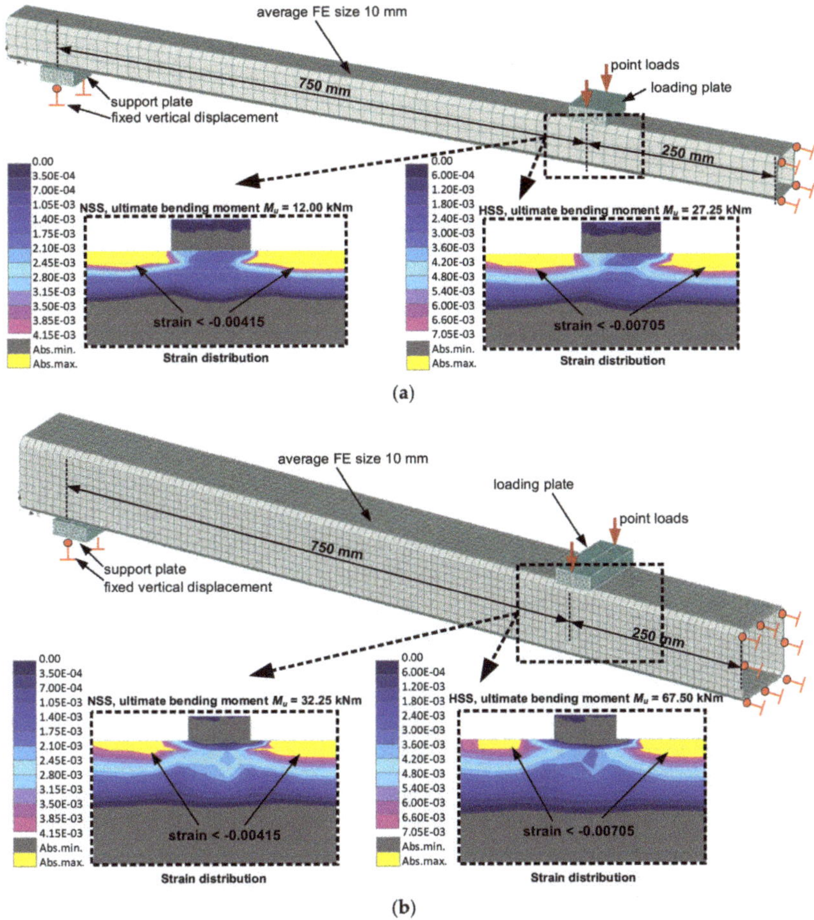

Figure 8. Finite element (FE) model and numerical simulation results (strain distribution) of the beams with different size of the cross-section: (**a**) $60 \times 60 \times 4$ mm; (**b**) $100 \times 100 \times 4$ mm.

The relationship between the relative increase of the load-bearing capacity of the beam specimen $\Delta M_{u,HSS/NSS}$ and the corresponding ratio of the steel strengths (grades) $\Delta\sigma_{0.2,HSS/NSS}$ determines the HSS utilisation efficiency. The following coefficient is used for that purpose:

$$c_{ef} = \frac{\Delta M_{u,HSS/NSS}}{\Delta\sigma_{0.2,HSS/NSS}}; \quad \Delta M_{u,HSS/NSS} = \frac{M_{u,HSS}}{M_{u,NSS}}; \quad \Delta\sigma_{0.2,HSS/NSS} = \frac{\sigma_{0.2,HSS}}{\sigma_{0.2,NSS}} \tag{4}$$

where $M_{u,HSS}$ and $M_{u,NSS}$ are the maximum bending moments, and $\sigma_{0.2,HSS}$ and $\sigma_{0.2,NSS}$ are the equivalent yielding stresses. The subscripts HSS and NSS correspond to the normal-strength steel and high-strength steel, respectively.

The manufacturing process of SHS profiles affects the mechanical properties of steel, inducing residual stresses in the corner regions [11] and enhancing the yield strength [7–11]. These mechanisms compensate each other and can be neglected in numerical simulations [11,21]. Thus, the FE analysis employs a simplified material modelling concept—the von Mises material model with bilinear stress-strain diagram and strain hardening was used. The experimental stress–strain relationships (Figure 2b) determined the material properties of the NSS and HSS profiles. The possible geometry imperfections were also neglected following the recommendation by the referred work [21]. The simulations were performed incrementally increasing the load (with 1 kN increment) until failure of the beam. The equivalent strain limit $\varepsilon_{0.2}$ was chosen as the failure criterion. As it can be observed in Table 2, the strain limits of 4.15‰ and 7.05‰ correspond to NSS and HSS profiles, respectively. Figure 8 shows the simulation results. For the $100 \times 100 \times 4$ mm cross-section, the estimated ultimate bending moment increased from 32.3 to 67.5 kNm (2.09 times) when NSS was replaced with HSS. For the alternative cross-section ($60 \times 60 \times 4$ mm), the corresponding increases reached 2.28 times (from 12.0 to 27.3 kNm). The comparison of the equivalent stresses $\sigma_{0.2}$, presented in Table 2, gave the ratio of 2.36 (= 1013/430). The coefficient c_{ef} (Equation 4) defined the 89% utilisation of HSS in the $100 \times 100 \times 4$ mm profile, whereas the 96% effectiveness was possessed in the $60 \times 60 \times 4$ mm cross-section. In other words, the latter profile enables the utilisation of the HSS in a more efficient way than the specimen used for the physical tests.

The actual efficiency of the utilisation of HSS in the $100 \times 100 \times 4$ mm profile, however, is less significant than that was observed in the physical tests. On average, the coefficient c_{ef} was equal to 85% for the data presented in Table 4. This is a consequence of the assumption of the simplified material models in the FE analysis; the deformation interaction mechanism between the flange and web was ignored as well. Thus, the identification of the optimum configuration of the cross-section is an object for further research.

5. Conclusions

Although the current trends in the construction industry are related to the use of high-strength steel (HSS), the increase of the load-bearing capacity of the member is often not proportional to the rise of the strength of the material. This study employed the flexural test results of two groups of HSS and normal-strength steel (NSS) tubular specimens with $100 \times 100 \times 4$ mm cross-section to illustrate this observation. A simplified finite element (FE) model was used to demonstrate the efficiency assessment principles. The ratio c_{ef} between the relative increase of the load-bearing capacity of the structural member and the respective increment of the steel strength was the parameter for defining the utilisation efficiency. The obtained results enabled the formulation of the following conclusions:

- An almost identical initial elastic modulus was characteristic of the considered steel grades S355 (NSS) and S900 (HSS). This parameter predominantly controlled the deformation behaviour of the elements.
- The failure of all beam specimens was a consequence of the deformation localisation process identified by using a digital image correlation system. Although the deformed shape of the profiles looked similar, the corresponding deformation mechanisms were different. The strain localisation in the NSS profile corresponded to the descending branch of the loading diagram, whereas the HSS element continuously demonstrated increased deformations until the ultimate loading stage.
- The actual bending resistance of the NSS profiles was comparable to the theoretical plastic bending moment M_{pl}. This meant that the strength term controlled the load-bearing capacity of the NSS elements. The M_u/M_{el} ratio gave the average increase of the theoretical load-bearing capacity at 18%. Alternatively, the stiffness condition governed the failure behaviour of the HSS specimens;

the load-bearing capacity was below the theoretical value M_{el} at 9%. The identification of the optimum configuration of the cross-section should be the subject for further research.

- The numerical analysis demonstrated that the HSS utilisation efficiency was dependent on the shape of the cross-section. The ratio c_{ef} was found as being equal to 89% for the $100 \times 100 \times 4$ mm profile, whereas it increased up to 96% for the $60 \times 60 \times 4$ mm cross-section. The actual efficiency of the utilisation of HSS in the $100 \times 100 \times 4$ mm specimen was less substantial than that obtained in the physical tests ($c_{ef} = 85\%$). This was a consequence of the idealised approximation of the deformation localisation process in the tubular profiles assumed in FE analysis.

Author Contributions: Conceptualization V.G.; methodology V.G. and I.M.; software R.J.; validation I.M. and A.R.; investigation, I.M., A.R.; resources V.G.; data curation A.R.; writing—original draft preparation I.M. and A.R.; writing—review and editing V.G.; visualization A.R.; supervision V.G.; project administration I.M. All authors have read and agreed to the published version of the manuscript.

Funding: The European Social Fund provided a financial support for this research within the framework of the project "Development of Competences of Scientists, other Researchers and Students through Practical Research Activities" (project no. 09.3.3-LMT-K-712).

Conflicts of Interest: The authors declare no conflict of interest.

References

1. Young, B.; Li, H.T. Behaviour of cold-formed high strength steel RHS under localized bearing forces. *Eng. Struct.* **2019**, *183*, 192–205.
2. Gkantou, M.; Theofanous, M.; Baniotopoulos, C. Plastic design of hot-finished high strength steel continuous beams. *Thin-Walled Struct.* **2018**, *133*, 85–95. [CrossRef]
3. Lan, X.; Chan, T.M.; Young, B. Structural behavior and design of chord plastification in high strength steel CHS X-joints. *Constr. Build. Mat.* **2018**, *191*, 1252–1267. [CrossRef]
4. Young, B.; Li, H.T. Design of cold-formed high strength steel tubular sections undergoing web crippling. *Thin-Walled Struct.* **2018**, *133*, 192–205.
5. Yan, J.J.; Chen, M.T.; Quach, W.M.; Yan, M.; Young, B. Mechanical properties and cross-sectional behavior of additively manufactured high strength steel tubular sections. *Thin-Walled Struct.* **2019**, *144*, 106–158. [CrossRef]
6. Misiunaite, I.; Rimkus, A.; Jakubovskis, R.; Sokolov, A.; Gribniak, V. Analysis of local deformation effects in cold-formed tubular profiles subjected to bending. *J. Construct. Steel Res.* **2019**, *160*, 598–612. [CrossRef]
7. Gardner, L.; Saari, N. Comparative experimental study of hot-rolled and cold-formed rectangular hollow sections. *Thin-Walled Struct.* **2010**, *48*, 495–507. [CrossRef]
8. Gardner, L.; Yun, X. Description of stress-strain curves for cold-formed steels. *Construct. Build. Mat.* **2018**, *189*, 527–538. [CrossRef]
9. Afshan, S.; Rossi, B.; Gardner, L. Strength enhancements in cold-formed structural sections – Part I: Material testing. *J. Constr. Steel Res.* **2013**, *83*, 177–188. [CrossRef]
10. Rossi, B.; Afshan, S.; Gardner, L. Strength enhancements in cold-formed structural sections – Part II: Predictive models. *J. Constr. Steel. Res.* **2013**, *83*, 189–196. [CrossRef]
11. Ma, J.L.; Chan, T.M.; Young, B. Material properties and residual stresses of cold-formed high strength steel hollow sections. *J. Constr. Steel Res.* **2015**, *109*, 152–165. [CrossRef]
12. Wang, J.; Afshan, S.; Gkantou, M.; Theofanous, M.; Baniotopoulus, C.; Gardner, L. Flexural behaviour of hot-finished high strength steel square and rectangular hollow sections. *J. Constr. Steel Res.* **2016**, *121*, 97–109. [CrossRef]
13. Ma, J.L.; Chan, T.M.; Young, B. Design of cold-formed high strength steel tubular beams. *Eng. Struct.* **2017**, *151*, 432–443. [CrossRef]
14. Ma, J.L.; Chan, T.M.; Young, B. Experimental investigation of cold-formed high strength steel tubular beams. *Eng. Struct.* **2016**, *126*, 200–209. [CrossRef]
15. Xiaoyi, L.; Chen, J.; Chan, T.M.; Young, B. The continuous strength method for the design of high strength steel tubular sections in compression. *Eng. Struct.* **2018**, *162*, 165–177.

16. Zhao, X.-L.; Fernando, D.; Al-Mahaidi, R. CFRP strengthened RHS subjected to transverse end bearing force. *Eng. Struct.* **2006**, *28*, 1555–1565. [CrossRef]

17. International Organization for Standardization (ISO). *Metallic Material – Tensile Testing – Part 1: Method of Test at Room Temperature*; EN ISO 6892-1: 2016; ISO: Geneva, Switzerland, 2016.

18. Standards Association of Australia (AS). *Metallic Material – Tensile Testing at Ambient Temperature*; AS 1391-2007; AS: Sydney, Australia, 2007.

19. American Society for Testing and Materials (ASTM). *Standard Test Methods for Tension Testing of Metallic Materials*; E8/E8M-13a; ASTM: West Conshohocken, PA, USA, 2013.

20. International Organization for Standardization (ISO). *Plastics-Determination of Tensile Properties - Part 1: General Principles*; EN ISO 527-1; CEN: Brussels, Belgium, 1996.

21. Schafer, B.W.; Pekoz, T. Computational modeling of cold-formed steel: characterizing geometric imperfections and residual stresses. *J. Constr. Steel. Res.* **1998**, *47*, 193–210. [CrossRef]

MDPI

St. Alban-Anlage 66

4052 Basel

Switzerland

Tel. +41 61 683 77 34

Fax +41 61 302 89 18

www.mdpi.com

Materials Editorial Office

E-mail: materials@mdpi.com

www.mdpi.com/journal/materials